Chapman & Hall/CRC
Statistics in the Social and Behavioral Sciences Series

Handbook of International Large-Scale Assessment

Background, Technical Issues, and Methods of Data Analysis

Edited by

Leslie Rutkowski
Indiana University
Bloomington, USA

Matthias von Davier
Educational Testing Service
Princeton, New Jersey, USA

David Rutkowski
Indiana University
Bloomington, USA

CRC Press is an imprint of the
Taylor & Francis Group, an **informa** business
A CHAPMAN & HALL BOOK

Chapman & Hall/CRC
Statistics in the Social and Behavioral Sciences Series

Aims and scope

Large and complex datasets are becoming prevalent in the social and behavioral sciences and statistical methods are crucial for the analysis and interpretation of such data. This series aims to capture new developments in statistical methodology with particular relevance to applications in the social and behavioral sciences. It seeks to promote appropriate use of statistical, econometric and psychometric methods in these applied sciences by publishing a broad range of reference works, textbooks and handbooks.

The scope of the series is wide, including applications of statistical methodology in sociology, psychology, economics, education, marketing research, political science, criminology, public policy, demography, survey methodology and official statistics. The titles included in the series are designed to appeal to applied statisticians, as well as students, researchers and practitioners from the above disciplines. The inclusion of real examples and case studies is therefore essential.

Published Titles

Analysis of Multivariate Social Science Data, Second Edition
David J. Bartholomew, Fiona Steele, Irini Moustaki, and Jane I. Galbraith

Latent Markov Models for Longitudinal Data
Francesco Bartolucci, Alessio Farcomeni, and Fulvia Pennoni

Statistical Test Theory for the Behavioral Sciences
Dato N. M. de Gruijter and Leo J. Th. van der Kamp

Multivariable Modeling and Multivariate Analysis for the Behavioral Sciences
Brian S. Everitt

Bayesian Methods: A Social and Behavioral Sciences Approach, Second Edition
Jeff Gill

Multiple Correspondence Analysis and Related Methods
Michael Greenacre and Jorg Blasius

Applied Survey Data Analysis
Steven G. Heeringa, Brady T. West, and Patricia A. Berglund

Informative Hypotheses: Theory and Practice for Behavioral and Social Scientists
Herbert Hoijtink

Foundations of Factor Analysis, Second Edition
Stanley A. Mulaik

Linear Causal Modeling with Structural Equations
Stanley A. Mulaik

Handbook of International Large-Scale Assessment: Background, Technical Issues, and Methods of Data Analysis
Leslie Rutkowski, Matthias von Davier, and David Rutkowski

Generalized Linear Models for Categorical and Continuous Limited Dependent Variables
Michael Smithson and Edgar C. Merkle

Incomplete Categorical Data Design: Non-Randomized Response Techniques for Sensitive Questions in Surveys
Guo-Liang Tian and Man-Lai Tang

CRC Press
Taylor & Francis Group
6000 Broken Sound Parkway NW, Suite 300
Boca Raton, FL 33487-2742

Printed on acid-free paper
Version Date: 20130923

International Standard Book Number-13: 978-1-4398-9512-2 (Hardback)

Library of Congress Cataloging-in-Publication Data

Handbook of International large-scale assessment : background, technical issues, and methods of data analysis / [edited by] Leslie Rutkowski, Matthias von Davier, David Rutkowski.
pages cm. -- (Statistics in the social and behavioral sciences series)
Includes bibliographical references and index.
ISBN 978-1-4398-9512-2 (hardback)
1. Educational tests and measurements--Cross-cultural studies. 2. Academic achievement--Cross-cultural studies. 3. Educational tests and measurements--Methodology. 4. Educational tests and measurements--Statistics. I. Rutkowski, Leslie, editor of compilation. II. Davier, Matthias von, editor of compilation. III. Rutkowski, David, editor of compilation.

LB3051.H31983 2013
371.26--dc23 2013025261

Visit the Taylor & Francis Web site at
http://www.taylorandfrancis.com

and the CRC Press Web site at
http://www.crcpress.com

Contents

Editors

Leslie Rutkowski, PhD, is an assistant professor of inquiry methodology in the Department of Counseling and Educational Psychology at Indiana University, Bloomington, Indiana, USA. Leslie joined the inquiry methodology program at Indiana University in January of 2010. She earned her PhD in educational psychology with a specialization in statistics and educational measurement from the University of Illinois at Urbana-Champaign. Leslie's research is focused in the area of international large-scale assessment from both methodological and applied perspectives. Her interests include the impact of background questionnaire quality on achievement results, latent variable models for achievement estimation, and examining methods for comparing heterogeneous populations in international surveys. Leslie teaches courses in quantitative methods, including covariance structure analysis, multilevel modeling, and categorical data analysis. As part of her work at the university, Leslie frequently collaborates with or consults for international organizations, including the International Association for the Evaluation of Educational Achievement (IEA) and the Organization for Economic Co-Operation and Development (OECD). Leslie is currently involved with the 2013 cycle of the OECD's Teaching and Learning International Survey (TALIS). She also conducts seminars and workshops in quantitative methods for researchers from around the world.

Matthias von Davier, PhD, is a research director in the Research & Development Division at the Educational Testing Service (ETS). He joined ETS, which is located in Princeton, New Jersey, USA, in 2000. He earned his PhD in psychology from the University of Kiel, Germany, in 1996, specializing in psychometrics. At ETS, Dr. von Davier manages a group of researchers concerned with methodological questions arising in large-scale international comparative studies in education. He is one of the editors of the periodical *Issues and Methodologies in Large-Scale Assessments,* which is jointly published by the International Association for the Evaluation of Educational Achievement (IEA) and ETS through the IEA-ETS Research Institute (IERI). His current work at ETS involves the psychometric methodologies used in analyzing cognitive skills data and background data from large-scale educational surveys, such as the Organization for Economic Co-Operation and Development's upcoming Program for the International Assessment of Adult Competencies (PIAAC) and the ongoing Programme for International Student Assessment (PISA), as well as the IEA's Trends in International Mathematics and Science Study (TIMSS) and Progress in International Reading Literacy Study (PIRLS). His work at ETS also involves the development of software for multidimensional models for item response data, and the improvement of models and estimation methods for the analysis of data from large-scale educational survey assessments.

Prior to joining ETS, Dr. von Davier led a research group on computer-assisted science learning, was codirector of the "Computer as a Tool for Learning" section at the Institute for Science Education (IPN) in Kiel, Germany, and was an associate member of the Psychometrics & Methodology Department of IPN. During his 10-year tenure at the IPN, he developed commercially available software for analyses with the Rasch model, with latent class analysis models, and with mixture distribution Rasch models. Dr. von Davier taught courses on foundations of neural networks and on psychometrics and educational psychology at the University of Kiel for the departments of psychology and education. He gave various invited workshops and mini-courses on psychometrics and recent developments in item response theory models. In 1997, he received a postdoctoral fellowship award from the ETS and an additional research award from the German Science Foundation. From 1993 to 1997 he was part of the research staff on the German Science Foundation-funded project, "Development and Validation of Psychometric Mixture Distribution Models" at the University of Kiel.

David Rutkowski, PhD, is an assistant professor of educational policy in the Department of Educational Leadership and Policy Studies at Indiana University, Bloomington, Indiana, USA. Dr. Rutkowski joined the program at Indiana University in January of 2011 and before that worked as a researcher at the International Association for the Evaluation of Educational Achievement (IEA) Data Processing and Research Center in Hamburg, Germany. He earned his PhD in educational policy with a research specialization in evaluation from the University of Illinois at Urbana-Champaign. Dr. Rutkowski's research is focused in the area of evaluation and policy and technical topics within international large-scale assessment. His interests include how large-scale assessments are used within policy debates, the impact of background questionnaire quality on achievement results, and topics concerning immigrant students at the international level. Dr. Rutkowski teaches courses in educational policy, evaluation, mixed methods, and statistics. As part of his work at the university, Dr. Rutkowski collaborates with or consults for national and international organizations, including the U.S. State Department, USAID, the IEA, and the Organization for Economic Co-Operation and Development (OECD). Dr. Rutkowski is currently involved in projects in Afghanistan, Austria, South Sudan, and the United States.

Contributors

Jim Allen
Research Centre for Education and the Labour Market (ROA)
Maastricht University
Maastricht, the Netherlands

Carolyn J. Anderson
Department of Educational Psychology
University of Illinois at Urbana-Champaign
Champaign, Illinois

Alka Arora
TIMSS and PIRLS International Study Center
Boston College
Chestnut Hill, Massachusetts

Jonas P. Bertling
Center for Academic and Workforce Readiness and Success
Educational Testing Service
Princeton, New Jersey

Cees Glas
Department of Research Methodology, Measurement and Data Analysis
University of Twente
Enschede, the Netherlands

Eugene Gonzalez
Educational Testing Service
Princeton, New Jersey

Stephen P. Heyneman
Department of Leadership, Policy and Organizations
Vanderbilt University Peabody College
Nashville, Tennessee

Khurrem Jehangir
Department of Research Methodology, Measurement and Data Analysis
University of Twente
Enschede, the Netherlands

Minjeong Jeon
Graduate School of Education
University of California, Berkeley
Berkeley, California

David Kaplan
Department of Educational Psychology
University of Wisconsin-Madison
Madison, Wisconsin

Bryan Keller
Department of Human Development
Teachers College
Columbia University
New York, New York

Jee-Seon Kim
Department of Educational Psychology
University of Wisconsin-Madison
Madison, Wisconsin

Patrick C. Kyllonen
Center for Academic and Workforce Readiness and Success
Educational Testing Service
Princeton, New Jersey

Bommi Lee
Vanderbilt University Peabody College
Nashville, Tennessee

Herbert W. Marsh
Centre for Positive Psychology and Education
University of Western Sydney
New South Wales, Australia

and

Department of Education
University of Oxford
Oxford, England

and

King Saud University
Riyadh, Saudi Arabia

Michael O. Martin
TIMSS and PIRLS International Study Center
Boston College
Chestnut Hill, Massachusetts

John Mazzeo
Statistical Analysis, Data Analysis and Psychometric Research
Educational Testing Service
Princeton, New Jersey

Ina V. S. Mullis
TIMSS and PIRLS International Study Center
Boston College
Chestnut Hill, Massachusetts

Benjamin Nagengast
Educational Psychology
University of Tübingen
Tübingen, Germany

Andreas Oranje
Statistical Analysis, Data Analysis and Psychometric Research
Educational Testing Service
Princeton, New Jersey

Soojin Park
Department of Educational Psychology
University of Wisconsin-Madison
Madison, Wisconsin

Corinna Preuschoff
Jones and Bartlett Learning
Burlington, Massachusetts

Sophia Rabe-Hesketh
Graduate School of Education
University of California, Berkeley
Berkeley, California

Frank Rijmen
CTB/McGraw-Hill
Monterey, California

Joseph P. Robinson
Department of Educational Psychology
University of Illinois at Urbana-Champaign
Champaign, Illinois

Keith Rust
Westat
Rockville, Maryland

and

Joint Program in Survey Methodology
University of Maryland
College Park, Maryland

David Rutkowski
Department of Educational Leadership and Policy Studies
Indiana University
Bloomington, Indiana

Leslie Rutkowski
Department of Counseling and Educational Psychology
Indiana University
Bloomington, Indiana

Martin Senkbeil
Department of Educational Science and Research Methodology
Leibniz Institute for Science and Mathematics Education
University of Kiel
Kiel, Germany

Yongyun Shin
Department of Biostatistics
Virginia Commonwealth University
Richmond, Virginia

Sandip Sinharay
CTB/McGraw-Hill
Monterey, California

Laura M. Stapleton
Department of Human Development and Quantitative Methodology
University of Maryland
College Park, Maryland

Rolf van der Velden
Research Centre for Education and the Labour Market (ROA)
Maastricht University
Maastricht, the Netherlands

Matthias von Davier
Research and Development Division
Educational Testing Service
Princeton, New Jersey

Hans Wagemaker
International Association for the Evaluation of Educational Achievement
Wellington, New Zealand

Jonathan P. Weeks
Research and Development Division
Educational Testing Service
Princeton, New Jersey

Jörg Wittwer
Department of Empirical Research on Instruction and Learning
University of Göttingen
Göttingen, Germany

Kentaro Yamamoto
Center for Global Assessment
Educational Testing Service
Princeton, New Jersey

Lei Ye
Educational Testing Service
Princeton, New Jersey

Yan Zhou
Department of Counseling and Educational Psychology
Indiana University
Bloomington, Indiana

Section I

Policy and Research Relevance of International Large-Scale Assessment Data

1

A Brief Introduction to Modern International Large-Scale Assessment

David Rutkowski
Indiana University

Leslie Rutkowski
Indiana University

Matthias von Davier
Educational Testing Service

CONTENTS

Introduction

The origins of modern-day international assessments of student skills are often traced back to the First International Mathematics Study (FIMS) conducted by the International Association for the Evaluation of Educational Achievement (IEA) in the early 1960s. The undertaking of an international project at that time, with a shoestring budget and few modern technological conveniences to speak of (no email, fax, or Internet, and only minimal access to international phone lines), speaks to the dedication and vision of the scholars that were willing to attempt such a feat. The first executive director of the IEA, T. Neville Postlethwaite (1933–2009), once recounted the story of sending off the first round of assessments and not knowing for months if the assessment booklets had even arrived at their destinations, let alone whether or not the assessment was actually being administered in the 12 countries that initially participated. In many ways the founders of this early study understood the obstacles of completing such a study, and they were equally interested in the process as well as the outcome. Alan Purves (1987), a former chair of the IEA, recounts, "the purpose of the [FIMS] study was twofold: to make inferences about intellectual functioning from multiple-choice items and to test the feasibility of a large-scale international study" (p. 11). Sweeping statements for and against international large-scale assessments (ILSAs) enjoy a prominent place in international educational discourse, particularly in modern educational research and policy debates (cf. Asia Society and Council of Chief State School Officers (CCSSO), 2010; Medrich and Griffith, 1992; Mislevy, 1995; Postlethwaite, 2005; Purves, 1987; Rizvi and Lingard, 2009; Schwippert, 2007), as well as in the public sphere (cf. Darling-Hammond 2011). Regardless of where one might fall on the ILSA ideological divide, 50 years later, the feasibility of such an endeavor is clear.

To that end, modern ILSAs such as the Trends in International Mathematics and Science Study (TIMSS), the Progress in International Reading Literacy Study (PIRLS), and the Programme for International Student Assessment (PISA) are an increasingly important part of the educational research and policy landscape internationally. Achievement rankings that compare performance of participating countries are fodder for newspaper headlines and politicians. For instance, in 2010, Andreas Schleicher, a senior education official at the Organization for Economic and Co-operative Development (OECD) who oversaw the development and implementation of the PISA, testified before the U.S. Senate Education Panel that if the United States could achieve gains on PISA similar to those of Poland, that "a long-term economic value of over 40 trillion dollars" was possible (Dillon 2010). Similar projections have been stated and restated based on research presented by economists (e.g., Hanushek and Woessmann 2011). Projections of this magnitude and long-term reach are obviously associated with a certain level of uncertainty, and it is the responsibility of researchers to point out that statements are likely based on a number of assumptions, some of which may be difficult to validate.

Along a related line, Shanghai participated in PISA for the first time in 2009. With high rankings across all tested domains, Shanghai's surprise results prompted Chester Finn, the former U.S. Assistant Secretary of Education and current president of the Thomas B. Fordham Institute, to react strongly, stating "on Pearl Harbor Day 2010, the United States (and much of the world) was attacked by China… will this be the wake-up call that America needs to get serious about educational achievement?" (Finn, 2010). The reforms that have been attributed to these studies continue to have a far and lasting reach both in the United States and around the world (Hopmann and Brinek, 2007; Pongratz, 2006). Further, the relatively recent and marked emphasis on educational accountability and system monitoring makes ILSAs, which are internationally representative studies at the system level, prime resources for researchers and policy makers interested in correlates of learning and achievement in a number of areas and across a number of primary and secondary levels of education around the world (von Davier et al., 2012). Before proceeding to a discussion of the aims and content of this volume, we very briefly outline some of the special methodological and design features of ILSAs in general. Several chapters dig deeper into these details (e.g., Chapters 4, 6 through 8, and 10), so our treatment is necessarily superficial.

In addition to cross-sectional estimates of achievement in a number of domains such as reading, science, and mathematics—often the only components of these studies that figure into policy debates—TIMSS, PIRLS, and PISA databases feature a wealth of auxiliary background information from participating students and their homes, teachers, and schools. Student background questionnaires solicit information regarding attitudes toward learning, home environment, study and leisure habits, and perceptions of school climate, among a host of other student background domains. Further, PIRLS and, in some countries, PISA include a questionnaire for parents or guardians regarding the home environment. At the teacher level, TIMSS and PIRLS collect information on pedagogical practices, perceptions of teachers' own preparation, and school climate. Finally, all three studies seek information from school principals on school resources, climate, and other school-level issues surrounding teaching and learning. These databases provide an excellent resource for researchers interested in the context and correlates of learning internationally.

Arguably, TIMSS, PIRLS, and PISA are the most widely discussed and used international assessments; however, the field continues to expand and includes (but is not limited to) assessments and surveys such as the International Computer and Information Literacy Study (ICLS); Teacher Education and Development Study in Mathematics (TEDS-M); Civic Education Study (CIVED); International Civic and Citizenship Education Study (ICCS); Programme for the International Assessment of Adult Competencies (PIAAC); and the Teaching and Learning International Survey (TALIS). In fact, Kamens and McNeely (2010) clearly demonstrate that there has been a large expansion of international educational testing and that there is no end in sight.

It is noteworthy that, typically, the data from ILSAs become part of the public domain. Organizations such as IEA and OECD maintain online repositories that allow easy access and provide help with downloading databases and documentation. Although these rich databases are useful and widely available resources for researchers, the complexity and structure of ILSA databases requires that specialized statistical analysis methods are used to appropriately account for these features (Rutkowski et al. 2010). To illustrate, we describe just a few of the complexities inherent in the data. For example, the clustered structure of the data (students nested in classes or schools, schools nested within countries) frequently necessitates a multilevel approach to analyses (see Chapters 18, 19, and 21 in this volume); the nature of the sampling framework often requires that sampling weights are used (see Chapters 6 and 17 in this volume) to produce unbiased parameter estimates; and the method used to estimate achievement demands its own careful treatment by data analysts (see Chapters 7 and 8 in this volume). These and other issues (and unique opportunities) are part and parcel with analyzing these types of data. Specific guidance on developing sensible models and analysis methods is fairly limited in nature and is generally relegated to the pages of the user guides of each study. Further, user guides are generally restricted to describing methods for generating unbiased descriptive statistics and simple tests of mean differences and correlations. It is in the context of this sort of policy and research landscape that the current edited volume is presented, the aims and content of which we subsequently introduce.

Our goal for this edited handbook is to address a gap in the literature about these studies by bringing together recognized scholars in the field of ILSA, behavioral statistics, and policy to develop a detailed guide for ILSA users that goes beyond ILSA user manuals. We acknowledge that this is a difficult task; one that requires a range of views and expertise that no single author can speak to comprehensively. Fortunately, we were able to collect contributions from some of the top researchers in their respective fields to explain (1) why ILSAs are important to educational policy and research; (2) how assessments are created and scaled and the future of these processes; and (3) how to properly analyze ILSA data using cutting-edge statistical methods. We also hope to make available a wide-ranging and instructional resource for analyzing ILSA data. Let us briefly look at an overview of each part in turn.

To highlight the importance of ILSA data in policy and research, the book opens with two important interrelated chapters discussing how ILSAs have been used in policy formation and academic research. The authors of these two chapters, Hans Wagemaker (Chapter 2) and Stephen Heyneman (Chapter 3), provide excellent discussions on the ways in which ILSAs have been used in various countries to influence and change educational policies. These chapters should be of interest to all users of ILSAs, from those that plan to use the data in their research to people working on technical or policy aspects of particular assessments. Wagemaker opens with a short overview of ILSAs and a firsthand account of how ILSAs have been used to influence educational policies around the world. As the executive director of the IEA, Wagemaker is uniquely positioned to provide exclusive insight into the overarching objectives of ILSAs, how and why they have grown over the past 50 years, and their impact on participants, from the system level down to the individual level. He concludes with the telling statement that "ILSAs have reached a stage of maturity that has made them a necessary condition for monitoring and understanding the outcomes of the significant investment that all nations make in education."

Nicely complementing and expanding on Wagemaker's chapter, Stephen Heyneman, a highly respected academic with a vast amount of policy knowledge from his many years working at the World Bank, and his coauthor Bommi Lee, provide a number of insights that should be of interest to both technical and applied researchers alike. Their perspective offers further understanding of how ILSAs have been used in a number of countries, while also discussing a number of strengths, challenges, and possible solutions to our policy makers' never-ending desire for more evidence in short time spans. Similar to Wagemaker, these authors suggest that ILSAs are an important component of our broader understanding of education and how national investment in education is working. The authors share the following in their conclusion: "Though the methods and measures have many caveats and problems, the fact is that these international surveys allow us to inquire as to the product and purposes of education itself and to either challenge or reinforce the reasons for the public investment." Our intent in including these two chapters is to provide the reader with an overview of why international assessments are important, how they are being used, and propose some possible ways forward.

In the two subsequent parts, the contributing authors take an approach of explaining methodologies and analytic procedures based on the specific design features of ILSAs. The level of prerequisites in terms of knowledge about quantitative methods makes these parts appropriate for advanced graduate students in quantitative psychology, educational measurement and statistics and related fields, as well as researchers and policy analysts who have a good grounding in quantitative methods. Section II describes methodological aspects and features of the studies based on operational considerations, analytics, and reporting. In other words, Section II emphasizes such ILSA issues as assessment design (Chapter 4), estimation methods (Chapter 5), sampling (Chapter 6), scale linking (Chapter 7), and other topics related to primary analysis and reporting. In Chapters 8 through 16, the methodological treatments of design features and analytic technologies are broadened to provide an overview of different perspectives on how to utilize the rich background questionnaires that are an important part of ILSAs. Questionnaires, in contrast to the cognitive tests, tap into the "softer" side of educational input and outcome variables. Chapters in this part answer questions such as: What types of variables are collected in the questionnaires and what are their interrelations? How do theoretical considerations guide the analysis of the questionnaire scales? What types of methodologies and assessment innovations are used to improve cross-cultural comparability of the questionnaires?

Section III describes methods that are of interest to the advanced analyst who is interested in more in-depth methods for analyzing released databases that all of the studies make available to the public. These databases reflect the complexity of the studies, and advanced statistical procedures should be used in order to arrive at results that adhere to the highest standards of defensibility.

Section III is likely to appeal to researchers who are looking for guidance on answering research questions by applying sophisticated statistical methods to ILSA data. For Section III, we asked authors to contribute practical insights for managing the complexities of ILSA data while offering an analytic perspective that goes beyond what is normally found in technical documentation and ILSA user guides. In particular, readers will find both a frequentist and Bayesian perspective on applying structural equation models to ILSA data (Rutkowski and Zhou, Chapter 19, and Kaplan and Park,

Chapter 23, respectively). Also to be found in these pages are specific insights into dealing with sampling weights (see Stapleton, Chapter 17) and missing data (Shin, Chapter 20). Throughout Section III, the presented examples also attend to these issues in an applied way. In particular, Anderson, Kim, and Keller's chapter (Chapter 21) on modeling discrete outcomes, and Kim, Anderson, and Keller's chapter (Chapter 18) on multilevel modeling both present innovative approaches for dealing with missing data and sampling weights in ways that complement the chapters dedicated to the topics. Finally, contributing authors have made available all data and syntax for interested researchers to replicate and extend their results. All supplementary materials can be found at the supporting website: http://indiana.edu/~ILSAHB. Despite the distinctions in content and focus between Sections II and III, both provide deep insights into how data from ILSAs are analyzed and reported, and how released data can be used with advanced methods in secondary analyses, regardless of the reader's specific research interest. We imagine that any analyst who hopes to take a model-based approach to understanding some ILSA-related phenomena would be served by understanding the complexities associated with the data. It is likely that measurement experts would find reading about the ways that these data are used after they are produced and reported a worthwhile way to spend time.

ILSAs have come a long way since the FIMS study. They have expanded in both reach and breadth. Technological and statistical advances along with a ubiquitous hunger for more information about the state of our educational systems have made it possible to assess more students, in more countries, more often, and in more subject domains. This rapid growth suggests that there is yet work to be done in the ILSA field from a number of perspectives. Our final contributors to Section III, Rijmen, Jeon, von Davier, and Rabe-Hesketh (Chapter 24), take up one stream of this discussion and offer a promising general and flexible approach for analyzing response data in international (and national) assessments.

Finally, we would be remiss if we did not acknowledge the many hands and minds that made this effort possible. In particular, we would like to thank John Kimmel and Marsha Pronin at CRC Press for their support during this process. We would also like to extend a heartfelt thanks to the many contributors and reviewers who gave their time and mental efforts to bring this volume to fruition. Also, we would like to acknowledge our ILSA ancestors—from the early pioneers that first tapped into the need to internationally compare educational systems to the many researchers who ushered in a new way to measure student achievement on the National Assessment of Educational Progress (NAEP) project in the 1970s and 1980s. We would finally like to thank the numerous researchers who continue to do important and innovative work on large-scale assessment projects such as NAEP, TIMSS, PIRLS, PISA, PIAAC, and others.

References

Asia Society and Council of Chief State School Officers (CCSSO). 2010. *International perspectives on US education policy and practice: What can we learn from high-performing countries?* 1–15. Retrieved from http://asiasociety.org/files/learningwiththeworld.pdf

Darling-Hammond, L. 2011, March 23. US vs. highest-achieving nations in education. *The Washington Post*. Retrieved from http://www.washingtonpost.com/blogs/answer-sheet/post/darling-hammond-us-vs-highest-achieving-nations-in-education/2011/03/22/ABkNeaCB_blog.html

Dillon. 2010, March 10. Many nations passing U.S. in education, expert says. *New York Times*. Retrieved from http://www.nytimes.com/2010/03/10/education/10educ.html?adxnnl=1&adxnnlx=1311174660-bh8MZ3pl/ZrA5XF7+fa0Ag

Finn, C. E. 2010, December 7. *Sputnik for the 21st Century*. Retrieved October 21, 2011, from http://www.educationgadfly.net/flypaper/2010/12/sputnik-for-the-21st-century/

Hanushek, E. A. and Woessmann, L. 2011. *GDP Projections*. Background paper prepared for the Education for All Global Monitoring Report 2012. Paris: UNESCO.

Hopmann, S. and Brinek, G. 2007. Introduction: PISA according to PISA—Does PISA keep what it promises? In: Hopmann, S. T., Brinek, G. and Retzl, M. (Eds.). *PISA According to PISA: Does PISA Keep What It Promises*? Vienna: LIT-Verlag.

Kamens, D. H. and McNeely, C. L. 2010. Globalization and the growth of international educational testing and national assessment. *Comparative Education Review*, 54(1), 5–25.

Medrich, E. A. and Griffith, J. E. 1992. International mathematics and science assessment: What have we learned? *National Center for Educational Statistics: Research and Development Report*, 92(011), 1–136.

Mislevy, R. J. 1995. What can we learn from international assessments. *Educational Evaluation and Policy Analysis*, 17(4), 419–437.

Postlethwaite, T. N. 2005. *What do international assessment studies tell us about the quality of school systems?* Background paper for EFA global monitoring report 2005. Retrieved from http://unesdoc.unesco.org/images/0014/001466/146692e.pdf

Pongratz, L. 2006. Voluntary self-control: Education reform as a governmental strategy. *Education Philosophy and Theory*, 38, 471–482.

Purves, A. C. 1987. The evolution of the IEA: A memoir. *Comparative Education Review*, 31(1), 10–28.

Rizvi, F. and Lingard, B. 2009. *Globalizing Education Policy*. London: Routledge.

Rutkowski, L., Gonzalez, E., Joncas, M., and von Davier, M. 2010. International large-scale assessment data: Recommendations for secondary analysis and reporting. *Educational Researcher*, 39(2), 142–151.

Schwippert, K. (Ed). 2007. *Progress in Reading Literacy: The Impact of PIRLS 2001 in 13 Countries*. Munster: Waxman.

von Davier, M., Gonzalez, E., Kirsch, I., and Yamamoto, K. (Eds.) 2012. *The Role of International Large-Scale Assessments: Perspectives from Technology, Economy, and Educational Research*. New York: Springer.

2

International Large-Scale Assessments: From Research to Policy

Hans Wagemaker

International Association for the Evaluation of Educational Achievement

CONTENTS

Introduction

The origins of ILSA can be traced back to early investigations into student performance conducted by the International Association for the Evaluation of Educational Achievement (IEA). The association began as the result of a meeting of scholars at the United Nations Educational, Scientific, and Cultural Organization (UNESCO) Institute for Education in Hamburg, Germany. The objective of their meeting was to develop a collaborative effort directed at gaining fuller understanding of relationships among the inputs to education and its outcomes. These early collaborators considered that the variety of ways in which countries address the challenges of providing mass education offered a "natural experiment," which, through close and careful study, might help unravel the complexity of input–output relationships not readily detectable within a single education system. They also considered that the resulting revelations could produce powerful insights into possible avenues for educational reform and improvement.

While the expansion of ILSA has been well documented (e.g., Kamens 2009), questions continue to be asked about the extent to which these investigations have an impact on the education systems of the participating countries. The question most often asked is whether ILSA contributes to educational reform and improvement in the respective countries. The asking of such questions is entirely reasonable given that some of the organizations primarily responsible for the conduct of these studies have clearly articulated aims with respect to study outcomes and impact. In this chapter, I consider the question of impact in broad terms as well as in terms of the expressed goals of some of the agencies, particularly IEA and the Organization for Economic Co-operation and Development (OECD), responsible for conducting the majority of the ILSA activities.

The Challenge of Assessing Impact

Numerous analysts and researchers have identified the challenge associated with judging the policy impact of a single piece of research. For example, when reflecting on why information produced by social scientists, particularly in the field of evaluation, was having little apparent impact on policy matters, Caplan (1979) and Wyckoff (2009) noted the "two communities" theory, which holds that social scientists and policymakers live in different worlds, divided by different values, reward systems, and (often) different languages. Weiss (1999) likewise argued that while evaluation—and, by extension, other areas of social science research—has much to offer policymakers, these individuals, because of competing demands, interests, and ideologies, rarely base policy decisions directly on evaluation outcomes.

Judging the impact of any piece of research is clearly not a simple task and often results in a rather pessimistic conclusion as to its efficacy (e.g., Burkhardt and Shoenfeld 2003). A common conclusion is that stakeholders need to recognize that major policy initiatives or reforms are more likely to result from a wide variety of inputs and influences rather than from a single piece of research. Research is also more likely to provide a heuristic for policy intervention or development rather than being directly linked, in a simple linear fashion, to a particular policy intervention. This outcome is due not only to the competing pressures of interests, ideologies, other information, and institutional constraints, but also to the fact that policies take shape over time through the actions of many policymakers.

Given these challenges to judging impact, how can efforts to determine the influence of ILSA be advanced? In this chapter, I argue that while the search for evidence of impact may be guided by considering the programs in terms of the stated objectives of the organizations conducting these studies, it should also consider other indicators of influence. It should, furthermore, take into account broader factors, such as the growth in overall participation in ILSA and the extent to which the information produced as an output of these assessments has entered educational research and policy discourse.

ILSAs and Their Goals

ILSA, as a strategy, has evolved over the last three decades beyond the initial efforts of the IEA. Organizations such as the OECD, UNESCO, and, to a certain extent, PASEC (CONFEMEN Programme for the Analysis of Education Systems; CONFEMEN stands for Conference of Ministers of Education of French-speaking Countries) have all developed assessments that extend the challenges of assessing educational achievement beyond national boundaries. Although the assessments conducted by these organizations share a common focus of measuring achievement outcomes, differences in their objectives and design, briefly outlined in the following descriptions, have significant implications for how they might potentially influence educational policy and the work of the wider research community. These differences also have implications with respect to the way in which these assessments might be expected to influence educational reform and improvement.

International Association for the Evaluation of Educational Achievement

IEA is recognized as the pioneer in the field of cross-national assessments of educational achievement. With more than five decades of research in the field, the association's aims have broadened beyond what was initially a focus on comparing student learning outcomes both within and across countries. Today, the association's aims, as stated on its website (www.iea.nl), are to

- Provide high-quality data that will increase policy-makers' understanding of key school- and non-school-based factors that influence teaching and learning.
- Provide high-quality data that will serve as a resource for identifying areas of concern and action, and for preparing and evaluating educational reforms.

- Develop and improve the capacity of education systems to engage in national strategies for educational monitoring and improvement.
- Contribute to the development of the worldwide community of researchers in educational evaluation.

These aims drive the design and approach that IEA takes when developing and conducting its studies. IEA's Progress in International Reading Literacy Study (PIRLS) and Trends in Mathematics and Science Study (TIMSS), for example, are designed to assess students' learning outcomes after a fixed period of schooling, and are fundamentally concerned with students' opportunity to learn and learning outcomes. The understanding sought, therefore, is the relationship between and across the following:

a. The knowledge, skills, attitudes, and dispositions described in the intended curriculum (i.e., the curriculum dictated by policy);
b. The implemented curriculum (i.e., the curriculum that is taught in schools); and
c. The achieved curriculum (what students learn).

Organization for Economic Cooperation and Development

The OECD's approach to assessing student outcomes is encapsulated in its Program for International Student Assessment (PISA). PISA began in 2000, and every 3 years since, the program has evaluated education systems worldwide by testing the skills and knowledge of 15-year-old students in participating countries/economies. PISA thus "assesses the extent to which students near the end of compulsory education have acquired some of the knowledge and skills essential for full participation in society" (www.pisa.oecd.org/pages/0,3417,en_32252351_32235918_1_1_1_1_1,00.html).

Each PISA cycle covers the domains of reading and *mathematical and scientific literacy* (my emphasis), not merely in terms of mastery of the school curriculum, but also in terms of important knowledge and skills needed in adult life. During the PISA 2003 cycle, the OECD introduced an additional domain—problem solving—to allow for ongoing examination of cross-curriculum competencies. The manner in which PISA evolved during the 2000s led the OECD to explain an apparent rejection of curricular influence by those engaged in shaping the assessments.

> Rather than test whether students have mastered essential knowledge and skills that might have been set out in the curricula that shaped their schooling, the [PISA] assessment focuses on the capacity of 15-year-olds to reflect on and use the skills they have developed.... [PISA furthermore assesses students'] ability to use their knowledge and skill to meet real-life challenges, rather than merely on the extent to which they have mastered a specific school curriculum. (OECD, 2007, p. 6)

The OECD's approach to assessing student outcomes is therefore distinctly different from that of the IEA's (see McGaw (2008) and OECD (2005)). The distinguishing features of PISA are its age-based sampling design and its noncurricular, skills-based focus on assessment. This design largely mitigates opportunities to inform policy at the classroom or curriculum level. As Klemencic (2010) observes, "Within the PISA framework there is nothing that could correspond to, for example, [the] TIMSS curriculum focus. Even more, PISA is not focused on a particular school/grade level; therefore it is difficult to identify the direct impact on national PM [policymaking]…" (p. 57).

United Nations Educational, Scientific, and Cultural Organization

Two organizations affiliated with UNESCO conduct ILSA:

- The Southern and Eastern Africa Consortium for Monitoring Educational Quality (SACMEQ)
- The Latin American Laboratory for Assessment of the Quality of Education (LLECE)

The Southern and Eastern Africa Consortium for Monitoring Educational Quality (SACMEQ) was initiated through the UNESCO International Institute for Educational Planning (IIEP) as an experimental research and training project intended to examine the conditions of schooling and the quality of educational outcomes in the Anglophone Africa region. Modeled on the work of IEA, SACMEQ consists of 15 ministries of education in southern and eastern Africa: Botswana, Kenya, Lesotho, Malawi, Mauritius, Mozambique, Namibia, Seychelles, South Africa, Swaziland, Tanzania, Uganda, Zambia, Zanzibar, and Zimbabwe. SACMEQ's mission is to

> undertake integrated research and training activities that will expand opportunities for educational planners to (a) receive training in the technical skills required to monitor, evaluate, and compare the general conditions of schooling and the quality of basic education; and (b) generate information that can be used by decision-makers to plan and improve the quality of education. (www.sacmeq.org/about.htm)

The consortium also provides opportunities for educational planners to work together in order to learn from one another and to share experience and expertise. The emphasis on capacity building expressed in the consortium's aims reflects one of the key features guiding SACMEQ's evolution over the last two decades and is one that, arguably, may be used in the search for impact.

Similar to the work conducted by IEA, SACMEQ employs a curriculum-focused, grade-based sampling design to assess Grade 6 students' reading and mathematical skills. Since its inception, SACMEQ has concluded three assessment cycles.

LLECE, the laboratory is the quality assessment system network for education in Latin America and the Caribbean. The UNESCO regional bureau, located in Santiago, Chile (Regional Bureau of Education for Latin America and the Caribbean (OREALC)/UNESCO Santiago), coordinates these assessment activities. The laboratory's goals are to

- Produce information on learning outcomes and analyze factors associated with educational achievement.
- Give support and advice to country measurement and evaluation units.
- Serve as a forum to debate, share, and reflect on new approaches to educational evaluation (www.llece.org/public/).

The studies conducted by the LLECE are similar in design to those conducted by the IEA and SACMEQ, but they assess different grade levels—Grades 3 and 6—in the Spanish-speaking education systems of Central and South America and the Caribbean. The subject areas assessed are mathematics, science, and reading. The LLECE conducted its First Regional Comparative and Explanatory Study (PERCE) in 13 countries in 1997. It was followed by the Second Regional Assessment (SERCE) in 16 countries in 2006. The third—and current—regional assessment, TERCE, commenced in 2010 and should be completed in 2013.

While operating on the basis of a somewhat different model, PASEC may also be considered to be one of the current examples of ILSA. As was the case with SACMEQ (the regional assessment for Anglophone Africa), the impetus for starting PASEC was to enhance educational quality in Francophone Africa. Although PASEC, which focuses on mathematics and French in Grades 2 and 5, shares some of the characteristics of other ILSAs, its primary objective is not to compare student achievement across a group of countries during a common period of time and via the same set of instruments. Rather, it is to analyze student achievement and identify key factors relevant to fostering educational quality in the Francophone Africa region on a country-by-country basis, using a more or less common assessment model with a focus on annual growth. PASEC currently acts as the central coordinating agency for this activity in the respective countries, and from 2014 will begin with the first truly multi-national concurrent assessment of mathematics and French in up to 10 countries from the French-speaking regions of Africa.

Measuring Impact

Growth in ILSA as a Measure of Impact

One might argue that the growth of international assessments and demand for participation in international assessments is evidence of the impact of

ILSA. Kamens and McNeely (Kamens, 2009) describe the "explosive growth" of international assessments and national testing, particularly among developed countries, which they attribute to an emerging consensus about not only the legitimacy but also the necessity of assessment. Earlier I argued that any measure of impact must take into account the broader context of the development of ILSA. While ILSAs are now regarded by many as a regular feature of the educational assessment landscape, they remain a relatively recent phenomenon that has evolved in line with increasing demand worldwide for greater accountability for educational inputs.

The history of large-scale international assessments dates back to the early 1960s. However, the report by the U.S. Commission on Excellence in Education (Gardner 1983) marked the beginning of a significant period of expansion for these assessments as described in more detail below. The commissioners expressed concerns in their report about educational performance not only in the United States but also in many OECD countries. Ambach (2011) notes that the meeting of state governors prompted by the release of the report produced a bipartisan consensus on the need for a statement of national goals for education in the United States. These goals, the governors concurred, would focus on improving the performance of elementary and secondary schools and include an expressed expectation that US student performance would exceed student performance in other countries. The governors' aims also tapped into a growing realization across OECD countries that education systems need to operate in a supra-national space, responding to demands to educate a citizenry capable of competing in a highly competitive, rapidly changing, globalized social, economic, and political world.

According to Tuijnman and Postlethwaite (1994), the Commission on Excellence Report, which had a high degree of visibility internationally, also marked a clear move away from expanding education systems in terms of enrollment and toward a more systematic expansion of national monitoring and questions of educational excellence, equity, and efficiency. Early reform efforts in education traditionally focus on concerns related to inputs and the challenges of ensuring equity in terms of school enrolments. Even today, under the aegis of Education for All, the Millennium Development Goals, and universal primary education, enrollment in education remains the central goal for developing nations. The focus on educational quality in terms of what students know and can do after a fixed period of schooling, and how these outcomes might relate to societal and curricular expectations, is a much more recent occurrence and has yet to achieve primary status in the development aspirations of many nations. For such countries, participation in ILSA can mark their transition to focusing on educational achievement couched in these terms.

These changes since the time of the report have also been both a product of and a reason for an increase in the numbers of countries participating in the various ILSAs, and the considerable growth in investment from donor agencies in support of both countries and organizations such as IEA, UNESCO, and the OECD. An example of this growth is evident in the participation in IEA's studies of mathematics and science. Since its first large-scale international assessment of mathematics, which dates back to IEA's Pilot 12 Country Study conducted in the early 1960s, the association has carried out more than 26 studies examining differences in student performance in and across the participating countries (Mullis et al. 2004). This same pattern of growth has been evident in OECD's PISA, launched in the late 1990s. Although the participating OECD member states continue to determine the design and content of the PISA assessments of 15-year-olds' achievement in mathematics, science, and reading, the program now attracts participants and interest beyond the original OECD membership (OECD 2006).

Today, the work of IEA and, in particular, its TIMSS and PIRLS assessments, along with OECD's PISA, are characterized by participation that is truly worldwide. The degree of interest in cross-national comparisons evidenced by this expansion has precipitated other cross-national assessment initiatives. The aforementioned PASEC, LLECE, and SACMEQ are obvious examples. PASEC was launched in 1991, and since then more than 15 individual country assessments have been carried out in Francophone Sub-Saharan Africa under the auspices of a secretariat in Dakar (CONFEMEN 2012). SACMEQ began in the early 1990s as a consortium for 15 ministers of education from countries in Southern and Eastern Africa, initially supported by the UNESCO IIEP. It has since developed into a standalone intergovernmental organization. Chile's collaboration in the mid-1990s with Latin American and Caribbean countries led to the establishment of LLECE, which acts as the umbrella agency charged with managing various assessments of education in the region.

As the number of organizations offering cross-national assessments has expanded, so too has participation in assessment activities. Participation in IEA studies increased from the 12 countries that participated in the first IEA assessment* to 20 systems for the Second International Mathematics Study (SIMS), 40 systems for TIMSS 1999, 66 systems for TIMSS 2007, and 79 participants for the TIMSS 2011 assessment, which includes a number of benchmarking US states and other subnational systems such as Dubai, one of the United Arab Emirates nations. Participation in the OECD PISA assessments has expanded beyond the members of the OECD countries to encompass 43 countries in 2000 and an expected 68 participants for the 2012 round.† What is also evident from an examination of participants in both the IEA and OECD assessments is the expansion beyond the developed nations of the world to less developed economies, as well as the growth of subnational

* Participation in IEA studies includes "benchmarking" participants or subnational administrative regions, including participants such as the Canadian provinces, U.S. states, and special administrative regions such as Hong Kong.

† This count includes subnational entities such as Macao and Dubai that are participating in their own right.

entities such as US states, which choose to benchmark their educational performance against the standards set by the rest of world.

Other developments related to ILSA include their extension into other areas of investigation related to education. In 2010, IEA, through the Teacher Education Development Study-Mathematics (TEDS-M), conducted a major investigation of the preparation (preservice training) of mathematics teachers in primary and lower-secondary schools in 17 countries (Tatto et al. 2012). TEDS-M explored such things as future teachers' content knowledge, pedagogical knowledge, and attitudes in the area of primary and lower secondary mathematics.

The OECD recently reprised interest in assessing adult competencies through its Program for the International Assessment of Adult Competencies (PIAAC), conducted across 25 OECD member countries. This investigation targets the adult population's (aged 15–65) literacy and basic numeracy skills.

Finally, the OECD now regularly surveys teachers at the secondary school level through its Teaching and Learning International Survey (TALIS), which focuses on issues surrounding the teaching and learning environment in schools and teachers' working conditions. Studies such as PIAAC, TEDS-M, and TALIS have taken ILSA outside their traditional domain of primary and secondary classrooms, and in so doing have expended their sphere of influence among policymakers and researchers alike.

The continued interest in ILSA as evidenced by the growth in participation, both quantitatively and qualitatively, suggests the emergence of a new culture of assessment. This culture is one that focuses on understanding educational outcomes in an increasingly global context, where the search for educational excellence is shaped not only by understanding the limits and possibilities of education as defined by the globalized educational community, but also by understanding the contexts within which the achievement of a country's economic competitors take place.

The concern with, and recognition of, a more globalized educational community is also shaping the shift from a tactical (one-off) approach to participation in ILSA to a more strategic investment that sees countries participating on a more regular basis in these assessment activities. In many cases, countries choose to participate in several different assessments because of the different perspectives that each provides. A majority of the OECD countries participate in both the IEA and OECD assessments, for example. Lockheed (2011) notes that a country's decision to participate in one ILSA is strongly related to its decision to participate in subsequent large-scale assessments. She gives, as an example, the fact that 76% of 17 developing countries participating in PISA 2006 had previously participated in at least one IEA study.

Discourse as a Measure of Impact

Evidence of ILSA having a marked impact on shaping discourse about education, particularly in terms of educational quality, is mounting. The work of IEA, perhaps more than the work of any other organization, has stimulated public and political concern about educational outcomes within the context of educational quality. A current Internet search for references to the TIMSS is likely to identify around 5.21 million listings. A search for listings on PIRLS will bring up some 3.25 million listings, and PISA typically gives around 4.37 million listings. These citations also reference more than 30 education systems, numerous professional organizations, governmental and nongovernmental bodies, and interested professionals.

The extent to which information resulting from IEA studies has entered public domain is prima facie evidence of impact not only in terms of its currency in the media and among the general public, but also in terms of the political debate that successive releases of IEA results have generated in many countries. Early in the history of the TIMSS assessments, results that were unpalatable were, as was in the case of South Africa, greeted by "outrage" (Howie 2011). The disquiet about educational performance in Israel was reflected in the following headline in an issue of *Education Week*: "Down in Rankings: Israel Seeks Changes in Education" (Goldstein 2004). In 2004, Alan Greenspan, then Chairman of the US Federal Reserve, used information from TIMSS in his address to the Committee on Education and the Workforce of the US House of Representatives. During his address Greenspan (2004) expressed concern about educational underachievement among sectors of the US population and the likely detrimental impact that this was having and would continue to have on the US national economy.

The political turmoil that can be stimulated by the release of educational achievement results is not, of course, limited to the work of the IEA (see, e.g., Figazzolo n.d.). Lehmann (2011) describes the "TIMSS shock" that occurred after release of the TIMSS resul-ts in Germany in the winter of 1996/1997. Evidence of poor mathematics and science performance and social disparity among the country's fourth- and eighth-graders caused considerable consternation. This event was followed several years later by the "PISA shock," which similarly raised concerns about educational achievement levels among German 15-year-olds. Gruber (2006) argues that the PISA shock, which occurred as the consequence of Germany scoring below the national average in the 2000 PISA survey, had much the same impact on educational discourse as the launch of the Russian satellite *Sputnik* and the publication of the *Nation at Risk* report had in the United States. According to one account, Gruber reports that the "Bundestag [Germany's federal parliament] ran a special 'PISA' session." He also observes that the event is now seen as demarcating a new era in German education known as "AP"—After PISA. At a policy level, PISA triggered significant educational reforms in Germany (Ertl, 2006), including a move toward implementation of national standards.

The impact of PISA was not confined to the countries of the OECD. As Shamatov (2010) notes, the PISA results "provided solid evidence of the terrible state of secondary education in Kyrgyzstan" (p. 156). More specifically, the study identified a shortage of school resources and highlighted issues of equity related to access. The fact that the results were disputed in some quarters precipitated a national sample-based assessment (NSBA) that served to validate the PISA results. A similar situation occurred in Denmark (Egelund 2008). There, serious questions were raised about the mediocre performance of Danish students despite the education system being relatively well funded.

Regional assessments such as SACMEQ have also stimulated significant policy debate. Murimba (2005) notes, "Results that were 'unexpected' frequently ignited serious policy debates" (p. 101). In both Mauritius and the Seychelles, heated national debates erupted when the extent of grade repetition and the use of private tutors were exposed. In Senegal, presentation of the first results to stakeholders, including the Minister of Education and cabinet members, engendered considerable discussion about grade repetition (Bernard and Michaelowa 2006).

Changes in Educational Policy as a Measure of Impact

As I noted earlier, initiatives in many countries directed at improving educational provisions have traditionally focused on inputs to education, particularly school resources. Quality was—and still is—often judged on the basis of input measures such as per-student expenditure on education. ILSAs attempt to capture these background (input) variables in order to aid understanding of how such factors relate to educational outcomes. The studies conducted by IEA, SACMEQ, LLECE, and PASEC, in particular, also attempt to secure information on factors relating to curriculum and instruction and their association with educational achievement. Interest in these factors tends to be high because they are deemed capable of policy intervention. There is compelling evidence that the insights provided by analysis of these data have influenced curricular and instructional reforms in many countries (Aggarwala 2004; Elley 2002; Mullis and Martin 2007; Robitaille 2000; Wagemaker 2011). There is also evidence that more and more countries are using TIMSS and PIRLS results as an integral means of informing their respective educational reform and improvement strategies. In Germany, for example, where participation in ILSA is mandated by law, Lehmann (2011) describes ILSAs such as TIMSS, PIRLS, and PISA as "a necessary condition for policy making" (p. 422).

Several evaluations of the impact of IEA studies (Aggarwala 2004; Elley 2002; Gilmore 2005; Lockheed 2009; Robitaille 2000; Schwippert 2007; Schwippert and Lenkeit 2012) have identified marked impacts on educational policy and reform. Wagemaker (2011, p. 261), in his review of impact, noted that

> In countries as different as Iceland, Kuwait, New Zealand, Norway, Romania and South Africa, TIMSS (1995) served as a catalyst for curricular review and change. In Iceland, the information collected during the TIMSS study resulted in a recommendation for increased teaching hours for mathematics and science instruction at the elementary level.

> In New Zealand, the TIMSS (1995) results precipitated the establishment of a taskforce with a focus on mathematics and science education.

Poon (2012) records the ongoing influence of ILSA in Singapore, where TIMSS and PIRLS results, along with other measures of performance, are used as international benchmarks that contribute to analysis of the data arising out of the nation's annual curriculum review over 6 years. Targeted secondary analysis of these data help guide the development of teaching resources that may be used, for example, to enhance student cognitive and meta-cognitive skills.

In Hong Kong, where students consistently achieve high average results on ILSA achievement scales for science and mathematics, a series of talks and workshops were used to disseminate reading results from PIRLS 2001 to schools. These activities were stimulated by the relatively average reading performance among Hong Kong students despite serious efforts since 1994 to improve language instruction. Ensuing political debates and dissemination strategies were augmented by the adoption of the PIRLS framework for the Chinese reading comprehension examinations in 2004. In 2006, Hong Kong was among the top-performing countries in PIRLS (Leung 2011).

Australia had a similar experience. As reported by Ainley et al. (2011), "...policy issues raised by IEA in its cross-national studies also applied to the making of educational policy at both the state and national level across Australia" (p. 330). More particularly, compensatory programs were introduced in Australia to address concerns arising out of significant gender and socioeconomic effects on student achievement in IEA studies (pp. 332–333). TIMSS 1995 in Australia was considered to have broken new ground in term of its scope, material, curriculum analysis, and the contextual information that was collected (Ainley et al. 2011).

In his recent review of the impact of PISA on policy, Breakspear (2012) concludes that PISA played a key role in monitoring educational standards and issues related to equity and accountability in a number of OECD and non-OECD countries. Breakspear considers the PISA frameworks, as well as other international assessment frameworks (such as those developed by the IEA), that have been particularly influential in terms of adaptation of assessment practices and curriculum standards determined at national and federal levels. Shamatov (2010) records a number of reforms arising out of the lessons of PISA 2006. These include the development of new standards and curriculums, a reduction of education load per teacher, the modernization of school infrastructure and equipment, improvements to teaching standards and performance, and the introduction of per-capita financing.

SACMEQ results have also served as an impetus in some of the participating countries to develop and implement policy focused on improving educational quality (Murimba 2005). In Kenya, for example, the results from the first SACMEQ assessment were used to monitor the impact of free universal primary education, particularly with respect to reassuring stakeholders that the quality of outcomes was being maintained.

Questions related to impact tend to reference TIMSS, PIRLS, and PISA. However, other IEA studies, including the IEA's Second Information Technology in Education Study Module 1 (SITES-M1) and the Civic Education Study (CIVED), have also played an important role in the educational policy debate. The SITES-M1 results highlighted several key policy issues when they were released (Plomp et al. 2009). The first major concern was that although a good number of participating countries had successfully managed to bring sufficient numbers of computers into schools, many teachers were ill equipped to use the technology as a pedagogical tool. As significant, perhaps, was the finding that despite expectations of computers transforming pedagogy and curriculum delivery, this transformation had not occurred. In Australia, for example, Ainley and colleagues (2011) reported that SITES exposed the need to enhance the use of technology for instructional purposes. The overall investigation across the participating countries (Ainley et al. 2011), also identified what was to become an increasing area of concern for educational institutions, namely the challenges of protecting young children from accessing inappropriate materials.

Conducted during a time of significant political reform and turmoil in Eastern and Central Europe, the IEA CIVED addressed key concerns related to the processes of civic education and education for citizenship in participatory democracies. Among its findings, this study revealed important differences among and within countries in terms of knowledge about and understanding of what is meant by democracy, student attitudes toward nation and government, and immigrants' and women's political rights. According to Cariola et al. (2011), CIVED played a role in supporting curricular reforms in civic and citizenship education in Chile after the country had elected a democratic government.

Curriculum Change as a Measure of Impact

With their curriculum focus, the IEA studies such as TIMSS and PIRLS along with the SACMEQ assessments, in particular, are designed not only to describe differences in student achievement within and across countries but also to address questions as to why such differences might exist, especially as they relate to curriculum and teaching practice. Even among the countries that achieve the highest scores on the international achievement scales, ILSAs appear to have had a substantial influence on the ways in which countries manage the process of educational reform and improvement. Singapore, for example, regularly analyzes its TIMSS results in order to identify specific student strengths and weaknesses. The findings of the analyses also serve as one of several references informing national curriculum reviews. Singapore furthermore uses the contextual variation within which TIMSS items are presented to gain insight into students' ability to apply their learning in the variety of contexts demanded by the Singapore curriculum.

When describing the longstanding involvement of the Russian Federation in ILSA, Kovaleva (2011) reported that assessments such as TIMSS and PIRLS have played a role in capacity building and developing educational standards in Russia. For example, specialists involved in standards development, having analyzed the TIMSS and PIRLS frameworks, developed recommendations for the new state educational standards for primary schools, which were introduced in 2011.

Klemencic (2010) describes the impact that TIMSS 1995 had on the syllabus plans for students in their ninth year of schooling in Slovenia. She also points out that the TIMSS 1999 results for Slovenia made clear that, relative to other countries, Slovenian teachers were not as burdened as they had believed they were. Pelgrum et al. (2011) state that, in Latvia, analysis of TIMSS 1995 findings identified gaps in skills related to application of mathematical knowledge in real-world situations, a finding that led to curricular change. The authors also note that Iran used the TIMSS curriculum framework to develop items for an assessment item bank for primary education.

In Chile, according to Cariola et al. (2011), the TIMSS 1999 results identified a curricular imbalance in the teaching of mathematics and science. This finding was instrumental, along with the results of the subsequent TIMSS study, in prompting curricular reform and the introduction of content standards in 2009. These changes were accompanied by the beginnings of reforms to the teacher education system, which had also been found wanting after release of the TIMSS 1995 results.

The potential for ILSA to impact national curriculum is not limited to PIRLS and TIMSS. Klemencic (2010), for example, states that although the IEA CIVED 1999 study appeared to have limited impact on national policymaking in Slovenia, the converse was true for the IEA ICCS 2009 survey. She predicted that this study would have a vigorous impact on national policymaking, given that the syllabus for Grades 7 and 8 were being revised and given that the Slovenian national research coordinator for ICCS had been appointed as the president of the national commission responsible for reviewing the syllabus.

Curricular impact is not confined to the highest-achieving countries. Qatar, which first participated in PIRLS 2006 and then in TIMSS 2007, recognized that the benefits of participation were not likely to be fully realized until future decades. However, the country's recently developed strategic plan was informed by insights gained from the secondary analysis of PIRLS and TIMSS data. The international standards provided by these assessments were used not only for curricular reforms but also as the basis of workshops for subject matter coordinators. The TIMSS and PIRLS frameworks in particular were used as a means of identifying the adequacy of the national curricula for mathematics, science, and reading.

PISA, too, has had some impact on curriculum. In Japan, one of the high-achieving countries in PISA, the findings of PISA 2009 stimulated a curriculum revision that included provision for PISA-type assessment tasks. According to Nakajima (2010), despite Japan's regular strong performance on ILSA, the country used TIMSS and PISA results to justify increasing the number of annual instructional days for schools and the amount of time given over to instruction.

Changes in Teaching as a Measure of Impact

The impact that ILSAs have beyond the classroom curriculum often extends to the preparation of teachers. Numerous examples now exist of countries using ILSA findings to make changes to their teaching and instruction policy and practice. The release of the TIMSS 1995 findings, for example, spotlighted issues associated with teaching in Australia, Canada, Spain, and Japan. In the Philippines, the government resolved to develop a program to train 100,000 teachers, while in England, teacher guidance materials were prepared, as were instructional materials in the province of Ontario, Canada (Robitaille 2000; Wagemaker 2010).

In her description of the influence of PIRLS 2006 on teaching in South Africa, Howie (2011) states that the PIRLS 2006 results spearheaded a learning campaign and a national reading strategy policy, implementation of which included a teachers' handbook, a teachers' toolkit, and parallel resources. Howie also observes that findings from IEA studies informed a national policy framework for teacher education and development. In Slovenia, Klemencic (2010) argued that TIMSS 2003 provided "expressive" sources for teacher training and professional development. Further, Klemencic notes, in particular, an increased focus in the primary school science curriculum on learning using experiments (problem solving), and material from the TIMSS assessment items contributing to new textbooks for primary school mathematics. In Singapore, insights gained from secondary analysis of TIMSS and PIRLS data were brought into the professional development of subject-matter specialists. The results also guided the development of teacher-training materials as well as resources directed at enhancing students' meta-cognition and self-monitoring skills (IEA 2012).

Capacity Building and Research Endeavors as a Measure of Impact

ILSAs, particularly TIMSS, PIRLS, SACMEC, and LLECE, need to be understood as both process and product. The agencies responsible for developing and coordinating these initiatives spend much time and effort to ensure that participants benefit from their experience of participation not only in terms of study outcomes but also in terms of the training they receive. The impact that this emphasis has on countries takes several forms.

Brassoi and Kadar-Fulop (2012), for example, point out that IEA studies inspired the development of a program of national assessments in Hungary. Macedonia, lacking experience of and capacity for conducting national assessments of student achievement before its first participation in TIMSS, was able, after the study, to develop a national assessment system, the nature of which was significantly influenced by the TIMSS design and methodologies. Describing Singapore's involvement in IEA studies and their impact, Dr. Chew Leng Poon (IEA 2012) recounted that the PIRLS and TIMSS studies gave specialists in curriculum, research, and evaluation considerable exposure to best practices through their involvement in review of the assessment frameworks, item development, scoring training, and data analysis. Furthermore, at Singapore's National Institute of Education and its Ministry of Education, PIRLS data showing, for example, changes in students' reading habits and attitudes stimulated new research that helped Singaporean policymakers better understand the impact of reading and library programs on reading behavior. In summarizing Singapore's experience, Dr. Chew Leng Poon concluded:

> The PIRLS and TIMSS findings are used in Singapore in a variety of ways to inform policy development, as well as enhance teaching and learning practices. ...The data collected over the three decades of participation have provided useful insights to inform policy decisions, research, and educational practice. (IEA 2012, p. 4)

In a similar account (IEA 2011), Dr. Galina Kovaleva, the director of the Center for Evaluating the Quality of Education in the Russian Academy of Education, noted that Russia was using TIMSS and PIRLS databases in seminars designed to train regional coordinators and specialists in educational measurement and data analysis. The two databases were also included, as a fundamental resource, in two new Master's degree programs established in 2009 (one at the Higher School of Economics and the other at the Academy of National Economy). Both programs require students to analyze data from the databases.

The enhancement of local capacity is also a fundamental concern for SACMEQ, with participating countries giving this outcome high priority. Every country, according to Murimba (2005), reported positively of the benefits of SACMEQ capacity-building initiatives in areas such as sampling, data collection, and data analysis. Countries also identified improved information systems (see also Saito and Capele, 2010) and a strengthening of policy-making processes as positive outcomes of participation: "In several countries, SACMEQ results served as input to key documents that were linked to developmental planning" (Murimba 2005, p. 99).

The influence that ILSAs have had on educational capacity-building and research endeavors in general in the participating countries' education systems can also be seen at the institutional level. In Australia, for example, Ainley and colleagues (2011, p. 365) attribute transformation of the Australian

Council for Educational Research (ACER) into a large and highly respected organization to "the involvement of the ACER in the wide range of IEA's studies and activities in educational research and the evaluation of educational achievement across a growing number of countries of the world." Moreover, in line with the research-driven interests of IEA and SACMEQ, in particular, ILSAs continue to have a substantial impact in the area of research and assessment methodology. Lockheed (2011, p. 12) reminds us that capacity building and development of expertise in the area of measurement and assessment is one of the factors currently motivating developing countries to take up participation in these studies.

In summarizing the impact of ILSA in developing countries, both Aggarwala (2004) and Elley (2002) offer conclusions on the impact TIMSS has had on educational capacity building in the developing nations that participate in TIMSS. According to Aggarwala (2004), the "comprehensive impact" of TIMSS led to

- Meaningful educational reforms in curriculum, evaluation and assessment standards, and tests and examinations, particularly in Egypt
- Reforms/changes in curriculum, testing, and scoring methodologies have been instituted
- Heightened awareness of the need for reforms in such areas as teaching methodologies, teachers' training, and licensing
- Adoption of new test formats and scoring methodologies
- Recognition of the urgent need for shifting the emphasis from rote learning to application of knowledge
- A renewed awareness of the essential need for educational reforms as a building block for a knowledge-based society (p. 4)

In her description of South Africa's experiences in ILSA, Howie (2011) states that prior to participating in IEA studies such as TIMSS and PIRLS, South Africa had very little objective data on educational outcomes. Participation in these studies and in the SACMEQ assessments helped South Africa develop local research capacity in assessment. Saito and Capele (2010) report similar positive outcomes for research capacity in Malawi from that country's participation in SACMEQ.

While ILSAs are now well entrenched in the educational assessment landscape of a great many countries and are therefore somewhat taken for granted, their evolution and expansion represents a significant development in terms of their contribution to research and measurement methodologies. The IEA studies such as TIMSS and PIRLS and, more recently, OECD's PISA are based on methodologies originally developed for the US National Assessment of Educational Progress (NAEP). The process of extending this model to allow comparison of educational achievement both within and across an ever larger number and an increasingly diverse range of countries has posed several major challenges.

For example, concern relating to the consequential validity of conducting these assessments has demanded adherence to the highest technical standards possible to ensure reliability and validity. Although ILSAs are considered to be low-stakes assessments in terms of the potential impact on individual life chances, they are potentially high risk with respect to the decisions countries make as a consequence of analysis of their findings. The IEA, in response to these concerns, published technical standards (Martin et al. 1999) that outline minimum technical requirements for the association's assessments. These standards and the published technical reports that accompany the release of each assessment cycle (see, e.g., Martin et al. 2007; Olson et al. 2007) provide those who analyze or read accounts of the data with the guidance necessary to appropriately interpret that information. The IEA technical standards and technical reports also make explicit the latest techniques for managing the challenges associated with developing sampling designs that meet technical demands for excellence, as well as the particular requirements driven by individual or groups of countries wanting to use the studies to collect data on matters of specific educational interest.

An examination of the TIMSS 2011 reports also reveals the challenge of working in linguistically diverse environments experienced by those designing and implementing ILSA. Unlike the SACMEQ and SERCE assessments, which operate in one language, the assessments conducted by IEA and OECD assess students in their language of instruction. Several countries, for example, South Africa and Spain, implement the assessments in several languages (11 for South Africa and 5 for Spain). Both IEA and OECD have had to develop methodologies robust enough to provide not only the highest-quality translations possible, but also translations that ensure comparability of test items.

One of the most significant endeavors to date to lessen the complexity of the environment in which ILSAs operate has been the use of digital technologies to deliver some of these assessments. Some background data for many studies are now routinely captured electronically. OECD's PISA and PIAAC, and, more recently, IEA's International Computer and Information Literacy Study (ICILS) have used or are using computer-based assessment technologies to deliver assessments in countries where infrastructure permits. The development needed to achieve this goal has been substantial in terms of overcoming operational challenges as well as challenges associated with translation into digital formats and with the technology itself.

The notion that educational reform and improvement rather than "assessment or the test *per se*" are the goals for ILSA utilization creates the imperative to ensure that the data gathered are readily accessible and used. Both IEA and the OECD encourage and facilitate analysis of ILSA data through targeted research conferences and access to online data analysis tools. Although training in data analysis is provided as part of the process of conducting ILSA, IEA in particular has taken a number of additional steps to ensure that the data from complex studies are not simply examined via commercially available statistical tools, but are accessible to a wide body of researchers and policy analysts through the International Database Analyzer. This data analysis tool, designed by the IEA, takes into account the complex sampling designs of and estimation techniques for both IEA studies and PISA. The electronic data analysis tools such as the Analyzer that are now available also help avoid serious calculation errors.

Use of these tools has been further supported through a regular series of training seminars offered by sponsoring organizations such as the US National Center for Educational Statistics and the World Bank. IEA, in keeping with its mission statement to develop the community of researchers and develop local capacity, committed to a strategy that included establishment of the IEA-ETS* Research Institute (IERI). In addition to providing training in both basic and advanced analytical techniques related to ILSA, IERI publishes an annual monograph called *Issues and Methodologies in Large-Scale Assessments*, which provides both a stimulus and a forum for research related to ILSA. The collaborative work conducted by IEA and ETS through IERI links into IEA's biennial international research conference (IRC), which offers an academic forum for the analysis, discussion, and dissemination of data related to all IEA studies.

ILSAs have contributed not only to the methodology and technology of large-scale assessments, but also to educational theory. Model building and testing have been features of the work of IEA since its inception (for a full description, see Keeves, 2011 and Keeves and Leitz, 2011). Over the course of its operation, the association has modeled theories of school/student learning in an attempt to enhance understanding of the impact of schooling on student achievement. A key feature of these efforts has been effort to move discourse about the quality of education from the input side to the output side of the equation, thus making learning outcomes the central focus of the assessments.

This process has been accompanied by attempts to disentangle the effects of the broader social context of student home-background experiences, the school environment and instructional practices, and their relationship to achievement. As Keeves and Leitz (2011) write:

> The theoretical perspectives of the IEA research studies during the early years of operation gave rise to six issues that involved the development of new approaches to inquiry in the field of education in a cross-national context.
>
> 1. There were issues concerned with the framing of specific hypotheses and the making of a decision to endorse or reject the hypothesis or, alternatively, to develop a more complex model that described and explained events occurring in the real world, together with the testing of the model.

* Educational Testing Service, Princeton, New Jersey, USA.

2. There were issues concerned with the collection of data from a sample that had a complex design, as well as the estimation of the associated parameters with their appropriate standard errors.
3. From the use of many observed variables that were interrelated in order to examine their influence on an identified outcome, the need arose to form latent variables that combined the effects of the observed variables, as well as to examine the indirect effects of the latent variables that operated through other latent variables.
4. Issues arose in the examination of variables from the situation that within most analyses of data in educational research studies, some cases under survey were nested within other cases, and the data had a multilevel structure.
5. In educational research there was primary concern for learning and development that involved the investigation of change over time. The examination of data in this context not only required the use of time series, longitudinal, or trend data, but also required that the data were measured on an interval scale across an extended scale that was not truncated at either the higher or lower ends.
6. The investigation of change over time required the use of data measured on an interval scale that was not truncated at the higher and lower ends. Such scales of measurement were based on count data or categorical data and employed Rasch scaling procedures. (p. 10)

Perhaps the most enduring feature of the IEA's work, attributable particularly to its early scholars, is the model of curriculum implementation depicted in broad terms in Figure 2.1. The feature that underpins this model

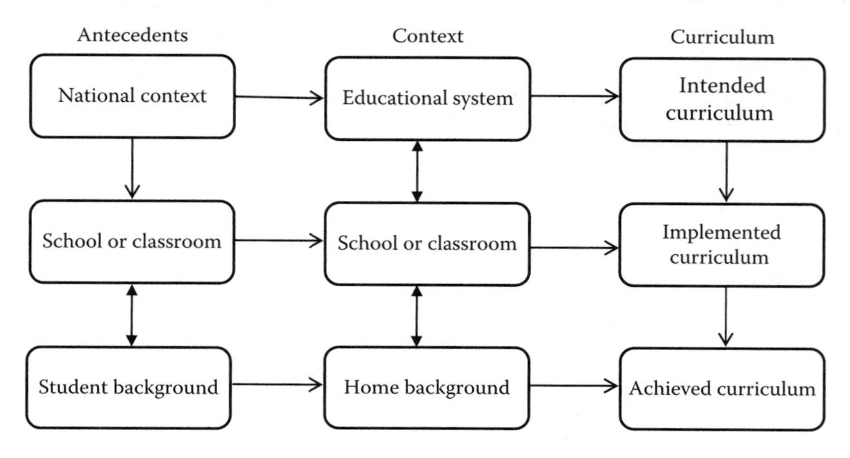

FIGURE 2.1
IEA general analytical model.

is the central component of much of the IEA's work and, in particular, the design, development, and implementation of its studies. That feature is the notion of opportunity to learn.

The three-level model of curriculum illustrated in Figure 2.1 specifies the intended curriculum (that which is dictated by policy), the implemented curriculum (what is taught), and the achieved curriculum (what students know and can do). The IEA's national background questionnaires, together with questionnaires directed at school principals, classroom teachers, students, and, in the case of PIRLS (and TIMSS in 2011), parents attempt to secure information related to the policy context, school resourcing and instructional practice, and student home backgrounds, respectively. This model is now deeply imbedded in the IEA's research design and in many other research investigations.

Global and Donor Responses as a Measure of Impact

One criticism that might be leveled by the educational reforms stimulated by the Millennium Development Goals in many countries around the world is the emphasis on achieving universal primary education at the expense of attention to a concern about educational quality as measured by educational achievement. This lack of attention to the quality of educational outcomes is somewhat surprising given the accelerating growth in participation in ILSA and the accompanying concerns about educational achievement.

While those who developed the Millennium Development Goals and Education for All were largely silent on the matter of educational outcomes, the World Bank, driven by information and accountability concerns, was an early advocate of a focus on educational quality and educational outcomes. This focus was most recently restated in its education sector strategy discussed in *Learning for All: Investing in People's Knowledge and Skills to Promote Development* (World Bank 2011a). What is evident in this document, as well as in the education strategies of major donor/development organizations such as the US Agency for International Development (USAID) and the UK's Department for International Development (DIFID), is their reliance on evidence from ILSAs such as PIRLS, TIMSS, and PISA to support the argument for a greater emphasis on learning outcomes and their exhortation for a larger number of developing countries to participate in ILSA. A similar emphasis on outcomes can be seen in the bilateral negotiations for Fast Track Initiative Countries under the auspices of the Global Partnership for Education (www.educationfasttrack.org/).

ILSAs have also attracted the interest of developmental economists (see, e.g., Lockheed 2011). This interest has undoubtedly helped secure funding for these assessments from donor organizations, including (among others) the World Bank, the United Nations Development Program, and the Inter-American

Development Bank. These organizations have also financially supported the participation of less-developed countries in ILSA.

The influence of ILSA on the policies and practice of development organizations such as the World Bank is nowhere more evident than in the recently established Arab Regional Agenda on Improving Education Quality (ARAIEQ), an organization committed to addressing educational quality in the Middle East and North Africa (MENA) region (World Bank, 2011b). Data from both PISA and, in particular, TIMSS, are heavily implicated in the proposal to establish a series of regional networks that have, as their primary focus, improving educational outcomes in the MENA region. This support and the ongoing work of organizations such as the IEA, OECD, UNESCO, and PASEC serve to ensure that concerns about educational quality remain at the forefront of the thinking of educational policymakers, planners, and researchers.

Conclusion

As nations and other bodies responsible for administering education systems have begun or continue to pursue educational excellence and to recognize that they now operate in a supra-national space, they have increasingly relied on ILSAs as one of the necessary sources of information on educational quality and its antecedents. There is now considerable evidence that participation in and the results of studies conducted by organizations such as the IEA, OECD, UNESCO, and PASEC provide powerful insights into the impact of policies and practices on student achievement in key subject-matter areas. The public discourse on educational outcomes, accountability, and equity that has been stimulated by the release and further analysis of ILSA data has been instrumental in surfacing fundamental issues about educational quality at both national and international levels.

Many factors, such as curriculum, instruction, instructional time, school resources, teacher preparation, and professional development are all amenable to policy intervention. For example, efforts by the research teams conducting PIRLS and, more recently, TIMSS 2011 to understand the role that parents play in the acquisition of early reading and mathematical skills are not only providing further insight into factors associated with learning differences among students, but also offering a potential source of information for new educational policy initiatives focused on the home. In addition, the expansion of ILSA beyond the boundaries of compulsory schooling is likely to provide additional insights for policymakers searching for ways to improve educational outcomes. Today, ILSAs have reached a stage of maturity that has made them a necessary condition for monitoring and understanding the outcomes of the significant investment that all nations make in education.

References

Aggarwala, N. 2004. *Quality Assessment of Primary and Middle Education in Mathematics and Science (TIMSS).* Takoma Park, MD: Eaton-Batson International.

Ainley, J., Keeves, J., Lokan, J. et al. 2011. The contribution of IEA research studies to Australian education. In: *IEA 1958–2008: 50 Years of Experiences and Memories*, ed. C. Papanastasiou, T. Plomp, and E. Papanstasiou, 317–372. Nicosia: Cultural Center of the Kykkos Monastery.

Ambach, G. 2011. The IEA of 1990–2001: A 50th year reflection. In: *IEA 1958–2008: 50 Years of Experiences and Memories*, ed. C. Papanastasiou, T. Plomp and E. Papanstasiou, 469–489. Nicosia: Cultural Center of the Kykkos Monastery.

Bernard, J.-M. and Michaelowa, K. 2006. Managing the impact of PASEC projects in Francophone Sub-Saharan Africa. In: *Cross-National Studies of the Quality of Education: Planning Their Design and Managing Their Impact*, eds. K. Ross and I. Jürgen-Genevois, 229–240. Paris: UNESCO/IIEP.

Brassoi, S. and Kadar-Fulop, J. 2012. How IEA influenced the education system in Hungary. In: *IEA 1958–2008: 50 Years of Experiences and Memories* eds. C. Papanastasiou, T. Plomp, and E. Papanstasiou, 431–446. Nicosia: Cultural Center of the Kykkos Monastery.

Breakspear, S. 2012. *OECD Education Working Papers No. 71: The Policy Impact of PISA.* Paris: OECD.

Burkhardt, H. and Shoenfeld, A. H. 2003. Improving educational research: toward a more useful, more influential, and better funded enterprise, *Educational Researcher* 32(9): 3–14.

Caplan, N. 1979. The two-communities theory of knowledge utilization. *American Behavioral Scientist 22*(3): 459–470.

Cariola, L., Covacevich, C., Gubler, J. et al. 2011. Chilean participation in IEA studies. In: *IEA 1958–2008: 50 Years of Experiences and Memories*, eds. C. Papanastasiou, T. Plomp and E. Papanstasiou, 373–388. Nicosia: Cultural Center of the Kykkos Monastery.

Confenmen PASEC 2012, www.confemen.org/spip.php?rubrique3 July 11

Egelund, N. 2008. The value of international comparative studies of achievement: A Danish perspective. *Assessment in Education: Principles, Policy, and Practice 15*(3): 245–251.

Elley, W. 2002. *Evaluating the impact of TIMSS-R (1999) in low- and middle-income countries: An independent report on the value of World Bank support for an international survey of achievement in mathematics and science.* Unpublished report, Christchurch, New Zealand.

Ertl, H. 2006. Educational standards and the changing discourse on education: The reception and consequences of the PISA study in Germany. *Oxford Review of Education* 32(5): 619–634.

Figazzolo, L. n. d. *Impact of PISA 2006 on the Education Policy Debate.* Brussels: Education International.

Gardner, D. C. 1983, April. *A Nation at Risk: The Imperative for Educational Reform.* Washington, DC: National Commission on Excellence in Education.

Gilmore, A. 2005. The impact of PIRLS (2001) and TIMSS (2003) in low- to middle-income countries: An evaluation for the value of World Bank support of international surveys of reading literacy (PIRLS) and mathematics and science (TIMSS). Unpublished report, Christchurch, New Zealand.

Goldstein, L. 2004, March 3. Down in rankings, Israel seeks changes in education. *Education Week* 23(25): 8.

Greenspan, A. 2004, March 11. *Address to the Committee on Education and the Workforce, U.S. House of Representatives*. Washington, DC: The Federal Reserve Board. http://federalreserve.gov/boarddocs/testimony/2004/20040311/default.htm

Gruber, K. H. 2006. The German 'PISA-Shock': some aspects of the extraordinary impact of the OECD's PISA study on the German education system. In: *Cross National Attraction in Education: Accounts from England and Germany*, ed. E. Ertl, Symposium Books.

Howie, S. 2011. The involvement of African countries. In: *IEA 1958–2008:50 Years of Experiences and Memories*, eds. C. Papanastasiou, T. Plomp, and E. Papanstasiou, 289–316. Nicosia: Cultural Center of the Kykkos Monastery.

International Association for the Evaluation of Educational Acheivement (IEA). 2011, September. Use and impact of TIMSS and PIRLS data in the Russian Federation. *IEA Newsletter* 38:5–6.

International Association for the Evaluation of Educational Achievement (IEA). 2012, April. Impact of IEA data: Bridging the gap between research, policy and practice. *IEA Newsletter* 39:3–6.

Kamens, D. H. 2009. Globalization and the growth of international educational testing and national assessment. *Comparative Education Review* 54(1): 5–25.

Keeves, J. P. 2011. IEA's contribution to research, methodology and measurement in education. In: *IEA 1958–2008: 50 Years of Experiences and Memories*, eds. C. Papanastasiou, T. Plomp, and E. Papanstasiou, Vol. 2, 217–253. Nicosia: Cultural Center of the Kykkos Monastery.

Klemencic, E. 2010. The impact of international achievement studies on national education policymaking: The case of Slovenia. How many watches do we need? In: *The Impact of International Achievement Studies on National Education Policymaking*, ed. C. Wiseman, 239–268. Bingley: Emerald.

Kovaleva, G. 2011. Use and Impact of TIMSS and PIRLS data in the Russian Federation, *IEA Newsletter*, Sept 2011

Lehmann, R. 2011. The impact of IEA on educational policy making in Germany. In: *IEA 1958–2008:50 Years of Experiences and Memories*, eds. C. Papanastasiou, T. Plomp, and E. Papanstasiou, 411–430. Nicosia: Cultural Center of the Kykkos Monastery.

Leung, F. K. 2011. The significance of IEA studies in education in East Asia. In: *IEA 1958–2008:50 Years of Experiences and Memories*, eds. C. Papanastasiou, T. Plomp, and E. Papanstasiou, 389–410. Nicosia: Cultural Center of the Kykkos Monastery.

Lockheed, M. 2009. The Effects of the Program of the Assessment of Student Achievement (PASA) on the Educational Assessment Capacity in Developing Countries: An Independent Evaluation. The World Bank.

Lockheed, M. 2011, December 2–3. *Assessing the assessors: Who benefits from international large-scale assessments in developing countries*. Paper presented at the SUNY Albany Conference "The Role of Benchmarking in Global Education: Who Succeeds in PISA and Why?" Albany, New York.

Martin, M. O., Rust, K., and Adams, R. 1999. *Technical Standards for IEA Studies*. Amsterdam: IEA.

Martin, M. O., Mullis, I. V., and Kennedy, A. M. 2007. *PIRLS Technical Report*. Chestnut Hill, MA: Boston College.

McGaw, B. 2008. The role of the OECD in international comparative studies of achievement. *Assessment in Education: Principles, Policy and Practice* 15(3): 223–243.

Mullis, I. V. and Martin, M. O. 2007. Lessons learned: What international assessments tell us about math achievement. In: *TIMSS in Perspective: Lessons Learned from IEA's Four Decades of International Mathematics Assessments*, ed. T. Lovelace, 9–36. Washington, DC: Brooking Institution Press.

Mullis, I. V., Martin, M. O., Gonzales, E. J. et al. 2004. *TIMSS International Mathematics Report. Findings from IEA's Trends in International Mathematics and Science Study at the fourth and eighth grades*. Chestnut Hill, MA: Boston College.

Murimba, S. 2005. The impact of the southern and eastern Africa consortium for monitoring educational quality (SACMEQ). *Prospects* 35(1), 91–208.

Nakajima, H. cited in Takayama, K. 2010. Politis of externalization in reflexive time: Reinventing Japanese education reform discourses through "Finnish PISA Success", comparative Education Review Vol. 54 No. 1, Feb 2010.

Organisation for Economic Co-operation and Development (OECD). 2005. *EDU/PISA/GB 200521* (long-term strategy document). Paris: Author.

Organisation for Economic Co-operation and Development (OECD). 2006. *Education at a Glance*. Paris: Author.

Organisation for Economic Co-operation and Development (OECD). 2007. *The Program for International Student Assessment*. Paris: Author.

Olson, J. F., Martin, M. O., and Mullis, I. V. 2007. *TIMSS 2007 Technical Report*. Chestnut Hill, MA: Boston College.

Pelgrum, H., Blahova, V., Dukynaite, R. et al. 2011. IEA experiences from Latvia, Lithuania, the Slovak Republic and Slovenia. In: *IEA 1958–2008:50 Years of Experiences and Memories*, eds. C. Papanastasiou, T. Plomp, and E. Papanstasiou, 447–468. Nicosia: Cultural Center of the Kykkos Monastery.

Plomp, T., Anderson, R. E., Law, N. et al. 2009. *Cross-national Information and Communication Techonlogy: Policies and Practices in Education*. Charlotte, NC: Information Age Publishing.

Poon, C. L. 2012. Singapore, *IEA Newsletter*, N39 April 2012.

Robitaille, D. B. 2000. *The Impact of TIMSS on the Teaching and Learning of Mathematics and Science*. Vancouver: Pacific Educational Press.

Saito, M. C. and Capele, F. 2010. Monitoring the quality of education: Exploration of concept, methodology, and the link between research and policy. In: *The Impact of International Achievement Studies on National Education Policymaking: International Perspectives on Education and Society*, ed. A. Wiseman, 3–34. Bingley: Emerald Publishing.

Schwippert, K. 2007. *Progress in Reading Literacy: The Impact of PIRLS 2001 in 13 Countries*. Hamburg: Waxman.

Schwippert, K. and Lenkeit, J. (Eds). 2012. Progress in reading literacy in national and international context: The impact of PIRLS 2006 in 12 countries. Münster: Waxmann.

Shamatov, S. A. 2010. The impact of sandardized testing on education quality in Kyrgzstan: The case of the Program for International Student Assessment (PISA) 2006. In: *The Impact of International Achievement Studies on National Education Policymaking: International Perspectives on Education and Society*, ed. A. Wiseman, 145–180. Bingley: Emerald Publishing.

Tatto, M. T., Schwillie, J., Senk, S. L., Ingvarson, L., Rowley, G., Peck, R., Bankov, K., Rodirguez, M., Rekase, and Rekase, M. 2012. Policy, Practice and Readiness to Teach Primary and Secondary Mathematics in 17 Countries: Findings from the IEA Teacher Education and Development Study in Mathematics (TEDS-M) IEA, Amsterdam, The Netherlands.

Tuijnman, A. C. and Postlethwaite, N. 1994. *Monitoring the Standards of Education.* Oxford: Pergamon.

United States National Commisson on Excellence in Education. 1983. *A nation at Risk: The Imperative for Educational Reform: A Report to the Nation and the Secretary of Education.* Washington, DC: United States Department of Education.

Wagemaker, H. 2010. IEA: Globalization and assessment. In: *International Encyclopaedia of Education*, ed. E. B. Penelope Peterson, Vol. 4, 663–668. Oxford: Elsevier.

Wagemaker, H. 2011. IEA: International studies, impact and transition. In: *IEA 1958–2008:50 Years of Experiences and Memories*, eds. C. Papanastasiou, T. Plomp and E. Papanstasiou, 253–273. Nicosia: Cultural Center of the Kykkos Monastery.

Weiss, C. H. 1999, October. The interface between evaluation and public policy. *Evaluation* 5(2): 468–486.

World Bank. 2011a. *Learning for all: Investing in People's Knowledge and Skills to Promote Development: Education Sector Strategy 2020.* Washington, DC: Author.

World Bank. 2011b, May 18. *The Arab Regional Agenda on Improving Education Quality (ARIEQ)* (working paper): Tunis: Author.

Wyckoff, P. 2009. *Policy and Evidence in a Partisan Age: The Great Disconnect.* Washington, DC: Urban Institute Press.

3

The Impact of International Studies of Academic Achievement on Policy and Research

Stephen P. Heyneman
Vanderbilt University Peabody College

Bommi Lee
Vanderbilt University Peabody College

CONTENTS

Introduction

It began as an educational experiment. In the late 1950s, Torsten Husen from the University of Stockholm was visiting his friends Benjamin Bloom and C. Arnold Anderson at the University of Chicago. "Why don't we test for academic achievement internationally," he was reported to have asked, "The world could be our laboratory" (Heyneman and Lykins 2008, p. 106). This was the origin of the International Association for the Evaluation of Education Achievement (IEA),* an organization that now includes 68 countries (see the chapter appendix) and assists in the testing of a half-dozen academic subjects including foreign languages, reading literacy, mathematics, science, civics, and writing. The IEA commenced as a loose association of university-based personalities interested in finding solutions to pedagogical and other problems. It sought solutions that could not be found locally and it expanded to include international agencies and a dozen different sources of funding. From the first "mom-and-pop shop" studies in the 1960s, international testing is now conducted not only by the IEA, which operates globally by regional organizations in the industrial democracies (the Organization for Economic Cooperation and Development, OECD), and others in Africa and Latin America. A list of these international testing efforts can be found in Table 3.1. As one can see, they have become more frequent and more global.

Beginning with the wealthier countries, the IEA now includes many middle- and even low-income countries. In the period from 1960 to 1989 there were 43 international surveys of academic achievement. In the period between 1990 and 1999 there were 66 international surveys, 49 regional surveys, and 205 local or national surveys. Between 2000 and 2009 there were 152 international surveys, 47 regional surveys, and 324 national or local surveys. The percentage of countries participating in at least one of the three types of academic achievement surveys includes 33% of countries in Europe and Central Asia, 50% in Sub-Saharan Africa and in the Arab States, 60% in East Asia and the Pacific, and 74% in Latin America and the Caribbean (Kamens 2013).

One observer concludes:

> The search for best practices is now an international mantra with extraordinary legitimacy and funding ... Comparison is viewed not only as possible but required for advancing knowledge (Kamens, forthcoming).

In the new education sector policy paper, the World Bank has called for funding to be used to support the systematic testing of children in all parts of the world (World Bank 2011). How did the world get from the tiny

* Ben Bloom was once asked: "Why was the long title reduced to just three letters—IEA." He responded, "Why not? Where is the rule that an acronym has to be the exact replica of an organization's title?"

TABLE 3.1

International Tests of Educational Achievement: Scope and Timing

Sponsor	Description	Countries	Year(s) Conducted
IEA	First International Mathematics Study (FIMS)	12 countries	1964
IEA	Six Subjects Study		1970–1971
	Science	19 systems	
	Reading	15 countries	
	Literature	10 countries	
	French as a foreign language	8 countries	
	English as a foreign language	10 countries	
	Civic Education	10 countries	
IEA	First International Science Study (FISS; part of Six Subjects Study)	19 countries	1970–1971
IEA	Second International Mathematics Study (SIMS)	10 countries	1982
IEA	Second International Science Study (SISS)	19 systems	1983–1984
ETS	First International Assessment of Educational Progress (IAEP-I, Mathematics Study and Science)	6 countries (12 systems)	1988
ETS	Second International Assessment of Educational Progress (IAEP-II, Mathematics and Science)	20 countries	1991
IEA	Reading Literacy (RL)	32 countries	1990–1991
IEA	Computers in Education	22 countries 12 countries	1988–1989
Statistics	International Adult Literacy Survey (IALS) Canada	7 countries	1994
IEA	Preprimary Project:		
	Phase I	11 countries	1989–1991
	Phase II	15 countries	1991–1993
	Phase III (longitudinal follow-up of Phase II sample)	15 countries	1994–1996
IEA	Language Education Study	25 interested countries	1997
IEA	Third International Mathematics and Science Study (TIMSS)	45 countries	1994–1995
	Phase I	About 40	1997–1998
	Phase II (TIMSS-R)		
IEA	Civics Education Study	28 countries	1999
OECD	Program for International Student Assessment (PISA)	43 countries	2000 (reading)
		41 countries	2003 (math)
		57 countries	2006 (science)
		65 countries	2009 (reading)
IEA	Progress in International Reading Literacy Study (PIRLS)	34 countries	2001
		41 countries	2006
		48 countries	2011
IEA	Trends in International Mathematics and Science Study (TIMSS)	45 countries	2003
		48 countries	2007
		63 countries	2011

Source: Adapted from Chromy, R.R. 2002. *Methodological Advances in Cross-National Surveys of Educational Achievement* (pp. 80–117). Washington, DC: National Academy Press.

experiment in the 1960s to today's "mantra"? What have the objections been to these surveys? To what extend have they had an effect on education policy and research? That is the subject to which we now turn.

Debates over Utility and Feasibility

In 1933, an argument broke out between M. Dottrens and Jean Piaget at a board meeting of the International Bureau of Education (IBE) in Geneva.* Dotterns proposed an international survey to record what countries were doing in education. Piaget was against it. He said: *"L'experience nous a montre qu'il est extremenent difficile d'estabir des tableaux statistiques comparables"* (In our experience it is extremely difficult to establish a table of comparable statistics; Smyth 1996, p. 4). Piaget had a point. At that time there was no common definition on what education meant, on how schooling might differ from ad hoc learning, or on how to distinguish education levels. The meaning of vocational and general education varied both between and within nations. There were 115 different ways to define literacy and 133 different ways to classify educational attainment by age group (Smyth 2005, p. 15). Dotterns, however, eventually won the argument. In spite of the many procedural complexities and the very real danger of receiving misleading results, the demand to know "what countries were doing" in the field of education was simply irresistible.

These same arguments have reoccurred with regularity in the interim. Though those who agree with Dottrens have lost many battles, they have all but won the war. The record of advancement in geographical coverage and qualitative depth in comparative education statistics has been unidirectional. From defining and counting schools in 1933 to videotaping teaching styles and capturing unit expenditures, the story of educational measurement and the resulting debates over its unprecedented findings is one true sign that there has been progress in education research. This extraordinary growth, both in the quantity and the quality of education data has brought fresh—and sometimes contentious—insights into perennial questions concerning educational achievement (Heyneman and Lykins 2008).

The modern-day equivalent of Piaget's doubt about the feasibility of counting schools in different countries is the criticism of international achievement testing. In the 1970s and 1980s objections included the characterization of cross-national achievement tests in developing countries as tantamount to a form of neocolonialism (Weiler 1994). The notion that developing countries were inherently inferior in education achievement helped stimulate regional efforts in Southern Africa and Latin America. Recent studies, however, have suggested that the results from some developing countries are better than

* The International Bureau of Education was established in 1925 as a nongovernmental education organization. In 1929, it allowed countries to join as members. Both Dotterns and Piaget were educational psychologists. Dotterns served on the Board of Directors; Piaget served as IBE's director for 40 years. Today, the IBE is a specialized agency within the United Nations Educational, Scientific, and Cultural Organization (UNESCO).

wealthier countries in both average achievement and in the efficiency of delivery (Heyneman 2004).

In the early 1990s, criticisms of international testing contained the argument that countries sent lower portions of their age cohorts to secondary school; hence higher scores in Germany, for instance, were based on biased samples (Rotberg 1990). This argument precipitated a formal reply from the members of the Board on International and Comparative Studies in Education of the National Academy of Sciences (BICSE). The members of the BICSE committee pointed out that while countries indeed select some to attend specialized secondary schools in preparation for university, sampling was purposefully random across all categories of schools; hence higher scores from countries such as Germany could not be attributed to a bias in sampling (Bradburn et al. 1991).

BICSE then laid out the virtues of international achievement testing:

- It may provide a wider laboratory on which to observe the consequences of different domestic policies and practices.
- That by studying the challenges and successes of other parts of the world, international test information helps define what is realistic in terms of domestic education policy.
- It may introduce concepts that have been overlooked domestically.
- It may raise important questions or challenge long-held assumptions that may not have been challenged using domestic sources of information alone.
- It may elicit results that were not anticipated but nevertheless have high value (Bradburn and Gilford 1990).

Once arguments over the technical methods of test administration declined, others quickly took their place. These included the probability that the public's expectations about what can be learned from international tests are unrealistic and the notion that other countries are better is based on a set of myths. These myths include whether tests provide valid measures of quality; whether the quality of education in the United States has declined; whether the problem in schools can be fixed with the introduction of new tests; and whether new testing can compensate for inadequate resources (Rotberg 1996, 2006, 2007, 2008; Ravitch 2010).

In addition to these doubts as to the utility of testing to inform policy, there are more specific objections to the nature of the test items, whether they reasonably guarantee equivalency across countries when students speak, read, write, and listen using entirely different systems of communication (Holliday and Holliday 2003); or whether local schooling situations can actually be compared (Fensham 2007). The assets and drawbacks of international assessments have been well summarized by Rowan (2002) and by Braun and Kanjee (2006).

Impact on Practice

Despite the debates over the utility of international assessments, experience suggests that they have often had an effect on domestic education policies, although the magnitude of impact differs. The range of these policies is wide, from pedagogy to national standards of assessment. In this section, we discuss some of these findings and how they influenced specific policies.

Pedagogy

The U.S. TIMSS 1999 Video Study, which examined instructional practices across countries, revealed that classroom instructions in Japan were significantly different from those in the United States (Stigler and Hiebert 1999; Wagemaker 2004). The findings suggested that Japan had the most skillful and purposeful teaching, with students being asked to solve challenging problems. In the United States, teaching tended to focus on learning terms and demonstrating procedures (Stigler and Hiebert 1999). Also, the curriculum content in Japan was more coherent. In the United States teachers had fewer opportunities to work with other teachers to improve teaching (Wagemaker 2004), whereas in Japan teachers were more collaborative and saw themselves as contributing to the knowledge of teaching practices and their own professional development (Stigler and Hiebert 1999). This result grabbed attention in the mathematics education community and the public at large (Park 2004). On the other hand, what the TIMSS videos did not reveal was the effect of Japanese Juku (private tutoring) schools either on achievement or on the pedagogy in public schools. Because the pedagogy of private tutoring is rigorously focused on the content of the university entrance examination, it is quite possible that, relieved of being solely responsible for student success on the examinations, public school teachers could "afford" to teach with more creativity. American teachers, by contrast, do not have such "luxury."

In England, the results of TIMSS 1995 showed that Grade 4 students used calculators in mathematics lessons more frequently than those in other countries. Although there was no association between frequency of calculator use and mean mathematics scores at Grade 4, this result from the teacher survey drew attention to the frequency of calculator use. Consequently, England began to emphasize the importance of mental calculation (Keys 2000).

Policy makers in some developing countries began to model teaching practices on developed countries after TIMSS. Eight among eighteen developing countries that participated in TIMSS-Repeat planned to change the direction of classroom instructions to emphasize activity-based learning, problem solving in mathematics, and more critical thinking in science (Elley 2005). In two countries, policy makers planned to reduce the amount of teacher lecturing and increase student engagement in lessons, with more discussion, questioning, experimenting in class, critical thinking, valuing of student opinion, and exploring students' curiosity (Elley 2005).

Not only teaching practices, but also national curricula were changed as results of TIMSS and PISA. In many countries, new topics were added because findings showed that higher-performing countries generally had greater levels of coverage of the intended curricula (Mullis and Martin 2007). Romania and the Slovak Republic added new topics in mathematics and science curricula as a result of TIMSS (Noveanu and Noveanu 2000; Berova and Matusova 2000). In Sweden, PISA results contributed to the introduction of national tests in biology, chemistry, and physics for 12- and 15-year-olds, as well as the development of diagnostic material in science for the younger ages (Breakspear 2012). In the Slovak Republic, key competencies characterized by PISA, which were not included in the previous curriculum, were incorporated into national curriculum standards (Breakspear 2012).

Teacher Training

TIMSS results led to changes in the existing preservice and in-service teacher training programs as well as development of instructional resources in some countries. In Canada (Ontario), instructional materials were developed as a direct outcome of TIMSS (Taylor 2000). New Zealand also developed resources materials and created professional development programs for mathematics and science teachers to address the areas of relative weakness (Chamberlain and Aalst 2000). In Macedonia, changes were made to in-service training, which encouraged the teachers to change from a teacher-centered to a more student-centered style (Elley 2005). In Israel, findings indicated that they should focus on teacher preparation programs instead of in-service training programs (Mevavech 2000). In Malaysia, the government paid more attention to teaching practices and teacher training after the TIMSS-R study (Elley 2005).

Class Size

Contrary to common belief, findings from ILSA have consistently shown that class size had little association with higher achievement. The IEA's First Mathematics Study and the Second Science Study showed that class size was not related to achievement (Medrich and Griffith 1992). More recent studies using PISA and TIMSS with school-fixed effects and instrumental variables also presented no statistically significant class size effect in most of the countries (Fertig and Wright 2005; Woessmann 2005a,b; Ammermueller et al. 2005; Woessmann and West 2006). While it is irrational to suggest that larger classes are better, these findings suggest that simply lowering class size without making other changes may not result in achievement gains.

Hours of Instruction

TIMSS results were not conclusive about whether more instruction time is related to higher achievement. Early IEA studies showed that the students had higher achievement when there was a greater amount of time spent on teaching and learning (Husen 1967; Anderson and Postlethwaite 1989; Keeves 1995). However, the Second Science Study showed that hours of school each week, hours of mathematics instruction each week, and hours of homework had virtually no relationship to the achievement (Medrich and Griffith 1992). In addition to this finding, TIMSS 2003 results showed that countries that provided more instruction time (the Philippines, Indonesia, and Chile) had relatively low achievement (Mullis and Martin 2007), a conclusion that can sometimes be a topic of local political interest (Heyneman 2009a). Related to the number of school days/year and hours/day is the issue of how much classroom time is actually devoted to the task of learning. Since the First Mathematics Study, the "opportunity to learn" variable emerged as an important indicator of performance, especially at the secondary level (Medrich and Griffith 1992). These findings provided insight into the relationship between the "intended curriculum," the "implemented curriculum," and the "achieved curriculum" (Wagemaker 2004).

Use of Technology

Both the IEA and OECD PISA study provide information on Information Communication Technology (ICT) use in education. The IEA has a long tradition of studying ICT in education (Pelgrum and Plomp 2008). The first one was the so-called Computer in Education (CompEd) study that was conducted in 1989 and 1992. The findings implied that greater attention should be given to how computers can be used in schools. The second wave of ICT large-scale international assessment was the IEA's Second Information Technology in Education Studies (SITES), which started with Module One in 1998 to 1999. The SITES Module One examined how ICT affects teaching and learning processes in different countries. Wagemaker (2004) notes that the first school survey, Module One of SITES, raised three issues—(i) the challenge of making teachers ready-to-use technology in their instruction; (ii) over-expectations suggesting that computers would transform the curriculum and pedagogy; and (iii) the difficulties that schools have in internet access and protecting children from inappropriate materials. Module One was followed by Module Two, which is a collection of qualitative case studies on ICT-supported pedagogical innovations, and findings were used in ongoing discussions in several countries (Kozma 2003; Wagemaker 2004). SITES 2006 is the most recent survey on ICT conducted by the IEA.

Nations such as Austria that introduced computer technologies in a systematic way with universal access, teacher training, and curriculum integration out-performed countries that added computers in an ad-hoc fashion without regard to standardization and planning (Keeves 1995). Perhaps as a result, the main area of interest in the studies of CompEd and three SITES studies shows a shift of focus from computer counts, access rates, and obstacles to how pedagogical practices are changing and how IT supports these practices (Pelgrum and Plomp 2008). TIMSS 2003 and PISA 2003 also have indicators of ICT availability and ICT use on core school subjects; however, these assessments cover only a small number of indicators (Pelgrum and Plomp 2008).

Despite the rising significance of using technology in education, research shows inconsistent but often negative findings on the use of technology and achievement. Although some IEA reports show mathematics achievement to be positively associated with computer usage (Mullis and Martin 2007), later investigations found a negative or no relationship (Fuchs and Woessmann 2004; Papanastasiou et al. 2004; Wittwer and Senkbeil 2008). Still, others found mixed results within the same study (Papanastasiou et al. 2003; Antonijevic 2007) and sometimes results differed according to the kind of software students used. More recent studies have suggested that it is not the computer that influences problem-solving skills, but the purpose to which the computer is put. Using computers to download games and videos, for instance, may adversely affect problem-solving skills. On the other hand, using computers to seek and analyze new information augments problem-solving skills (de Boer 2012).

Subject-Specific Findings: Math

As fluency in Latin was the mark of an educated person and a prerequisite for employment throughout Europe in the past, mathematics is in many key respects the new Latin (Plank and Johnson 2011). Mathematics has become the key indicator for educational success (Plank and Johnson 2011), as a country's best students in mathematics have implications for the role that country will play in the future advanced technology sector and for its overall international competitiveness (OECD, 2004a). Mathematics is also straightforward in comparing across countries, while comparing performance in other subject matters (literacy, history, science) is complicated by cross-national differences in language and culture. Mathematics provides a ready yardstick for international comparisons (Plank and Johnson 2011). Most findings from international studies use mathematics achievement as a dependent variable.

The focus of earlier IEA studies was observing whether differences in selective and comprehensive systems lead to different outcomes (Walker 1976). The Six Subjects Study found that students in academic courses in selective schools attained higher levels in mathematics than students following similar courses in comprehensive schools (Walker 1976). This suggests that selective school systems may have higher outcomes than comprehensive school systems.

Results from the TIMSS 2003 and 2007 supported some common beliefs about factors for high achievement. Findings indicated that high-performing students had parents with higher levels of education, and had more books at home. High achievers also had computers at home. Findings from TIMSS 2007 also indicated that most high-performing students spoke at home the same language used for the test, and generally had positive attitudes and self-confidence. School factors were also important for high achievers. Generally, high-achieving students were in schools with fewer economically disadvantaged students and the school climate was better. These schools had fewer problems with absenteeism, missing classes, and students arriving late. Both TIMSS 2003 and 2007 findings show that being safe in school was positively associated with math achievement.

The findings from PISA are similar to the findings from TIMSS. In terms of learning behavior, the general learning outcomes from PISA 2000 presented that those who process and elaborate what they learn do better than those who memorize information (OECD and the United Nations Educational, Scientific, and Cultural Organization [UNESCO] 2003). The results also showed that cooperative learning is not necessarily superior to competitive learning, and evidence suggests that these strategies are complementary. For achievement among subgroups, PISA 2000 results showed that there was a significant number of minority students in all countries who displayed negative attitudes toward learning and lack of engagement with school, which is associated with poor performance (OECD and UNESCO 2003). PISA 2000 results also supported the fact that spending is positively associated with mean student performance. However, spending alone is not sufficient to achieve high levels of outcomes and other factors play a crucial role (OECD and UNESCO 2003).

From the results of PISA 2003, two major findings emerged. Some of the best-performing countries showed only a modest gap between high and low performers, suggesting that wide disparities are not a necessary condition for a high level of overall performance (OECD 2004a). Additionally, student performance varied widely between different areas of mathematical content (OECD 2004a), implying that balance and sequence of the national curriculum may be independently important to high achievement.

Subject-Specific Findings: Science

Science is another important area of concern for policy makers, as it is a subject associated with national economic growth in an information, communication, and knowledge-based society. Early studies showed the need for highly qualified teachers in science (Comber and Keeves 1973; Walker 1976). The findings from the first IEA study showed that when teachers specialized in science, had received more post-secondary education, or had participated in science curricular reforms, their students tended to perform better on science achievement tests (Comber and Keeves 1973; Walker 1976). Implications from TIMSS studies were that competent and committed science teachers are needed for successful teaching of science in secondary schools (Walker 1976; Vlaardingerbroek and Taylor 2003).

Another finding from the IEA Six Subject Study was that achievement was closely linked to opportunity to learn, because students in developing countries found the tests to be much more difficult than their peers elsewhere (Walker 1976). In fact, PISA 2000 results also showed that students' performance in low- and middle-income countries was lower than that of high-income countries (OECD and UNESCO 2003). This raises attention to the gross differences in educational quality between OECD and developing countries (Heyneman 2004).

The gender gap in science was reduced over time. TIMSS 1995 results showed that, in most countries, boys performed significantly higher in science, particularly in physical science, than girls in both seventh and eighth grades (Mullis et al. 1996). Not only did boys perform better, but they also expressed a liking for this content area more often than girls (Mullis et al. 1996). However, TIMSS 2003 results showed that in the eighth grade there was greater improvement for girls than boys since 1999 (Mullis et al. 2004). This suggests a closed gender gap in science achievement. Average science achievement also improved for both boys and girls since 1995 (Mullis et al. 2004).

TIMSS 2003 results also presented some factors that were associated with high performance across countries. Higher science achievement was associated with having at least moderate coverage of the science topics in the curriculum, although high coverage in the intended curriculum itself does not necessarily lead to high student achievement. School climate and safety was also strongly associated with higher science achievement. Principal's perception of school climate was also highly related to average science achievement (Mullis et al. 2004).

Access to computers in science classrooms remained a challenge in many countries. Teachers reported that, on average, computers were not available for 62% of eighth grade students and 54% of fourth grade students internationally (Mullis et al. 2004). Even in countries with computers available in schools, using computers in science class was rare in eighth grade (Mullis et al. 2004). This finding raises concerns about how well schools and teachers are prepared in using computers for pedagogical purposes, particularly in science.

Subject-Specific Findings: Reading and Writing

Perhaps because families influence their children through language, the findings from studies of reading achievement showed stronger home effects. There is a lack of explanatory power of school and classroom-based measures in accounting for the differences in reading achievement between students (Keeves 1995). Mother-tongue instruction beyond an early grade did not seem to advance reading comprehension skills.

There were some significant changes in PISA results since 2000. First, reading scores rose significantly in 13 countries and fell in four others, indicating that countries, in general, are increasing students' achievement. Second, improvement in results is largely driven by improvements at the bottom end of distribution, indicating that there was progress in improving equity (OECD 2010). The percentage of immigrant students increased, but the reading performance gap between students with and without an immigrant background narrowed (OECD 2010). On the other hand, gender differences stayed or widened (OECD 2010). One interesting finding among the countries that improved fastest—Chile, Peru, Albania, and Indonesia—is that the relationship between socioeconomic background and learning outcomes has weakened (OECD 2010).

Gender differences were larger in reading than in mathematics achievement. The PISA 2000 findings indicated a growing problem for males, particularly in reading literacy. In mathematics, females on average remained at a disadvantage in many countries, but this was due to high levels of performance of a comparatively small number of males (OECD and UNESCO 2003). Education systems have made significant efforts toward closing the gender gap; however, much remains to be done. Improving reading skills for males and stimulating self-concept among females in mathematics need to be major policy objectives if gender equality in educational outcomes is to be achieved (OECD and UNESCO 2003).

There are not many large-scale assessments of writing, and perhaps the IEA study of written composition is the only one. The study found that there was differential treatment in most school systems between boys and girls in the provision of writing instruction and in the rating of writing performance. Girls profited from differential treatment in most school systems, where mostly women provide instruction (Purves 1992). This, again, shows that there is a gender gap in writing performance as in reading performance. The problem with international writing tests is that the definition of excellence is heavily influenced by culture. This makes it difficult to compare achievement across cultures.

Subject-Specific Finding: Civic Education

The IEA conducted its first study on civic education in 1971. The civic education measures civic knowledge, support for democratic values, support for the national and local government, and participation in political activities. The Second Civic Education Study (CIVED) was carried out in 1999, when many countries were experiencing political, economic, and social transitions. An additional survey of upper secondary students was undertaken in 2000. The findings indicated that schools appear to be an important factor responsible for civic knowledge and engagement. Schools are also responsible for the content of the curriculum and offer places for students to practice democracy as much as they offer places for learning facts (Amadeo et al. 2002).

In 2009 the IEA conducted a study of civic and citizenship education (ICCS) in 38 countries that built upon previous IEA studies of civic education. The study took place because of significant societal changes including rapid development of new communication technologies, increased mobility of people across countries, and the growth of supranational organizations (Schulz et al. 2010). There was a decline in civic content knowledge since 1999, students were more interested in domestic political issues rather than foreign issues, and the levels of interest were greatly influenced by parents (Schulz et al. 2010). The findings from the ICCS also provide evidence that although students' home backgrounds largely influence students' civic knowledge, schools also can contribute to civic knowledge and intentions to vote in adulthood (Schulz et al. 2010).

Impact on Specific Regions

Sub-Saharan Africa

The Southern and Eastern Africa Consortium for Monitoring Educational Quality (SACMEQ) began in 1995 with a meeting of 15 Ministers of Education.* Coordinated by the International Institute of Education Planning (UNESCO/Paris) data on mathematics and reading were collected in 2000 and 2007. Publications have included national policy briefs, national reports, working documents, working papers, policy issues, and technical reports. The Programme d'analyse des systems educatifs des pays del la CONFEMEN (PASEC) is roughly the equivalent in 17 French-speaking countries.† Tests have been applied in French, mathematics, and occasionally in the national language at the beginning and end of the school year so as to capture value-added information. More than 500 reports are available online, which detail the methods and the results.

Latin America

UNESCO and UNICEF help coordinate the regional Laboratorio Latin americano de Evaluation de la Calidad de la Educación and its Monitoring of Learning Achievement (MLA) studies, an illustration of how much the region has changed. In the 1960s and 1970s international tests of academic achievement were considered to be "northern" threats to Latin American autonomy. Today, Latin American plays a leading role in the use and interpretation of

* Members include Botswana, Lesotho, Kenya, Malawi, Mauritius, Seychelles, South Africa, Swaziland, Tanzania, Tanzania-Zanzibar, Uganda, Zambia, and Zimbabwe.

† Members include Mauritania, Cape Verde, Senegal, Guinea Bissau, Guinea, the Ivory Coast, Togo, Benin, Burkina-Faso, Niger, Central African Republic, Cameroon, Congo (Brazzaville), Gabon, Madagascar, Comoros, the Seychelles, Mauritius, Djibouti, and Burundi.

international tests. Perhaps the most important stimulant has been the influence of the Partnership for Educational Revitalization in the Americas (PREAL). Since 1995, PREAL has worked to improve the quality and equity of public education in the region. It explains itself as

> A network of private institutions and individuals dedicated to changing the way public and private leaders in Latin America think about schools and education. It seeks to improve learning and equity by helping to promote informed debate on education policy, identify and disseminate best practice, and monitor progress toward improvement. PREAL encourages business and civil society to work with governments in a common effort to improve schools and strengthen their capacity to do so. (PREAL 2011)

PREAL has been responsible for publishing "Report Cards," which give grades to countries for their educational performance. As of 2011, 4 regional, 21 national, and 6 sub-national report cards have been issued; each has stimulated alterations in education policy and practice. Initiatives have stimulated the minister of education in Peru and Honduras to establish national education standards and have been used as a guide to new standards in Chile. Recommendations from PREAL's report cards and "Policy Audits" have been incorporated into education policy in El Salvador, Panama, and the Dominican Republic. PREAL's demand for accountability has directly influenced the orientations of education ministers in Jamaica, Brazil, and Mexico City. To promote informed debate on education policy, PREAL published a report on how Latin America and the Caribbean performed on the Second Regional Student Achievement Test (SERCE; Ganimian 2009) and the regional results from PISA 2009 (Ganimian and Rocha 2011). The results of PISA 2009 were used in spurring public discussion of education challenges in the region, and informed policy makers of specific issues that require attention. PREAL also strongly urged countries to participate in global achievement tests. As a result, El Salvador participated in the TIMSS 2007 for the first time, and the Panamanian minister of education was also convinced to participate in the 2009 PISA exam for the first time (PREAL 2011). Thus, the PREAL used the PISA results to bring educational issues to policy makers and also urged them to participate in the global tests. Since 1995, PREAL has been responsible for publishing three books on testing, 58 working papers, 28 educational synopses (short policy summaries), 29 policy series, and 32 best-practice documents in English, Spanish, or Portuguese. The Working Group on Standards and Assessment (GTEE) alone has been responsible for 16 publications. No organization in any other region seems to have had the impact of PREAL on both policy and practice. Today the Latin American region is in an advanced state of assessment use, innovation, and implementation. This use of education data to underpin transparency and accountability has occurred in spite of the fact that students in the region tend to learn less and its systems appear more inefficient than many other parts of the world. It is not irrelevant to note that the emphasis on educational accountability has occurred in parallel fashion with the shift from authoritarian dictatorships to democracy. Latin America, and the influence of PREAL specifically, is an illustration of the power voters have over the existence of information on which to judge their school systems. Without assessments, political leaders can claim excellence without reference to the facts; with assessments, claims of excellence require proof based on international standards of evidence.

Impact on Specific Countries

In England, the first TIMSS results were announced at a time when there were concerns about standards in education, and it was in 1997 that the policy makers had the incentives to make change because of the new government. Shortly after the results were announced, what was then the Schools Curriculum and Assessment Authority (SCAA), former Qualifications and Curriculum Development Agency (QCDA), investigated the TIMSS data to provide more information about students' strengths and weaknesses in relation to the national curriculum and its assessment (Shorrocks-Taylor et al. 1998).

Whereas TIMSS results stimulated changes in assessment in England, PIRLS 2001 results confirmed that their national education strategy worked well. England introduced the National Literacy Project in 1996, which was a large-scale pilot of a new approach to literacy teaching in the first 6 years of schooling (Twist 2007). One example of this new project was whole-class and highly interactive teaching (Twist 2007). However, this national strategy was not evaluated internationally until PIRLS 2001 was conducted. PIRLS 2001 was the first international survey to include children in England who spent most of their primary education being taught using the approaches detailed in the National Literacy Strategy (Twist 2007). Results showed that their performance was significantly better than other participating English-speaking countries (Twist 2007). England's relative success in PIRLS 2001 was seen as endorsement of an innovative national strategy (Twist 2007).

Both the government and the unions in England used PISA 2006 to support their arguments. The unions lobbied against the government's argument favoring the expansion of different types of schooling by using PISA findings, which suggested that countries with a school system segmented by vocational and academic institutions do not perform better. The government also used the PISA findings to create the Masters in Teaching and Learning (MTL) program, which is to be tested in three universities in the North West of England (Figazzolo 2009). Today, interpreting the results of international tests are a daily activity in the National Education Ministry (Heyneman 2009b).

In Iceland, it was already decided to increase the number of national examinations at the elementary level, but after the first results of TIMSS were made public, they decided to test for science achievement in Grade 10 (Gudmundsson 2000).

In Switzerland, prior to PISA there were efforts to harmonize the different education standards in each canton and test student performance on a regular basis (Bieber and Martens 2011). However, only after PISA results came out did they start to concretize the reform. PISA revealed unexpected shortcomings in reading competencies and also showed that Switzerland was one of the OECD countries where the influence of students' socioeconomic backgrounds on their reading skills was the most pronounced (OECD 2002). Education became top of the political agenda and reform pressures increased. PISA played a vital role in the proceeding high degree of convergence of policies (Bieber and Martens 2011).

In Ireland, PISA 2003 demonstrated that the national standard of mathematics education had to improve. With its focus on science, PISA 2006 served to expedite changes in the curriculum for lower secondary education that had been initiated in 2003 (Figazzolo 2009).

In Iran, changes were initiated by the Mathematics Curriculum Committee on the basis of the TIMSS results. TIMSS have been used in the development of the first national item bank for primary education in Iran. TIMSS items formed the framework for the development of tables of specification and items. Another new initiative was that they sponsored a national research project where they used TIMSS curriculum framework in developing test items (Kiamanesh and Kheirieh 2000).

Compared to TIMSS, the PIRLS 2001 results were more quickly disseminated and acted on, as the students did not perform well overall in PIRLS. The policy makers emphasized the ongoing effort, which should be directed toward sound language learning, especially for those whose mother tongue is not Farsi. Consequently, several long-term programs were developed for primary education. Also, the Organization for Research and Educational Planning has incorporated suggestions from the analysis of the PIRLS 2001 results for the future policies and programs (Karimi and Daeipour 2007).

In Latvia, the TIMSS result stimulated a significant step to improve their educational curriculum. The TIMSS results showed that Latvian students did not know how to apply acquired knowledge to real-life situations, which was a rather surprising result to them as the existing school examinations did not inform them of this fact. Thus, in 1997 they formed the new national compulsory education standard and established a centralized examination system for secondary education graduation examinations (Geske and Kangro 2000).

In Russia, TIMSS results contributed to the development of new educational standards in mathematics and science. Russia had no tradition in using standardized assessments in school practice; however, as a consequence of TIMSS, Russia began to use standardized tests for assessing student achievement (Kovalyova 2000).

In Germany, the mediocre performance in the 1995 TIMSS study came as a shock to German teachers, researchers, and policy makers. The results were confirmed by the PISA 2000, which revealed average performance and a large disparity between the federal states within Germany (Neumann et al.

2010). Before the PISA study, the German education system did not traditionally rely on standardized testing. However, after the low performance, particularly in science achievement, German policy makers decided for a major reform, which included the introduction of national education standards (NES). The NES were framed by the PISA framework and the particular deficits of German science education identified by PISA (Neumann et al. 2010).

PISA attracted more interest than PIRLS in Germany because the results were unexpected. However, PIRLS also indicated a clear picture of what is going on in German schools (Schwippert 2007). For example, the PIRLS data showed that one in every five of the German students participating in the study was from a family where at least one parent was not born in Germany, and they often performed poorly in school. This raised some concerns about the immigrant populations that had been growing steadily since the late 1960s due to Germany's booming economy and demands additional labor. Thus, researchers argued the need for remedial programs for the children from immigrant families (Schwippert 2007). The most significant change after PISA and PIRLS results in Germany is the shift of academic discourse on education in Germany toward a more empirical and practice-focused framework of research from general and didactic research (Bohl 2004). This might have been possible with German teacher unions who began to support large-scale national and international surveys, and began to build strong relationships with educational researchers since the first presentations of PIRLS results (Schwippert 2007).

In Mexico, the OECD was directly involved in the formulation and implementation of reform. The poor results of Mexico in PISA 2006 were more than once used as a justification for the implementation of the reform by the government, as well as by the media and by the OECD itself. The reform especially focused on "improving teacher performance" and "consolidating evaluation" (Figazzolo 2009).

In Denmark, the result of PISA 2000 raised doubts about the efficiency of the Danish education system. Compared to their well-funded education system, the results showed only middle-range outcomes. The results also questioned why social equity continued to be a problem despite the significant investment in social welfare programs (Egelund, 2008). After the PISA 2000 result, Denmark subsequently implemented a range of reform policies, including increased national assessment and evaluation, and strategies to target socio-economically disadvantaged and immigrant students (Egelund 2008).

In Japan, the national curriculum was revised after the release of the PISA 2003 results in 2004 to incorporate the PISA-type competencies. Implementation began in 2007 (Breakspear 2012).

In the Czech Republic, there was a long-standing belief that its system was homogenous and provided an equal opportunity to all students. However, findings from TIMSS 1995 raised an important debate over social equity. TIMSS also accelerated changes in strengthening technical and vocational schools, which had already been planned (Strakova et al. 2000). In terms of an assessment system, TIMSS played another important role in the Czech Republic, as it was the first time that Czech students had encountered multiple-choice items, and the first time that standardized tests had been administered in Czech schools (Strakova et al. 2000). As a result, the ministry of education created national standardized examinations for students in their final year of secondary school. Also, in order to improve the quality of tests developed within the Czech Republic, they used TIMSS as an exemplar and as a methodological guide. TIMSS also gave Czech scholars their first opportunity to explore the relationship between home background and achievement (Strakova et al. 2000).

In Italy, the government used PISA results to advocate a rather neo-liberal reform of the country's education system. The government justified reductions in spending and the promotion of evaluation systems to increase efficiency by referring to PISA results. The reform introduced a yearly bonus for "good" teachers in order to foster the quality of the Italian school system. Interestingly, PISA results were employed by unions and opposition parties to criticize this very same reform. They also rejected the government's recommendations by arguing that PISA did not take into account the peculiarity of the Italian system (Figazzolo 2009).

In Australia, PISA results have been used to support preexisting positions. For example, the government has used it to sustain an increasing focus on testing and evaluation. The Federal Education Minister has quoted Australia's position on the PISA ranking in support for her own already existing agenda (Figazzolo 2009).

In New Zealand, high performance on PISA seemed to reinforce existing policies and thus there was no impetus for substantial change (Dobbins 2010).

Some countries emphasized that PISA results complement national data derived from national/federal assessments. For high-performing countries, PISA results are used to compare and validate data from their own national assessments (Breakspear 2012).

In Canada, PISA is used as a complementary indicator to monitor overall system performance. They even redesigned the national assessment program in 2007 to better complement PISA in terms of schedule, target population, and subject areas covered (Breakspear 2012).

In Singapore, PISA data complements national assessment data to inform about the effectiveness of the education system (Breakspear 2012).

In Spain, PISA is used as a means of comparing and contrasting the country's own data generated by national and regional assessment studies. They mention that PISA is an important referent (Breakspear 2012).

In Kuwait, the principal goal of participation in TIMSS was to evaluate the standard of the curriculum with reference to international benchmarks. As a direct result of TIMSS 1995, committees were formed to revise the mathematics and science curricula. In addition, some aspects of the teacher preparation programs were being questioned. Also, examination regulations were revised, particularly at the secondary school level. The impact of TIMSS was not limited to only Kuwait, but throughout a number of Arab countries. As a result, the Kuwait Society for the Advancement of Arab children has begun another study similar to TIMSS for the Gulf Arab countries (Hussein and Hussain 2000).

In Romania, the TIMSS-R results came as a "shock," because they fell below the international average. This finding was surprising because the past success of Romania's top students in Olympiad competitions left educators and the public with the impression that Romanian standards of achievement were high.* With this "wake'up call," Romania made important changes in curriculum guides and the new textbooks reflecting the impact of TIMSS studies. The curriculum guides referred to the TIMSS findings and presented examples of new test items following the TIMSS models. Other changes have occurred in the time devoted to mathematics and science, sequence of topics, emphasis given to the nature of science, experimentation in science, and statistics and probability (Noveanu and Noveanu 2000). Another effect of international studies was on Romanian national assessments. As a result of international assessments, local assessments began to use a variety of item types, particularly multiple-choice and performance items, and to emphasize links between mental processes and test outcomes in teacher seminars. In conclusion, there was a widespread consensus that participation in TIMSS-R was "extremely important" for Romania (Elley, 2005).

In Macedonia, the results of TIMSS-R were a surprise, as they were in Romania, since elite students performed well in Olympiad competitions. Changes were made in curricula, national assessments, and in-service training. New topics have been added to the curriculum. For in-service training, the findings of TIMSS-R were used as a lever to get teachers to take seriously the need to change their teaching styles from predominantly lecturing style to one of interaction (Elley 2005).

Findings from PIRLS also had significant influence in Macedonia. The finding that early literacy activities are very important for further student performance in reading contributed to the government's decision to make the school starting age 6 instead of 7 years old (Naceva and Mickovska 2007). Another main impact of both PIRLS and TIMSS is that educational policy makers are now aware that empirical data provide very important indicators of the quality of education systems. This awareness led to the government establishing a National Center for Assessment and Examinations, which aims to monitor and raise standards in pre-university education by employing large-scale assessment and formal examinations (Naceva and Mickovska 2007). TIMSS-R methodology was used when the national "Matura" examinations began (Elley 2005).

* In the former Soviet Union and throughout Eastern and Central Europe, it was common to assume that the quality of a school system was defined by performance in the international academic Olympics. One important impact of international surveys of academic achievement was to instill the notion of having a result that represents a representative sample of students at large.

Use of Surveys to Generalize about School System Structures

ILSA results show that, although achievement was higher in countries with selective school systems, the effect of socioeconomic status was stronger than the factors associated with achievement. The Six Subject Study found that students in academic courses in countries with a selective education system exhibited higher levels of mathematics achievement than students who had similar courses in countries with a comprehensive school system (Walker 1976). This finding suggests that selective school systems may have higher outcomes than comprehensive school systems. However, the First IEA study also found that family background strongly influences the selection when it occurs early in age (Husen 1967).

Noting these observations, recent studies using TIMSS and PISA data examined the relationship between tracking systems and social inequalities. Studies found that there was a significant effect of early tracking on inequality, whereas there was no clear effect of tracking on mean performance (e.g., Hanushek and Woessmann 2006). Studies also found that later tracking systems are associated with the reduced effect of family background (Brunello and Checchi 2007; Schuetz et al. 2008; Woessmann et al. 2009), although one study by Waldinger (2006) found that tracking does not increase the impact of family background after controlling for pretracking differences.

Thus, one of the recommendations for equity from the OECD is that school structures should be dissuaded from selective models in favor of integrated ones, because selection and tracking reinforce socioeconomic disparities (OECD 2004b). Germany used the PISA 2006 findings to advocate against early tracking in school and for a more socially equitable school system (Figazzolo 2009). To accommodate the differences in abilities of students in comprehensive school systems, the OECD requires teacher training be combined with an integrated school system, as teachers should individually promote students from different backgrounds (OECD 2009).

Use of Surveys to Uncover More Specific Areas for Investigation

Jimenez and Lockheed's (1995) study provides a good example of how surveys can be used as a follow-up of findings from large-scale data. Their earlier work (1986, 1988, 1989) used SIMS data for analysis at the national level. Recognizing that simple findings from large-scale data cannot easily explain what occurs within a classroom, they conducted a "mini-survey," in addition to using SIMS data. The study (1995) shows how student characteristics are different in public and private schools, and what the private school effect is on achievement. Students in private schools come from more privileged families than those in public school, but, with student background and selection bias held constant, students in private schools outperform students in public schools on a variety of achievement tests.

Having found that private schools are more effective, Jimenez and Lockheed wanted to find out why. They explored what characteristics of private schools contribute to the effectiveness. Using the same large-scale data, they further examined the contribution of student selectivity, peer effects, and some school inputs, such as teacher training and experience and student–teacher ratios. However, they were particularly interested in how private schools allocate their resources and inputs and manage school systems more efficiently. Their hypothesis is that private schools may be more effective than public schools because they invest more of their resources in instructional materials and in improving teacher performance than do public schools. Another hypothesis is that private schools are more flexible in terms of organizational structure because they are more responsive to the demands of parents, students, and educational professionals than are public schools.

Therefore, using a mini-survey they examined the two aspects of public and private schools that they were unable to explore with the large-scale data: school-level resources and school-level management (Jimenez and Lockheed 1995). They invited a senior researcher in each country to gather systematic data about a variety of institutional practices in public and private schools, using a survey instrument that Lockheed and her colleagues provided (Jimenez and Lockheed 1995). The results from the mini-survey confirm what they had observed with the large-scale data. More importantly, it also supports their hypotheses and provides evidence on why private schools are more effective and efficient than public schools. In addition, it shows that simple resource availability cannot explain the differences in effectiveness because the public and private schools in the sample were very similar with respect to their overall resources (Jimenez and Lockheed 1995). Furthermore, the results support the fact that it is resource allocation that makes private schools different. The lesson from this innovative methodology that Lockheed and her colleagues adopted is that the large-scale survey was used not to find out what works in terms of predicting academic achievement, but as a question filtering mechanism for generating feasible hypotheses to be followed-up by small-scale and qualitative work.

Use of Surveys to Develop New Techniques and Discover New Variables

One of the largest limitations of large-scale international data from IEA and OECD is that the data are available only for the countries that participate in the achievement tests. Barro and Lee (2001) have covered a broad group of countries by using census and survey observations to measure educational attainment. Barro and Lee (2001) have recognized the importance of gathering data on educational attainment across the world as an important indicator of human capital. For measuring quality of education, they compiled test scores on examinations in science, mathematics, and reading that have been conducted in various years for up to 58 countries by the IEA and the International Assessment of Educational Progress (IAEP). At the same time, these measures were restricted by the limited sample, which consists mostly of OECD countries (Barro and Lee 2000). Thus, Barro and Lee explain that educational attainment still provides the best available information to cover a broad number of countries (Barro and Lee 2000). They expanded their data set on educational attainment from 1950 to 2010 and constructed a new panel data set that is disaggregated by sex and age. The data are broken down into 5-year age intervals, and by adding former Soviet republics the coverage has now expanded to 146 countries. The data are more accurate than before because it incorporates recently available census/survey observations (Barro and Lee 2010).

Smaller, Quicker, and Cheaper: Improving Learning Assessments for Developing Countries

LSEAs increasingly became a key tool for meeting the demand of accountability and systematic evaluation in Least Developed Countries (LDCs), particularly after the "Education for All" initiative. However, the complexity and expense of LSEAs have led some to question the utility of conducting LSEAs in LDCs (Wagner 2011). LSEAs are complex data because the number of participating countries and population samples are large and testing instruments must be vetted by experts. A more serious limitation of LSEAs is that they are also costly in terms of money and time. It often costs millions of dollars and takes multiple years to achieve closure, which is a long turnaround time (Wagner 2011).

Recently, there is a new hybrid methodology termed the "smaller, quicker, and cheaper" (SQC) approach, which Wagner (2011) describes. This new approach seeks to focus more directly on the needs of LDC contexts. SQC pays close attention to a variety of factors such as population diversity, linguistic and orthographic diversity, individual differences in learning, and timeliness of analysis (Wagner 2011). One well-known current example of recent hybrid assessment in reading is the Early Grading Reading Assessment (EGRA). The largest advantage of this method is its modest size in assessment. Because the scale is small, one can implement it with a much smaller budget and more quickly assess achievement than with LSEAs. Because of its frequency, this approach can also be more useful for policy makers if they are interested in early detection and near-term impact. Another advantage of this method is that researchers can explore more deeply the factors that affect learning outcomes, such as language of instruction, language of assessment, and opportunity to learn (Wagner 2011). In sum, SQC assessments can better track learning over time, can better adapt to local linguistic contexts, and can be better designed to understand poor-performing children (Wagner 2011).

Impact on Research Findings

The first study of American schools using a representative national sample generated a debate that, 4 decades later, continues to dominate discussions of education policy. This was the first attempt to summarize the many different influences on learning—neighborhood, socioeconomic status, curriculum, student attitudes, teacher training, and the like. Once summarized, the Equality of Educational Opportunity Report (EEOR; Coleman et al. 1966) concluded that the predominant influence on academic achievement came not from the school but from the characteristics of the home, a set of influences over which education policy was either ineffective or immaterial. Reanalyses of the EEOR (Coleman et al. 1966) data challenged some of the findings, but the main conclusions were reenforced (Mosteller and Moynihan 1972).* The implication was that because the influence of school quality on school achievement in comparison to home background was so weak, equality of opportunity could not occur through changes in education policy alone; it required changes in social policy affecting neighborhoods and the patterns of school attendance.†

In the spring of 1971, Stephen Heyneman was offered the opportunity to do a similar study of the primary schools in Uganda as had been done in the United States. The "Coleman format" was used. This format included separate questionnaires for students, teachers, and headmasters, a lengthy counting of the physical equipment in each school, and achievement tests of math, general knowledge, and English from the primary school leaving examination, the test used to determine entry to secondary school. In 1972, Coleman himself had moved from Johns Hopkins to the University of Chicago and served on the committee of the resulting dissertation (Heyneman 1975a). The irony was that the results from Uganda diverged from the results of the original Coleman Report. Little correlation could be found between a student's

* The Coleman report was cited in 132 publications in 1975; an average of 71/year during the 1980s; 48/year during the 1980s and 50/year after 2000. It is one of the most influential studies in the history of the social sciences.

† Coleman used ordinary least squares (OLS) as the analytical tool. Today OLS has largely been replaced by HLM, which is able to more accurately capture different levels of institutional effects at the classroom, school, and school district level. Using HLM to analyze Coleman's data, school quality appears to have a greater effect than student background (Borman and Dowling 2010).

socioeconomic status and school achievement, and the predominant influence on achievement was the quality of the school, not the home background of the student (Heyneman 1976, 1977a, 1979).

The Uganda study was the first survey of primary schools in Sub-Saharan Africa and led to better understanding of the school system in many ways. It helped isolate the importance of particular teacher characteristics, particularly what they actually know about the subject they are teaching (Heyneman 1975b). It helped to better understand the influences on student learning from school community (Heyneman 1977b); school construction, facilities, and equipment (Heyneman 1977c); language of instruction (Heyneman 1980c); and school administration (Heyneman 1975c). In fact, because the data were representative of the nation's schools, it allowed a comparison of the methods of distribution of school supplies in which it was found that under the government monopoly, supplies were distributed more inequitably than when they were purchased on the open market by the school's parent/teacher committee (Heyneman 1975c). However, perhaps the most important utility from this new category of research came with the identification of textbooks as the single most important contributor to academic achievement, an influence that was all but ignored by the original Coleman Report (Heyneman 1980a). This led the way to a line of investigation on textbook effects, beginning with a meta-analysis (Heyneman et al. 1978) and moving on to experiments comparing the intervention effects of textbooks and education radio (Jamison et al. 1981), and to a nationwide experiment in which the ratio of textbooks-to-child was reduced from 10:1 to 2:1. This generated national gains in academic achievement of unprecedented magnitude (Heyneman et al. 1984). The original Uganda survey was also useful as a baseline to compare changes in the patterns of academic achievement before and after the political catastrophe of Idi Amin (Heyneman and Loxley 1983a).

The Coleman Report's findings, however, had been so powerful that many were convinced that they were universal. If so, it would imply that investments in schools and in school quality would only reinforce preexisting patterns of social stratification; that schooling was in fact harmful to the social opportunities of the poor. This interpretation became popular with the generalization of the Coleman findings outside the United States (Simmons and Alexander 1978). This interpretation, however, had used only those low-income countries that had participated in the IEA studies (Chile and Hungary). They had not included the results from Uganda and elsewhere. When these other studies were included, the pattern of achievement seemed to diverge from the "norm" typical of high-income countries (Heyneman 1980b; Heyneman and Loxley, 1982). The result of a 6-year meta-analysis on this question, sponsored by the World Bank, resulted in the comparison of achievement data from 29 countries, including 15 low- and middle-income countries, which suggested that the pattern was in fact linear: the lower the income of the country, the greater influence school quality

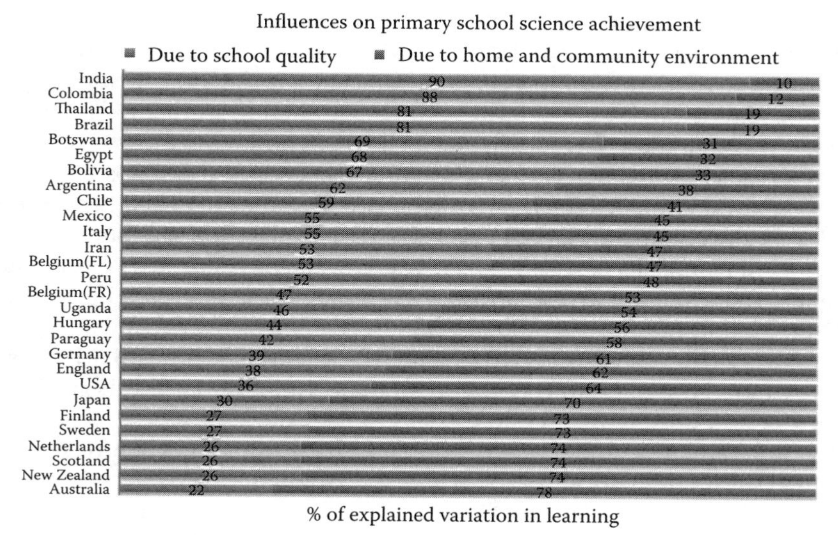

* Technical details in "The effect of primary school quality on academic achievement across 29 high and low income countries," *American Journal of Sociology*, (May 1983).

* Correlation between the influence of school quality and national GNP per capita [$R = -0.72$ ($p < 0.001$)].

Source: Heyneman and Loxely (1983).

FIGURE 3.1
Influences on academic achievement in 29 countries.

had on academic achievement (Heyneman and Loxley, 1983a).* The pattern looked like this (Figure 3.1).

Opposition to the findings and generally to the use of achievement tests across countries rose very quickly in the 1980s. The first set of challenges concerned sampling strategies, test reliability, and measures of socioeconomic status, which, it was alleged, were not sufficiently sensitive to local culture (Theisen et al. 1983). A separate set of issues arose with the use of new regression technologies. Hierarchical linear modeling (HLM) was said to be more accurate in capturing the effects when variables—classroom, school, system—were nested within one another. At first, it was alleged that HLM made the use of OLS simply outdated and the findings derived from OLS unreliable (Riddell 1989a, 1997). While HLM did seem to provide a new look at the model, it appeared unfair to criticize previous analyses as being inadequate. It was as though one criticized Charles Lindberg for flying across the Atlantic without using radar, when radar had not been invented at the time of his flight (Heyneman 1989; Riddell 1989b).

* The pattern of distribution of school quality within countries did not differ between high- and low-income countries, suggesting that the pattern of distribution was not responsible for the differences in the achievement findings (Heyneman and Loxley 1983b).

The real question of whether the Heyneman/Loxley effect was valid came once the TIMSS results were released.* TIMSS allowed the comparison of many new countries and increasingly accurate information. The first attempt to retest the Heyneman/Loxley effect used both OLS and HLM as methods. The conclusion was that there were no greater school effects in low-income countries, either because Heyneman and Loxley were inaccurate or perhaps because in the 20-year interim between their work and TIMSS the effect had been attenuated with development (Baker et al. 2002). This conclusion was supported by the work of Hanushek and Luque (2003) and by Harris (2007), who found insufficient school effects or evidence of diminishing marginal returns (Table 3.2).

On the other hand, the countries that participated in TIMSS were quite a bit different from those that participated in the earlier studies. TIMSS countries included the Russian Federation, Romania, Lithuania, Slovakia, Greece, Portugal, and middle-income countries (Table 3.3).

The per-capita income of the countries in the sample used by Baker et al. was 300% over the world mean, while the Heyneman/Loxley sample was only 50% over the world mean. Did that mean that the two samples were systematically different, and did this difference explain the differences in the results between Heyneman and Loxley versus Baker et al.? Work by Gameron and Long tested this by taking the countries used in the Baker et al. sample and adding 10 countries from Latin America, including Honduras, Bolivia, the Dominican Republic, Paraguay, and Mexico. The per capita income of their sample was $3409, comparable to that of Heyneman and Loxley ($2896) and considerably less than that of Baker et al. ($17,429; Table 3.4).

TABLE 3.2

Summary of Re-Analyses of the Heyneman/Loxley Effect

Source	Findings
Baker et al.	No greater school effect in low-income countries. Hypothesis: due to (i) economic development or (ii) threshold effect.
Hanushek and Luque	No effect of school resources (anywhere) after controlling for SES.
Harris	Search for diminishing marginal returns (DMR) to school quality (SQ): no solid conclusion.
Gameron and Long	School effects greater in low-income countries perhaps because of differences in gross national income (GNI) and differences in the distribution of SQ.
Chudgar and Luschei	School effects greater in low-income countries perhaps due to the inequality in income and the distribution of SQ.
Chiu and Chow	Poorer countries have higher classroom discipline (due to high value of education), hence good teachers have more effect in high-income countries.

* The theory of higher school effects in low-income countries was given the label of the Heyneman/Loxley effect by Baker et al. (2002).

TABLE 3.3

Countries in Different Samples

Heyneman/Loxley (1983)	Baker et al. (2002)	Long (2006)
Uganda	Russia	Honduras
Bolivia	Romania	Bolivia
Egypt	Thailand	Dominican Republic
Iran	Columbia	Paraguay
El Salvador	Latvia	Columbia
Thailand	Lithuania	Brazil
Peru	Slovakia	Venezuela
Paraguay	Hungary	Chile
Columbia	Czech Republic	Mexico
Brazil	Korea	Argentina
Botswana	Slovenia	(plus all countries from Baker et al.)
Chile	Greece	
Mexico	Portugal	
Hungary	Cyprus	
Argentina	New Zealand	
New Zealand	Spain	
Australia	Israel	
Italy	Australia	
United Kingdom	Canada	
Belgium	Hong Kong	
Singapore	France	
Netherlands	United Kingdom	
Finland	Belgium	
Germany	Singapore	
Sweden	Netherlands	
United States	Ireland	
Japan	Austria	
	Germany	
	Iceland	
	Denmark	
	United States	
	Norway	
	Switzerland	

Their conclusions support the original Heyneman/Loxley effect suggesting that the theory is supportable (including the use of HLM) and true over time (Gameron and Long 2007). Their results were also supported by the reanalyses of Chudgar and Luschei, who find that school effects are greater in low-income countries (Chudgar and Luschei, 2009), and the work of Chiu and Chow who find that in low-income countries teachers have more effect, perhaps on grounds that school discipline is higher due to the high value placed on education (Chiu and Chow 2010). Furthermore, a recent reanalysis of PISA data suggests that poor school performance does not necessarily follow from a student's disadvantaged background (Schleicher 2009).

TABLE 3.4

Reanalyses of the Heyneman–Loxley Effect: Comparison of Samples Per-Capita Income and School Effects

Sample	PCI (Per-Capita Income)	School Effects (% of Variance Explained)
Heyneman/Loxley (1983a)	2896	50.5
Baker et al. (2002)	17,429	34.4
Gameron and Long (2007)	3409	56.7

The sum result of this important debate would not have been possible without the ability to compare the influences on academic achievement across nations. Though the methods and measures have many caveats and problems, the fact is that these international surveys allow us to inquire as to the product and purposes of education itself and to either challenge or reinforce the reasons for the public investment. As to whether the Heyneman/Loxley effect is, in the end, supported, the jury is still out. The theory has helped stimulate debate for three decades, and may continue to stimulate debate for three more decades. What is clear is that the patterns of influences on academic achievement are not uniform around the world; they vary by gender, dependent variable (math is more subject to school effects; reading more influenced by home background), and by student age. What we do know is that reasons for disappointing results in the United States, particularly in urban areas, may be informed by the results from these studies conducted in very different environments, in which all children—including all poor children—have a high desire to learn.

Appendix

List of the Members of the International Association for the Evaluation of Education Achievement

Africa

Botswana
Kenya
Nigeria
South Africa

Asia

Armenia
China, People's Republic of
Chinese Taipei
Georgia
Hong Kong SAR
Indonesia
Japan

Europe

Austria
Belgium (Flemish)
Belgium (French)
Bosnia and Herzegovina
Bulgaria
Croatia
Cyprus
Czech Republic
Denmark
England
Estonia
Finland
France
Germany

North Africa and Middle East

Egypt
Iran
Israel
Jordan
Kuwait
Morocco
Palestinian National Authority
Qatar
United Arab Emirates

The Americas

Brazil
Canada

Kazakhstan	Greece	Chile
Korea, Republic of	Hungary	Colombia
Malaysia	Iceland	Mexico
Philippines	Ireland	United States
Singapore	Italy	
Thailand	Latvia	
	Lithuania	
Australasia	Luxembourg	
Australia	Macedonia	
New Zealand	Netherlands	
	Norway	
	Portugal	
	Romania	
	Russian Federation	
	Scotland	
	Slovak Republic	
	Slovenia	
	Spain	
	Sweden	
	Turkey	

References

Amadeo, J., Purta, J.T., Lehmann, R., Husfeldt, V., and Nikolova, R. (2002). *Civic Knowledge and Engagement: An IEA Study of Upper Secondary Students in Sixteen Countries.* Amsterdam: International Association for the Evaluation of Educational Achievement (IEA).

Ammermueller, A., Heijke, H., and Woessmann, L. 2005. Schooling quality in Eastern Europe: Educational production during transition. *Economics of Education Review*, 24(5), 579–599.

Anderson, L.W. and Postlethwaite, T.N. 1989. What IEA studies say about teachers and teaching. In: Purves, A.C. (Ed.), *International Comparisons and Educational Reform.* Alexandria, VA: Association for Supervision and Curriculum Development. (ERIC Document Reproduction Service No. ED 316 494).

Antonijevic, R. 2007. *Usage of computers and calculators and students' achievement: Results from TIMSS 2003.* Paper presented at the International Conference on Informatics, Educational Technology and New Media in Education, March 31–April 1, Sombor, Serbia.

Baker, D.P., Goesling, B., and Letendre, G.K. 2002. Socioeconomic status, school quality, and national economic development: A cross-national analysis of the 'Heyneman–Loxley Effect' on mathematics and science achievement. *Comparative Education Review*, 46(3), 291–312.

Barro, R.J. and Lee, J.W. 2000. International data on educational attainment: Updates and implications. NBER Working Paper N. 7911.

Barro, R.J. and Lee, J.W. 2001. International data on educational attainment: Updates and implications. *Oxford Economic Papers*, 53(3), 541–563.

Barro, R.J. and Lee, J.W. 2010. A new data set of educational attainment in the world, 1950–2010. NBER Working Paper N° 15902.

Berova, M. and Matusova, S. 2000. Slovak Republic. In: Robitaille, D.F., Beaton, A.E., and Plomp, T. (Eds.), *The Impact of TIMSS on the Teaching & Learning of Mathematics & Science* (pp. 133–138). Vancouver, Canada: Pacific Educational Press.

Bieber, T. and Martens, K. 2011. The OECD PISA study as a soft power in education? Lessons from Switzerland and the U.S. *European Journal of Education*, 46(1), 101–116.

Bohl, T. 2004. Empirische Unterrichcsforschung und Allgemeine Didaktik, Entstehung, Situation und Konsequenzen eines prekaren Spannungsverhaltnisses im Kontext der PISA-Studie. *Die Deutsche Schule*, 96, Ig/Heft 4, 414–425.

Borman, G. and Dowling, M. 2010. Schools and inequality: A multilevel analysis of Coleman's equality of educational opportunity data. *Teachers College Record* (on line version) January 28, 2010.

Bradburn, N. and Gilford, D.M (Eds.) 1990. *A Framework and Principles for International Comparative Studies in Education*. Washington, DC: National Academy Press.

Bradburn, N., Haertel, E. Schwille, J., and Torney-Purta, J. 1991. A rejoinder to 'I never promised you first place.' *Phi Delta Kappan*, 72(10), 774–777.

Braun, H. and Kanjee, A. 2006 Using assessment to improve education in developing nations. In: Cohen, J.E., Bloom, D.E., and Malin, M.B. (Eds.), *Educating All Children: A Global Agenda* (pp. 303–53). Cambridge, MA: American Academy Press.

Breakspear, S. 2012. The policy impact of PISA: An exploration of the normative effects of international benchmarking in school system performance. *OECD Education Working Papers*, No. 71, OECD Publishing. http://dx.doi.org/10.1787/5k9fdfqffr28-en.

Brunello, G. and Checchi, D. 2007. Does school tracking affect equality of opportunity? New international evidence. *Economic Policy*, 22(52), 781–861.

Chamberlain, M. and van Aalst, I. 2000. New Zealand. In: Robitaille, D.F., Beaton, A.E., and Plomp, T. (Eds.), *The Impact of TIMSS on the Teaching & Learning of Mathematics & Science* (pp. 98–103). Vancouver, Canada: Pacific Educational Press.

Chiu, M.M. and Chow, B.W. 2010. Classroom discipline across 41 countries: School, economic and cultural differences. *Journal of Cross-Cultural Psychology*, 20(10), 1–18.

Chromy, R.R. 2002. Sampling issues in design, conduct and interpretation of international comparative studies of school achievement. In: National Research Council (Ed.), *Methodological Advances in Cross-National Surveys of Educational Achievement* (pp. 80–117). Washington, DC: National Academy Press.

Chudgar, A. and Luschei, T.F. 2009. National income, income inequality, and the importance of schools: A hierarchical cross-national comparison. *American Education Research Journal*, 46(3), 626–658.

Coleman, J.S., Campbell, E.Q., Hobson, C.J., McPartland, J., Mood, A.M., Weinfall, F.D., and York, R.L. 1966. *The Equality of Educational Opportunity* (two volumes). Washington, DC: United States Department of Health, Education and Welfare.

Comber, L.C. and Keeves, J.P. 1973. *Science Education in Nineteen Countries*. Stockholm: Almqvist and Wiksell, and New York: Wiley.

De Boer, J. 2012. The role of engineering skills in development. Doctoral dissertation. Vanderbilt University, Nashville, Tennessee.

Dobbins, M. 2010. Education policy in New Zealand—Successfully navigating the international market for education. In: Martens, K., Kenneth, N.A., Windzio, M., and Weymann, A. (Eds.), *Transformation of Education Policy*. Basingstoke: Palgrave.

Egelund, N. 2008. The value of international comparative studies of achievement—A Danish perspective. *Assessment in Education: Principles, Policy & Practice*, 15(3), 245–251.

Fensham, P.J. 2007. Context or culture: Can TIMSS and PISA teach us about what determines educational achievement in science?. In: Atweh, B., Barton, A.C., Borba, M.C., Gough, N., Keitel, C., Vistro-Yu, C., and Vithal, R. (Eds.), *Internationalization and Globalization in Mathematics and Science Education* (pp. 151–172). Dordrecht, Netherlands: Springer.

Fertig, M. and Wright, R.E. 2005. School quality, educational attainment and aggregation bias. *Economics Letters*, 88(1), 109–114.

Figazzolo, L. 2009. Impact of PISA 2006 on the education policy debate. *Education International*. Retrieved from http://www.ei-ie.org/research/en/documentation.php.

Fuchs, T. and Woessmann, L. 2004. *Computers and Student Learning: Bivariate and Multivariate Evidence on the Availability and Use of Computers at Home and at School*. Munich, Germany: IFO Institute for Economic Research at the University of Munich.

Gameron, A. and Long, D. 2007. Equality of educational opportunity: A 40 year retrospective. In: Teese, R., Lamb, S., and Duru-Bellat, M. (Eds.), *International Studies in Educational Inequality, Theory and Policy* (pp. 23–48). Dordrecht, Netherlands: Springer.

Ganimian, A. 2009. *How Much Are Latin American Children Learning?: Highlights from the Second Regional Student Achievement Test (SERCE)*. Washington, DC: PREAL.

Ganimian, A. and Rocha, A.S. 2011. *Measuring Up? How did American and the Caribbean perform on the 2009 Programme for International Student Assessment (PISA)?* Programa de Promocion de la Reforma Educativa en America Latina y el Caribe (PREAL) Retrieved from Inter American Dialogue website: http://www.thedialogue.org/PublicationFiles/Preal_PISA_ENGLowres.pdf.

Geske, A. and Kangro, A. 2000. Latvia. In: Robitaille, D.F., Beaton, A.E., and Plomp, T. (Eds.), *The Impact of TIMSS on the Teaching & Learning of Mathematics & Science* (pp. 85–88). Vancouver, Canada: Pacific Educational Press.

Gudmundsson, E. 2000. Iceland. In: Robitaille, D.F., Beaton, A.E., and Plomp, T. (Eds.), *The Impact of TIMSS on the Teaching & Learning of Mathematics & Science* (pp. 56–60). Vancouver, Canada: Pacific Educational Press.

Hanushek, E.A. and Luque, J.A. 2003. Efficiency and equity in schools around the world. *Economics of Education Review*, 22, 481–502.

Hanushek, E.A. and Woessmann, L. 2006. Does educational tracking affect performance and inequality? Differences-in-differences evidence across countries. *Economic Journal*, 116(510), 63–76.

Harris, D.N. 2007. Diminishing marginal returns and the production of education: An international analysis. *Education Economics*, 15(1), 31–53.

Heyneman, S.P. 1975a. Influences on academic achievement in Uganda: A "Coleman Report" from a non-industrial society. Doctoral dissertation. University of Chicago, Chicago, IL.

Heyneman, S.P. 1975b. Changes in efficiency and in equity accruing from government involvement in Ugandan primary education. *African Studies Review*, 18(1), 51–60.

Heyneman, S.P. 1975c. Relationships between teachers' characteristics and differences in academic achievement among Ugandan primary schools. *Education in Eastern Africa*, 6(1), 41–51.

Heyneman, S.P. 1976a. A brief note on the relationship between socioeconomic status and test performance among Ugandan primary school children. *Comparative Education Review*, 20(1), 42–47.

Heyneman, S.P. 1977a. Influences on academic achievement: A comparison of results from Uganda and more industrial societies. *Sociology of Education*, 11(2), 245–259.

Heyneman, S.P. 1977b. Relationships between the primary school community and academic achievement in Uganda. *Journal of Developing Areas*, 11(2), 245–259.

Heyneman, S.P. 1977c. Differences in construction, facilities, equipment and academic achievement among Ugandan primary schools. *International Review of Education*, 23, 35–46.

Heyneman, S.P. 1979. Why impoverished children do well in Ugandan schools. *Comparative Education*, 15(2), 175–185.

Heyneman, S.P. 1980a. Differences between developed and developing countries: Comment on Simmons and Alexander's determinants of school achievement. *Economic Development and Cultural Change*, 28(2), 403–406.

Heyneman, S.P. 1980b. Student learning in Uganda: textbook availability and other determinants. *Comparative Education Review*, 24(2) (June) 108–118 (coauthored with Dean Jamison).

Heyneman, S.P. 1980c. Instruction in the mother tongue: The question of logistics. *Journal of Canadian and International Education*, 9(2), 88–94.

Heyneman, S.P. 1983. Education during a period of austerity: Uganda, 1971–1981. *Comparative Education Review*, 27(3), 403–413.

Heyneman, S.P. 1989. Multilevel methods for analyzing school effects in developing countries. *Comparative Education Review*, 33(4), 498–504.

Heyneman, S.P. 2004. International education quality. *Economics of Education Review*, 23, 441–52.

Heyneman, S.P. 2009a. "Should school time be extended each day and in the summer?" *The Tennessean* (October 6).

Heyneman, S.P. 2009b. An education bureaucracy that works. *Education Next* (November) http://educationnext.org/an-education-bureaucracy-that-works/

Heyneman, S.P., Farrell, J.P., and Sepulveda-Stuardo. 1978. *Textbooks and Achievement: What We Know.* Washington, DC: World Bank Staff Working Paper No. 298 (October) (available in English, French, and Spanish).

Heyneman, S.P. and Loxley, W. 1982. Influences on academic achievement across high and low-income countries: A re-analysis of IEA data. *Sociology of Education*, 55(1), 13–21.

Heyneman, S.P. and Loxley, W. 1983a. The effect of primary school quality on academic achievement across twenty-nine high- and low-income countries, *American Journal of Sociology*, 88(6), 1162–1194.

Heyneman, S.P. and Loxley, W. 1983b. The distribution of primary school quality within high- and low-income countries. *Comparative Education Review*, 27(1), 108–118.

Heyneman, S.P., Jamison, D., and Montenegro, X. 1984. Textbooks in the Philippines: Evaluation of the pedagogical impact of a nationwide investment. *Educational Evaluation and Policy Analysis*, 6(2), 139–150.

Heyneman, S.P. and Lykins, C. 2008. The evolution of comparative and international education statistics. In: Ladd, H.F. and Fiske E.B. (Eds.), *Handbook of Research in Education Finance and Policy* (pp. 105–127). New York: Routledge.

Holliday, W.G. and Holliday, B.W. 2003. Why using international comparative math and science achievement data from TIMSS is not helpful. *The Education Forum* 63(3), 250–257.

Husen, T., (Ed). 1967. *International Study of Achievement in Mathematics: A Comparison of Twelve Countries*. New York: John Wiley and Sons.

Hussein, M.G.A. and Hussain, A.A. 2000. Kuwait. In: Robitaille, D.F., Beaton, A.E., and Plomp, T. (Eds.), *The Impact of TIMSS on the Teaching & Learning of Mathematics & Science* (pp. 82–84). Vancouver, Canada: Pacific Educational Press.

Jamison, D.T., Heyneman, S.P., Searle, B., and Galda, K. 1981. Improving elementary mathematics education in Nicaragua: An experimental study of the impact of textbooks and radio on achievement. *Journal of Educational Psychology*, 73(4), 556–567.

Jimenez, E. and Lockheed, M.E. 1995. Public and private secondary education in developing countries. Discussion Paper No. 309. The World Bank. Washington, DC.

Jimmenez, E., Lockheed, M., and Wattanawaha, N. 1988. The relative efficiency of public and private school: The case of Thailand. The World Bank.

Kamens, D.H. 2013. Globalization and the emergence of an audit culture: PISA and the search for 'best practice' and magic bullets. In: Benavot, A. and M. Heinz-Dieter (Eds.), *PISA, Power and Policy: the Emergence of Global Educational Governance* (Oxford Studies in Comparative Education). Oxford, UK: Symposium Books.

Karimi, A. and Daeipour, P. 2007. The impact of PIRLS in the Islamic Republic of Iran. In: Schwippert, K. (Ed.), *Progress in Reading Literacy: The Impact of PIRLS 2001 in 13 Countries*. New York, NY: Munster Waxmann.

Keeves, J.P. 1995. *The World of School Learning: Selected Key Findings from 35 Years of IEA Research*. The Hague: International Association for the Evaluation of Educational Achievement.

Keys, W. 2000. England. In: Robitaille, D.F., Beaton, A.E., and Plomp, T. (Eds.), *The Impact of TIMSS on the Teaching & Learning of Mathematics & Science*. Vancouver, Canada: Pacific Educational Press.

Kiamanesh, A.R. and Kheirieh, M. 2000. Iran. In Robitaille, D.F., Beaton, A.E., and Plomp, T. (Eds.), *The Impact of TIMSS on the Teaching & Learning of Mathematics & Science* (pp. 61–65). Vancouver, Canada: Pacific Educational Press.

Kovalyova, G. 2000. Russia. In: Robitaille, D.F., Beaton, A.E., and Plomp, T. (Eds.), *The Impact of TIMSS on the Teaching & Learning of Mathematics & Science* (pp. 120–124). Vancouver, Canada: Pacific Educational Press.

Kozma, R.B. (Ed.) 2003. *Technology, Innovation and Educational Change: A Global Perspective*. Eugene, OR: International Society for Technology in Education.

Lockheed, M.E., Vail, S.C., and Fuller, B. 1986. How textbooks affect achievement in developing countries: Evidence from Thailand. *Educational Evaluation and Policy Analysis Winter*, 8(4), 379–392.

Lockheed, M.E., Fuller, B., and Nyrongo, R. 1989. Family effects on students' achievement in Thailand and Malawi. *Sociology of Education*, 62, 239–256.

Medrich, E.A. and Griffith, J.E. 1992. International Mathematics and Science Assessments: What Have We Learned? U.S. Department of Education, Office of Educational Research and Improvement, NCES-92–011, January.

Mevavech, Z.R. 2000. Israel. In: Robitaille, D.F., Beaton, A.E., and Plomp, T. (Eds.), *The Impact of TIMSS on the Teaching & Learning of Mathematics & Science* (pp. 66–70). Vancouver, Canada: Pacific Educational Press.

Mosteller, F. and Moynihan, D. (Eds.) 1972. *On Equality of Educational Opportunity: Papers Deriving from the Harvard Faculty Seminar on the Coleman Report*. New York: Vintage Books.

Mullis, I.V.S. and Martin, M.O. 2007. Lessons learned: What international assessments tell us about math achievement. In: Loveless, T. (Ed.), *TIMSS in Perspective: Lessons Learned from IEA's Four Decades of International Mathematics Assessments.* Washington, DC: Brookings Institution Press.

Mullis, I.V., Martin, M.O., Gonzalez, E.J., and Chrostowski, S.J. 2004. TIMSS 2003 International Mathematics Report.

Mullis, I.V., Martin, M.O., Gonzales, E.J., Kelly, D.L., and Smith, T.A. 1996. *Mathematics Achievement in the Middle School Years: IEA's Third International Mathematics and Science Study (TIMSS).* Chestnut Hill, MA: TIMSS International Study Center, Boston College.

Naceva, B. and Mickovska, G. 2007. The impact of PIRLS in the Republic of Macedonia. In: Schwippert, K. (Ed.), *Progress in Reading Literacy: The Impact of PIRLS 2001 in 13 Countries.* New York, NY: Munster Waxmann.

Neumann, K., Fischer, H.E., and Kauertz, A. 2010. From PISA to educational standards: The impact of large-scale assessments on science education in Germany. *International Journal of Science and Mathematics Education*, 8, 545–563.

Noveanu, G. and Noveanu, D. 2000. Romania. In Robitaille, D.F., Beaton, A.E., and Plomp, T. (Eds.), *The Impact of TIMSS on the Teaching & Learning of Mathematics & Science* (pp. 117–119). Vancouver, Canada: Pacific Educational Press.

OECD. 2002. *Lesen Kann Die Welt verandern: Leistung und Engagement im Landervergleich.* Paris: OECD.

OECD. 2004a. *Learning for Tomorrow's World. First Results from PISA 2003.* Paris: OECD.

OECD. 2004b. *What Makes School Systems Perform. Seeing School Systems Through the Prism of PISA.* Paris: OECD.

OECD. 2009. *Education Today. The OECD Perspective.* Paris: OECD.

OECD and UNESCO. 2003. *Literacy Skills for the World of Tomorrow—Further Results from PISA 2000.* Paris and Montréal: OECD and UNESCO Institute for Statistics.

OECD. 2010. *PISA 2009 Results: Executive Summary.* Paris: OECD.

Papanastasiou, E.C., Zembylas, M., and Vrasidas, C. 2003. Can computer use hurt science achievement? The U.S.A results from PISA. *Journal of Science Education and Technology*, 12(3), 325–332.

Papanastasiou, E.C., Zembylas, M., and Vrasidas, C. 2004. Reexamining patterns of negative computer-use and achievement relationships. Where and why do they exist? In: Papanastasiou, C. (Ed.), *Proceedings of the IRC-2004. TIMSS. Volume 1.* Lefkosia, Cyprus: Cyprus University Press.

Park, K. 2004. Factors contributing to East Asian students' high achievement: Focusing on East Asian teachers and their teaching. Paper presented at the APEC Educational Reform Summit, 12, January 2004.

Pelgrum, W. and Plomp, T. 2008. Methods for large-scale international studies on ICT in education. In: Voogt, J., Knezek, G. (Eds.), *International Handbook of Information Technology in Primary and Secondary Education* (pp. 1053–1066). Springer Science + Business Media, LLC, New York, 2008.

Plank, D.N. and Johnson, Jr. B.L. 2011. Curriculum policy and educational productivity. In: Mitchell, D.E., Crowson, R.L. and Shipps, D. (Eds.), *Shaping Education Policy: Power and Process* (pp. 167–188). New York, NY: Routledge Taylor & Francis Group.

PREAL. 2011. *Better Schools through Better Policy.* Washington, DC: PREAL.

Purves, A.C. (Ed.). 1992. *The IEA Study of Written Composition II: Education and Performance in Fourteen Countries*. Oxford: Pergamon Press.
Ravitch, D. 2010. *The Death and Life of the Great American School System: How Testing and Choice are Undermining Education*. New York: Basic Books.
Riddell, A.R. 1989a. An alternative approach to the study of school effectiveness in third world countries. *Comparative Education Review*, 33(4), 481–497.
Riddell, A.R. 1989b. Response to Heyneman. *Comparative Education Review*, 33(4), 505–506.
Riddell, A.R. 1997. Assessing designs for school effectiveness research in developing countries. *Comparative Education Review*, 41(2), 178–204.
Rotberg I.C. 1990. I never promised you first place. *Phi Delta Kappan*, 72, 296–303.
Rotberg I.C. 1996. Five myths about test score comparisons. *School Administrator*, 53, 30–31.
Rotberg I.C. 2006. Assessment around the world. *Educational Leadership*, 64(3), 58–63.
Rotberg I.C. 2007. Why do our myths matter? *School Administrator*, 64(4), 6.
Rotberg I.C. 2008. Quick fixes, test scores and the global economy: Myths that continue to confound us. *Education Week*, 27(41), 32.
Rowan, B. 2002. Large scale cross-national surveys of educational achievement: Pitfalls and possibilities. In: Porter, A.C. and Gamoran, A. (Eds.), *Methodological Advances in Cross-National Surveys of Educational Achievement* (pp. 321–349). *Board on Comparative Studies in Education*. Washington, DC: National Academy Press.
Schleicher, A. 2009. Securing quality and equity in education: Lessons from PISA. *UNESCO Prospects*, 39, 251–263.
Schuetz, G., Ursprung, H.W., and Woessmann, L. 2008. Education policy and equality of opportunity. *Kyklos*, 61(2), 279–308.
Schulz, W., Ainley, J., Fraillon, J., Kerr, D., and Losito, B. 2010. *ICCS 2009 International Report: Civic Knowledge, Attitudes, and Engagement among Lower-Secondary School Students in 38 Countries*. Amsterdam: International Association for the Evaluation of Educational Achievement.
Schwippert, K. 2007. The impact of PIRLS in Germany. In Schwippert, K. (Ed.), *Progress in Reading Literacy: The Impact of PIRLS 2001 in 13 Countries*. New York, NY: Munster Waxmann.
Shorrocks-Taylor, D., Jenkins, E., Curry, J., Swinnerton, B., Laws, P., Hargreaves, M., and Nelson, N. 1998. *An Investigation of the Performance of English Pupils in the Third International Mathematics and Science Study (TIMSS)*. Leeds: Leeds University Press.
Simmons, J. and Alexander, L. 1978. The determinants of school achievement in developing countries: A review of research. *Economic Development and Cultural Change* (January), 341–358.
Smyth, J. 1996. The origins, purposes and scope of the international standard classification of education. Paper submitted to the ISCED revision task force. Paris: UNESCO (February) (mimeographed).
Smyth, J. 2005. *International Literacy Statistics 1950–2000*. Paris: UNESCO (mimeographed).
Stigler, J.W. and Hiebert, J. 1999. *The Teaching Gap*. New York: Free Press.
Strakova, J., Paleckova, J., and Tomasek, V. 2000. Czech Republic. In Robitaille, D.F., Beaton, A.E., and Plomp, T. (Eds.), *The Impact of TIMSS on the Teaching & Learning of Mathematics & Science* (pp. 41–44). Vancouver, Canada: Pacific Educational Press.
Taylor, A.R. 2000. Canada. In: Robitaille, D.F., Beaton, A.E., and Plomp, T. (Eds.), *The Impact of TIMSS on the Teaching & Learning of Mathematics & Science* (pp. 50–55). Vancouver, Canada: Pacific Educational Press.

Theisen, G.L., Achola, P.W., and Boakari, F.M. 1983. The underachievement of cross-national studies of academic achievement. *Comparative Education Review*, 27(1), 46–68.

Twist, L. 2007. The impact of PIRLS in England. In: Schwippert, K. (Ed.), *Progress in Reading Literacy: The Impact of PIRLS 2001 in 13 Countries*. New York, NY: Munster Waxmann.

Vlaardingerbroek, B. and Taylor, T.G.N. 2003. Teacher education variables as correlates of primary science ratings in thirteen TIMSS systems. *International Journal of Educational Development*, 23, 429–438.

Wagemaker, H. 2004. IEA: International studies, impact and transition. Speech given on the occasion of the *1st IEA International Research Conference*, University of Cyprus, Lefkosia, Cyprus, 11–13, May 2004.

Wagner, D.A. 2011. *Smaller, Quicker, Cheaper. Improving Learning Assessments for Developing Countries*. Paris, France: International Institute for Educational Planning. UNESCO.

Waldinger, F. 2006. *Does Tracking Affect the Importance of Family Background on Students' Test Scores*? Mimeo: London School of Economics.

Walker, D.A. (Ed.). 1976. *The IEA Six-Subject Survey: An Empirical Study of Education in Twenty-One Countries*. New York: Wiley.

Weiler, H. 1994. The failure of reform and the macro-politics of education: Notes on a theoretical challenge. In: Val Rust (Ed.), *Educational Reform in International Perspective* (pp. 43–54). Greenwich, CT: JAI Press.

Wittwer, J. and Senkbeil, M. 2008. Is students' computer use at home related to their mathematical performance at school? *Computers & Education*, 50(4), 1558–1571.

Woessmann, L. 2005a. Educational production in East Asia: The impact of family background and schooling policies on student performance. *German Economic Review*, 6(3), 331–353.

Woessmann, L. 2005b. Educational production in Europe. *Economic Policy*, 20(43), 446–504.

Woessmann, L. and West, M.R. 2006. Class-size effects in school systems around the world: Evidence from between-grade variation in TIMSS. *European Economic Review*, 50(3), 695–736.

Woessmann, L., Luedemann, E., Schuetz, G., and West, M.R. 2009. *School Accountability, Autonomy, and Choice Around the World*. Cheltenham, UK: Edward Elgar.

World Bank. 2011. *Learning for All: Investing in People's Knowledge and Skills to Promote Development. Education Sector Strategy, 2020*. Washington, DC: The World Bank.

Section II

Analytic Processes and Technical Issues Around International Large-Scale Assessment Data

4

Assessment Design for International Large-Scale Assessments

Leslie Rutkowski
Indiana University

Eugene Gonzalez
IEA-ETS Research Institute and Educational Testing Service

Matthias von Davier
Educational Testing Service

Yan Zhou
Indiana University

CONTENTS

Introduction

As an invariant constraint to producing high-quality measurement instruments, questionnaire length is a significant factor in the development process. Owing to issues of fatigue, attrition, or the logistics of available study participants, instrument (referred to hereafter as *test*) developers must constantly weigh measuring the construct(s) of interest in a valid and reliable way against the real limitations associated with administration. To that end, a method referred to as multiple matrix sampling (MMS) is used in large-scale educational surveys such as the Programme for International Student Assessment (PISA) and the Trends in International Mathematics and Science Study (TIMSS) to balance the demands of test length against the need for sound measurement. Although methods for estimating latent traits from multiple matrix designs are the subject of much current or recent methodological research (e.g., Thomas and Gan 1997; von Davier and Sinharay 2007, 2009), advances in MMS theory, outside of mathematically optimizing designs (e.g., van der Linden and Carlson 1999; van der Linden et al. 2004) have largely come to a slowdown. Further, much of the primary literature on MMS has lain dormant for many years or, excepting a few recent papers (e.g., Childs and Jaciw 2003; Frey et al. 2009; Gonzalez and Rutkowski 2010), ascetic discussions around the matter have been relegated to the pages of technical manuals. Therefore, we pick up this discussion by detailing the origins of MMS, describing where it has been applied in educational assessments and who currently uses it (and why), and we suggest how this particular assessment design might be applied in new ways in the future. We focus many of our views for the future on international, national, and state-level large-scale assessments; however, the potential of MMS is not limited to these contexts.

In this chapter, we trace the historical antecedents and representative contributors to the development of closely related methods. We then describe early development of what we know today as MMS and some of the breakthroughs that resulted in its wide application in educational surveys. We also overview the predominant international studies that currently utilize multiple matrix assessment designs by way of describing the most recently completed cycles of TIMSS (2011), PIRLS (2011), and PISA (2009). We propose future directions for the continued development and application of MMS to large-scale surveys. As part of our discussion, we provide numerous references on a topic that goes by many pseudonyms—item sampling, item–examinee sampling, matrix sampling, split questionnaire survey design, and partial survey design, to name but a few. Finally, we suggest areas for further research and other possible applications, especially in large-scale educational surveys. First, however, we briefly introduce MMS and explain why it is useful for a particular class of measurement settings.

Background

To facilitate our discussion of MMS, we introduce some basic terms and concepts from the sampling field. The process of sampling, in general, is concerned with drawing a representative subset from a population of interest in such a way that defensible inferences about the sampled population can be made (see Chapter 6 in the current volume for more information). In sampling parlance, the individual sampled units for which a measure is taken are typically referred to as *observation units*, which are generally human subjects in the social sciences. Subject to certain constraints, the population of interest is referred to as the *sampling frame*, from which the observation units are drawn. In the social sciences, the sampling frame is typically some population of human subjects, for example, all eighth grade students in the United States, female alcoholics admitted to a recovery program, or registered voters in a particular state. Usually, a study design has just one sampling frame from which to draw observation units; however, this is not a necessary condition for drawing a study sample, as we subsequently explain in the case of MMS.

What Is MMS?

To better understand MMS, we contrast it with the more familiar sampling, where the sampling frame is some population of human subjects. For ease and to avoid confusion later, we use the term examinees or respondents instead of observation units to refer to human subjects selected for participation in some study. Typically, examinee sampling involves selecting participants from a larger population according to a sampling framework, measuring the participants with an instrument and drawing inferences about the population of interest. MMS differs in one critical respect from examinee sampling: there are essentially two mutually exclusive sampling frames from which to draw observation units. That is, in addition to taking a sample of examinees from the population of interested, a sample of measures is also drawn from the universe of interest, assuming that the total assessment comprises the universe of interest. In practice, MMS is a method of assembling and administering a survey or assessment where each respondent is measured on a sample of the total assessment. To illustrate, we compare traditional sampling with MMS. Figure 4.1 illustrates the traditional approach to sampling with a hypothetical sampling frame from which a sample of n examinees (shaded in gray) is selected for measurement from a population of N students. The n sampled examinees are measured on every possible item in the universe K. We contrast traditional sampling with an example of MMS, illustrated in Figure 4.2, where we see that three groups of examinees (n_1, n_2, and n_3) are administered group k_1, k_2, and k_3 items, respectively, sampled from the total universe of K possible items. In this particular design there is no item overlap across the item–unit samples; however, each sampled examinee is administered a common core of items. It is important to note that this example is just one of a multitude of possible MMS designs. Besides a number of practical and theoretical constraints associated with choosing a multiple matrix design (Frey et al. 2009), combinatoric or other mathematical optimization routines exist for choosing an assessment design subject

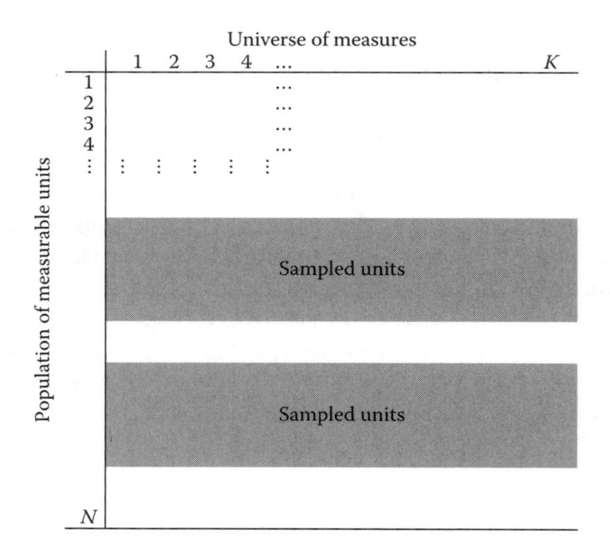

FIGURE 4.1
Examinee sampling.

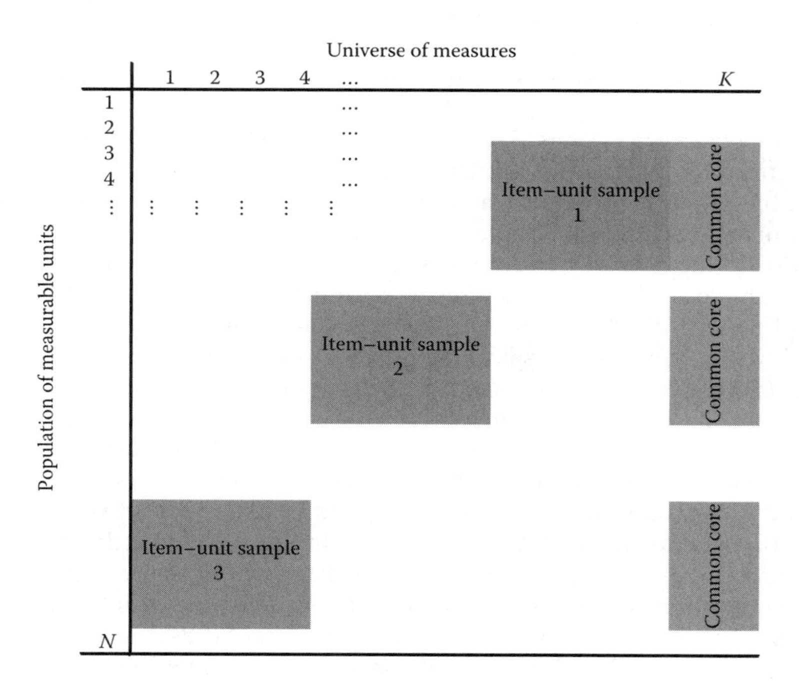

FIGURE 4.2
A multiple matrix sampling design with a common core of items.

to a number of constraints (e.g., van der Linden and Carlson 1999; van der Linden et al. 2004). We discuss a number of currently implemented designs later in this chapter.

MMS offers a number of advantages and disadvantages. Although an assessment design that uses MMS allows for coverage of a broad content domain in a reasonable amount of time for the individual examinees, this approach poses challenges for scaling the items and estimating individual proficiency, since only a fraction of the sample of examinees respond to a given item, and any examinee responds to only a fraction of the total assessment. The sparseness of the matrix and student sample size will influence the accuracy of the item parameters and the examinees' proficiency estimates (e.g., Mislevy 1991; Gonzalez and Rutkowski 2010). Although stable item parameter estimates can be obtained with standard item response theory models (assuming that a sufficient sample of responses to each of the items is available), making inferences about student proficiencies requires additional structure in the form of a population model.

The population models used in large-scale assessment are based on an integration of item response theory and latent-regression methods, and are sometimes referred to as plausible value methods (Mislevy et al. 1992a, b; Chapter 7 of this volume). The idea is to estimate the proficiency distribution based on the matrix sample of item responses and additional information available about the students taking the test when a multiple matrix design is used. Because students only see a fraction of the total item pool, individual student ability estimates are associated with a large error of measurement. Rather than a method for estimating individual student achievement, this method efficiently and precisely estimates *population* characteristics using individual student responses and other background variables collected in the assessment. For an accessible primer on plausible value methods, interested readers are referred to von Davier et al. (2009).

Why Is MMS Useful?

Generally, large-scale assessment programs feature broad content and cognitive domains that involve a number of processes, concepts, and objectives. For example, TIMSS 2007 included in its eighth grade mathematics assessment four content domains: *number, algebra, geometry,* and *data and chance.* Within these broadly defined content domains, students were expected to draw on the cognitive domains of *knowing, applying,* and *reasoning.* In spite of TIMSS's broad domains, it was necessary to summarize performance in a sensible way. Historically, TIMSS achievement reporting was limited to only the content domains; however, the 2007 administration also reported on the cognitive domains (Mullis et al. 2005). This approach brought the total number of reporting scales in mathematics to eight (seven subscales plus overall mathematics). Reporting in this fashion implies that an ample number of items in each of the reporting categories should have been administered such that sufficiently precise estimates of proficiency distribution in reporting subgroups were possible.

In total, more than 10 hours of testing time was covered in the TIMSS 2007 assessment, which made it impossible to test all examinees on every item. Instead, test developers used an assessment design that distributed 429 total mathematics and science items across 14 mathematics blocks and 14 science blocks. The blocks were subsequently arranged into 14 booklets with four blocks each. This design, sometimes referred to as a partially balanced incomplete block design (discussed later), ensured sufficient linkage across booklets since each block (and therefore each item) appeared in two different booklets. Booklets were then administered such that the total assessment material was divided into more reasonable 90-minute periods of testing time for each student. With this design, students were administered just a small subset of the total available items. See Mullis et al. (2005) for complete details on the TIMSS 2007 assessment design. Later in this chapter we review a few of the common designs currently used in the most prominent large-scale educational surveys. First, we provide a short history of MMS and its evolution over time.

Evolution of a Method

Item Sampling

Traditional approaches to sampling for surveys generally rely on sampling respondents from the population; however, a number of initial investigations (Johnson and Lord 1958; Lord 1962; Pugh 1971) suggested that item sampling, where small subsets of the total items are administered in groups to the entire population, could be an efficient and cost effective way of assessing examinees. One of the earliest empirical examinations of item sampling by Johnson and Lord, who simply referred to the method as Procedure B, compared an 11-item assessment that administered every item to every respondent to an assessment that administered the same 11 items, albeit individually, to each respondent. The design was such that each item was attempted by three examinees. Findings from this study surprisingly indicated that estimates of group means were more consistent in Procedure B than in the traditional examinee sample design, where a subset of examinees (and their responses to all of the items) was used to estimate group means.

In 1960, Lord coined the term "item sampling" in a study that extended Johnson and Lord's (1958) earlier success with estimating group means to estimating the entire frequency distribution of a 70-item vocabulary test score for a norm population of 1000 individuals. In Lord's study, 7 items were drawn 10 times at random *with replacement* to create 10 possibly overlapping 7-item tests. Each of these 10 tests was administered to 1 of 10 groups of 100 examinees (randomly assigned to each group from the population of 1000). Findings from this investigation indicated that the generating norm distribution was adequately represented by the results from the item-sampled design, except for a few groups at the highest end of the distribution, even though 18 (of 70 total) items were drawn more than once and 25 items were never drawn at all.

Several subsequent empirical studies continued the technique of sampling items at random for assignment to subtests (*booklets* in large-scale assessment terminology; e.g., Lord 1962; Plumlee 1964; Shoemaker 1970a, b). During approximately the same period, researchers seeking to innovate and extend the idea of item sampling began to use stratification techniques to assign items to subtests based on item difficulty (Kleinke 1972; Scheetz and Forsyth 1977) or item discrimination (Myerber 1975; Scheetz and Forsyth 1977). Lord (1965) formalized the methods of item sampling as it related to test theory using formulas developed by Hooke (1956a). Further, he described a number of reasonably straightforward formulas for estimating means and variances for item-sampled assessments under varied conditions. To that end, many of the early studies depended on percent-correct methods for estimating test scores on multiple-matrix sampled assessments (cf. Johnson and Lord 1958; Lord 1962; Plumlee 1964; and Pugh 1971) with general formulas given by Hooke (1956a, b).

Balanced Incomplete Block Designs

A significant advance in the field came when Lord (1965) suggested that for item sampling, balanced incomplete block (BIB) designs (from experimental design and combinatorial mathematics) might fulfill the conditions that every item should appear an equal number of times and that many subtests should be administered, with one of these subtests to each student. In the same manuscript, Lord first referred to item sampling as "matrix sampling" (Wilks 1962), both of which names persist today. The concept of applying experimental designs to matrix sampling was first realized when Knapp (1968) incorporated Lord's suggestion and found that a BIB design was an extremely efficient means of estimating the means, variances, and reliability coefficients of several assessments. Although the current chapter makes mention of several different experimental designs that are applicable to the context of large-scale assessment, readers are directed to Frey et al. (2009) for a more detailed overview of this topic.

A short digression to introduce the details of a BIB design is in order. In early testing literature, the parameters used to specify a BIB design were associated with slightly different components of the assessment than are currently in common use. In early applications of BIB designs for testing (e.g., Knapp 1968), the parameters of concern shared a one-to-one correspondence with experimental design (the typical nomenclature of which is denoted in parentheses), where v is the total number of items in an item pool (*treatments*); b is the total number of examinees (*blocks*); k is the number of items to be administered to any one individual (*block size*); r is the number of times each item is administered (*replications*); and λ corresponds to the number of times each pair of items (*pair of treatments*) is administered. Parameters specifying older assessment applications of a BIB design, thus, are v, b, k, r, and λ. Limits on Knapp's specific approach were that the number of items must be prime and the number of examinees must be $v(v-1)/2$. Practical difficulties might arise from these constraints, including when the number of items is very large. For example, with 199 items, 19,701 examinees are required (Knapp).

Current testing approaches for specifying a BIB design differ in that instead of assigning items to examinees, items are assigned to clusters, which are assigned to booklets, which are in turn assigned to examinees. Specifically, the total item pool is divided into t clusters of items (adding an advantage that the number of items per cluster can vary). The t clusters are then assigned to b booklets. Each booklet is of equal length with k clusters. Every cluster t appears only once in any booklet and r times across all booklets, with each pair of clusters appearing λ times. Parameters then specifying the design are t, b, r, k, and λ. This implies that $rt = bk$. One important constraint on this design is $t \leq b$. That is, the number of clusters should not exceed the number of booklets. In practical applications, discussed subsequently, some of these conditions are not met. Namely, λ is less than 1 in at least one international large-scale assessment, rendering this design not strictly a BIB design but a partially balanced incomplete block (PBIB) design (Giesbrecht and Gumpertz 2004). A disadvantage of the condition $\lambda < 1$ is that not every cluster appears in a booklet with every other cluster, making the possibility of a pair-wise, cluster-specific context effect difficult to evaluate and control. In addition, item linking is heterogeneous as item pairs are consequently unbalanced.

Several researchers pursued Knapp's (1968) original line of inquiry to further establish the utility of experimental designs to the field of testing. In particular, Shoemaker (1973) investigated the utility of a PBIB design, acknowledging the difficulties of a strict BIB design when the number of items was large. His findings suggested that a PBIB design was effective at reproducing known means under a number of conditions. In the same analysis, Shoemaker used the jackknife to estimate standard errors with limited success. This line of inquiry continued (Pandey and Shoemaker 1975; Sirotnik and Wellington 1974) with little resolution regarding sampling distributions for the test mean and variance.

The rise of item response theory (IRT) methods in assessment was associated with calls for advances that integrated the MMS schemes with better measurement models (Bock et al. 1982). In response, Mislevy (1983, 1984), Reiser (1983), and Mislevy and Sheehan (1987), among others, developed group-level models that estimated the underlying latent traits measured by assessments that followed a multiple-matrix design. This approach provided, among other innovations, a tractable solution to the historical problem of estimating standard errors. The dominant approach, applied to the US National Assessment of Educational Progress (NAEP) since 1983/1984 (Beaton 1987;

von Davier et al. 2006) and clearly laid out in Mislevy (1991) and Sinharay and von Davier (Chapter 7, the current volume), uses a population model to integrate the advantages of MMS with BIB or PBIB designs with IRT models. This approach is commonly referred to as conditioning, population modeling, or latent regression, but sometimes also called plausible value methodology. The naming confusion stems from the fact that the main data products of the population model are known as *plausible values,* which are essentially multiple imputations of achievement based on the posterior distribution of abilities, given item responses and the expected distribution of skills and given background data. The extraction of plausible values from the population model draws on Rubin's (1978, 1987) multiple imputation methods to impute examinee proficiency values valid for population-level inference. This approach to estimating traits from multiple-matrix designs is frequently used in national and international large-scale assessment applications today. Von Davier et al. describe this methodology and discuss variations and extensions of the methods used in national and international assessments. Adams et al. (2007) describe the application of this methodology in PISA.

Current Applications of MMS in International Assessment

Currently, only PISA 2009 uses a pure BIB design approach to matrix sampling when assembling assessment booklets. The most recent IEA studies, including TIMSS 2011 and PIRLS 2011, use a variation of a PBIB design, where each cluster appears a set number of times but does not appear with every other cluster. We briefly review the designs used in the most recently completed TIMSS, PISA (at the time of writing; PISA 2012 will be released at the end of 2013), and PIRLS cycles, beginning with the most recently released PISA cycle. A side-by-side comparison of these three studies and their design features can be found in Table 4.1.

PISA 2009

PISA 2009 featured one major domain, *reading,* and two minor domains, *science* and *mathematics* (OECD 2009). The total reading item pool consisted of 131 cognitive reading items arranged around 37 reading units,* 26 of which have been in use for linking since 2000. An additional 11 items from 2000, not used since that cycle, were also included. The remaining items were newly developed. The mathematics portion of the assessment comprised 34 items in 18

* PISA items are grouped around common stimuli including text, tables, graphs, and diagrams. These groups, containing as many as five items, are referred to as *units*.

TABLE 4.1

Comparison of Most Recently Available ILSA Designs

	PISA 2009	PIRLS 2011	TIMSS 2011
Target population	15-year-old students in grade 7 and higher participating in: • full or part-time in educational institutions; • vocational training programs or other related types of educational programs; • foreign schools within countries, no matter what nationality students hold.	Students at their fourth formal schooling year defined by ISCED (International Standard Classification of Education developed by the UNESCO Institute for Statistics)	Students at their fourth or eighth formal schooling year defined by ISCED
Test contents	Major domain: Reading Minor domains: Science and Mathematics	Reading	Mathematics and Science
Test design	Core paper and pencil test design: • Standard test design: 13 booklets, 4 clusters in every booklet; each cluster appears in four booklets in different positions. • Easier test design: 13 booklets, 4 clusters in every booklet; each cluster appears in four booklets in different positions. • UH booklet: half of the normal booklet items, designed for special-needs students. Digital reading test design: • 6 forms, 3 total clusters, 2 clusters in each form.	PIRLS 2011 test design: • 13 booklets, 2 blocks in each booklet; • 10 blocks, 5 literary passages, and 5 information texts. • Each block, apart from the 5th literary block and the 5th informational block, appears in 3 booklets excepting for the 13th booklet; • The 5th literary block and the 5th informational block only appear in the 13th booklet. Pre-PIRLS 2011 test design: • 6 blocks, 3 literary passages, and 3 information texts; • 9 booklets, 2 blocks in each booklet; • each block appears in three booklets.	• 14 booklets, 4 blocks in each booklet; • 28 blocks, 14 mathematics blocks, and 14 science blocks. • Each block appears in two booklets at different positions, the first position paired with the fourth position, and the second position paired with the third position.

TABLE 4.1 (continued)

Comparison of Most Recently Available ILSA Designs

	PISA 2009	PIRLS 2011	TIMSS 2011
Context questionnaires	• Student questionnaire: administrated to students; • School questionnaire: administrated to school principals; • ICT familiarity questionnaire (international option): administrated to students; • Parent questionnaire (international option): administrated to parents; • Educational career questionnaire (international option): administrated to students.	Student questionnaire: administrated to students; Learning to read survey (home questionnaire): administrated to parents or primary caregivers; Teacher questionnaire: administrated to teachers; School questionnaire: administrated to school principals; Curriculum questionnaire: administrated to national research coordinator.	Student questionnaire: administrated to students; Teacher questionnaire: administrated to all teachers of tested, subject for each sampled student; School questionnaire: administrated to school principals; Curriculum questionnaire: administrated to nationally selected curriculum specialist.
Total testable material	510 min	400 min	510 min at grade 4 600 min at grade 8
Actual testing time	Paper and pencil assessment: • Standard and easier test design: 2 hours with a break after one hour; • UH booklet design: 1 hour. Digital test design: 40 min Context questionnaires: • Student questionnaire: 30 min; • School questionnaires: 30 min; • ICT familiarity questionnaire: 5 min; • Parent questionnaire: 20 min; • Educational career questionnaire: consisted of 7 simple questions.	PIRLS 2011 assessment: 80 min Context questionnaires: • Student questionnaire: 15–30 min; • Home questionnaire: 10–15 min; • Teacher questionnaire: 30 min; • School questionnaire: 30 min; • Curriculum questionnaire: not stated.	TIMSS 2011 cognitive assessment: • The fourth grade: 72 min; • The eighth grade: 90 min. Context questionnaire: • Student questionnaire: 15–30 min; • Teacher questionnaire: 30 min; • School questionnaire: 30 min; • Curriculum questionnaire: not stated.

units, while the science portion included 53 items in 18 units. For both minor domains, selected items were a subset of a larger group of items used in 2006.

The content-specific units were divided into 13 clusters: 7 reading, 3 science, and 3 mathematics clusters, where each cluster represents about 30 minutes of testable material. Table 4.2 details the contents of each of the 20* PISA booklets, where each of the component clusters are indicated by *R* for reading, *S* for science, and *M* for mathematics. Educational systems with previously or expected low performance were offered the option of an easier set of test booklets. This is reflected in Table 4.1 by the addition of seven booklets (booklets 21–27) and two designs: standard (booklets 1–13) or easier (booklets 8–13; 21–27). In the easier booklets, two of the standard reading clusters (e.g., R3A and R4A) were substituted with easier reading clusters (R3B and R4B). Notice that regardless of the booklet set, under the PISA design, booklets may be missing (by design) entire content areas. For example, students who

TABLE 4.2

PISA 2009 Booklet Design

Booklet ID	Cluster				Booklet Set	
					Standard	Easier
1	M1	R1	R3A	M3	X	
2	R1	S1	R4A	R7	X	
3	S1	R3A	M2	S3	X	
4	R3A	R4A	S2	R2	X	
5	R4A	M2	R5	M1	X	
6	R5	R6	R7	R3A	X	
7	R6	M3	S3	R4A	X	
8	R2	M1	S1	R6	X	X
9	M2	S2	R6	R1	X	X
10	S2	R5	M3	S1	X	X
11	M3	R7	R2	M2	X	X
12	R7	S3	M1	S2	X	X
13	S3	R2	R1	R5	X	X
21	M1	R1	R3B	M3		X
22	R1	S1	R4B	R7		X
23	S1	R3B	M2	S3		X
24	R3B	R4B	S2	R2		X
25	R4B	M2	R5	M1		X
26	R5	R6	R7	R3B		X
27	R6	M3	S3	R4B		X

* There was also an additional one hour booklet (UH—*Une Heure*) for students with special needs. As it is not part of the primary study design, we do not discuss it here. Interested readers are referred to OECD (2009) for details.

TABLE 4.3

PISA 2009 Digital Reading Assessment Design

Form	Cluster 1	Cluster 2
1	A	B
2	B	A
3	B	C
4	C	B
5	C	A
6	A	C

receive booklet 1 answer no science items, while students who receive booklet 4 answer no math items; however, study participants receive proficiency values on each content scale. This is a straightforward BIB design under either condition with parameters $t = 13$, $b = 13$, $k = 4$, $r = 4$, $\lambda = 4$, and indeed $rt = 4 * 13 = bk = 13 * 4$. An advantage of this approach is that it accounts for position effects, since each cluster appears in every position; however, it does not account for context effects (Sirotnik 1970) or what Frey et al. (2009) refer to as *carry-over effects*. In other words, the PISA 2009 design does not protect against the possibility that a given block will be perceived by examinees as differentially easier or more difficult because of the block that precedes it.

We also note the addition of a digital reading assessment (DRA) that was available to participating countries in 2009. This optional component was intended to measure reading skills associated with electronic media such as websites and emails. With 29 items arranged around nine digital reading units, the DRA represented an additional 60 minutes of testable material. Available items were allocated to three clusters (A, B, or C), which appeared in one of two positions. Thus, each DRA booklet contained two of three clusters, representing 40 minutes of testing time. The DRA design located in Table 4.3 can be considered, in the experimental design parlance, to be a fully factorial cross-over design with three treatment levels since each cluster appeared with every other cluster in all possible combinations. Under this design, the six assessment forms serve as *blocks*, the cluster positions serve as *periods*, and the clusters serve as *treatment levels* (Kirk 1995).

TIMSS 2011

The TIMSS 2011 design differs from the core PISA 2009 design in that it is a partial BIB design (λ is not an integer). To explain, we detail the eighth-grade portion of the TIMSS 2011 design. The fourth-grade design is generally similar, albeit with a shorter actual testing time (Mullis et al. 2009). The eighth-grade TIMSS assessment included mathematics and science items distributed across 14 mathematics clusters (M01–M14) and 14 science clusters

TABLE 4.4
TIMSS 2011 Booklet Design

Booklet	Part 1		Part 2	
ID	Cluster 1	Cluster 2	Cluster 3	Cluster 4
1	M01	M02	S01	S02
2	S02	S03	M02	M03
3	M03	M04	S03	S04
4	S04	S05	M04	M05
5	M05	M06	S05	S06
6	S06	S07	M06	M07
7	M07	M08	S07	S08
8	S08	S09	M08	M09
9	M09	M10	S09	S10
10	S10	S11	M10	M11
11	M11	M12	S11	S12
12	S12	S13	M12	M13
13	M13	M14	S13	S14
14	S14	S01	M14	M01

(S01–S14), with 10–15 items per cluster. Of the 28 math and science clusters, eight each were carried forward from 2007 and 12 clusters (six in math and six in science) were newly developed for 2011. The 28 clusters were arranged into 14 booklets with four clusters each. Under this design, each cluster (and therefore each item) appears in two booklets. The parameters specifying this PBIB design are $t = 28$, $b = 14$, $k = 4$, and $r = 2$. A representation of the TIMSS 2011 assessment design can be seen in Table 4.4.

The TIMSS 2011 administration is such that the first two clusters are administered, examinees take a short break after 45 minutes (36 minutes in fourth grade), and then the last two clusters are administered. While each pair of clusters does not appear in this design—thereby not controlling for variation that might result from particular cluster pairs—each cluster appears in the first position (in either part 1 or part 2) and the last position (again, in either part 1 or part 2). Further, each assessment part is consistent in content. That is, both clusters within a part feature the same content area. A weakness of each type of these designs is that it neither fully accounts for possible context nor position effects, since not every cluster appears with every other cluster, and every cluster does not appear in every possible position. As an example of possible context effects, imagine that a feature of math cluster 1 (M01) causes students to perform at lower than expected levels on any cluster that follows M01. A plausible position effect that would be desirable to control is differentially poor performance on the fourth cluster of items due to fatigue. Under a PBIB design, it would be difficult to completely control for these types of context and position effects.

PIRLS 2011

PIRLS 2011 utilizes 10 *regular* (distinguished from *special*, described subsequently) item clusters in total, with 5 literary and 5 informational clusters distributed across 12 booklets (Mullis et al. 2009), which represents a total of 6 hours of testing time. Six of the 10 clusters were carried forward from previous PIRLS assessments for linking purposes, while the remaining clusters were newly developed for PIRLS 2011. In addition, there are two special-item clusters (one informational cluster and one literary cluster) that appear together in booklet 13. This 13th booklet is called a *reader*, the discussion of which is omitted here. Suffice to say, this booklet, presented in a magazine-type format, is intended to create a more natural, authentic reading experience

Although the PIRLS approach generally follows that of TIMSS and PISA, a critical difference is that all items are directly linked to an item stem that delivers a reading passage, either literary or informational in nature. The PBIB design used in TIMSS 2011 is illustrated in Table 4.4. The design is specified by the parameters $t = 8$, $b = 12$, $k = 2$, and $r = 3$, omitting the special reader booklet. Note that there is no restriction that a literary cluster appears with an informational cluster. Further, every cluster appears in the first and last position in the booklet, controlling for position effects associated with a given cluster. Context effects, on the other hand, are not controlled in several booklets that are mixed across reading purpose, namely booklets 8 through 12. In booklets 1 through 7, context effects are only controlled to the degree that the two clusters are like in content. The PIRLS 2011 booklet design can be seen in Table 4.5.

TABLE 4.5

PIRLS 2011 Booklet Design

Booklet ID	Cluster 1	Cluster 2
1	L1	L2
2	L2	L3
3	L3	L4
4	L4	I1
5	I1	I2
6	I2	I3
7	I3	I4
8	I4	L1
9	L1	I1
10	I2	L2
11	L3	I3
12	I4	L4

How Can MMS Be Used in the Future?

Given the long history and established uses of MMS, we would argue that several areas in international large-scale assessment might benefit from the innovative application of MMS in future assessment cycles. As one example, all of the studies discussed so far are susceptible to context effects (Sirotnik 1970). That is, due to the study designs of PISA, TIMSS, and PIRLS, there is no provision to control for the possible influence of preceding clusters on a given cluster. In contrast, NAEP controls for context effects, at least to some degree, by using a focused-BIB or focused-PBIB design, where each booklet is focused on a single content area, rather than using the mixed-content designs of TIMSS and PISA. We detail just one of these designs given that the tested content in NAEP varies from year to year. In particular, we discuss a booklet design, described in a recent NAEP technical report, used to test mathematics in 2000 (National Center for Educational Statistics 2008). The NAEP design in Table 4.6 features 13 mathematics clusters assembled into 26 mathematics booklets with three clusters each. This design is a BIB design with the parameters $t = 13$, $b = 26$, $k = 3$, $r = 6$, and $\lambda = 1$ ($rt = 6 * 13 = bk = 26 * 3 = 78$). Notable in this design is that every cluster appears in each of the three booklet positions and the content is the same within each cluster. Both of these attributes offer some control over context and position effects. A disadvantage of focused designs in general is that when multiple content domains are assessed (e.g., NAEP assessed reading, math, and science at the eighth grade in 2009), it is not possible to measure correlations between these domains.

In addition to 90 minutes or more of tested material, ILSAs also collect extensive background information from students, teachers, and principals. Given the vast number of theories that relate student and school contexts to achievement, there is a desire to expand the background data that is collected on continuing cycles of studies such as TIMSS, PIRLS, and PISA. Multiple-matrix sampling might well be an option for expanding the covered content of background questionnaires; however, it is important to investigate the impact that extending multiple matrix designs to the background questionnaires will have on estimating proficiencies scores on these tests, and what obtaining more contextual information might cost the precision of population achievement estimates.

As noted, nearly three decades ago, by Bock et al. (1982), "Educational assessment was born of a desire to estimate attainment in culturally or educationally defined groups, with results detailed enough to be useful, but at costs low enough to be practical" (p. 11). Side-by-side with advances in measurement theory, MMS provides a viable method for balancing useful results and low costs, especially in terms of examinee-response costs. A number of important factors must be weighed by the test designer and other stakeholders when considering the implementation of an MMS-designed assessment (e.g., Pandey and Carlson 1976; Childs and Jaciw 2003; Gonzalez

TABLE 4.6

NAEP 2000 Mathematics Assessment Booklet Design

Booklet	Cluster 1	Cluster 2	Cluster 3
1	M03	M04	M07
2	M04	M05	M08
3	M05	M06	M09
4	M06	M07	M10
5	M07	M08	M11
6	M08	M09	M12
7	M09	M10	M13
8	M10	M11	M14
9	M11	M12	M15
10	M12	M13	M03
11	M13	M14	M04
12	M14	M15	M05
13	M15	M03	M06
14	M03	M05	M10
15	M04	M06	M11
16	M05	M07	M12
17	M06	M08	M13
18	M07	M09	M14
19	M08	M10	M15
20	M09	M11	M03
21	M10	M12	M04
22	M11	M13	M05
23	M12	M14	M06
24	M13	M15	M07
25	M14	M03	M08
26	M15	M04	M09

and Rutkowski 2010). While established guidelines exist for many design variables and practical considerations, at least one area that deserves more attention includes the assignment of items to blocks (and subsequently to booklets). In general, technical documentation related to this issue is sparse and the literature on the topic has not kept pace with developments in the models used to estimate achievement. Most articles examining related issues (assigning items to booklets) tapered off around the 1970s (e.g., Barcikowski 1974; Scheetz and Forsythe 1977), with a few resurgent efforts (e.g., Gressard and Loyd 1991; Zeger and Thomas 1997) and generally equivocal results. What remains to be explored in this regard is how best to assign items to blocks, given that population models that produce plausible values are the predominant way of estimating distributions of proficiencies in populations and policy-relevant sub-populations.

Finally, international assessments, such as TIMSS, PIRLS, and PISA, continue to grow in terms of the number of countries participating from cycle to cycle. This necessarily leads to a better representation of countries; however, this growth also implies a more diverse set of assessed populations using translated versions of the same assessment. Currently, participating countries range widely from economically developing countries (with huge gradients between rural and urban living environments) to industrialized countries with a more evenly distributed standard of living. This heterogeneity across assessed populations leads to increased variance in achievement results. Important advances in this regard were undertaken in PISA 2009 and PIRLS 2011, described previously. In addition to these recent developments, the challenge of assessing such disparate populations might also be ameliorated to some degree by integrating more elaborate matrix designs, with further deviations from the assignment of the same booklets to all subgroups of the population. Statistical linking procedures based on extensions of IRT linking methods (e.g., von Davier and von Davier 2007; Olivieri and von Davier 2010) as well as a careful design of the link across booklets will enable the organizations conducting these assessments to design a system that generated comparable results and is appropriate for all participating countries.

References

Adams, R., Wu, M., and Carstensen, C. 2007. Application of multivariate Rasch models in international large scale educational assessment. In M. von Davier and C. H. Carstensen (Eds.) *Multivariate and Mixture Distribution Rasch Models: Extensions and Applications* (pp. 271–280). New York: Springer.

Barcikowski, R. S. 1974. The effects of item discrimination on the standard errors of estimate associated with item-examinee sampling procedures. *Educational and Psychological Measurement, 34*, 231–237.

Beaton, A. E. 1987. *Implementing the new design: The NAEP 1983-84 technical report (No. 15-TR-20).* Princeton, NJ: Educational Testing Service, National Assessment of Educational Progress.

Bock, R. D., Mislevy, R., and Woodson, C. 1982. The next stage in educational assessment. *Educational Researcher, 11*(3), 4–11 + 16.

Childs, R. and Jaciw, A. 2003. Matrix sampling of items in large-scale assessments. *Practical Assessment, Research and Evaluation, 8(*16). Retrieved January 25, 2010 from http://PAREonline.net/getvn.asp?v=8&n=16.

Frey, A., Hartig, J., and Rupp, A. 2009. An NCME instructional module on booklet designs in large-scale assessments of student achievement: Theory and practice. *Educational Measurement Issues and Practice, 28*(3), 39–53.

Giesbrecht, F. G. and Gumpertz, M. L., 2004. *Planning, Construction, and Statistical Analysis of Comparative Experiments*. Hoboken, New Jersey: John Wiley and Sons.

Gressard, R. and Loyd, B. 1991. A comparison of sampling plans in the application of multiple matrix sampling. *Journal of Educational Measurement, 28*(2), 119–130.

Gonzalez, E. and Rutkowski, L. 2010. Principles of multiple matrix booklet designs and parameter recovery in large-scale assessments. *IEA-ETS Research Institute Monograph, 3*, 125–156.

Hooke, R. 1956a. Symmetric functions of a two-way array. *The Annals of Mathematical Statistics, 27*(1), 55–79.

Hooke, R. 1956b. Some applications of bipolykays to the estimation of variance components and their moments. *The Annals of Mathematical Statistics, 27*(1), 80–98.

Johnson, C. and Lord, F. 1958. An empirical study of the stability of a group mean in relation to the distribution of test items among students. *Educational and Psychological Measurement, 18*(2) 325–329.

Kirk, R. E. 1995. *Experimental Design: Procedures for the Behavioral Sciences*. Pacific Grove, CA: Brooks/Cole.

Kleinke, D. 1972. *The Accuracy of Estimated Total Test Statistics*. Washington, DC: National Center for Educational Research and Development (ERIC Document Reproduction Service No. ED064356).

Knapp, T. R. 1968. An application of balanced incomplete block designs to the estimation of test norms. *Educational and Psychological Measurement, 28*, 265–272.

Lord, F. 1960. Use of true-score theory to predict moments of univariate and bivariate observed-score distributions. *Psychometrika, 25*, 325–342.

Lord, F. 1962. Estimating norms by item-sampling. *Educational and Psychological Measurement, 22*(2), 259–267.

Lord, F. 1965. *Item Sampling in Test Theory and in Research Design*. Princeton: Educational Testing Service.

Mislevy, R. J. 1983. Item response models for grouped data. *Journal of Educational Statistics, 8*, 271–288.

Mislevy, R. J. 1984. Estimating latent distributions. *Psychometrika, 49*, 359–381.

Mislevy, R. J. 1991. Randomization-based inference about latent variables from complex samples. *Psychometrika, 56*, 177–196.

Mislevy, R. J., Beaton, A. E., Kaplan, B., and Sheehan, K. M. 1992a. Estimating population characteristics from sparse matrix samples of item responses. *Journal of Educational Measurement, 29*(2), 133–161.

Mislevy, R. J., Johnson, E. G., and Muraki, E. 1992b. Scaling procedures in NAEP. *Journal of Educational Statistics, (17)*2, 131–154.

Mislevy, R. J. and Sheehan, K. M. 1987. Marginal estimation procedures. In A. E. Beaton (Ed.), *The NAEP 1983-84 Technical Report (No. 15-TR-20)*. Princeton, NJ: Educational Testing Service, National Assessment of Educational Progress.

Mullis, I., Martin, M., Ruddock, G., O'Sullivan, C., Arora, A., and Erberber, E. 2005. *TIMSS 2007 Assessment Frameworks*. Chestnut Hill, MA: TIMSS and PIRLS International Study Center, Boston College.

Mullis, I., Martin, M., Ruddock, G., O'Sullivan, C., and Preuschoff, C. 2009. *TIMSS 2011 Assessment Frameworks*. Chestnut Hill, MA: TIMSS and PIRLS International Study Center, Boston College.

Myerberg, N. J. (1975, April). *The Effect of Item Stratification in Multiple Matrix Sampling*. Paper presented at the Annual Meeting of the American Educational Research Association, Washington, DC.

National Center for Educational Statistics. 2008. *Student Booklets for the Mathematics Assessment*. Retrieved from http://nces.ed.gov/nationsreportcard/tdw/instruments/booklet_math.asp.

Olivieri, M. and von Davier, M. 2010, May. *Investigation of Model Fit and Score Scale Comparability in International Assessments*. Paper presented at the National Council on Measurement in Education annual meeting, Denver, CO.

Organisation for Economic Co-operation and Development. 2009. *PISA 2009 Assessment Framework: Key Competencies in Reading, Mathematics, and Science*. Paris: OECD.

Pandey, T. N. and Carlson, D. 1976. Assessing payoffs in the estimation of the mean using multiple matrix sampling designs. In D. DeGruijter and L. J. van der Kamp (Eds.), *Advances in Psychological and Educational Measurement*. London: Wiley.

Pandey, T. N. and Shoemaker, D. M. 1975. Estimating moments of universe scores and associated standard errors in multiple matrix sampling for all item-scoring procedures. *Educational and Psychological Measurement, 35*, 567–581.

Plumlee, L. 1964. Estimating means and standard deviations from partial data: An empirical check on Lord's item sampling technique. *Educational and Psychological Measurement, 24*(3), 623–630.

Pugh, R. C. 1971. Empirical evidence on the application of Lord's sampling technique to Likert items. *Journal of Experimental Education, 39*(3), 54–56.

Reiser, M. 1983. An item response model for the estimation of demographic effects. *Journal of Educational Statistics, 8*(3), 165–186.

Rubin, D. B. 1978. Bayesian inference for causal effects: The role of randomization. *Annals of Statistics, 6*, 34–58.

Rubin, D. B. 1987. *Multiple Imputation for Nonresponse in Surveys*. New York: Wiley.

Scheetz, J. and Forsythe, R. 1977, April. *A Comparison of Simple Random Sampling Versus Stratification for Allocating Items to Subtests in Multiple Matrix Sampling*. Paper presented at the Annual Meeting of the National Council on Measurement in Education, New York, New York.

Shoemaker, D. M. 1970a. Allocation of items and examinees in estimating a norm distribution by item sampling. *Journal of Educational Measurement, 7*(2), 123–128.

Shoemaker, D. M. 1970b. Item-examinee sampling procedures and associated standard errors in estimating test parameters. *Journal of Educational Measurement, 7*(4), 255–262.

Shoemaker, D. M. 1973. *Principles and Procedures of Multiple Matrix Sampling*. Cambridge, MA: Ballinger Publishing Company.

Sirotnik, K. 1970. An investigation of the context effect in matrix sampling. *Journal of Educational Measurement, 7*(3), 199–207.

Sirotnik, K. and Wellington, R. 1974. Incidence sampling: An integrated theory for matrix sampling. *Journal of Educational Measurement, 14*(4) 243–299.

Thomas, N. and Gan, N. 1997. Generating multiple imputations for matrix sampling data analyzed with item response models. *Journal of Educational and Behavioral Statistics*, 22(4), 425–445.

van der Linden, W. J. and Carlson, J. E. 1999. *Calculating Balanced Incomplete Block Design for Educational Assessments.* Enschede, the Netherlands: Faculty of Educational Science and Technology, University of Twente.

van der Linden, W. J., Veldkamp, B. P., and Carlson, J. E. 2004. Optimizing balanced incomplete block designs for educational assessments. *Applied Psycho-Logical Measurement, 28*, 317–331.

von Davier, M., Gonzalez, E., and Mislevy, R. 2009. What are plausible values and why are they useful? *IERI Monograph Series: Issues and Methodologies in Large Scale Assessments, 2*, 9–36.

von Davier, M. Sinharay, S., Oranje, A., and Beaton, A. 2006. Statistical Procedures used in the National Assessment of Educational Progress (NAEP): Recent Developments and Future Directions. In C.R. Rao and S. Sinharay (Eds.), *Handbook of Statistics (Vol. 26): Psychometrics*. Amsterdam: Elsevier.

von Davier, M. and Sinharay, S. 2007. An importance sampling EM algorithm for latent regression models. *Journal of Educational and Behavioral Statistics, 32*(3), 233–251.

von Davier, M. and Sinharay, S. 2009. *Stochastic Approximation Methods for Latent Regression Item Response Models* (ETS RR-09-09). Princeton, NJ: Educational Testing Service.

von Davier, M. and von Davier, A. 2007. A unified approach to IRT scale linkage and scale transformations. *Methodology, 3*, 115–124.

Wilks, S. 1962. *Mathematical Statistics*. New York: Wiley and Sons, Inc.

Zeger, L. and Thomas, N. 1997. Efficient matrix sampling instrument for correlated latent traits: examples from the National Assessment of Educational Progress. *Journal of the American Statistical Association, 92*, 416–425.

5

Modeling Country-Specific Differential Item Functioning

Cees Glas
University of Twente

Khurrem Jehangir
University of Twente

CONTENTS

Introduction

The growing awareness of the importance of education for knowledge economies has led to even greater emphasis on improving educational systems. Educational surveys play a prominent role in taking stock of the state of educational systems. Educational surveys not only depict the current state of the educational system but also help identify weaknesses and handicaps that can be addressed with proper policy planning. Large-scale educational surveys enable comparisons of large groups of students within and across countries. They allow countries to gauge the performance of their populations on a comparative scale, to evaluate their global position, and to gain insights into the factors that determine the effectiveness of their educational systems.

However, large-scale surveys are a complex undertaking and present many challenges, especially with respect to insuring that the results are comparable across diverse target groups. An especially important problem has to do with cultural bias. Modern educational surveys not only measure the cognitive abilities of students in areas of interest, but also include a set of context or background questionnaires. Culture-specific differential item functioning (CDIF) may occur both in cognitive items and in items of background questionnaires, but the background questionnaires may be more vulnerable. CDIF in achievement tests may, for example, occur through the content of the context stories in a math or language achievement test. Still, though the framing of a question may influence the response behavior, it is reasonable to assume that the underlying construct, say math achievement or language comprehension, is stable over countries and cultures. In background questionnaires, cultural bias may be more prominent. First, it is no minor task to define constructs such as the socioeconomic status or the pedagogical climate in such a way that they allow for comparisons over countries and cultures and, second, culture-related response tendencies may bias the comparability between countries and cultures.

Both educational achievement and most of the explanatory variables on the student-, parent-, teacher-, classroom-, and school-level are viewed as latent variables. The data from tests of educational achievement and background questionnaires are usually analyzed using item response theory (IRT) models (see, for instance, Lord 1980). In this chapter, we present a statistical methodology to identify and account for CDIF. This methodology will be applied to the background questionnaires that were used in the 2009 cycle of the Program for International Student Assessment (PISA) survey. In the PISA study, the data of the background questionnaires were modeled using an exponential family IRT model, that is, the partial credit model (PCM; see Masters 1982). The statistical methodology presented here will be developed in the framework of the PCM, but also in the framework of a more general model, the generalized partial credit model (GPCM; see Muraki 1992). To assess the impact of modeling CDIF using the PCM and GPCM, the rank order of the participating countries on the constructs measured by the PISA background questionnaires will be evaluated.

Item Response Theory

The background questionnaires in the PISA project consist mostly of polytomously scored items, that is, the scores on an item indexed i ($i = 1,2,\ldots,K$) are integers between 0 and M_i, where M_i is the maximum score on item i. In the GPCM, the probability of a student n ($n = 1,\ldots,N$) scoring in category j on item i (denoted by $X_{nij} = 1$) is given by

$$P(X_{nij} = 1 \mid \theta_n) = P_{ij}(\theta_n) = \frac{\exp(j\alpha_i\theta_n - \beta_{ij})}{1 + \sum_{h=1}^{M_i} \exp(h\alpha_i\theta_n - \beta_{ih})}, \tag{5.1}$$

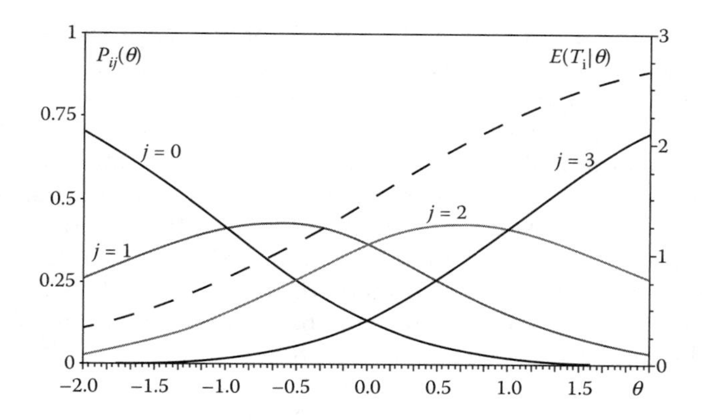

FIGURE 5.1
Response functions and ITF under the GPCM.

for $j = 1,\ldots,M_i$. Note that the probability of a response in category $j = 0$ is thus given by

$$P(X_{ni0} = 1 \mid \theta_n) = P_{i0}(\theta_n) = \frac{1}{1 + \sum_{h=1}^{M_i} \exp(h\alpha_i\theta_n - \beta_{ih})}. \tag{5.2}$$

An example of the category response functions $P_{ij}(\theta)$ for an item with four response categories is given in Figure 5.1. The graph also shows the item-total score function (ITF)

$$E(T_i \mid \theta) = \sum_{j=1}^{M_i} jE(X_{ij} \mid \theta) = \sum_{j=1}^{M_i} jP_{ij}(\theta), \tag{5.3}$$

where the item-total score is defined as $T_i = \Sigma_j jX_{ij}$. Note that the ITF increases as a function of θ. The location of the response curves are related to location parameters defined by $\eta_{i1} = \beta_{i1}$ and $\eta_{ij} = \beta_{ij} - \beta_{i(j-1)}$, for $j = 2,\ldots,m_i$. The location parameter η_{ij} is the position on the θ scale, where the curves $P_{i(j-1)}(\theta)$ and $P_{ij}(\theta)$ intersect. Finally, the so-called discrimination parameter α_i gauges the steepness of the curves. If the discrimination parameters for all items are constrained to be equal, the GPCM specializes to the PCM.

The GPCM and the PCM are not the only IRT models giving rise to sets of response curves where a higher level on the latent scale, that is, the θ scale, is associated with a tendency to score in a higher response category. The sequential model by Tutz (1990) and the graded response model by Samejima (1969) have response curves that can hardly be distinguished on the basis of empirical data (Verhelst et al. 1997). Therefore, the choice between the GPCM and these two alternatives is not essential.

Estimating all the parameters in the GPCM concurrently has both practical and theoretical drawbacks. The practical problem is the sheer amount of parameters (the sample size of the PISA project approaches half a million, with the analogous number of θ parameters), which renders standard computational methods such as the Newton–Raphson method infeasible. Theoretical problems have to do with the consistency of such concurrent estimates (Haberman 1977). Depending on the model and the psychometrician's preferences, various alternative estimation methods may solve the problem. One of the most-used methods, and the method used in the present chapter, is the maximum marginal likelihood (MML; Bock and Aitkin, 1981) estimation method. To apply this method, it is assumed that the θ parameters have one or more common normal distributions. So we consider populations indexed g ($g = 1,\ldots,G$) and assume that

$$\theta_n \sim N(\mu_{g(n)}, \sigma^2_{g(n)})$$

where $g(n)$ is the population to which respondent n belongs. Populations may, for instance, be the countries in an educational survey, gender, or countries crossed with gender, and so on. In MML, the likelihood function is marginalized over the θ parameters, that is, the likelihood function of all item parameters α, β and all means and variances μ,σ given all response patterns x_n ($n = 1,\ldots,N$) is given by

$$L(\alpha,\beta,\mu,\sigma)=\prod_{n}^{N}\int p(x_n \mid \theta_n,\alpha,\beta)p(\theta_n;\mu_{g(n)},\sigma^2_{g(n)})\,d\theta_n$$

where $p(x_n \mid \theta_n, \alpha, \beta)$ is the probability of response pattern x_n and $p(\theta_n;\mu_{g(n)},\sigma^2_{g(n)})$ is the normal density related to population $g(n)$.

The likelihood equations are derived by identifying the likelihood equations as if the θ parameters were observed, and then taking the posterior expectation of both sides of the equation with respect to the posterior of the θ parameters. For instance, if the θ parameters were observed, the likelihood equation for the mean μ_g would be

$$\mu_g = \sum_{n|n(g)=g} \theta_n$$

and taking posterior expectations of both sides gives the MML estimation equation

$$\begin{aligned}\mu_g &= \sum_{n|n(g)=g} E\left(\theta_n \mid x_n;\alpha,\beta,\mu_g,\sigma_g^2\right)\\ &= \sum_{n|n(g)=g} \int \theta_n p(\theta_n \mid x_n;\alpha,\beta,\mu_g,\sigma_g^2)\,d\theta_n \qquad (5.4)\end{aligned}$$

In the next section, the MML framework will be applied to detection and modeling of differential item functioning (DIF).

Detection and Modeling of DIF

Part of the process of establishing the construct validity of a scale may consist of showing that the scale fits a one-dimensional IRT model. This means that the observed responses can be attributed to item and person parameters that are related to a one-dimensional latent dimension. Construct validity implies that the construct to be measured is the same for all respondents. Item bias, or DIF, violates this assumption. An item displays DIF if the probabilities of responding in different categories vary across subpopulations (say countries or genders) given equivalent levels of the underlying attribute (Holland and Wainer 1993; Camilli and Shepard 1994). Or, equivalently, an item is biased if the manifest item score, conditional on the latent dimension, differs between subpopulations (Chang and Mazzeo 1994).

Several techniques for detecting DIF have been proposed. Most of them are based on evaluating differences in response probabilities between groups, conditional on some measure of the latent dimension. The most generally used technique is based on the Mantel–Haenszel statistic (Holland and Thayer 1988), while others are based on loglinear models (Kok et al. 1985), or on IRT models (Hambleton and Rogers 1989). The advantage of IRT-based methods over the other two approaches is that IRT offers the possibility of modeling DIF for making inferences about differences regarding the average scale level of subpopulations.

In this chapter, the logic for the detection of DIF will be based on the logic of the Lagrange multiplier (LM) test (Rao 1948; Aitchison and Silvey 1958). Applications of LM tests to the framework of IRT have been described by Glas (1998, 1999), Glas and Falćon (2003), and Glas and Dagohoy (2007). In this chapter, our primary interest is not in the actual outcome of the LM test, because due to the very large sample sizes in educational surveys such as PISA, even the smallest model violation, that is, the smallest amount of DIF, will be significant. The reason for adopting the framework of the LM test is that it clarifies the connection between the model violations and observations and expectations used to detect DIF. Further, it produces comprehensible and well-founded expressions for model expectations, so that the value of the LM test statistic can be used as an effect size of DIF. Finally, the procedure can be easily generalized to a broad class of IRT models. Before the general approach to detect DIF is outlined, a special case is presented to clarify the method.

Consider two groups labeled the reference group and the focal group. For instance, the reference group may be girls and the focal group may be boys. Define a background variable:

$$y_n = \begin{cases} 1 & \text{if person } n \text{ belongs to the focal group,} \\ 0 & \text{if person } n \text{ belongs to the reference group.} \end{cases}$$

For clarity, the method will be introduced in the framework of the two-parameter logistic model (2PLM), which is the special case of the GPCM pertaining to dichotomously scored items. Consider a model where the probability of a positive response is given by

$$P_i(\theta_n) = \frac{\exp(\alpha_i\theta_n - \beta_i + y_n\delta_i)}{1 + \exp(\alpha_i\theta_n - \beta_i + y_n\delta_i)}. \tag{5.5}$$

For the reference population, $y_i = 0$ and the model is analogous to the 2PLM. For the focal population, $y_i = 1$, so in that case, the model is also the 2PLM, but the item location parameter β_i is shifted by δ_i.

The LM test targets the null hypothesis of no DIF, that is, the null hypothesis $\delta_i = 0$. The LM test statistic is computed using the MML estimates of the null model, that is, δ_i is not estimated. The test is based on evaluation of the first-order derivatives of the marginal likelihood with respect to δ_i evaluated at $\delta_i = 0$ (see Glas 1999). If the first-order derivative in this point is large, the MML estimate of δ_i is far removed from zero, and the test is significant. If the first-order derivative in this point is small, the MML estimate of δ_i is probably close to zero and the test is not significant. The actual LM statistic is the squared first-order derivative divided by its estimated variance, and it has an asymptotic Chi-square distribution with one degree of freedom. However, as already mentioned above, the primary interest is not so much in the test itself, but on the information it provides regarding the fit between the data and the model. Analogous to the reasoning leading to likelihood Equation 5.4, we first derive the first-order derivative assuming that θ_n is observed, and equate it to zero. This results in the likelihood equation

$$\sum_{n=1}^{N} y_n x_{ni} = \sum_{n=1}^{N} y_n P_i(\theta_n).$$

Note that the left-hand side is the number of positive responses given in the focal group and the right-hand side is its expectation if θ_n were observed. Taking expectations with respect to the posterior distribution of θ_n results in

$$\sum_{n=1}^{N} y_n x_{ni} = \sum_{n=1}^{N} y_n E\left(P_i(\theta_n) \mid x_n; \alpha, \beta, \mu_{g(n)}, \sigma^2_{g(n)}\right).$$

So the statistic is based on the difference between the number-correct score in the focal group and its posterior expected value. Note that the difference between the two sides of the likelihood equation can be seen as a residual. Further, if we divide the two sides by the number of respondents in the focal group, that is, with $\Sigma_n y_n$, the expressions become the observed and expected average item score in the focal group. This interpretation provides guidance in judging the size of the DIF, that is, it provides a framework for judging whether the misfit is substantive or not, referenced to the observed score scale.

For a general definition of the approach, which also pertains to polytomously scored items, define covariates y_{nc} ($c = 1,\ldots,C$). Special cases leading to specific DIF statistics will be given below. The covariates may be separately observed person characteristics, but they may also depend on the observed response pattern, though without the response to the item i targeted. The probability of a response is given by a generalization of the GPCM, that is

$$P_{ij}(\theta_n) = \frac{\exp(j\alpha_i\theta_n - \beta_{ij} + j\sum_c y_{nc}\delta_{ic})}{1 + \sum_{h=1}^{M_i}\exp(h\alpha_i\theta_n - \beta_{ih} + h\sum_c y_{nc}\delta_{ic})}. \tag{5.6}$$

For one or more reference populations, the covariates y_{nc} ($c = 1,\ldots,C$) will be equal to zero. These populations serve as a baseline where the GPCM with item parameters α and β holds. In the other populations, one or more covariates y_{nc} are nonzero. The LM statistic for the null hypothesis $\delta_{ic} = 0$ ($c = 1,\ldots,C$) is a quadratic form in the C-dimensional vector of first-order derivatives and the inverse of its covariance matrix (for details, see Glas 1999). It has an asymptotic Chi-square distribution with C degrees of freedom. This general formulation can be translated into many special cases. Three are outlined here and will also be used in the example presented below.

For the first special case, one population serves as the focal population; all other populations serve as reference. The GPCM has only one additional parameter, δ_i, that is, $C = 1$. This leads to the residual

$$r_i = \sum_{n=1}^{N}\sum_{j=1}^{M_i} y_n j X_{ij} - \sum_{n=1}^{N}\sum_{j=1}^{M_i} y_n j E\left(P_{ij}(\theta_n) \mid x_n; \alpha, \beta, \mu_{g(n)}, \sigma^2_{g(n)}\right) \tag{5.7}$$

Dividing this residual by the number of respondents in the focal group produces a residual which is the difference of the observed and expected average item-total score in the focal group. The residual gauges the so-called uniform DIF, that is, the residual indicates whether the ITF $\Sigma_j jP_{ij}$ of the focal group is uniformly shifted over the θ scale relative to the reference group or not. The associated LM statistic has an asymptotic Chi-square distribution with one degree of freedom.

A second version of the statistic emerges when y_n is a dummy code for a country. The residuals defined by formula 5.7 then become country-specific, say r_{ic} ($c = 1,\ldots,C$). To assess CDIF, C is equal to the number of countries minus one, because one country must serve as a reference group or baseline. The associated LM statistic has an asymptotic Chi-square distribution with a degree of freedom equal to the number of countries minus one.

Besides uniform DIF, nonuniform DIF may also occur. In this case, the ITFs of focal and reference groups may not be just shifted, but they may also cross. That is, in some locations on the θ scale, the ITF of one group is higher, while the reverse is true in other locations. Since θ cannot be directly observed, detection of nonuniform DIF must be based on a proxy for θ. The proxy is a respondent's rest score, which is the test score on all items except the targeted item i, that is, $\Sigma_n\Sigma_{k\neq i}\Sigma_j jX_{kj}$. The range of these scores is divided into C nonoverlapping subranges. Usually, C is between 3 and 6. Residuals are used to evaluate whether the ITF of the focal population and reference populations are different given different rest scores. So, y_{nc} is equal to one if n belongs to the focal population and obtained a rest score in subrange c ($c = 1,\ldots,C$), and zero otherwise. This leads to the third version of the test based on the residual

$$r_{ic} = \sum_{n=1}^{N}\sum_{j=1}^{M_i} y_{nc} jX_{ij} - \sum_{n=1}^{N}\sum_{j=1}^{M_i} y_{nc} jE\left(P_{ij}(\theta_n) \mid x_n; \alpha, \beta, \mu_{g(n)}, \sigma^2_{g(n)}\right) \tag{5.8}$$

These are only three examples of the general approach of identifying DIF, with Equation 5.6 serving as an alternative model to the GPCM. The residuals may also be based on the frequencies in the response categories rather than on the ITF.

Identification and modeling of DIF is an iterative process where the item with the worst misfit in terms of its value of the LM statistic and its residual is given country-specific item parameters, followed by a new concurrent MML estimation procedure and a new DIF analysis. So, DIF items are treated one-at-a-time. From a practical point of view, defining country-specific item parameters is equivalent to defining an incomplete design where the DIF item is split into a number of virtual items, and each virtual item is considered as administered in a specific country. The resulting design can be analyzed using IRT software that supports the analysis of data collected in an incomplete design. Below, items with country-specific parameters will also be referred to as split items.

The method is motivated by the assumption that a substantial part of the items function the same in all countries and a limited number of items have CDIF. In the IRT model, it is assumed that all items do pertain to the same latent variable θ. Items without CDIF have the same item parameters in all countries. The items with CDIF have item parameters that are different across countries. This is, these items refer to the same latent variable θ as all the other items, but their location on the scale is different across countries. For instance, the number of cars in the family may be a good indicator of wealth, but the actual number of cars at a certain level of wealth may vary across countries. Or even within countries. Having a car in the inner city of Amsterdam is clearly a sign of wealth, but in the rural eastern part of the Netherlands, an equivalent level of wealth will probably result in ownership of three cars.

The number of items with country-specific item parameters is a matter of choice, where two considerations are relevant. First, there should remain a sufficient number of anchor items in the scale. Second, the model, including the split items, should fit the data. DIF statistics no longer apply to the split items. However, the fit of the item response curve of an individual item, say item i, can be evaluated using the test for nonuniform DIF described above, but evaluated using a model including country-specific items parameters. Also in this application, ranges of the rest score are used as proxies for locations on the θ scale, and the test evaluates whether the model with the country-specific items parameters can properly predict the ITF.

Examples

Two examples will be given. The objective of the first one is to give the flavor of the model, and the second one is meant to show how the approach works in a large-scale international survey. Starting with the first example, the data stem from the field trial of the PISA 2009 cycle. The data pertained to 20 countries and the total sample size was 9522 students. The scale analyzed was "Online Reading Activities" and consisted of 11 items, all scored 0–4. Table 5.1 shows the results of an analysis where one of the countries served as a focal group, while the rest of the countries served as a reference group. Using the GPCM, concurrent MML estimates were obtained for all item parameters and a separate population distribution was used for each country. The focal group consisted of 586 students; the reference group consisted of 8936 students.

The column labeled "LM" gives the values of the LM statistics based on the residuals r_i defined by formula 5.7. In this case, the LM has one degree of freedom. The significance probabilities are not given: as expected, all tests were significant due to the sample sizes. However, the values indicate that item 8 had the largest misfit in this country. The following four columns give the observed and expected values on which the test is based, for the focal and reference group, respectively. The values are average item scores. It can be seen that for item 1, the observed average in the focal group was 2.88, while the expected value was 2.94. So, the focal group scores lower than expected. Since the observed score ranges from 0 and 4, the difference is quite small. Note further that the difference for the reference group is 0.01, which is very small. This is, however, due to the fact that the reference group was much larger and has put far more weight on the estimate of the item parameters. The last column gives the value of r_{ic} as defined by formula 5.7. Again, it can be concluded that item 8 had the worst fit: the focal group scored far too low.

Besides information about the interaction between items and countries, an overall assessment of DIF is also of interest. Table 5.2 presents such

TABLE 5.1

Tests for Differential Item Functioning Targeted at Items within a Country

		Focal Group		Reference Group		
Item	**LM**	**Obs**	**Exp**	**Obs**	**Exp**	r_i
1	39.5	2.88	2.94	2.44	2.43	−0.05
2	86.3	3.57	3.33	2.78	2.79	0.24
3	49.6	2.78	2.59	2.09	2.10	0.19
4	54.2	2.54	2.80	2.39	2.38	−0.26
5	42.6	1.27	1.45	1.30	1.29	−0.18
6	21.7	2.42	2.34	1.97	1.97	0.08
7	14.3	2.70	2.73	2.33	2.33	−0.03
8	136.2	2.80	3.02	2.77	2.76	−0.22
9	3.4	2.04	2.05	1.66	1.66	−0.01
10	62.3	1.17	1.37	1.24	1.23	−0.20
11	31.4	2.39	2.24	1.89	1.90	0.15

information. This information is obtained in the same MML estimation run as used for Table 5.1. The second and third columns, labeled "LM" and "Av. Dif," give information aggregated across countries. The LM statistic has 19 degrees of freedom. Again, significance probabilities are not given: all tests were significant due to the large sample size. Further, again, item 8 has the largest misfit. The column labeled Av. Dif gives an effect size of the DIF

TABLE 5.2

Tests for Differential Item Functioning Targeted at Items within a Country

	No Item Split		2 Items Split		4 Items Split	
Item	**LM**	**Av. Dif**	**LM**	**Av. Dif**	**LM**	**Av. Dif**
1	1107.7	0.20	915.4	0.19		
2	831.0	0.21	581.3	0.21	606.0	0.21
3	664.4	0.18	589.9	0.18	475.8	0.16
4	779.9	0.19	679.9	0.20		
5	1541.2	0.25				
6	414.7	0.14	330.7	0.14	271.5	0.12
7	520.9	0.13	402.9	0.14	355.9	0.12
8	1672.7	0.42				
9	422.1	0.16	396.1	0.16	380.1	0.15
10	384.0	0.10	354.4	0.11	366.1	0.11
11	314.9	0.11	250.7	0.10	232.2	0.10

aggregated across countries: it is the mean over the countries of the absolute residuals, that is, the absolute differences between the observed and the expected as defined in formula 5.7.

Next, in an iterative process of splitting items into virtual items, MML estimation and evaluation of LM tests, items 8, 5, 1, and 4 were split, in that order. The columns labeled "2 Items Split" give the values after splitting items 8 and 5, the columns labeled "4 Items Split" give the values after all four items were split. Note that the first analysis does not always determine the order in which the items are split: item 2 seems to have more bias than item 4 at first, but their order is reversed in the process. The reason is that the presence of DIF items can also bias the estimates of the parameters of items that are not biased.

What is also needed to justify the procedure is evidence that the complete concurrent model, including the link and the split items, fits the data for every country. Information that can contribute to such evidence is given in Table 5.3 for the same country, as used for Tables 5.1 and 5.2. The table gives information regarding the fit of the ITF within a country after items are split. For every item, the rest score range is divided into three subranges and the observed and expected average item scores in the resulting student subgroups are given. The last column gives the means over these subgroups of the residuals defined in formula 5.8, that is, of the absolute difference between observed and expected values in subgroups. The split items are marked with an asterisk. It can be seen that the split items fit the model well. For the items that were not split, the table gives information regarding non-uniform DIF. The reason is that the expected values are computed using the assumption that the same item parameters apply in all countries, while the

TABLE 5.3

Fit of the ITF within a Country

			Group 1		Group 2		Group 3		
Item	**LM**	**Prob**	**Obs**	**Exp**	**Obs**	**Exp**	**Obs**	**Exp**	**Av. Dif**
1*	0.1	0.94	2.43	2.41	2.90	2.89	3.27	3.28	0.01
2	29.8	0.00	3.19	2.95	3.67	3.47	3.79	3.72	0.17
3	5.3	0.07	2.14	2.11	2.82	2.70	3.29	3.20	0.08
4*	1.4	0.48	1.95	1.92	2.45	2.51	3.08	3.05	0.04
5*	1.1	0.56	1.11	1.07	1.26	1.22	1.44	1.47	0.03
6	3.8	0.14	1.85	1.96	2.40	2.38	2.86	2.79	0.07
7	19.7	0.00	2.16	2.36	2.70	2.80	3.16	3.17	0.10
8*	0.4	0.82	2.54	2.50	2.77	2.78	3.04	3.03	0.02
9	7.1	0.03	1.42	1.58	2.04	2.11	2.59	2.69	0.11
10	61.2	0.00	1.01	1.20	1.14	1.37	1.35	1.63	0.23
11	5.0	0.08	1.95	1.91	2.44	2.32	2.82	2.74	0.08

observations may reveal differences in the regression of the item scores on the rest scores.

The LM statistics have two degrees of freedom. The sample sizes within the country (586 students) are now such that significance probabilities of the LM tests become informative. Items 2 and 10 show the largest misfit. Consistent with the results in Table 5.1, the ITF of item 2 is too high, while the ITF of item 10 is too low. This is an indication of uniform rather than nonuniform DIF. So it might be worthwhile to also split these items. On the other hand, the link must also remain substantial. There is some trade-off between these two considerations and some element of arbitrariness cannot be avoided.

The second example pertains to the main study of the PISA 2009 cycle. The data consisted of samples of 1000 students from 31 OECD countries. The analyses consisted of two steps. First, the data of all countries were analyzed simultaneously to identify items with country-specific DIF. This was done in an iterative process. In each iteration, MML estimates were obtained and the item with the worst misfit was identified. In the next iteration, this item was given country-specific item parameters and a new MML estimation run was made. This was repeated between two and four times depending on the scale analyzed. Finally, the fit of the resulting model with country-specific item parameters for the DIF items and the parameters of the non-DIF items, which were fixed over the countries, was evaluated. In the second step, the impact of DIF was evaluated by computing the correlations of the countries' mean latent trait values estimated with and without country-specific item parameters. Analyses were done using the PCM and the GPCM. Finally, to evaluate the impact of the choice of the model, the correlations of the countries' mean latent trait values estimated using the PCM and GPCM were computed. The reason is that the PISA project uses the PCM, so as a side line we will make an unassuming comparison between the results obtained using these two models.

Table 5.4 gives the codes and names of the scales that were investigated and the number of items in each scale. Labels starting with ST refer to scales from the student questionnaire and labels starting with IC refer to scales from the ICT questionnaire. To compute the results in Table 5.4, MML analyses using the GPCM were made for every scale, with all available OECD countries entered in an analysis simultaneously. The number of countries was 31 for the student questionnaire and 26 for the ICT questionnaire. Absolute values of the residuals as defined in formula 5.7 were counted and the percentages of values above 0.25 and 0.20 are displayed in the two last columns of Table 5.4, respectively. Note that the scales ST25 ("Like Reading") and IC04 ("Home Usage of ICT") displayed the most DIF. The scales ST27(a) and ST27(b), ST34, ST37, IC02, IC05, IC8, and IC10 were relatively free of DIF.

In Table 5.5, the results are further broken down to the item level. Items with effect sizes above 0.20 are highlighted. The items causing the DIF can be easily identified. Further breaking down these residuals can

TABLE 5.4
Overview of CDIF in the Student Questionnaire and the ICT Questionnaire

Label	Scale	Number of Items	Percentage Item by Country Interaction	
			Residual > 0.25	Residual > 0.20
ST24	Reading Attitude	11	7	12
ST25	Like Reading	5	60	66
ST26	Online Reading Activities	7	18	22
ST27(a)	Use of Control Strategies	4	6	7
ST27(b)	Use of Elaboration Strategies	4	1	3
ST27(c)	Use of Memorisation Strategies	5	12	16
ST34	Classroom Climate	5	2	4
ST36	Disciplinary Climate	5	2	3
ST37	Stimulate Reading Engagement	7	6	10
ST38	Teacher Structuring Strategies	9	10	12
ST39	Use of Libraries	7	15	22
IC02	ICT Availability at School	5	3	4
IC04	Home Usage of ICT	8	24	30
IC05	ICT for School-Related Tasks	5	9	14
IC06	Use of ICT at School	9	11	18
IC08	ICT Competency in Different Contexts	5	7	9
IC10	Attitude Toward Computers	4	3	5

lead to interesting insights. It is beyond the scope of this chapter to discuss all item-by-country interactions in detail, so one example must do. As already mentioned, ST25 has the largest bias. ST25 consists of the stem overall question, "How often do you read these materials because you want to?" followed by the items in reference: magazines, comic books, fiction (novels, narratives), nonfiction books, and newspapers. Response categories indexed from 0 to 4 are "never or almost never," "a few times a year," "about once a month," "several times a month," and "several times a week." It turns out that in Finland reading of comic books is much more salient than in other countries. The average observed and expected score over all countries except Finland is 1.25. The average item score in Finland is 2.58, compared to an expected value of 1.78, resulting in a residual of 0.87. The conclusion is that the Finnish students like to read more than the average OECD student, but they are especially fond of comic books. Within this example, country-specific item parameters solved the problem for Finland in the sense that the absolute values of all residuals as defined by formula 5.8 dropped below 0.10.

The impact of DIF was assessed using both the PCM and GPCM. The countries were rank ordered on their mean value on the latent variable, for

TABLE 5.5

Size of Residuals on the Item Level

Scale	Q1	Q2	Q3	Q4	Q5	Q6	Q7	Q8	Q9	Q10	Q11
ST24	0.10	0.07	0.07	0.13	0.09	0.06	0.09	0.11	0.15	0.16	0.12
ST25	0.24	0.41	0.20	0.07	0.35						
ST26	0.20	0.20	0.19	0.09	0.13	0.15	0.11				
ST27(E)	0.08	0.06	0.08	0.05							
ST27(M)	0.10	0.12	0.17	0.17							
ST27(C)	0.15	0.07	0.06	0.09	0.10						
ST34	0.06	0.09	0.04	0.06	0.08						
ST36	0.07	0.07	0.04	0.09	0.06						
ST37	0.10	0.09	0.09	0.17	0.08	0.10	0.07				
ST38	0.16	0.08	0.14	0.13	0.12	0.16	0.07	0.07	0.13		
ST39	0.14	0.17	0.13	0.10	0.09	0.06	0.16				
IC01	0.13	0.16	0.04	0.18	0.04	0.05	0.06	0.10			
IC02	0.07	0.13	0.02	0.10	0.13						
IC04	0.13	0.16	0.15	0.11	0.09	0.15	0.23	0.46			
IC05	0.12	0.20	0.10	0.08	0.13						
IC06	0.15	0.16	0.11	0.07	0.12	0.06	0.13	0.15	0.08		
IC08	0.09	0.19	0.15	0.11	0.05						
IC10	0.09	0.12	0.08	0.14							

both models and with and without items with country-specific parameters. An example pertaining to scale ST26 is given in Table 5.6. The first two items of the scale were split. The values in the four last columns are the MML estimated means of the latent trait distributions. Note that at first sight, the rank order of the countries looks quite stable.

Table 5.7 gives the correlations between the estimates obtained using 2 or 4 split items. The iterative process of splitting items was stopped when either 4 items were split, or 95% of the residuals defined by formula 5.7 were under 0.25. The first two columns of Table 5.7 give the rank correlation between order of countries obtained with and without split items using either the PCM or the GPCM as the measurement models. The last two columns present the analogous correlations. It can be seen that many of the correlations between the country means are quite high except for the scales "Like Reading," "Online Reading Strategies," "Use of Memorization Strategies," and "Use of Control Strategies," which had substantial DIF. Also, "Use of Libraries" seems affected by DIF. There are no clear differences between the correlations obtained using the PCM and GPCM.

The impact of using either the PCM or the GPCM was further evaluated by assessing differences in the estimated means of the countries on the latent scale and also the rank ordering obtained using the two models. These results are given in Table 5.8. The last two columns give the rank correlation and

TABLE 5.6

Rank Order and Mean Scale Level of Countries on the Scale ST26 for the GPCM and PCM with and without 2 Split Items

	Rank Order				Mean on Latent Scale			
	PCM	GPCM	PCM	GPCM	PCM	GPCM	PCM	GPCM
Country	**With**	**With**	**Without**	**Without**	**With**	**With**	**Without**	**Without**
AUS	8	11	9	10	−0.131	−0.074	−0.109	−0.076
AUT	20	21	21	21	0.067	0.069	0.069	0.066
BEL	4	7	3	5	−0.281	−0.128	−0.260	−0.165
CAN	6	9	6	7	−0.203	−0.103	−0.164	−0.106
CZE	30	30	30	30	0.605	0.546	0.486	0.472
DNK	24	25	24	26	0.149	0.211	0.142	0.183
FIN	11	13	7	9	−0.105	−0.059	−0.134	−0.095
FRA	7	10	8	8	−0.153	−0.094	−0.130	−0.099
DEU	23	23	22	24	0.143	0.167	0.121	0.136
GRC	21	17	20	18	0.067	−0.018	0.054	0.003
HUN	29	29	29	29	0.455	0.437	0.326	0.339
ISL	28	28	25	25	0.237	0.251	0.149	0.173
IRL	2	2	2	2	−0.576	−0.539	−0.486	−0.480
ITA	17	14	17	15	−0.024	−0.053	−0.004	−0.032
JPN	1	1	1	1	−0.673	−0.618	−0.509	−0.500
KOR	25	5	26	14	0.167	0.168	0.158	−0.046
LUX	16	19	15	17	−0.035	0.002	−0.036	−0.017
MEX	3	3	4	3	−0.376	−0.491	−0.248	−0.350
NLD	14	24	13	22	−0.060	0.176	−0.058	0.089
NZL	5	4	5	4	−0.234	−0.267	−0.196	−0.222
NOR	27	27	27	27	0.217	0.223	0.171	0.187
POL	31	31	31	31	0.744	0.562	0.624	0.528
PRT	26	26	28	28	0.217	0.215	0.184	0.190
SVK	13	15	11	11	−0.072	−0.040	−0.098	−0.074
ESP	10	12	12	13	−0.109	−0.073	−0.066	−0.049
SWE	18	20	18	19	0.040	0.024	0.027	0.023
CHE	15	18	14	16	−0.058	−0.007	−0.055	−0.023
TUR	22	16	19	23	0.132	−0.029	0.134	0.026
QUK	19	22	19	23	0.067	0.138	0.049	0.098
USA	9	8	10	6	−0.119	−0.126	−0.104	−0.107
CHL	12	6	16	12	−0.099	−0.135	−0.036	−0.072

product moment correlation of the latent-scale means of countries obtained using PCM and GPCM when no items were split. The two previous columns give the analogous correlations for the number of split items given in the column labeled "Item Split." In general, the correlations are high. The main exception is ST25. Therefore, given our criteria for comparing model fit, it can

TABLE 5.7

Correlations between Country Means of Latent Distributions Estimated with and without Split Items

Label	Scale	Items Split	Rank Correlation		Correlation	
			PCM	GPCM	PCM	GPCM
ST24	Reading Attitude	2	0.847	0.964	0.978	0.991
ST25	Like Reading	2	0.589	0.861	0.610	0.968
ST26	Online Reading Activities	2	0.616	0.819	0.936	0.962
ST27(a)	Use of Control Strategies	2	0.646	0.706	0.914	0.934
ST27(b)	Use of Elaboration Strategies	2	0.838	0.919	0.969	0.973
St27(c)	Use of Memorization Strategies	2	0.510	0.616	0.784	0.922
ST34	Classroom Climate	2	0.870	0.870	0.973	0.967
ST36	Disciplinary Climate	2	0.885	0.906	0.979	0.979
ST37	Stimulate Reading Engagement	2	0.933	0.966	0.982	0.991
ST38	Teacher Structuring Strategies	2	0.951	0.958	0.979	0.989
ST39	Use of Libraries	2	0.883	0.880	0.954	0.954
IC02	ICT Availability at School	2	0.851	0.823	0.923	0.901
IC04	Home Usage of ICT	2	0.876	0.894	0.980	0.981
IC05	ICT for School-Related Tasks	2	0.850	0.844	0.969	0.969
IC06	Use of ICT at School	2	0.969	0.969	0.995	0.995
IC08	ICT Competency	2	0.829	0.822	0.959	0.953
IC10	Attitude Toward Computers	2	0.801	0.743	0.985	0.960
ST24	Reading Attitude	4	0.804	0.919	0.996	0.984
ST26	Online Reading Activities	4	0.606	0.798	0.857	0.905
ST37	Stimulate Reading Engagement	4	0.767	0.829	0.922	0.996
ST38	Teacher Structuring Strategies	4	0.888	0.889	0.956	0.966
ST39	Use of Libraries	4	0.788	0.853	0.927	0.945
IC04	Home Usage of ICT	4	0.879	0.894	0.980	0.981
IC06	Use of ICT at School	4	0.976	0.920	0.995	0.862

be concluded that there is little support for preferring the GPCM over the PCM as an analysis model.

Conclusions

Large-scale educational surveys often give rise to an overwhelming amount of data. Simple unequivocal statistical methods for assessing the quality and structure of the data are hard to design. The present chapter presents diagnostic tools to tackle at least one of the problems that emerge in educational surveys, the problem of differential item functioning. Given the

TABLE 5.8

Correlations between Country Means of Latent Distributions Estimated Using the PCM and GPCM

			With Split Items		Without Split Items	
Label	**Scale**	**Items Split**	**Rank Corre-lation**	**Corre-lation**	**Rank Corre-lation**	**Corre-lation**
ST24	Reading Attitude	2	0.940	0.993	0.913	0.988
ST25	Like Reading	2	0.643	0.897	0.574	0.666
ST26	Online Reading Activities	2	0.962	0.994	0.879	0.988
ST27(a)	Use of Control Strategies	2	0.960	0.993	0.959	0.992
ST27(b)	Use of Elaboration Strategies	2	0.954	0.994	0.968	0.997
St27(c)	Use of Memorization Strategies	2	0.988	0.998	0.805	0.966
ST34	Classroom Climate	2	0.983	0.996	0.987	0.998
ST36	Disciplinary Climate	2	0.976	0.997	0.986	0.997
ST37	Stimulate Reading Engagement	2	0.993	0.998	0.981	0.996
ST38	Teacher Structuring Strategies	2	0.977	0.998	0.978	0.996
ST39	Use of Libraries	2	0.981	0.993	0.990	0.998
IC02	ICT Availability at School	2	0.998	0.997	0.968	0.987
IC04	Home Usage of ICT	2	0.959	0.980	0.941	0.978
IC05	ICT for School-Related Tasks	2	0.974	0.993	0.980	0.996
IC06	Use of ICT at School	2	0.992	0.998	0.994	0.998
IC08	ICT Competency	2	0.942	0.990	0.972	0.994
IC10	Attitude Toward Computers	2	1.000	0.983	0.980	0.997
ST24	Reading Attitude	4	0.968	0.995		
ST26	Online Reading Activities	4	0.964	0.996		
ST37	Stimulate Reading Engagement	4	0.978	0.994		
ST38	Teacher Structuring Strategies	4	0.985	0.997		
ST39	Use of Libraries	4	0.972	0.988		
IC04	Home Usage of ICT	4	0.959	0.980		
IC06	Use of ICT at School	4	0.936	0.880		

complicated and large data, it comes as no surprise that the tools presented here have both advantages and drawbacks. On the credit side, concurrent MML estimation is well founded, practical, and quick. Further, in combination with LM statistics, few analyses are needed to gain insight into the data. Above, searching for DIF was presented as an iterative procedure, but this procedure can be easily implemented as one automated procedure. On the other hand, the advantage that, contrary to most test statistics for IRT, the LM statistics have a known asymptotic distribution loses much of its impact, because of the power problem in large data sets. What remains is a procedure that is transparent with respect to which model violations are exactly targeted and the importance of the model violation in terms of the actual observations. Further, the procedure is not confined to specific IRT models, but can be generally applied. Finally, the procedure supports the use of group-specific item parameters. The decision of whether group-specific item parameters should actually be used depends on the inferences that are to be made next. In that sense, the example where the order of countries on a latent scale is evaluated is just an example. Often, other inferences are made using the outcomes of the IRT analyses, such as multilevel analyses relating background variables to educational outcomes. Also in these cases, the impact of using country-specific item parameters can be assessed by comparing different analyses.

The present chapter was mainly written to present statistical methodology and not to draw ultimate conclusions regarding the PISA project. Still, some preliminary conclusions can be drawn. The analyses showed that certain scales of the student background questionnaire and the ICT questionnaire are indeed affected by the presence of CDIF. The scale most affected by CDIF was ST25 "Like Reading." Other scales where DIF was evident were ST26 "Online Reading Activities," ST27c "Memorization Strategies," ST39 "Use of Libraries," and IC04 "Home Usage of ICT." Correlations between ordering of countries showed that the detected CDIF did indeed have an impact. However, other criteria for impact may be more relevant.

Finally, using either the PCM or GPCM had little impact. Overall, the discrimination parameters were quite high and differences between these indices within the scales probably canceled out when evaluating the order of the countries. Also, the conclusions regarding CDIF items were not substantially affected by the model used.

References

Aitchison, J. and Silvey, S.D. 1958. Maximum likelihood estimation of parameters subject to restraints. *Annals of Mathematical Statistics 29,* 813–828.

Bock, R.D. and Aitkin, M. 1981. Marginal maximum likelihood estimation of item parameters: An application of an EM-algorithm. *Psychometrika, 46,* 443–459.

Camilli, G. and Shepard, L.A. 1994. *Methods for Identifying Biased Test Items.* Thousand Oaks, CA: Sage.

Chang, H. and Mazzeo, J. 1994. The unique correspondence of the item response function and item category response function in polytomously scored item response models. *Psychometrika, 59,* 391–404.

Glas, C.A.W. 1998 Detection of differential item functioning using Lagrange multiplier tests. *Statistica Sinica, 8,* 647–667.

Glas, C.A.W. 1999. Modification indices for the 2-pl and the nominal response model. *Psychometrika, 64,* 273–294.

Glas, C.A.W. and Dagohoy, A.V. 2007. A person fit test for IRT models for polytomous items. *Psychometrika, 72,* 159–180.

Glas, C.A.W. and Suárez-Falćon, J.C. 2003. A comparison of item-fit statistics for the three-parameter logistic model. *Applied Psychological Measurement*, *27*, 87–106.

Hambleton, R.K. and Rogers, H.J. 1989. Detecting potentially biased test items: Comparison of IRT area and Mantel-Haenszel methods. *Applied Measurement in Education*, *2*, 313–334.

Haberman, S.J. 1977. Maximum likelihood estimates in exponential family response models. *The Annals of Statistics*, *5*, 815–841.

Holland, P.W. and Thayer, D.T. 1988. Differential item functioning and the Mantel-Haenszel procedure. In: H. Wainer and H.I. Braun (Eds.), *Test Validity*. Hillsdale, NJ: Lawrence Erlbaum Associates Inc.

Holland, P.W. and Wainer, H. 1993. *Differential Item Functioning*. Hillsdale, NJ: Erlbaum.

Kok, F.G., Mellenbergh, G.J., and van der Flier, H. 1985. Detecting experimentally induced item bias using the iterative logit method. *Journal of Educational Measurement*, *22*, 295–303.

Lord, F.M. 1980. *Applications of Item Response Theory to Practical Testing Problems*. Hillsdale, NJ, Erlbaum.

Masters, G.N. 1982. A Rasch model for partial credit scoring. *Psychometrika*, *47*, 149–174.

Muraki, E. 1992. A generalized partial credit model: Application of an EM algorithm. *Applied Psychological Measurement*, *16*, 159–176.

Rao, C.R. 1948. Large sample tests of statistical hypothesis concerning several parameters with applications to problems of estimation. *Proceedings of the Cambridge Philosophical Society*, *44*, 50–57.

Samejima, F. 1969. Estimation of latent ability using a pattern of graded scores. *Psychometrika, Monograph Supplement, No. 17*.

Tutz, G. 1990. Sequential item response models with an ordered response. *British Journal of Mathematical and Statistical Psychology*, *43*, 39–55.

Verhelst, N.D., Glas, C.A.W., and de Vries, H.H. 1997. A steps model to analyze partial credit. In: W.J. van der Linden and R.K. Hambleton (Eds.), *Handbook of Modern Item Response Theory*. (pp. 123–138). New York, NJ: Springer.

6

Sampling, Weighting, and Variance Estimation in International Large-Scale Assessments

Keith Rust

Westat

Joint Program in Survey Methodology

CONTENTS

Introduction

Many educational assessment programs are conducted with the aim of obtaining results at the student, school, and administrative unit level. Such "high stakes" assessments may be used to make decisions about individual student progress through the education system, or as a tool used in the evaluation of teachers and schools. In such cases, generally every student in the population of interest is assessed, as results are desired for each student. In these circumstances there are no issues of sample design and selection involved, and no issues related to the need to provide analysis weights. In large-scale international assessments, however, almost invariably the goals of the study do not include the provision of individual student achievement results for all the individuals in the population. Rather, as is clearly demonstrated in the rest of this book, the purpose is to make inference about a whole population. This extends to interest in providing results for a wide variety of population subgroups, examining the distribution of achievement within and across these subgroups, and exploring the relationship of student and school characteristics to achievement.

Given these goals, it is not a specific requirement of the study to obtain achievement data for each student in the population. The inferences of interest can be obtained from a suitably designed and executed sample of students. This, of course, offers the potential to greatly reduce the cost and burden of these assessments, rendering them possible in cases where it would be infeasible to assess all students. This also permits such studies to simultaneously assess multiple subjects, or cognitive domains, while not unduly burdening any individual student.

While sampling methods provide the means to carry out assessments in an affordable manner, considerable attention to detail is required in designing and selecting samples from the participating countries. Correspondingly, care is required in the analysis of data collected from these samples. Analytic techniques are required that address the fact that, as a result of the way in which the sample is selected, the distribution of characteristics in the sample is likely to be different from that of the population from which the sample is drawn. The most obvious and best-known technique for addressing this issue is to apply sampling weights in the analysis of the data. This can be reasonably straightforward for some analyses, but is less so for others, particularly in cases where the model used to analyze the data has features that parallel the design of the sample. The estimation of sampling variances that account for the covariance structure of the data induced by the sample design is also a key feature of the analysis of data obtained from the sample designs typically used in large-scale assessments.

The use of probability-based, or scientific, methods for selecting samples for large-scale assessments has been practiced for many decades. Since the 1960s the International Association for the Evaluation of Educational

Achievement (IEA) has conducted many of these studies, and over the intervening period a number of other organizations have also conducted them. Over time those conducting these studies have increasingly seen the need to promote and, to the extent feasible, enforce a standardized approach to sample design and selection, with common standards applied. Perhaps these efforts can best be seen coming to fore in two studies conducted in the early 1990s. The IEA conducted the International Study of Reading Literacy from 1989 to 1991 (Elley 1992), while the Educational Testing Service conducted the International Assessment of Educational Progress (IAEP) in mathematics and science (Lapointe et al. 1988, 1992). These studies featured sampling manuals, with procedures described in detail, and considerable efforts to ensure the quality of the resulting samples in each country. Subsequent IEA studies in a variety of academic subjects, and the Programme for International Student Assessment (PISA), conducted by the Organisation for Economic Co-operation and Development in several cycles since 2000, have continued the promotion of adopting high-quality sampling procedures, well-documented and subject to substantial quality-control procedures.

Thus, the use of scientific sampling methods has become standard in these assessment programs. In this chapter we describe the most common sampling techniques that are applied in these studies, indicating the goals that the methods are designed to achieve. We then discuss the implications for analysis, specifically through the use of survey weights, and variance estimation procedures appropriate for the design. We also describe how missing data resulting from nonresponse can impact the analysis, with a discussion of the way in which the survey weights can be modified so as to reduce the potential for bias that is introduced by such nonresponse. We include a brief discussion of software available for analyzing data from these studies that addresses the need to incorporate weights and estimate sampling errors appropriately. However, experience has taught us that any detailed treatment of currently available software soon becomes dated, and so that will not be attempted. Users who undertake, or anticipate undertaking, regular analyses of complex survey data, whether from large-scale assessments or other surveys, are recommended to consult the book on this topic by Heeringa et al. (2010).

Sample Design

It has become routine over the past 25 years (and there were certainly earlier precedents) that the sample designs for international large-scale assessments follow the principles and practices of scientific, probability-based, sampling methods. A full introduction to such methods is given in Lohr (2010). In this section the key elements of scientific probability sampling that apply in the large-scale assessment context are described.

Population Definition

Strictly speaking, of course, the need to define the population of inference for an assessment study, and to ensure that this is done comparably across participating countries, is not related to whether one uses a sample or surveys the entire population. However, when a sample is selected in each country, it may not be as readily evident whether the population coverage is comparable across countries as might be the case if all students in the population were selected. For example, suppose that there is a school included in the sample that has 300 full-time students in the population, and 15 part-time students. If all the students in the school are to be assessed, it will be fairly readily apparent if those administering the assessment decide not to include any of the part-time students. But if a sample of 25 students is selected, and part-time students are omitted from the sampling frame, the fact that no part-time students end up being selected in the sample might not be noted. Issues with population definition and coverage tend to revolve around relatively small groups in the population, but ones whose distribution of achievement may be quite different from the rest of the population. Thus, on the one hand, their absence from the sample might not be noticed, while on the other hand, failure to cover them in the sampling procedure might induce a substantial bias in certain analyses of the data.

Large-scale assessments typically define some cohort of the student population as the population of interest. The two basic approaches are to define a single grade of interest, or a particular birth-cohort, typically a 12-month span. Each of these approaches has its advantages and limitations.

In the past two decades, the IEA studies have generally defined a common grade to be surveyed in each country. This has two main advantages: (1) within a given country, the population is meaningful to policy makers. There is interest in knowing "how our eighth-graders compare to the rest of the world"; and (2) by surveying a single grade, it is relatively straightforward to define the population, sample it, and administer the assessment. The major drawback is being able to define grades that are comparable across countries. "Eighth grade" is a very arbitrary label, affected by national policies about the age of starting school, and the labels given to early years of schools. Typically this is addressed by defining the population as, for example, the grade containing the largest proportion of 14-year-old students (i.e., students born during a specific 12-month period, with this same birth cohort used in every country). This is somewhat effective, but has three drawbacks. First, even if every country surveys the correct grade, there can legitimately be almost a 12-month spread in the mean age of students across the participating countries. Comparing students from a grade in one country where the mean age is 8.6 years with another where the mean age is 9.4 years can be problematic. Second, often when countries participate for the first time, they do not actually know which is the correct grade in cases where there are two grades each with a large proportion of the specified age cohort. As more and more countries have come to participate in these studies, this issue has diminished as a problem. A pilot test can also indicate if there is a problem with age distribution of the chosen grade. The third problem is related to the first two; a slight change in the definition of the cohort can lead to a change in the target grade, especially if the age cohort definition is tied to the time of the assessment. That is, it is possible that in March the greatest number of 14-year-olds are in grade 9, but by May (shifting the cohort definition by 2 months) the greatest number are in grade 8. Problems can also arise when comparisons are made over time within a country, if there is a policy change about the age of schooling. In the late 1990s, Norway made such a policy change, and over a very short span of years the average age of students in each grade dropped by about 1 year. After this change, it was difficult to know which was the most appropriate grade to survey when an important goal of the study was to assess the trend over time. The IEA Trends in International Mathematics and Science Study was forced to confront this issue. A further complication is that different provinces within a country may have different policies relating to age of schooling. Thus, in the first TIMSS study in 1995, different grades were assessed in different states in Australia. In each state the two adjacent grades with the most 13-year-olds were surveyed. In some states the upper of these two grades was grade 9, but in others it was grade 8. In federal systems this type of complication seems inevitable.

PISA and the IAEP studies, on the other hand, have defined the population using a 12-month birth cohort. This approach too has its benefits and deficiencies; not surprisingly these complement those of the grade-based approach. One does not need to study, or understand in depth, the educational system of a country to define the target population, and it is relatively easy to conduct quality control procedures to ensure that the correct population has been surveyed. Diverse countries can be compared, but they will each have a population that is uniformly distributed in age over a common 1-year age span (setting aside the problem of students in the later teenage years dropping out of school differentially across countries). The use of an age definition also avoids issues that can arise because of artifactual distortions of a grade population. In addition to the presence of students in ungraded programs, in some cases there is a particular grade at the top of one level of school, where students may tend to disproportionately be "held back." Thus if there is a key exit exam at the end of grade 8, for example, the grade 8 cohort may contain relatively many students who are repeating the grade (and these will be among the lower achievers). Thus, a form of length-biased sampling may occur. This was a particular issue in the 1995 TIMSS study of grade 12 students, and the population was therefore defined not as students in grade 12, but as students who were in grade 12 for the first time. Such problems are avoided when an age-based population definition is used.

There are four main drawbacks to the use of an age-based definition. First, the population is often not very interesting at a national level. It might be of interest to know how U.S. 15-year-olds do compared with 15-year-olds in Finland. But if 15-year-olds in the United States are half in grade 10 and half in grade 9, internal comparisons within the United States are not very enlightening, and the implications of the results for instructional policy are not as clear as when a grade-based approach is used. Second, administering the assessments is often somewhat unwieldy, since within a school students have to be pulled from classes in two or three different grades for the assessment. The third issue is that obtaining a frame of schools that comprehensively covers the population can be difficult. Often there are many schools in a given country that might have just a few eligible students enrolled, because the grades taught in the school generally have very few students of the required age. National study centers are loath to include such schools in the sampling frame, as they know that these schools will be a burden and relatively costly on a per-student basis to include. However, the students in such schools tend to be very much skewed toward one end or the other of the ability distribution. Consider the case of PISA in the United States. About 1–2% of the PISA age cohort is in eighth grade, with the rest in higher grades. Most schools that have grade 8 do not have higher grades, and there are many more schools with eighth grades than there are schools with higher grades (as middle schools tend to be smaller than high schools). This means that the sampling frame for the United States contains a large number of schools that have grade 8 as the highest grade. When such a school is sampled, the procedure is likely to be to include the handful of eligible students enrolled in the school. These will by definition be the oldest students in the school, which in many cases means that they are not the most able. Thus, there is a strong disincentive to include these schools, both on the part of the national center (because of the operational cost and burden), and also on the part of the school, which cannot understand the desire to assess these few atypical students, and question the representativeness of the resulting sample. Fourthly, the analysis and interpretation of school- and system-level factors, and their relationship to achievement, are problematic in cases where the population of students is divided across different levels of schooling.

There are other groups of students whose inclusion (or otherwise) must be carefully addressed whether an age- or grade-based population definition is used: special needs students, recent immigrants with a different first language from the test language, vocational students, apprenticeship, and other part-time students. Increasingly it seems likely that online students may be an issue at higher grades, although no study seems to have been as yet noticeably impacted by this phenomenon.

Multistage Sampling

It is entirely natural that one obtains a sample of students within a country by first selecting a sample of schools, and then selecting students within schools. Such a procedure is known in the survey sampling literature as two-stage sampling, and it is a commonly applied one. It is primarily used for reasons of cost and operational feasibility. It is just not practicable to assess thousands of students within a country by testing one or two students from each of thousands of schools (although perhaps the rapid growth of computer-based testing may render this statement inaccurate within a matter of several years). In many cases also, there is no centralized data source listing all of the students in the population (and even if there is it is likely to be somewhat outdated). So the best approach to getting high-quality coverage of the student population is to sample schools and ask the schools directly for a current list of eligible students. However, in the case of achievement studies, there is an additional reason for the use of this type of design. Often analysis of the data involves consideration of school effects through a multilevel (mixed) model, and such models cannot be effectively analyzed if the sample size of students from any one school is very small.

Thus, the use of the two-stage approach means that there are two distinct elements to sample selection; the selection of schools and the selection of students. We will address each of these in the following. It is important to note that the two are closely related. In particular, the goal of the school sample design is primarily to lead to a sound sample of students, and the approach used to designing the school sample is driven by this consideration. Thus, the school sample that results does not necessarily have the characteristics that one would design for if the goal were solely to select a sample of schools to analyze school characteristics.

In certain countries a third, prior, stage of sampling may be called for. This is the selection of geographic domains, from within which schools will be sampled. There are two reasons for such an approach. The first is that there may be no up-to-date list of schools available centrally for the entire country. By selecting only a sample of a few dozen geographic regions, the construction of school lists of adequate quality may be made feasible. This is not required in many countries, as even developing countries generally have a centralized list of schools, or can easily compile one. The prime example that we are aware of where this was an issue was in Russia in the early 1990s, at the time of the first TIMSS study. A more common reason for using a geographic stage of sampling is to save on administrative costs, when the assessments are administered separately. Obviously this is an issue only for larger countries, and Russia and (at times, but not recently), the United States have adopted a three-stage design for this reason.

The use of a multistage design, and specifically a two-stage design of selecting schools and then students, can (and generally does) have profound implications for the precision of the resulting estimates. If one selects a simple random sample of schools, all of which have the same size enrollment of eligible students, and then draws a simple random sample of students from within each selected school, the sampling variance is given in formula 6.1. This formula ignores issues related to sampling from a finite population. This has been for simplicity of explication, but can often be justified on a combination of the following conditions: (1) the sample is selected using a with-replacement procedure; (2) the sample sizes are small in comparison with the population size (this is likely to be true of the school sample, but not the student sample within each school); and (3) inferentially, strict reference to a finite population is not warranted (this is the case in most other areas of statistical analysis, and can particularly apply to the student sample in a large-scale assessment program, where we do not wish to limit our inference to the particular student population that happened to exist at the time the study was conducted).

$$Var\left(\bar{\bar{y}}\right) = \frac{1}{m}\left\{\sigma_1^2 + \frac{1}{M}\sum_{j=1}^{M}\sigma_{2j}^2/n_j\right\} \tag{6.1}$$

where

$\bar{\bar{y}}$ is the estimate of the per-student mean of y, $\bar{\bar{Y}}$;
m is the sample size of schools;
n_j is the sample size of students in school j;
M is the population size of schools;

$\sigma_1^2 = \frac{1}{M}\sum_{j=1}^{M}\left(\bar{Y}_j - \bar{\bar{Y}}\right)^2$ is the between-school variance of y;

$\sigma_{2j}^2 = \frac{1}{N}\sum_{i=1}^{N}\left(y_{ij} - \bar{Y}_j\right)^2$ is the within school variance of y;

N is the number of students in each school; and
$\bar{Y}_j$ is the mean of y for school j.

This compares with the variance of a simple random sample of students of the same size:

$$Var\left(\bar{\bar{y}}\right) = \left(\sigma_1^2 + \frac{1}{M}\sum_{j=1}^{M}\sigma_{2j}^2\right) / n_o \tag{6.2}$$

where $n_o = \sum_{j=1}^{m} n_j$ is the total sample size of students.

The ratio of these two quantities is referred to as the design effect for this particular design (a two-stage sample with simple random sampling at each stage):

$$Deff = \frac{n_o}{m}\frac{\left(\sigma_1^2 + \frac{1}{M}\sum_{j=1}^{M}\sigma_{2j}^2/n_j\right)}{\left(\sigma_1^2 + \frac{1}{M}\sum_{j=1}^{M}\sigma_{2j}^2\right)} \tag{6.3}$$

One can see that unless the ratio of the between-school variance to the average within-school variance is substantially smaller than the mean of the n_j, this design effect is likely to be noticeably greater than 1.0. This indicates that the level of sampling error from such a two-stage design is likely to be much higher than for a simple random sample. Under the simple case that all σ_{2j}^2 squared are equal, and all n_j are equal (to n say), this equation reduces to

$$Deff = \frac{n\sigma_1^2 + \sigma_2^2}{\sigma_1^2 + \sigma_2^2} \tag{6.4}$$

where $\sigma_2^2 = \frac{1}{M}\sum_{j=1}^{M} \sigma_{2j}^2$.

Defining the intraclass correlation, *Rho,* as

$$Rho = \sigma_1^2 / \left(\sigma_1^2 + \sigma_2^2\right) \tag{6.5}$$

the Design Effect can be expressed as

$$\begin{aligned} Deff &= \frac{n\sigma_1^2 + \sigma_2^2}{\left(\sigma_1^2 + \sigma_2^2\right)} \\ &= 1 + (n-1)Rho \end{aligned} \tag{6.6}$$

To mitigate the negative effects of this design on sampling variances, we use techniques of stratification and systematic selection to effectively reduce the sizes of σ_1^2 and σ_2^2 for the two-stage design, thereby lowering the design effect. Also, it is important to keep in mind that the much lower per-student cost of administering the assessment to a two-stage sample means that while the amount of sampling variance per student is likely to be considerably higher for a two-stage design, the amount of sampling variance per unit cost will be considerably lower for a two-stage design, and this is the reason that two-stage designs are universally adopted in large-scale assessments of school students.

Sample Size Determination

Before turning to the sample design features of stratification, systematic selection, and probability-proportional-to-size selection, we will consider the question of sample size determination for international studies.

The topic is usually approached by considering what level of precision is required for the national student mean on an achievement scale, measured in standard deviation units, and what level of precision is required for an estimate of a proportion of a certain size, again with the students as the units of analyses. IEA studies have specified the target to be that the standard error for a measure of student mean performance on an assessment should be 0.05 standard deviations, and that the standard error on an estimate proportion of about 0.5 should be 0.025. Both of these considerations dictate what is referred to as an effective sample size of 400 students. That is the sample size that would be required if, in fact, a simple random sample of students were selected. The actual sample size required for a given sample design is the product of the required effective sample size and the design effect.

Using the simplified model for the design effect above, this means that the student sample size required to meet the precision requirement is: $400(1 + (n-1)\ Rho)$. Thus for a value of *Rho* equal to 0.2 (a typical value in practice), with a sample of 30 students selected from each school (also not atypical), the required student sample size is 2720 assessed students, which would require a sample of about 90 schools.

However, other considerations also come into play in determining the desired sample size of students and schools. First, it is desirable to ensure that a certain minimum number of schools are included in the sample. This is to ensure that analyses involving school level variables have good precision, and also guards against undesirable consequences that can arise when one or two unexpectedly atypical schools are included in the sample. Also, there is generally interest in subgroup analyses for even moderately small population subgroups, and also the interest is not just in the mean student performance, but also in the overall distribution, including relatively extreme values, such as the 5th or 95th percentiles.

The combination of these considerations, and the establishment of historical precedent over the past 20 years, mean that a requirement to have a minimum of 150 participating schools, with 30 assessed students (or one classroom, in the case of IEA studies), has become an accepted norm. Of course individual countries often elect to adopt larger samples, almost always because they wish to obtain adequate precision for results at the region level—this is particularly the case for countries with federal governing systems, such as Australia, Canada, and Spain, where responsibility for education policy largely lies at the provincial level.

Stratification

Stratification involves forming a partition of the population, and selecting separate samples from each of the resulting groups (called strata). This has the effect of eliminating the variance between strata as a contributor to the sampling variance of any estimate, since all possible samples selected represent each stratum in exact proportion to the stratum population size (after weighting). Stratification also allows the sampling rates to differ across strata, so that some groups of particular interest can be sampled more heavily than others.

Stratification can be applied at both the school and student levels of sampling. In all large-scale assessment programs it is applied at the school level. It is also on occasion applied at the student level. This is generally only done when there are rare subgroups of special interest, which are not restricted to particular schools. Two examples from PISA are indigenous students in Australia (who constitute about 1–2% of the population) and recent immigrant students in Denmark.

As noted in Section "Multistage Sampling," when the intraclass correlation of schools is even moderately high, the design effect can be substantial, leading to considerable loss in the precision of estimates. Fortunately, in most countries there are school-level variables available in readily accessible sources, for all schools in the country, that are substantially correlated with school mean achievement. Thus, using these variables to form strata significantly reduces the impact of between-school variance on sampling variance, and thus effectively lowers the intraclass correlation and thus the design effect. An obvious variable that applies in many countries in the case of age-based studies is the grade level of the school. When surveying 15-year-olds, as in the case of PISA, in many countries some of these students are enrolled in upper-level schools, while the remainder are enrolled in lower-level schools. When surveying students in the upper levels of secondary education, in many countries good strata can be created by distinguishing between academic, vocational, and general high schools. In some countries the distinction between public and private schools is an important one. And in most countries some sort of geographic stratification is beneficial, as often schools in metropolitan areas have higher levels of achievement than those in more rural areas.

Table 6.1 shows the example of stratification for Sweden from PISA 2012. This illustrates the fact that a variety of school characteristics can be used to create school-level strata.

Under proportional allocation, in which the sample of schools allocated to each stratum is proportional to the population size of schools, Equation 6.4 for the design effect of the mean is replaced by

$$Deff = \frac{n\sigma_{s1}^2 + \sigma_2^2}{\left(\sigma_1^2 + \sigma_2^2\right)} \tag{6.7}$$

where

$$\sigma_{s1}^2 = \frac{1}{M}\sum_{l=1}^{S}\sum_{j\in l}^{M_l}\left(\overline{Y}_j - \overline{\overline{Y}}_l\right)^2$$

M_l denotes the school population size of stratum l;

$\overline{\overline{Y}}_l$ is the per student mean of y in stratum l;

S is the number of strata.

It can be seen that σ_{s1}^2 is generally less than σ_1^2, and often relatively considerably less, with the reduction being proportional to the squared correlation

TABLE 6.1

School Stratification and Sample Allocation for Sweden, PISA 2012

Stratification				Population Figures			Sample Sizes		
Type of Control	School Level	Community Type	Stratum ID	Number of Schools	Enrollment of 15-Year-Olds	Percentage of Total 15-Year-Old Enrollment	Number of Schools	Expected Number of Students	Percentage of Student Sample
Public	Lower secondary	Metropolitan	01	459	30,784	30.5	57	1710	30.0
		Other large city	02	331	22,166	22.0	42	1260	22.1
		Pop. at least 12,500	03	268	18,800	18.6	34	1020	17.9
		Manufacturing	04	85	6265	6.2	12	360	6.3
		Sparse population	05	117	5501	5.5	10	293	5.1
	Upper secondary		06	525	1953	1.9	19	82	1.4
Independent	Lower secondary	Metropolitan	07	234	7579	7.5	17	436	7.7
		Other large city	08	133	5199	5.2	14	371	6.5
		Pop. at least 12,500	09	82	1770	1.8	4	88	1.5
		Manufacturing	10	10	107	0.1	2	30	0.5
		Sparse population	11	12	124	0.1	2	25	0.4
	Upper secondary		12	488	613	0.6	12	19	0.3
			Total	2744	100,861	100.0	225	5694	100.0

between the stratification variables and the school mean achievement. Thus, the design effect is reduced in this case. However, a reduction in design effect may not result if, rather than using stratification with proportional allocation, stratification is used as a means to oversample certain types of schools. Since disproportionate allocation acts to increase the design effect, this may outweigh the gains that result from stratification and in aggregate lead to an overall increase in the design effect, but with the benefit of being able to produce reliable estimates for subgroups of interest.

Systematic Sampling

Systematic equal-probability sampling is used in place of simple random sampling. The technique involves ordering the population list (referred to as the sampling frame) in some systematic order. A sampling interval, I, is then obtained by dividing the population size, N, by the desired sample size, n. A random start, R, is then selected, uniformly between 0 and I. The selected units are those corresponding to the number R, $R + I$, $R + 2I$, and so on up until $R + (n - 1)I$. These numbers are in general not integers; rounding them up to the next exact integers identifies the sampled units.

Systematic selection has two advantages. First, it reduces sampling variance in much the same way that stratification does. The sampling variance from a systematic sample is very similar to that of a stratified design in which n equal-sized strata are created and one unit is sampled from each stratum. The second advantage is that it is easy to implement, and easy to check, on a large scale, such as is needed in international assessments in which many countries participate. Its disadvantages are that there is no unbiased variance estimator available, even for a simple mean, and, in certain rare instances, it can actually lead to an increase in design effect rather than a decrease as occurs with stratification.

Systematic sampling can be used in conjunction with stratification, and also with probability-proportional-to-size selection, discussed in the next section. When used with stratification, the typical application is to use one set of variables for stratification, and other variables for systematic selection within strata. The variables used for stratification would tend to be categorical variables (especially those with no natural ordering), and variables most strongly related to achievement, as well of course as variables defining groups to be oversampled. Variables used for sorting for systematic selection are often continuous in nature, and perhaps less highly correlated with achievement. Thus, for example, one might stratify a sample of schools by province, public/private, and level (upper/lower), and then within each resulting stratum, sort schools by size prior to systematic selection. This ensures that the resulting sample is well-balanced with respect to school size, even though that is not a stratification variable.

Systematic sampling can also be used effectively to select student samples in cases where no oversampling is required. For example, when selecting students within each school, sorting the sample by grade (in the case of age-based samples) and gender, prior to systematic selection, can be very effective in reducing sampling errors. In this case it is σ_2^2 that is being effectively reduced, much in the way that school stratification reduces σ_1^2. It is generally much easier from a practical standpoint to obtain a sorted list of students from each sampled school and to select a systematic sample from the list, than it is to stratify the student list and select independent samples within each stratum.

Probability-Proportional-to-Size Selection

In the discussion to this point, it has been assumed that all units within a stratum receive the same probability of selection. However, in large-scale assessment surveys, almost always the school sample selection procedure selects larger schools with larger probabilities than smaller schools. This method provides a robust approach, when the primary goal of school sampling is to select a sample of students. The reason why this approach is effective is discussed later in this section.

In probability-proportional-to-size (PPS) selection, each unit is assigned a measure of size. In the case of assessment surveys, the measure of size for a school is generally a nondecreasing function of the estimated number of eligible students enrolled in the school. The probability that the unit (i.e., school) is included in the sample is directly proportional to this measure of size. There are many ways to select such a sample, although if more than two units are selected from a given stratum, all but one method are very complicated. Thus the approach that is generally adopted is to use systematic sampling to select a PPS sample. Let the measure of size for school j be denoted as Z_j, and let the sum of Z_j across all M schools be Z. Schools are sorted in the desired systematic order, and the Z_j values are cumulated. Let Q_j denoted the cumulated measure of size associated with school j. Thus $Q_1 = Z_1$, $Q_2 = Z_1 + Z_2$, and $Q_M = Z$. An interval I is calculated as $I = Z/M$, and a random start R is chosen, uniformly between 0 and I. The schools selected are those whose values of Q_j correspond to the values R, $R + I$, $R + 2I$, and so on up until $R + (n - 1)I$. That is, school j is selected if Q_j equals or exceeds one of the selection values, while $Q_{(j-1)}$ is less than that same selection value. The result is that school j is selected with probability equal to nZ_j/Z. If any school has a value for Z_j that exceeds mZ/M, then it is included with certainty.

The PPS selection of schools is paired with a student sampling procedure that selects an equal sample size of students from each school. That is (setting aside any oversampling at the student level), the sample size in each selected school is set equal to a constant, n, whenever $N_j > n$, and is set to N_j otherwise. Provided that $\max(N_j, n)$ is highly correlated with Z_j, this two-stage sampling procedure leads to a "self-representing" student sample, that is, one in which each student throughout the whole population has close to the same chance of selection. All else being equal, samples in which each student has close to the same selection probability have a lower design effect (i.e., lower sampling variance for the same sample size), than ones in which the student sampling probabilities vary considerably.

A self-representing sample could also be obtained by selecting an equal-probability sample of schools, and then selecting a fixed fraction of students within each school. But while selecting, say, 30 students out of 150 eligible students in a school provides for an operationally feasible assessment, selecting 200 students from 1000 might be inconvenient, and selecting 2 students out of 10 seems like folly.

Furthermore, having variable sample sizes within schools acts to increase somewhat the design effect due to the clustering of the sample. If schools are of unequal size, and a simple random sample is selected at each stage, with students having equal probability of selection, the variance for the mean of a student characteristic, y, is given as

$$Var\left(\bar{y}\right)=\frac{1}{m}\left[\frac{1}{M}\sum_{j=1}^{M}\left(\frac{MN_j}{N_o}\bar{Y}_j-\bar{\bar{Y}}\right)^2+\frac{1}{N_o\bar{n}}\sum_{j=1}^{M}N_j\sigma_{2j}^2\right] \tag{6.8}$$

where

$$N_o=\sum_{j=1}^{M}N_j$$

and

$$\bar{n}=\frac{1}{m}\sum_{j=1}^{m}n_j$$

However, if the sample is chosen PPS with the selection probabilities exactly proportional to the true school size, N_j, with a constant sample size of n students selected in each school, the variance becomes

$$Var\left(\bar{y}\right)=\frac{1}{m}\left[\frac{1}{N_o}\sum_{j=1}^{M}N_j\left(\bar{Y}_j-\bar{\bar{Y}}\right)^2+\frac{1}{N_o\bar{n}}\sum_{j=1}^{M}N_j\sigma_{2j}^2\right] \tag{6.9}$$

Thus, the within-school variance contribution is the same in Equations 6.8 and 6.9, but the between-school variance is different. The between-school variance is likely to be higher in Equation 6.8, because not only the variation in school means for y contributes to this variance, but also the variation of the schools sizes, N_j, about the average school size, N_o/M.

Thus, for both practical and design efficiency reasons, the preferred design is one in which the school sample is selected with PPS, and a fixed number of students is selected from each school (or all students if the school is smaller than the target student sample size).

It is important to note that the whole emphasis of this design discussion has been aimed at obtaining a student sample that will provide reliable estimates of student characteristics. It is sometimes also of interest to make statements about the population of schools. If each school in the *population* is given equal weight in such analyses, then a PPS sample of schools generally provides a rather poor basis for analysis, since the sample is heavily skewed toward larger schools. Population estimates can be rendered unbiased through weighting, but the resulting sampling errors will generally be much larger than would be the case if an equal probability sample of schools were selected.

However, even when analyzing school characteristics, it is often the case that, at least implicitly, inference is about the student population in the schools. Thus if one asks the question: "What proportion of schools have a principal who has a postgraduate degree?," it is likely that the real question of interest is: "What proportion of students attend a school where the principal has an advanced degree?" For this second question, a PPS sample of schools is substantially more efficient than an equal probability sample of schools. It is effective to include more large schools in the sample, because large schools each affect a greater number of students. Thus the use of a PPS school sample, with an equal probability student sample, is effective for most kinds of analyses that are likely to be of interest in large-scale assessment studies.

Classroom Samples

The above discussion of student sampling assumed that students are sampled directly from lists within each sampled school. This is the case in studies such as PISA and IAEP, for this is a very natural approach for age-based samples, where students may be drawn from several different grades, and in any one class in the school it is quite likely that not all students are eligible for inclusion in the population. However, as mentioned previously, IEA studies such as TIMSS select whole classrooms of students. This is a natural procedure when the population is grade based.

The selection of classrooms is a method of selecting students within a school, but by sampling clusters of students (classrooms) rather than selecting individual students via systematic sampling. However, similar principles apply as with direct student sampling. It is good to select schools with PPS, and then to select a fixed number of classrooms in each school (usually one or two), with equal-probability. Classroom characteristics can be used to sort the list of classrooms, and then systematic sampling used to select the sample.

Depending upon the type of classrooms used to define the list of classrooms to be provided for sampling, it is quite likely to be the case that a classroom-based sample has a much higher level of sampling error per student than does a direct student sample. This is because classrooms within a school tend to have students of somewhat different mean ability. Especially in middle and upper levels of schooling, this can be an intentional part of school organization.

For example, the IEA TIMSS assessments select one or two mathematics classes from grade 8 (or equivalent). At that level of education, mathematics classes are very often arranged on the basis of mathematics proficiency of the students. Thus, students from a single mathematics classroom are much more likely to be similar to one another than students drawn at random from throughout the school. Thus, this intraclass correlation at the classroom level acts to increase the design effect of the resulting student sample, and often very substantially, since generally all students within the selected classrooms are included.

Classroom-based designs continue to be used despite their higher design effects. Not only do they facilitate the selection of the student sample, but this design also offers the possibility to analyze classroom effects, which would be unwieldy with a direct student sample (perhaps every teacher in the school needs to be surveyed, and carefully linked to individual students), and may greatly weaken the ability to analyze the data using multilevel models, using the classroom as the second level unit, with students as the first level.

The Overall Sample Design

Putting all of these design attributes together, we see that the typical sample design for a national sample in an international large-scale assessment most often has the following characteristics:

a. A two-stage sample of schools, and students within schools.
b. The sample size is at least 150 schools and 4000 students, and can often be much larger.
c. The school sample is stratified by a variety of characteristics that are related to mean school achievement, defined groups of special interest, or both.
d. The school sample is selected with probability proportional to a measure of size (PPS).
e. The student sample within each school is selected with equal probability, and an equal-size sample (of students or classrooms) is selected from within each school.
f. The school sample is selected using systematic sampling, sorted using additional characteristics likely to be related to school mean achievement.
g. The student sample is selected using systematic sampling, sorted using characteristics likely to be related to student achievement.

The use of designs such as these has implications for estimating population parameters from the sample, quantifying the precision of those estimates, and conducting multivariate analyses of the data. Compounding these issues is the presence of nonresponse, at the school level and at the student level. In the next section, we discuss the implications of the design for deriving estimates, which is addressed through weighting. Then we discuss the issues of nonresponse, and how the weights are modified to account for that, and finally how we estimate sampling variances appropriately for a wide variety of estimators, accounting for the effects of both the design and nonresponse adjustments.

Weighting

The Purpose of Weighting

The provision of survey weights on the data files for large-scale assessments and other surveys is intended to assist the user in obtaining estimates of population parameters that: (1) do not suffer from bias resulting from the use of a complex sample design, and (2) have minimal bias due to the presence of nonresponse, to the extent possible. In many applications there is a third objective: (3) to reduce sampling errors in estimates by utilizing auxiliary information about the population; but this objective is seldom, if ever, present for international large-scale assessments of student populations.

Weighting is, in effect, a part of the estimation procedure. The intention is to incorporate features that would ideally be included in many analyses into a single weight (or, on occasion, into several different weights) and place this on the data file. This has two benefits. It obviates the need of the analyst to repeat this part of the estimator for every analysis, and it ensures consistency across analyses, if all use the same weight variable.

Base Weights

The term "base weight" refers to the weight component that reflects the impact of the sample design, and specifically the inclusion probability. The base weight is also referred to as the design weight. The base weight is given as the reciprocal of the inclusion probability of the unit. It can be separated into two components. The first is the school-level base weight, which is the reciprocal of the inclusion probability of the school to which the student belongs. The second is the reciprocal of the inclusion probability of the student, given that the school is included in the sample. The overall student base weight is given as the product of these two components, and it is this weight that is required in order to obtain unbiased (or consistent, in the case of more complex estimators) estimators of student population characteristics (in the absence of nonresponse). However, the two components are generally needed when conducting multilevel (hierarchical) linear model analyses, as the overall student base weight by itself is not sufficient for estimating parameters of such models correctly.

If we let d_i denote the overall base weight for student i (the "d" is used to denote "design weight" since this weight reflects the features of the sample design, but no other weight components), then for characteristic y, the mean of y for the student population is estimated from sample s as

$$\hat{\bar{y}} = \left(\sum_{i \in s} d_i y_i\right) \Big/ \left(\sum_{i \in s} d_i\right) \tag{6.10}$$

More generally, for a more complex statistic, such as regression coefficient, for example, each student's contribution to the estimator is multiplied by the base weight. Thus the estimator for a regression coefficient, where y is the dependent variable and x is the independent variable, is given as

$$\hat{B}_{yx} = \frac{\sum_{i \in s} d_i y_i \left(x_i - \hat{\bar{x}}\right)}{\sum_{i \in s} d_i \left(x_i - \hat{\bar{x}}\right)^2} \tag{6.11}$$

Nonresponse Adjustments

Weights are adjusted for nonresponse by creating nonresponse adjustment factors by which the base weights are multiplied. For nonresponse adjustments of this type to be effective in reducing nonresponse bias in estimates from the survey, they must be a function of a variable, available for both nonrespondents and respondents, that has two properties: It is correlated with the response status, and it is correlated with the outcome variables of interest. There are two basic approaches to establishing variables for use in creating nonresponse adjustments: The first is to find variables that are related to response, and hope that they are related to the outcome variables (perhaps looking for evidence that this is the case). The second is to find variables that are related to key outcome variables and response status. Whereas the first method generally uses data from the survey to be adjusted, the second method is generally based on historical data from a similar survey. Because over time the relationship with the outcome is likely to be more stable than the relationship to the response (which can be affected by policies to try to increase response rates, or current events that affect response rates at a given time), generally with this approach the emphasis is concentrated on finding variables that are correlated with the outcome, perhaps confirming that they are also correlated with response status.

Either approach is viable for international large-scale achievement studies. But in fact, for several reasons, the approach has generally been to focus on historical data with the aim of finding variables that are related to both the outcomes of interest, and response status. First, in these studies, unlike many surveys, there are generally a few outcome variables that are clearly the most important—namely, achievement on the assessments conducted. Second, as the weights for many countries must be processed in a short time, a fairly standard approach to creating nonresponse adjustments is needed. This is in contrast to the case with stratification, where time permits the tailoring of the choice of stratification variables for each individual country. Since these studies are often repeated over time, there is good opportunity to use historical data to determine which variables to use in creating nonresponse adjustments.

Generally nonresponse adjustments are implemented by creating groups of schools, and students, and applying a common weight-adjustment factor to all the units in the same group. Adjustments are made to the school base weight component to account for nonresponse by entire schools. Then adjustments are made to the student weights, to adjust for nonresponse by individual schools for which some students respond, but not all.

School Nonresponse Adjustments

Generally school nonresponse adjustment classes are created using stratification variables. These are a natural choice, since they are generally chosen for stratification for either or both of two reasons: (1) They are related to the outcomes of interest; a necessary condition for effective nonresponse adjustment, or (2) they represent subgroups of special interest, and it is important that such groups be represented in the correct proportions in the final weighted data. It is also the case that often the different strata experience different school response rates. This is particularly the case when strata represent different school systems (public versus private), different tracks (academic versus vocational), different levels (upper secondary schools versus lower secondary schools), or different provinces, which may differ in their level of support for the study. For schools grouped in school nonresponse class C, the nonresponse adjustment factor that is applied to the school component of the base weight, is given as

$$f_{1C} = \left(\sum_{j \in C} d_j e_j\right) \bigg/ \left(\sum_{j \in C_R} d_j e_j\right) \tag{6.12}$$

where

C_R denotes the set of responding schools in class C; and

e_j denotes the anticipated enrollment of eligible students in school, based on frame data.

Note that the adjustment factor utilizes not only the base weight of each school, but also its enrollment of eligible students, as estimated from the data on the sampling frame. This estimated enrollment is used, rather than the actual enrollment, because the actual enrollment is not generally available for the nonresponding schools. If there is a systematic difference between the estimated and actual enrollments, and the actual enrollments were used for the respondents, this would lead to a bias being incorporated into the nonresponse adjustments.

The enrollment data are incorporated into the adjustment, because the purpose of the adjustment is to reduce, as much as possible, bias in estimates concerning the student population that is induced by the school nonresponse. The number of students in the population that are "represented" by each school in the sample is approximated by $d_j e_j$, and thus this is the contribution of each school to the school nonresponse adjustment. Therefore these adjustments are not suitable if one wishes to use the data to make inference about the population of schools (with each school in the population counted as a single unit, regardless of enrollment size), since in that case each school should be weighted just by d_j.

Recall that, with probability proportional to size sampling, except perhaps in the case of very small and very large schools, d_j is proportional to $1/e_j$, which means that often the school nonresponse adjustment is close in value to the ratio of the number of schools sampled in class C to the number of responding schools in class C_R. In some applications, in fact, this ratio can adequately serve as the nonresponse adjustment factor, without having to apply Equation 6.12. This would generally only be the case if each class C falls entirely within a single stratum.

It is typical in international large-scale assessments that school response rates vary considerable from country to country, or across strata within countries. Thus, it is not uncommon for a country to achieve a very high rate of response for public schools, but a lower rate for private schools. Because of these variations within and across countries, school nonresponse adjustments often serve an important role in bias reduction of survey estimates, but the importance of these adjustments varies considerably across countries.

Student Nonresponse Adjustments

Whereas school response varies considerably across countries, and typically many countries achieve 100% school participation or very close to it, it is usually more the case that student response rates are less variable across countries. There is almost always some student nonresponse, as a result of day-to-day absences of students from school, but the amount of student nonresponse is seldom high. High student nonresponse generally has one of two causes: (1) poor operational procedures in organizing the assessments within schools; and (2) a requirement that written parental consent be obtained prior to assessing the student (but situations where parents are notified of the assessment in advance, and may exercise an option to opt their child out of the assessment, do not generally result in a high degree of nonresponse).

One might think that these circumstances would mean that student nonresponse is of less concern for the validity of the data. However, experience shows that, while school nonresponse is often quite high, it is also often not strongly related to the average achievement of students within the school. All sorts of local political factors can come into play in determining whether a school will participate, and often these have little to do with the level of student achievement within the school. But for student nonresponse, experience has shown that this is very often related to student achievement. Generally this is in the direction of a potential upward bias, as less capable students are more likely to be absent from the assessment. This relationship may be causal—less able students may seek to avoid the assessment—but it is also likely that the two share a common cause—lack of interest in school, or chronic poor health, for example. Whatever the case, the application of student nonresponse adjustments can substantially reduce student nonresponse bias.

In contrast to the case of school nonresponse adjustments, generally the information available at the student level for making student nonresponse adjustments is very similar across countries. Furthermore, available student characteristics are often limited—gender, age (relevant for grade-based studies), and grade (relevant for age-based studies) are often the only variables available for nonresponding students. However, often *school-level* variables are very strong predictors of *student* response. Thus, it is common to observe differential student response in upper- compared to lower-level schools, public compared to private schools, vocational compared to academic schools, and so on. Therefore, effective student nonresponse adjustments are generally accomplished by forming classes of students from similar schools (or even using a single school as the basis of a class), and perhaps adding gender and age or grade.

Student nonresponse adjustments are calculated using both the student base weight and the school nonresponse adjustment factor that applies to the student's school. Thus the student nonresponse adjustment factor is given as

$$f_{2C^*} = \left(\sum_{i \in C^*} d_{ij} f_{1C(j)} \right) \Big/ \left(\sum_{i \in C_R^*} d_{ij} f_{1C(j)} \right) \tag{6.13}$$

where

C_R^* denotes the set of responding students in class C^*;

d_{ij} denotes the base weight for student i in school j (reflecting both school and student-within-school inclusion probabilities); and

$f_{1C(j)}$ denotes the school nonresponse factor for the class of schools which contains the school (j) to which student i belongs.

While the absolute and relative importance of school, gender, and age/grade in determining student nonresponse bias obviously varies across studies, findings from research into an effective method of forming student nonresponse classes in PISA, reported in Rust et al. (2012), are informative.

In the 2003 cycle of PISA, each individual school formed a student nonresponse adjustment class. Further subclassification within school was done on the basis of grade where feasible. However, in most cases, the sample size did not permit this finer classification. It was noted that, in PISA, gender is strongly related to achievement (girls substantially outperform boys in reading in almost every country, while boys somewhat outperform girls in mathematics in almost every country). Furthermore, in many countries gender was related to response, with typically girls responding at a slightly higher rate than boys. Thus, there was evidence that the method used to create the student nonresponse adjustment classes for PISA in 2003 might not have been fully effective in removing nonresponse bias. In an attempt to improve the effectiveness of these adjustments, for the 2006 cycle of PISA an alternative approach was proposed. Instead of using the individual school as the highest level of classification for forming adjustment cells, school nonresponse adjustment classes were used. These classes were based on the school stratification variables. Such classes were formed even for countries with 100% school response. By creating classes based on school characteristics that were several-fold larger than those used in 2003, much scope was provided for further subdividing the cells based on gender and grade.

As an evaluation of the change in procedure, student nonresponse adjustments were applied to the 2006 PISA data using the method proposed for 2006, and that used in 2003. As expected, in those countries with either a sizeable response rate difference between boys and girls, or a response rate difference across grades, the two methods produced somewhat different results.

Table 6.2 shows the student response rates and the mean achievement in each of reading, mathematics, and science, for a selection of countries that participated in PISA in 2006. The countries listed are those for which the two methods of nonresponse adjustment led to the greatest changes in results, or where there were noticeable response rate differences, either by gender or grade. Many of them have over a 2-percentage point difference in response rates by gender. Australia is unusual in that the response rate was higher for males than for females. Note that in every country females had much higher achievement than males in reading, but noticeably lower achievement in mathematics. Thus, absent an adequate adjustment for nonresponse, results in many of these countries are likely to be biased upward for reading, and downward for mathematics (exceptions being Belgium, Ireland, and the United States, with little difference in the response rates by gender, and Australia where the directions of the biases are likely to be reversed).

Table 6.3 shows response rates by grade for the subset of countries listed in Table 6.2 for which a substantial proportion of the sample was drawn from more than one grade. Note that in every case, the response rate for the lowest

TABLE 6.2

Student Response Rates and Mean Achievement Scores, by Gender, for Selected Countries PISA 2006

	Females				Males				Difference			
Country	Response Rate (%)	Reading Mean	Mathematics Mean	Science Mean	Response Rate (%)	Reading Mean	Mathematics Mean	Science Mean	Response Rate (%)	Reading Mean	Mathematics Mean	Science Mean
Australia	85.2	532	513	527	86.4	495	527	527	–1.2	37	–14	0
Austria	92.9	513	494	507	88.2	468	517	515	4.8	45	–23	– 8
Belgium	93.1	522	517	510	92.9	482	524	511	0.2	40	–7	–1
Denmark	90.7	509	508	491	87.5	480	518	500	3.2	30	–10	–9
Iceland	84.4	509	508	494	81.8	460	503	488	2.6	48	4	6
Ireland	83.8	534	496	509	83.7	500	507	508	0.0	34	–11	0
Italy	93.2	489	453	474	90.9	448	470	477	2.2	41	–17	–3
Poland	92.8	528	491	496	90.6	487	500	500	2.1	40	–9	–3
Portugal	87.2	488	459	472	85.6	455	474	477	1.7	33	–15	–5
Spain	90.3	479	476	486	86.7	443	484	491	3.7	35	–9	–4
Tunisia	95.5	398	358	388	93.5	361	373	383	2.1	38	–15	5
USA	90.9	*	470	489	91.0	*	479	489	0.0	*	–9	–1

Note: * denotes reading results are not available for the United States.

TABLE 6.3

Student Response Rates, by Grade*, for Selected Countries PISA 2006

	Response Rate				
Country	**Grade 7**	**Grade 8**	**Grade 9**	**Grade 10**	**Grade 11**
Australia			81.6%	86.9%	83.9%
Belgium		81.8%	90.7%	95.5%	87.4%
Denmark		84.9%	89.8%		
Ireland		65.9%	85.8%	83.1%	81.4%
Italy		82.8%	83.1%	94.2%	87.8%
Poland		77.6%	92.5%	95.1%	
Portugal	76.0%	82.2%	86.8%	88.9%	
Spain		64.1%	80.7%	96.2%	
USA			84.1%	92.5%	88.2%

Note: * denotes only grades with atleast 100 students in sample are given.

grade shown is less than for the other grades. However, in many of these countries the response rate for the highest grade shown is below that of the next lowest grade. So absent an adequate adjustment, there is a potential for bias in the overall results for these countries, but the likely direction is not always clear.

Table 6.4 shows the results for this set of countries from each of the two methods of nonresponse adjustment. Note that in several countries there is little difference between the methods (Australia, Austria, Belgium, Italy, and Tunisia). This reflects the fact that either the 2003 method was adequate to remove the nonresponse bias, or the biases from differential response rates across grades have a canceling effect, since the lowest response rates occurred in the highest and lowest grades, with higher response rates in the middle grades (this is seen in Table 6.3 for Australia and Belgium in particular). However, in a few countries there was a noticeable difference in the results of the two methods, which strongly suggests that the 2006 method at least partially reduced nonresponse bias that the 2003 method failed to remove. The 2006 method gave higher scores in Ireland, and lower scores in Poland, Portugal, and Spain. In all cases the differences were less than one standard error of the mean (and less than half a standard error in all cases except that of Spain). The fact that in most cases the differences were the same for all subjects suggests that the differences in response rate by grade were contributing more bias than were the differences by gender.

This study, while hardly providing a definitive approach to performing nonresponse adjustments in international large-scale assessments, does illustrate the potential importance of these methods for reducing the biases that are very likely to be introduced through differential response to the assessments.

TABLE 6.4

Mean Achievement Scores, PISA 2006, Using Two Different Student Nonresponse Adjustment Methods, for Selected Countries

	Reading					Mathematics					Science				
	2003 Adjustment Method		2006 Adjustment Method			2003 Adjustment Method		2006 Adjustment Method			2003 Adjustment Method		2006 Adjustment Method		
Country	Mean	S.E.	Mean	S.E.	Difference	Mean	S.E.	Mean	S.E.	Difference	Mean	S.E.	Mean	S.E.	Difference
Australia	512.9	2.1	512.9	2.1	0.0	520.0	2.2	519.9	2.2	0.1	527.0	2.2	526.9	2.3	0.1
Austria	490.3	4.1	490.2	4.1	0.1	505.4	3.7	505.5	3.7	–0.1	510.9	3.9	510.8	3.9	0.0
Belgium	500.9	3.0	500.9	3.0	0.0	520.5	2.9	520.3	3.0	0.1	510.4	2.5	510.4	2.5	0.0
Denmark	495.0	3.1	494.5	3.2	0.5	513.1	2.6	513.0	2.6	0.1	496.0	3.1	495.9	3.1	0.1
Iceland	484.9	1.9	484.4	1.9	0.4	505.9	1.8	505.5	1.8	0.3	491.0	1.7	490.8	1.6	0.2
Ireland	516.4	3.6	517.3	3.5	–1.0	500.6	2.8	501.5	2.8	–0.9	507.4	3.2	508.3	3.2	–1.0
Italy	468.5	2.4	468.5	2.4	0.0	461.3	2.3	461.7	2.3	–0.4	475.2	2.0	475.4	2.0	–0.2
Poland	509.0	2.8	507.6	2.8	1.4	496.3	2.4	495.4	2.4	0.8	498.7	2.3	497.8	2.3	0.9
Portugal	473.5	3.6	472.3	3.6	1.2	467.2	3.1	466.2	3.1	1.0	475.5	3.0	474.3	3.0	1.2
Spain	462.4	2.2	460.8	2.2	1.6	481.5	2.3	480.0	2.3	1.6	490.0	2.6	488.4	2.6	1.5
Tunisia	380.5	4.0	380.3	4.0	0.2	365.3	4.0	365.5	4.0	–0.2	385.3	3.0	385.5	3.0	–0.2
USA	*	*	*	*	*	474.9	4.0	474.4	4.0	0.5	489.5	4.2	488.9	4.2	0.6

Note: * denotes reading results are not available for the United States.

Variance Estimation

Because large-scale survey assessments invariably involve complex sample designs, special procedures are needed to estimate the variances of the resulting estimates. An examination of the formula for the actual sampling variance of a mean achievement score, shown in Section "Probability-Proportional-to-Size Selection" (e.g., Equation 6.9) reveals that sampling variance from these designs is not the same as in the case of a simple random sample. To obtain unbiased (or consistent) estimates of sampling variance therefore requires the application of specialized variance estimation procedures.

As is well-documented in the literature (see, e.g., Wolter [2007]), there are two aspects to this problem. The first is to find a variance estimator that is unbiased (preferably) or consistent, for very simple estimators. By "very simple" we essentially mean estimated totals of population characteristics, calculated using the base weights. This estimator, known as the Horvitz-Thompson estimator, is given by the numerator in Equation 6.10. Formulae for variance of the Horvitz–Thompson estimator for a variety of complex designs are given in well-known sampling texts such as those by Särndal et al. (1992) and Lohr (2010). The second aspect arises from the fact that, especially in large-scale assessments, there is generally very little interest in anything that can be estimated via the Horvitz–Thompson estimator. As Equation 6.10 demonstrates, even something as simple as an estimator of mean achievement, obtained without any adjustments being made to the base weights, is not of this form.

There are two approaches to dealing with these joint issues of a complex sample and complex estimators (meaning estimators other than the Horvitz–Thompson estimator). Wolter (2007) provides a detailed treatment of both of these approaches. The first is to use Taylor series linearization to approximate the complex estimator by a linear combination of Horvitz–Thompson estimators, and then use the variance estimation formulae appropriate for the particular design involved to estimate the variance of this linear combination. The second general approach is to use a replication (or resampling) procedure, to estimate the complex variance via a fairly simple formula, but one that involves a large number of estimates of the parameter of interest, each obtained from a subsample of the full sample. In more general applications the best known of such method these days is the bootstrap (Efron, 1993), but in typical survey sampling applications such as large-scale assessment surveys, other methods such as the jackknife and balanced repeated replication are available and are more efficient than the bootstrap in most practical applications. For a discussion of the use of replicated variance estimation methods in surveys generally, see Rust and Rao (1996).

For a variety of reasons, both practical and historical, in large-scale assessment surveys the replication approach has tended to be used more frequently than the linearization approach for the estimation of sampling variance. This has not been a result of any technical superiority of the resulting variance estimators. Studies of the asymptotic properties of linearization and replication estimators have shown them to be of similar consistency, and empirical simulations have supported these findings of similarity. The reasons for the popularity of replication methods can be traced to two practical aspects. First, many of those analyzing the data are not familiar with the special formulae needed to estimate the variances of Horvitz–Thompson estimators directly, and are not particularly anxious to master them. Second, many different types of complex estimators are employed in the analysis of assessment data. With the linearization approach a different variance formula is needed for each type of estimator. But for a given replication approach (the jackknife, say), a common formula can be used to estimate the variance of a wide variety of different kinds of parameter estimators, incorporating the effects of both the survey design and the survey weighting process in each case. Consequently, in the remainder of this chapter we focus on replication variance estimators.

Replication Variance Estimators in International Large-Scale Assessments

There are two replication methods used commonly in large-scale assessments: The jackknife, or jackknife repeated replication (JRR), and balanced repeated replication (BRR). In application they are quite similar, and have similar properties, with some slight differences. Traditionally IEA studies such as TIMSS and PIRLS have used the jackknife, while for PISA a modified version of BRR is used.

In most, if not all cases, the approach is to "model" the sample design as being a stratified (explicitly) two-stage sample, with two primary sampling units selected from each stratum with replacement (three units for some strata, in cases of an odd number of units). We use the expression "model the sample design," because generally the design does not actually fit the description. It is the case that two-stage sampling is generally involved, with schools as the first-stage units and students or classes as the second-stage units. However, most often more than two schools are selected per explicit stratum, and the schools are sorted within strata, based on some characteristic likely to be related to mean school achievement on the assessment in question. Following this the schools are sampled using systematic selection—a without replacement sampling method, rather than with replacement.

By approximating the design as a two-per-stratum design selected with replacement, the construction of appropriate replicate subsamples and the ensuing variance estimation formula is greatly simplified. Yet there is little bias introduced into the variance estimation. Systematic sampling very closely approximates a highly stratified design with independent selections of two (or three) units within each stratum. Treating the school sampling as being with-replacement when in fact it is not, leads to some overestimation of variance, but in most applications it is minor. In cases where the effect of assuming without replacement sampling results in a noticeable overestimation of variance, more complex replication procedures are available (see Rizzo and Rust (2011), Lin et al. (2013)), and have been implemented in practice (Kali et al. 2011), but not in an international context.

We first consider the jackknife and BRR methods ignoring the complications of having an odd number of schools in the sample. We will also set aside the possibility that any school has a school-level selection probability of 1.0 (which can certainly happen in smaller countries). We also ignore in this discussion the effect of imputation error when estimating the variance of achievement scale scores and other latent characteristics—this is addressed in Chapter 7 (von Davier and Sinharay, 2013).

The Jackknife

Suppose that there are $2K$ schools in the sample for a given country, paired into K "strata" (hereafter called "variance strata") so as to reflect both the actual school stratification and the systematic sampling used within strata. The first replicate is formed by "dropping" from the sample one of the two schools in variance stratum 1, identified at random. For the dropped school, the design weights for this first replicate are set to zero. The design weights are doubled for all students in the remaining school for variance stratum 1. The design weights for all other students (i.e., those whose schools are not in variance stratum 1) are set equal to those used for the full sample. This process is repeated K times, once for each variance stratum. This results in each student on the data file having not only a full-sample design weight, d_i, but also a set of K replicate weights, where these are given by Equation 6.14.

$$\begin{aligned} d_i^{(k)} &= 0 \text{ if student } i \text{ is from the school in variance stratum } k \text{ that is} \\ &\qquad \text{dropped for replicate } k \\ &= 2d_i \text{ if student } i \text{ is from the other school in variance stratum } k \qquad (6.14) \\ &= d_i \text{ otherwise} \\ k &= 1,\ldots,K. \end{aligned}$$

Ideally, nonresponse adjustments are applied to each set of replicate weights in turn, resulting in a set of K nonresponse adjusted replicate weights. Sometimes in practice this step is omitted, and the replication is applied to the nonresponse adjusted weights for the full sample. That is, nonresponse adjusted weights take the place of the design weights in Equation 6.14. This procedure fails to replicate the effect on sampling variance of the nonresponse adjustments, but in most applications these are relatively small.

To obtain the estimated variance of a particular estimate, a set of K replicate estimates is then generated by using each of the replicate weights in turn in place of the original design weight. Thus, for example, for an estimate of mean achievement, we obtain the following replicate estimates:

$$\hat{\bar{y}}_{(k)} = \left(\sum_{i \in s} d_i^{(k)} y_i\right) \Big/ \left(\sum_{i \in s} d_i^{(k)}\right) \quad k = 1, \ldots, K. \tag{6.15}$$

The variance estimate for the mean achievement is then estimated via Equation 6.16 (again, ignoring the imputation variance, which must be added in the case of a mean scale score and other estimated latent characteristics).

$$var\left(\hat{\bar{y}}\right) = \sum_{k=1}^{K} \left(\hat{\bar{y}}_{(k)} - \hat{\bar{y}}\right)^2 \tag{6.16}$$

Note the simplicity of this formula. But not only is this formula simple, the same variance estimation formula applies to every parameter estimator for which the variance is estimated (e.g., the proportion of students above a cut score, or a regression coefficient in a multiple regression analysis).

With this approach of creating a set of jackknife replicate weights that are included in the data file, the analyst can derive valid inferences from the data without having to have any detailed understanding of either the sample design or the jackknife procedure. All that is needed is the application of formula 6.16. As discussed in Section "Software for Variance Estimation," many software packages available now utilize the replicate weights and this formula routinely, making this method even more straightforward for the analysts to apply. One cautionary note of application is that there some variants on the jackknife procedure that result in some variation of formula 6.16 being required. Thus, caution is needed to ensure that the correct form of variance estimator is being used, corresponding to the specific implementation of the jackknife replicate formation.

A theoretical limitation of the jackknife is that, in a single-stage sample, it has been demonstrated that the method does not provide consistent variance estimation for "nonsmooth" estimators such as the median and other population quantities (see Wolter 2007, Section 4.2.4). Roughly speaking, this means that, while the method may not result in a substantial bias, the variance estimator remains very unstable no matter how large the sample size or how many jackknife replicates are created. While the empirical evidence is strong that this problem is essentially completely mitigated in most applications with two-stage sampling, theoretical justification for this finding is lacking. Thus, when possible, it is probably best to avoid using the jackknife for estimating the variance of the median, and similar statistics, when analyzing large-scale assessment surveys.

Balanced Repeated Replication

Like the jackknife, BRR, or balanced half-sampling, proceeds by creating parallel sets of weights that vary systematically from the full sample weights. With BRR, the weights for a single replicate are created by increasing the weights for students in one school from each variance stratum, while simultaneously decreasing the weights of the remaining schools. In "classical" BRR, the increased weights are obtained by doubling the base weights, while the decreased weights are obtained by setting the replicate base weights to zero. However, it has become common to use a different approach to varying the base weights, for reasons discussed below.

If the choice of schools for which the weights were to be increased were both random and independent across variance strata and across replicates, this method would constitute a form of bootstrap variance estimator. Efficiency is gained by imposing a correlation structure on the assignment across variance strata and replicates, so that far fewer replicates are required than is generally the case with the bootstrap.

The determination as to which primary units will have their weights increased for a given replicate, and which decreased, is determined via an orthogonal array, known as a Hadamard matrix. A Hadamard matrix is a square matrix, with entries that have value +1 or –1, with the property that the vector product of any row, with any column of a different number is zero, while the product of a row and column of the same number is equal to the rank of the matrix. Hadamard matrices can be readily constructed of arbitrarily large size, and exist for most dimensions that are a multiple of 4. To construct the BRR replicates for a set of H pairs of primary units, one uses the Hadamard matrix of size H^*, where H^* is the smallest multiple of 4 that exceeds H. For further discussion of Hadamard matrices and their use with BRR, see Wolter (2007).

Let a_{ij} denote the (i, j) entry of Hadamard matrix A, of dimension K^*. Each value of a_{ij} is either +1 or –1. The properties of the matrix are such that

$$\sum_{k=1}^{K^*} a_{hk}a_{kl} = 0 \quad \text{for all } h \neq l$$
$$\sum_{k=1}^{K^*} a_{lk}a_{kl} = K^* \quad \text{for all } l \tag{6.17}$$

We create one set of replicate weights corresponding to each row of H, for a total of K^* replicate weights. In each primary unit pair, one is arbitrarily numbered 1 and the other 2, with the numbering assignment being independent across pairs. Then for pair l in replicate k, if $a_{lk} = +1$, the weight for unit 1 is increased and the weight for unit 2 is decreased, and vice versa when $a_{lk} = -1$.

Thus

$$
\begin{aligned}
d_i^{(k)} &= (1+\delta)d_i \text{ if student } i \text{ is from school labelled } (1.5-0.5a_{lk}) \text{ in pair } l \\
&= (1-\delta)d_i \text{ if student } i \text{ is from school labelled } (1.5+0.5a_{lk}) \text{ in pair } l \\
k &= 1,\ldots,K^*
\end{aligned}
\tag{6.18}
$$

where δ is a factor greater than zero, but no greater than 1. In classical BRR, $\delta = 1$, but in PISA and many other surveys, $\delta = 0.5$.

The variance estimator is somewhat similar to that for the jackknife but with sufficient crucial differences that it is disastrous if one uses the BRR formula with jackknife replicate weights, and vice versa (only if $K^* = \delta^{-2}$ do the jackknife and BRR variance formulae coincide).

$$
var_{BRR}\left(\hat{\bar{y}}\right) = \frac{1}{K^*\delta^2}\sum_{k=1}^{K^*}\left(\hat{\bar{y}}_{(k)} - \hat{\bar{y}}\right)^2 \tag{6.19}
$$

As with the jackknife, the same variance estimation formula can be used for a wide variety of complex estimators. Nonresponse weighting adjustments are applied to each set of replicate base weights. The BRR method is consistent for "nonsmooth" estimators such as the median and other quantiles, unlike the jackknife, and so is a preferable method when these quantities are of particular interest.

Some Practical Considerations in the Use of Replication Variance Estimators

Odd Number of Primary Units (Schools)

The discussion in section "Replication Variance Estimators in International Large-Scale Assessments" assumed that there is an even sample size of primary units (i.e., schools in most large-scale assessment surveys), but this is not always the case. To deal with an odd number of schools, we form one (or a few) triples of schools, rather than pairs. The approach then used with BRR is given in Adams and Wu (2002), Appendix 12. For two possible approaches when using the jackknife, see Rizzo and Rust (2011), and Martin and Kelly (1997, Chapter 5).

Large Numbers of Primary Units

Large numbers of primary units can arise in two situations in large-scale assessments. The first case is when a participating country wishes to obtain reliable estimates at a region level. Thus, for example, in PISA Canada typically has a sample of about 1000 schools, since they wish to obtain reliable estimates for each language by province combination. The second case occurs in small countries, where individual schools may be included with certainty. In these cases the students are the primary units. Thus in PISA Iceland typically includes all schools, and the resulting sample of over 3000 students is thus composed of over 3000 primary units. In these cases, some kind of grouping of primary units is required, since it is not feasible to produce a set of replicate weights that number in the same order as the number of primary units. However, considerable care is required in the way in which units are grouped. An inappropriate approach can result in biased variance estimators (either positively or negatively biased), unstable variance estimators, or both. For a detailed discussion of the best approaches to reducing the number of replicate weights required, while maintaining the precision of variance estimation and introducing little bias, see Lu, Brick, and Sitter (2006), Rust (1986), and Wolter (2007).

Software for Variance Estimation

Software suitable for analyzing complex survey data is widely available, and is frequently being updated. For this reason it is of little value to give a detailed treatment of the currently available software in a volume such as this, since such a discussion will be outdated shortly after publication.

There are a few key features that analysts of large-scale assessment surveys should be alert to when determining which software to use that should be kept in mind when considering whether to use a particular program or package.

A. *The ability of the program to handle the kind of sample design that is used in large-scale surveys.* Not all software is suitable for the multistage designs used in large-scale assessments, although increasingly these designs are covered. The software must be able to deal with, for example, an analysis that compares the United States, with a two-stage sample of schools and students, with Iceland, where all schools are included, and perhaps all students.

B. *The ability to use sampling weights appropriately.* Generally software that is capable of dealing with complex sample designs can also deal with weights appropriately. But, for example, if one uses the default option with SAS to obtain a weighted estimate of a population standard deviation, unless the weights are normalized in advance the result will be completely spurious. See Chapter 17 (Stapleton, 2013) for a discussion of weight normalization.

C. The ability to take advantage of the replicate weights, and other variables related to variance estimation, that are included in the data files. It is much more foolproof if the user is able to utilize replicate weights that are already provided, rather than attempting to use software to create replicate weights from scratch (sometimes confidentiality provisions and other considerations mean that the data needed to estimate variances, without using the replicate weights provided, are not available).

D. *The ability of the software to do other kinds of analyses where the complex design can reasonably be ignored.* It is a nuisance to have to use one type of software to estimate the variance of a mean appropriately, and something else to implement a complex modeling procedure. This is the strength of systems such as STATA, SAS, and R, which provide such "one-stop shopping."

E. *The ability to include measurement error as well as sampling error.* Generally when analyzing large-scale achievement data, valid inferences can be drawn only if the analyst incorporates a measure of uncertainty related to the individual estimates of the latent constructs—that is, the student's scale score. This can be achieved either through the multiple imputation ("plausible value") approach, or through an explicit maximum likelihood-based approach. It is very convenient if the same software can be used to incorporate both sources of error simultaneously, rather than requiring the user to estimate the components separately and then write additional software to combine them.

F. *Price, user support, and documentation.* Some suitable software is freely available, and this can be very attractive, especially for a one-off application. The user should beware, however, as to the availability of sound documentation and user support in cases where the software is free.

To see a comprehensive comparative review of current software for analyzing complex survey data, the reader is encouraged to visit the Survey Research Methods Section of the American Statistical Association website devoted to this issue: www.hcp.med.harvard.edu/statistics/survey-soft/.

Without specifically endorsing any particular set of software, listed below are some candidates that warrant consideration by anyone wishing to analyze large-scale assessment data. Indicated for each are those aspects from the list above that are particular strengths of the software. First-time users are likely to want to pay particular attention to those programs that support the analysis of latent variables (E), since student achievement scales are latent. All packages listed below have the properties A and B. Again, the reader is reminded that both the capabilities and availability of specific software packages vary over time, often changing rapidly.

i. AM Software (E, F)
ii. R (C, D, F)
iii. Mplus (C, E, F)
iv. SAS (D)
v. SPSS (D)

vi. Stata (D)
vii. SUDAAN (C)
viii. WesVar (C, E, F)

Conclusion

International Large Scale Assessments invariably involve the collection of data on student achievement via the conduct of a sample survey with a complex design. A variety of standard survey sampling techniques can be applied so that data can be collected at a reasonable cost, yet can provide estimates of adequate reliability for useful inference. The estimation procedures applied generally also call for adjustments to account for nonresponse at both the school and student levels.

Because the data are collected via a complex design, methods of inference must be used that reflect the sampling variance that actually results from the design, rather than relying on methods that assume a much simpler design. This applies both to parameter estimators, and estimators of sampling variance. In the case of parameter estimators, unbiased estimates will result only if survey weights are incorporated into the analyses. For variance estimation, techniques must be employed that give unbiased (or nearly unbiased) and reliable estimators of the true sampling variance engendered by the design. This is further complicated by the fact that this approach needs to be incorporated with methods used to reflect the measurement uncertainty of key latent characteristics measured in the survey, most notably student achievement placed on a scale.

Despite these challenges, well-established methods exist for the appropriate analysis of ILSA data and are routinely used both by those who produce the initial reports from each study, but also by secondary analysts of the data. Research continues into the best methods to analyze large-scale assessment data obtained from surveys.

In the future it seems likely that there will be interest in incorporating auxiliary data into the estimation procedures for these studies. It is fair to say that, with the exception of adjustments for nonresponse, current practice is that auxiliary data are frequently used at the design stage, but seldom at the analysis stage, at least for initial reporting. There are several reasons for this. Because the design information varies in nature and quality across countries, it is difficult to meet the time requirements for reporting the results if additional steps are added to the estimation, especially when these must be tailored to each country. A second consideration is that, generally speaking, errors in auxiliary data used at the design phase lead to some increase in variance, but errors in auxiliary data used in analysis can often lead to bias in the parameter estimates. Nevertheless, it seems likely that there will be interest in the future in modifying estimation procedures to use external data to increase the reliability and usefulness of the results, as this is happening in other fields of survey research, and statistical analysis more generally.

References

Adams, R. and Wu, M., eds., 2002. *PISA 2000 Technical Report*. OECD: Paris.

Efron, B. 1993. *Introduction to the Bootstrap*. Chapman & Hall: New York.

Elley, W.B. 1992. *How in the World do Students Read? IEA Study of Reading Literacy*. IEA: The Hague, Netherlands.

Heeringa, S.G., West, B.T., and Berglund, P.A. 2010. *Applied Survey Data Analysis*. CRC Press: Boca Raton, FL.

Kali, J., Burke, J., Hicks, L., Rizzo, L., and Rust, K.F. 2011. Incorporating a first-stage finite population correction (FPC) in variance estimation for a two-stage design in the National Assessment of Educational Progress (NAEP). *Proceedings of the Section on Survey Research Methods, American Statistical Association*, 2576–2583.

Lapointe, A.E., Mead, N.A., and Phillips, G.W. 1988. *A World of Differences: An International Assessment of Mathematics and Science*. Educational Testing Service: Princeton, NJ.

Lapointe, A.E., Mead, N.A., and Askew, J.A. 1992. *Learning Mathematics*. Educational Testing Service: Princeton, NJ.

Lin, C.D., Lu, W., Rust, K., and Sitter, R.R. 2013. Replication variance estimation in unequal probability sampling without replacement: One-stage and two-stage. *Canadian Journal of Statistics*. To appear.

Lohr, S. 2010. *Sampling: Design and Analysis*, 2nd Edition. Duxbury Press: Pacific Grove, CA.

Lu, W.W., Brick, J.M., and Sitter, R.R. 2006. Algorithms for constructing combined strata variance estimators. *Journal of the American Statistical Association* **101**, 1680–1692.

Martin, M.O. and Kelly, D.L., eds., 1997. *Third International Mathematics and Science Study Technical Report. Volume II: Implementation and Analysis*. Boston College: Chestnut Hill, MA.

Rizzo, L. and Rust, K. 2011. Finite population correction (FPC) for NAEP variance estimation. *Proceedings of the Section on Survey Research Methods*, American Statistical Association, 2501–2515.

Rust, K., Krawchuk, S., and Monseur, C. 2013. PISA nonresponse adjustment procedures. In Prenzel, M., Kobarg, M., Schöps, K., and Rönnebeck, S., eds., *Research on PISA*. Springer: Heidelberg, Germany.

Rust, K.F. 1986. Efficient replicated variance estimation. *Proceedings of the American Statistical Association, Survey Research Methods Section*, 81–87.

Rust, K.F. and Rao, J.N.K. 1996. Variance estimation for complex surveys using replication techniques. *Statistical Methods in Medical Research* **5**, 281–310.

Särndal, C-E., Swensson, B., and Wretman, J. 1992. *Model Assisted Survey Sampling*. Springer-Verlag: New York.

Stapleton, L.M. 2013. Incorporating sampling weights into linear and non-linear models. Ch. XII in Rutkowski, L., von Davier, M., and Rutkowski, D., eds., *Handbook of International Large-Scale Assessment Data Analysis.* CRC Press: Boca Raton, FL.

von Davier, M. and Sinharay, S. 2013. Item response theory extensions and population models. Ch. VII in Rutkowski, L., von Davier, M., and Rutkowski, D., eds., *Handbook of International Large-Scale Assessment Data Analysis.* CRC Press: Boca Raton, FL.

Wolter, K.M. 2007. *Introduction to Variance Estimation*, 2nd Edition. Springer: New York.

7

Analytics in International Large-Scale Assessments: Item Response Theory and Population Models

Matthias von Davier
Educational Testing Service

Sandip Sinharay
CTB/McGraw-Hill

CONTENTS

Introduction

International large-scale assessments (ILSAs) report results in terms of estimated average scale score in a subject area for all examinees in the target population in participating countries, and also as the percentage of examinees who attain predefined levels of proficiency. In addition, ILSA results include estimates for subgroups within countries based on characteristics such as gender, race/ethnicity, and school type. ILSAs do not report scores for individual students, however, and are typically of no consequence to the students participating in these assessments.

ILSAs are faced with a number of challenges. One is the conflicting goal of maintaining a short testing time (typically less than 2 hours, sometimes less than 1 hour of testing time) while trying to validly and reliably assess a broadly defined subject domain. ILSAs are kept short because fatigue affects performance and because examinees' motivation may be low due to the test having low stakes for them. Therefore, in ILSAs, each examinee is assessed in only a portion of the contents of the framework. However, the end result is an assessment of the total framework across examinees by systematically exposing each examinee to one booklet (out of many possible ones) with two or three blocks of test items.* This methodology makes raw results obtained from the different booklets on the examinees' performance not suitable for individual reporting. However, ILSAs are not intended to produce and disseminate individual results at the respondent or even the classroom or school level. They are targeted toward reporting results for well-defined subgroups, such as gender. A subgroup-based analysis approach is conducted on the results to overcome the lack of accuracy on the individual test-taker level.

Furthermore, response behaviors of ILSA test takers are modeled using item response theory (IRT), analyzing response patterns on the item level rather than aggregated raw scores. In addition, blocks of items are systematically linked across booklets (i.e., each block appears in most cases at least once in each block position and paired with several or all other blocks in the test).

A test booklet usually contains not much more than 20 items per content domain (scale). Thus, it is difficult to make claims about individual performance with great accuracy. However, in ILSAs, there is usually enough information to warrant a point estimate of the average proficiency of each of several examinee subgroups of interest. Because the proficiency distribution is not directly observed, a latent regression (Mislevy 1984, 1985) is carried out to estimate the average proficiencies of examinee subgroups given evidence about the distribution and associations of collateral variables in the data. In summary, the ILSA method exploits the several advantages of IRT paired with an explanatory approach based on collateral information to produce accurate subgroup results, even though the testing time is short and the data sparse. This chapter provides a description of the technical details involved in estimating the latent regression model.

The next section provides an introduction to IRT with a focus on its use in ILSAs as well as an overview of the ILSA latent regression model and the estimation method currently used. The section Current Research on Alternative Approaches discusses some suggested model extensions, and the Outlook section lays out conclusions and future directions.

* This design of exposing each examinee to one of many possible booklets is referred to as matrix sampling.

Latent Regression Item Response Theory

ILSAs fit a statistical model that integrates a variety of sources of information about the examinees. Besides responding to test questions, examinees provide information about themselves by filling out a background questionnaire. The questionnaire collects information about academic and nonacademic activities, attitudes, and other subjective variables, as well as demographic variables. These variables are viewed as covariates and are therefore integrated into the model differently from the test responses. The analytic procedures used to integrate these sources of information are divided into a three-step procedure.

The first stage, *Item Calibration*, fits an IRT model (Lord and Novick 1968; Rasch 1960) for a mixed format test, consisting of binary and ordered polytomous responses to the examinee test response data and produces estimates of item parameters. The section on Item Response Theory will provide an overview of IRT, focusing on its use in ILSAs.

The second stage, *Population Modeling*, assumes that the item parameters are fixed at the estimates found in the calibration stage. The first part of this stage consists of fitting the latent regression model to the data, that is, estimation of the regression weights, referred to as Γ in the following, and Σ, which denotes the residual variance–covariance matrix of the latent regression. In the second part of this stage, *plausible values* (Mislevy and Sheehan 1987; von Davier et al. 2009) for all examinees are obtained using the estimates of Γ and Σ as well as the estimates obtained in the item calibration stage. The plausible values are basically multiple imputations obtained using the latent regression IRT model as an imputation model (Little and Rubin 2002) and utilized form estimation of averages for various examinee subgroups.

The third stage, *variance estimation*, estimates the variances corresponding to the examinee subgroup averages, using a jackknife approach (see, e.g., Johnson 1989; Johnson and Rust 1992) and the plausible value generated in the second stage.

Item Response Theory

Assume $Y_1,\ldots,Y_I$ categorical-observed random variables with realizations on these variables sampled from N examinees. Let $(\boldsymbol{y}_v) = (y_{v1},\ldots,y_{vI})$ denote the realization for examinee v. Let θ_v be a real-valued variable and let

$$P(\boldsymbol{y}_v \mid \theta_v) = \prod_{i=1}^{I} P(y_{vi} \mid \theta_v) \tag{7.1}$$

be a model describing the probability of the observed categorical variables given the parameter θ_v, referred to as ability or proficiency variable in the subsequent text. For the product terms above, assume that

$$P(y_{vi} \mid \theta_v) = P(y_{vi} \mid \theta_v; \zeta_i)$$

with some vector valued parameter ζ_i. The parameters ζ_i are quantities that describe the items and do not vary across respondents, while θ_v refers to a quantity that varies by respondent, and in terms of modeling is considered a random effect.

In ILSAs, the model $P(y_{vi} \mid \theta_v; \zeta_i)$ is an IRT model (Lord and Novick 1968). The Program for International Student Assessment (PISA) has used the partial credit model (PCM; Masters 1982)

$$P\left(y_{vi} \mid \theta_v; \zeta_i\right) = \frac{\exp(y\theta_v - b_{iy})}{1 + \sum_{z=1}^{m_i} \exp(z\theta_v - b_{iz})} \tag{7.2}$$

as a basis with item parameter $\zeta_i = (b_{i1}, \ldots, b_{im})$ and an observed response $y_{vi} \in \{0, \ldots, m\}$ of person v on item i. This model is a generalization of the dichotomous Rasch model (Rasch 1960) for polytomous, ordinal responses and reduces to the dichotomous Rasch model in the case of binary data. Figure 7.1 shows the trace lines of nine different values for b_i, ranging from −3 to +5 plotted for values of θ ranging from −4 to +4.

The following exhibits why this approach, the Rasch model, and related approaches of IRT are regarded as suitable choices for modeling test data. When looking at test performance by age (a proxy of developmental state), Thurstone (1925) found that the proportion of respondents who are successfully mastering a problem is monotonically related to age. Figure 7.2 shows this relationship.

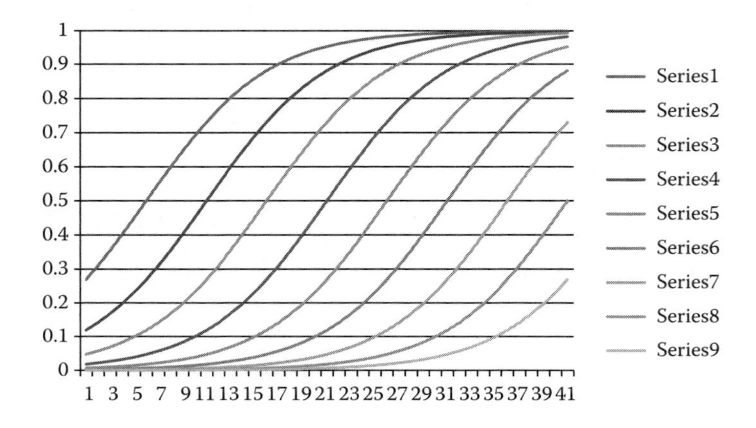

FIGURE 7.1
Trace lines, in IRT referred to as item characteristic curves, for nine different values of b_i ranging from −3 to +5. The x-axis represents values for the parameter θ ranging from −4.0 to 4.0, the y-axis represents the probability of a correct response.

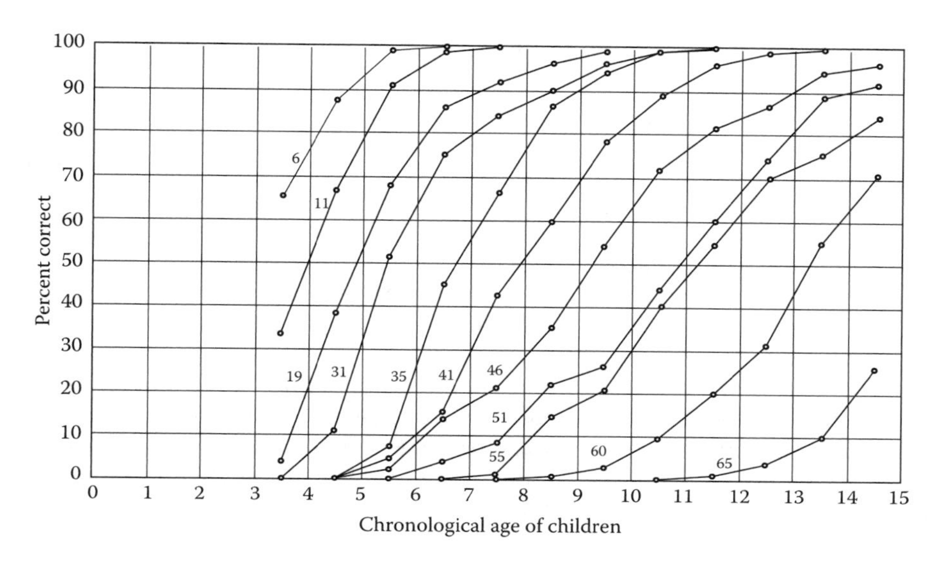

FIGURE 7.2
Trace lines obtained by plotting percent correct against age from a series of tasks (Figure 5 in Thurstone, 1925).

When, instead of developmental age, the total number correct on a longer test is used, similar graphs are obtained (Lord 1980). Natural choices for a parametric model suitable for these types of relationships are the probit and the logit (e.g., Cramer 2004; Fischer 1974).

While the Rasch model specifies a single-item parameter b_i in the form of a negative intercept, more general models can be defined that allow for variation of the trace lines in terms of slopes and asymptotes. PISA currently uses a multidimensional version of the Rasch model (e.g., Adams and Wu 2007) as the basic analytical model. To model dichotomous items, adult assessments such as the Adult Literacy and Lifeskills (ALL) survey and the Programme for the International Assessment of Adult Competencies (PIAAC) use the two-parameter logistic (2PL) IRT model, while the Trends in International Mathematics and Science Study (TIMSS), Progress in International Reading Literacy Study (PIRLS), and the National Assessment of Educational Progress (NAEP) use the 2PL and 3PL IRT models. The latter model is given by

$$P(y_{vi}=1|\,\theta_v;\zeta_i)=c_i+(1-c_i)\frac{\exp(a_i(\theta_v-b_i))}{1+\exp(a_i(\theta_v-b_i))} \tag{7.3}$$

for binary items. In the equation above, c_i denotes the guessing parameter (which is set to 0.0 for the 2PL), b_i denotes the item difficulty parameter, and a_i is the slope parameter. The model used for polytomous ordinal items, for example in PIRLS and TIMSS, and NAEP, is the generalized PCM (Muraki 1992), given by

$$P(y_{vi} \mid \theta_v; \zeta_i) = \frac{\exp(ya_i(\theta_v - b_{iy}))}{1 + \sum_{z=1}^{m_i} \exp(za_i(\theta_v - b_{iz}))} \tag{7.4}$$

assuming a response variable with $(m_i + 1)$ ordered categories.

For the proficiency variable θ_v it is sometimes assumed that $\theta_v \sim N(\mu,\sigma)$. Alternatively, other types of distributions can be chosen (Haberman et al. 2008; Heinen 1996; Xu and von Davier 2008a,b).

When there are several abilities as, for example, in TIMSS Mathematics and Science or in PISA Mathematics, Science, and Reading, these are represented in a d-dimensional vector $\theta_v = (\theta_{v1}, \ldots, \theta_{vd})$. In that case, one may assume $\theta_v \sim MVN\ (\mu, \Sigma)$. For the IRT models used in ILSAs, these d dimensions are assumed to be measured by separate sets of items, so that

$$\boldsymbol{y}_v = ((y_{v11}, \ldots, y_{vI_1 1}), \ldots, (y_{v1d}, \ldots, y_{vI_d d}))$$

represents d sets of I_1 to I_d responses. This concatenation of scales does produce a vector of observed responses $\boldsymbol{y}_v$ that contains all item response data put together from the d dimensions. For example, an assessment of reading, math, and science could administer 30 items each in reading and math, while 120 science items are administered. Each respondent would be asked to provide answers on a subset of these 30, 30, and 120 items to minimize the burden. A d-dimensional version of the model in Equation 7.1 is defined as

$$P(\boldsymbol{y}_v \mid \theta_v; \zeta_{i^*}) = \prod_{k=1}^{d} \prod_{i=1}^{I_k} P\left(y_{vik} \mid \theta_{vk}; \zeta_{ik}\right) \tag{7.5}$$

with item-level IRT models (2) through (4) plugged in for $P(y_{ik} \mid \theta_k; \zeta_{ik})$ as appropriate. The model given in Equation 7.5 is a multidimensional IRT model for items that show between-item multidimensionality (Adams et al. 1997; Adams and Wu 2007). This approach to multidimensional modeling is what is commonly found in ILSAs.

Population Modeling Using Latent Regressions

In latent regression models, the distribution of the proficiency variable θ is assumed to depend on a number of predictors $X_1, \ldots, X_K$ in addition to the item responses $Y_1, \ldots, Y_I$. Let $(\boldsymbol{y}_v, \boldsymbol{x}_v) = (y_{v1}, \ldots, y_{vi}, x_{v1}, \ldots, x_{vK})$ denote the realization of these variables for examinee v. The X variables are typically real-valued random variables observed for each examinee $v = 1, \ldots, N$. In practice, ILSAs often collect information on hundreds of background variables on each examinee, and a principal component analysis is used to determine a number of principal components ($x_{v1},\ldots,x_{vK}$) that explain 90% of the variation for further analysis. For the regression of these predictors on ability variable θ_v, we assume that

$$\boldsymbol{\theta}_v \sim N(\boldsymbol{x}_v\Gamma, \Sigma)$$

with $d \times d$-variance–covariance matrix Σ and K-dimensional regression parameter Γ. Let $\phi(\theta_v \mid \boldsymbol{x}_v\Gamma, \Sigma)$ denote the multivariate normal density with mean $\boldsymbol{x}_v\Gamma$ and covariance matrix Σ.

For brevity, we will assume that $d = 1$; for a multidimensional version, plug Equation 7.5 into the following equations. The likelihood of an observation ($\boldsymbol{y}_v,\boldsymbol{x}_v$), expressed as a function of Γ and Σ, can then be written as

$$L(\zeta,\Gamma,\Sigma;\boldsymbol{y}_v,\boldsymbol{x}_v) = \int_{\theta}\left(\prod_{i=1}^{I} P(y_{vi} \mid \theta;\zeta)\right)\phi(\theta \mid \boldsymbol{x}_v\Gamma,\Sigma)d\theta,$$

and, for the collective likelihood function involving all N examinees, we have

$$L(\zeta,\Gamma,\Sigma;\boldsymbol{Y},\boldsymbol{X}) = \prod_{v=1}^{N}\int_{\theta}\left(\prod_{i=1}^{I} P(y_{vi} \mid \theta;\zeta)\right)\phi(\boldsymbol{\theta} \mid x_v\Gamma,\Sigma)d\theta. \tag{7.6}$$

This log-likelihood function is the objective function to be maximized with respect to parameters Σ, Γ, and ζ.

In ILSA operational analysis, the item parameters ζ associated with the $\prod_i P(y_i \mid \theta;\zeta)$ are often determined in the item calibration stage, prior to the estimation of the latent regression $\phi(\theta \mid \boldsymbol{x}_v\Gamma,\Sigma)$, as described earlier. In the later population-modeling stage, the latent regression parameters Σ and Γ are estimated conditionally on the previously determined item parameter estimates $\hat{\zeta}$. von Davier et al. (2006) describe this procedure, which is the current method of choice in analysis of many national and ILSAs, including NAEP, TIMSS, PIRLS, and PISA.

The evaluation of the likelihood function in Equation 7.6 requires the calculation or approximation of multidimensional integrals over the range of $\boldsymbol{\theta}$. Mislevy et al. (1992) suggested use of the EM algorithm (Dempster et al. 1977) for maximization of L (Equation 7.6) with respect to Γ and Σ. The EM algorithm treats estimation problems with latent variables as an incomplete data problem and "completes" the required statistics in one phase (the E-Step, where E stands for expectation), and then maximizes the resulting complete-data likelihood with respect to the parameters of interest (the M-Step, where M stands for maximization). In the case of the latent regression, the EM algorithm as suggested by Mislevy et al. (1992) operates as follows:

E-Step: The integral over the unobserved latent variable θ is evaluated in the (t)th E-Step using the provisional estimates $\Sigma^{(t)}$ and $\Gamma^{(t)}$. As in the previous section, let $P(y_{vi} \mid \theta; \zeta_i)$ denote the probability of response y_{vi} given ability θ and item parameters ζ_i. Then the likelihood of an examinee's response pattern conditional on these provisional estimates is given by

$$\int_{\theta} \left(\prod_{i=1}^{I} P(y_{vi} \mid \theta; \zeta) \right) \phi(\theta \mid x_v \Gamma^{(t)}, \Sigma^{(t)}) d\theta.$$

The evaluation of this integral is replaced by the evaluation of the function $(\Pi_{i=1}^{I} P(x_{vi} \mid \theta; \zeta)) \phi(\theta \mid x_v \Gamma^{(t)}, \Sigma^{(t)})$ at the posterior mean of θ given observed item responses y_v and covariates x_v. Let this quantity be denoted by $\tilde{\theta}_v^{(t)} = E_{(\Sigma,\Gamma,\zeta)^{(t)}}(\theta \mid y_v, x_v)$. Mislevy et al. (1992) argue that the posterior expectation $\tilde{\theta}_v^{(t)}$ can subsequently be plugged into the ordinary least-squares (OLS) estimation equations. Note that the calculation of the $\tilde{\theta}_v^{(t)}$ also requires the evaluation or, at least, the approximation of an integral, because calculation of the expectation

$$E_{(\Sigma,\Gamma,\zeta)^{(t)}}(\boldsymbol{\theta} \mid y_v, x_v) = \int_{\theta} \boldsymbol{\theta} \left(\prod_{i=1}^{I} P(y_{vi} \mid \theta; \zeta) \right) \phi\left(\theta \mid x_v \Gamma^{(t)}, \Sigma^{(t)}\right) d\theta$$

is involved.

M-Step: Following the arguments put forward by Mislevy et al. (1992), the maximization of the likelihood for step $(t + 1)$, based on provisional estimates from the preceding E-Step, is accomplished by means of using OLS estimation equations with the posterior means generated in the E-Step, as outlined above. This yields

$$\tilde{\Gamma}^{(t+1)} = (X^T X)^{-1} X(\tilde{\theta}_{1t}, \ldots, \tilde{\theta}_{nt}),$$

where $\tilde{\theta}_{vt} = E(\theta \mid X_v, Y_v, \Gamma_t, \Sigma_t)$ is the posterior mean for examinee v given the parameter estimates at iteration t.

An estimate of the conditional variance–covariance matrix is given by

$$\Sigma^{(t+1)} = V(\tilde{\theta} - X\Gamma^{(t+1)}) + E(V_{(\Sigma,\Gamma,\zeta)^{(t)}}(\theta \mid y_v, x_v)).$$

The determination of estimates of $E_{(\Sigma,\Gamma,\zeta)^{(t)}}(\theta \mid y_v, x_v)$ and $V_{(\Sigma,\Gamma,\zeta)^{(t)}}(\theta \mid y_v, x_v)$ for each of the N observations makes the computation time consuming.

The complete ILSA model obtained by combining IRT and a linear regression in two levels is referred to either as a conditioning model, a multilevel/hierarchical IRT model (Adams et al. 1997), or a latent regression model.

Besides reliable reporting of ability distributions for most subgroups, there are other advantages associated with this approach. First, the underlying item response model facilitates the use of advanced IRT linking techniques across assessment years based on common items and/or randomly equivalent populations. Also, this model provides a straightforward approach to modeling multidimensional content domains. To do this, the latent regression approach essentially assumes simple structure of the items, which is sometimes referred to as between-item multidimensionality (Adams and Wu 2007), and thus resolves some of the disadvantages of exploratory multidimensional IRT, such as indeterminacy of parameters. Finally, the uncertainty in the parameters of the latent regression can be used to quantify measurement error. Assuming that the group membership information is precise, the uncertainty of the regression parameters can contribute to the uncertainty reflected in the latent distribution. Based on this, an imputation model can be constructed, imputing proficiency for all examinees and deriving *plausible values* (explained in the section Current Estimation Approaches and the MGROUP Programs) for each respondent as part of his or her vector of group membership. The variance between these plausible values can be regarded as an estimate of measurement error.

Current Estimation Approaches and the MGROUP Programs

The remaining part of this section focuses on the population modeling step that involves the EM algorithm discussed earlier. The DGROUP set of programs developed at Educational Testing Service (Rogers et al. 2006) consists of a number of different implementations of this EM algorithm. The main differences between these can be roughly described as variations on how the numerical integration is carried out in the E-Step (Mislevy and Sheehan 1992; von Davier and Sinharay 2007; Thomas 1993; von Davier and Yu 2003; von Davier and Yon 2004; von Davier and Sinharay 2010).

The M-step requires the estimates of the posterior means $E(\theta_v \mid X, Y, \Gamma_t, \Sigma_t)$ and the posterior variances $\mathrm{var}(\theta_v \mid X, Y, \Gamma_t, \Sigma_t)$ for the examinees. Correspondingly, the $(t + 1)$th E-Step computes the two required quantities for all examinees.

Different versions of the MGROUP program are used, depending on the statistical method used to perform the E-Step. For example, BGROUP uses numerical integration, NGROUP (e.g., Beaton 1988) uses Bayesian normal theory, and CGROUP (Thomas 1993) uses Laplace approximation, while MCEM-GROUP uses Monte-Carlo integration (von Davier and Sinharay 2007) and SAEM-GROUP uses stochastic approximation (von Davier and Sinharay 2010).

The BGROUP and CGROUP programs are used operationally in ILSAs. When the dimension of θ_v is less than or equal to two, BGROUP is used. When the dimension of θ_v is larger than two, CGROUP may be used. NGROUP is not used for operational purposes in ILSAs because it assumes normality of both the likelihood and the prior distribution and may produce biased results if these assumptions are not met (Thomas 1993).

CGROUP, the approach used operationally in TIMSS, PIRLS, NAEP, and several other large-scale assessments, uses the Laplace approximation of the posterior mean and variance to approximate the means and covariances of the components of θ_v as

$$E(\theta_j \mid X, Y, \Gamma_t, \Sigma_t) \approx \hat{\theta}_{j,\text{mode}} - \frac{1}{2}\sum_{r=1}^{p} G_{jr} G_{rr} \hat{h}_r^{(3)}, \quad j = 1, 2, \ldots, p, \tag{7.7}$$

$$\begin{aligned} cov\left(\theta_j, \theta_k \mid X, Y, \Gamma_t, \Sigma_t\right) \approx G_{jk} &- \frac{1}{2}\sum_{r=1}^{p}\left(G_{jr}G_{kr}G_{rr}\right)\hat{h}_r^{(4)} \\ &+ \frac{1}{2}\sum_{r=1}^{p}\sum_{s=1}^{r}\{[1 - \frac{1}{2}I(r = s)]\hat{h}_r^{(3)}\hat{h}_s^{(3)}G_{rs} \\ &\times (G_{js}G_{ks}G_{rr} + G_{js}G_{kr}G_{rs} + G_{jr}G_{ks}G_{rs} \\ &+ G_{jr}G_{kr}G_{ss})\}, \quad j, k = 1, 2, \ldots, p, \end{aligned} \tag{7.8}$$

where

$\hat{\theta}_{j,mode} = j$th dimension of the posterior mode of θ,

$h = -\log\left[f(y \mid \theta)\phi(\theta \mid \Gamma_t' x, \Sigma_t)\right]$,

$$G_{jk} = \left(\frac{\partial h}{\partial\theta_r \partial\theta_s}\Big|_{\hat{\theta}mode}\right)^{-1}_{jk}, \tag{7.9}$$

$\hat{h}_r^n = n$th pure partial derivative of h with respect to θ_r, evaluated at $\hat{\theta}_{mode}$.

Details about formulas 7.7 through 7.9 can be found in Thomas (1993). The Laplace method does not provide an unbiased estimate of the integral it is approximating. It may provide inaccurate results if p is large, and the number of items measuring each skill is small, which is not rare in ILSAs. Further, if the posterior distribution of an θ_v is multimodal (which can occur, especially for a small number of items), the method may perform poorly. Therefore, the CGROUP method is not entirely satisfactory. Figure 7.2 in Thomas (1993), plotting the posterior variance estimates of 500 randomly selected examinees using the Laplace method and exact numerical integration for two-dimensional θ_v, shows that the Laplace method provides inflated variance estimates for some examinees (see also von Davier and Sinharay 2004, 2010). The departure may be more severe for models with higher dimensions.

Estimating Group Results Using Plausible Values

After completing the EM algorithm in the conditioning step, multiple imputations—in this context called plausible values—for each examinee's proficiency are drawn from an approximation of the conditional posterior distribution of proficiency, using the following algorithm:

- Draw $\Gamma^{(m)} \sim N(\hat{\Gamma}, \hat{V}(\hat{\Gamma}))$, where $\hat{V}(\hat{\Gamma})$ is the estimated variance of the maximum likelihood estimate $\hat{\Gamma}$ obtained from the EM algorithm.
- Conditional on the generated value $\Gamma^{(m)}$ and fixed variance $\Sigma = \hat{\Sigma}$ (obtained from the EM algorithm), calculate the mean $\bar{\theta}_v$ and variance Σ_v for the conditional posterior of each examinee v in the sample.
- Draw a single $\theta_v \sim N(\bar{\theta}_v, \Sigma_v)$ independently for each examinee $v = 1,2,\ldots,N$.

The above three steps are repeated M times, producing M sets of imputations (plausible values) for each examinee in the sample. The use of an approximation of the posterior by a multivariate normal can be justified by the following rationale: The latent regression model typically involves a large number of predictors, so that the posterior variance is considerably smaller than the total variance of abilities. Therefore, any deviations from multivariate normality of ability distribution can be well approximated by a mixture of normals. The plausible values are used to approximate posterior expected values and variance in subsequent computations of different reporting statistics, such as percentiles and percentages above achievement levels. The multiple imputed plausible values are made publicly available in files for secondary analyses, so that researchers can utilize these files in analytic routines suitable for analyses with multiple imputations.

One of the major goals of ILSAs is estimating the performance of examinee subgroups by examining the average value of some function $g(\theta)$ over all individuals in that group. Some examples of $g(\theta)$s are (i) $g(\theta) = \theta_j$, when the interest is in the average of one subscale; (ii) $g(\theta) = \Sigma_j c_j \theta_j$, when the interest is an average of an overall/composite ability (where the scale-weights c_js are usually determined by experts or may reflect the percentage of subscale items in the assessment framework); and (iii) $g(\theta) = I_{\Sigma_j w_j \theta_j > K}$, when the interest is in the proportion of examinees with overall/composite ability above a specific threshold.

To estimate the average value of $g(.)$ for all individuals in group G, the ILSA operational analysis calculates the posterior expected value of

$$\hat{g}_G(\theta) = \frac{1}{n_G} \sum_{v \in G_s} w_v g(\theta_v),$$

with $n_G = \sum_{v \in G_s} w_v$ and G_s denoting the members of group G who are in the sample s. Then the above expected value is estimated, using the plausible values, as

$$\tilde{g}_G = \frac{1}{M}\sum_{m=1}^{M}\frac{1}{n_G}\sum_{v \in G_s} w_v g(\theta_v^{(m)}).$$

Variance Estimation

The third stage in the three-stage ILSA operational analysis is to find the variance of the group estimators $\tilde{g}_G$. There are two primary sources of uncertainty associated with the estimates. First, ILSAs sample only a portion of the entire population of examinees. Second, the true values of the sampled examinees' proficiencies θ_vs are not directly observed; examinees simply provide multiple discrete indicators (item responses) of their proficiency by responding to the assessment items.

The operational analysis, therefore, splits the total variation into two additive pieces: the variation due to sampling, and the variance due to latency of proficiencies. ILSAs use a jackknife repeated-sampling estimate of the sampling variance (Efron 1982) and the multiple imputation or plausible-value methodology (Rubin 1987) to estimate the variance due to latency.

Variance due to Latency of the Proficiency θ_v

The variance due to the latency of θ_v, or measurement error of θ_v, is the variance of the latent proficiency given the item responses. If the abilities θ_v were observed directly the measurement error would be zero. However, because we estimate the proficiencies based on a finite number of item responses for each examinee, the measurement error will be nonzero.

The operational analysis estimates the variance of $\tilde{g}_G$ due to latency of θ_v for subgroup G, denoted U_G, by calculating the variance of the posterior expected values over the plausible values. That is,

$$U_G = \frac{1}{M-1}\sum_{m=1}^{M}(\hat{g}_G^{(m)} - \tilde{g}_G)^2.$$

A more detailed explanation of why imputation-based plausible values are required in secondary analyses and aggregated in this way can be found in Little and Rubin (2002). A didactic on the rationale of the plausible values methodology was provided by von Davier et al. (2009).

Variance due to Sampling of Examinees

The variance due to sampling of examinees arises because ILSAs use only a sample of examinees from the total examinee population, and different samples would lead to different estimates. Most ILSA estimation procedures use a jackknife estimator or a balanced repeated replication (BRR) approach (Efron 1982; Rust 1985; Rust and Rao 1996) to estimate the variance due to sampling. The BRR approach forms pairs of schools, which are then utilized in the resampling process as the basis for replicate weights that drop one school from the pair (by setting the weight of students in that school to zero) and double the weight of the observations from the other school. In contrast, the grouped jackknife approach forms new samples by producing replicate weights that drop single clusters (schools) from the sample while leaving the remaining weights unchanged. NAEP employs a BRR approach to obtain the variance estimate of a statistic. TIMSS and PIRLS use the jackknife technique (e.g., Efron 1982) for variance estimation. PISA uses a modified BRR developed by Fay (Fay 1989; Judkins 1990). While the BRR approach selects one school for removal and doubles the weights for the other school, the Fay modification uses weights of –0.5 for the school selected for removal, and 1.5 for the school selected for inclusion. For further discussion of weighting and the variance estimation procedures used in ILSAs, interested readers can refer to Chapter 6 of this volume.

It is important to note that the estimates aggregated using resampling schemes use a set of plausible values obtained from an IRT latent regression model estimated using the total sample. This means that the jackknife and the BRR use the plausible values as they are, instead of reestimating the latent regression and the IRT item parameters for each subsample. Hsieh et al. (2009) present a more comprehensive alternative approach in which the plausible values used in each jackknife sample are estimated based on reestimating both the latent regression and recalibrating of the IRT item parameters for each jackknife sample.

Current Research on Alternative Approaches

A number of modifications to the current ILSA methodology have been suggested in the literature. These evolved out of criticisms of (i) the complex nature of the ILSA model, and (ii) the approximations made at different stages of the ILSA estimation process.

Application of an Importance Sampling EM Method

Von Davier and Sinharay (2004) suggested an approach to fit the ILSA latent regression model, in which one approximates the posterior expectation and variance of the examinee proficiencies θ_i in the above-mentioned M-step using importance sampling (e.g., Gelman et al. 2003).

The posterior distribution of θ_i, denoted as $p(\theta_i \mid X, Y, \Gamma_t, \Sigma_t)$, is given by

$$p(\theta_i \mid X, Y, \Gamma_t, \Sigma_t) \propto f(y_{i1} \mid \theta_{i1})...f(y_{ip} \mid \theta_{ip})\phi(\theta \mid \Gamma_t, \mathbf{X}_i, \Sigma_t) \tag{7.10}$$

using Equation 7.2. The proportionality constant in Equation 7.10 is a function of y_i, Γ_t, and Σ_t. Let us denote

$$q(\theta_i \mid X, Y, \Gamma_t, \Sigma_t) \equiv f(y_{i1} \mid \theta_{i1})...f(y_{ip} \mid \theta_{ip})\phi(\theta \mid \Gamma_t, \mathbf{X}_i, \Sigma_t). \tag{7.11}$$

We drop the subscript i for convenience for the rest of the section and let θ denote the proficiency of an examinee.

One has to compute the conditional mean and variance of θ having the posterior distribution $p(\theta \mid X, Y, \Gamma_t, \Sigma_t)$. Given a random sample $\theta^1, \theta^2, \ldots, \theta^n$, from a distribution $h(\theta)$ approximating $p(\theta \mid X, Y, \Gamma_t, \Sigma_t)$ reasonably, one can approximate the posterior mean using importance sampling by A/B, where

$$A = \frac{1}{n}\sum_{j-1}^{n} \frac{\theta^j q(\theta^j \mid X, Y, \Gamma_t, \Sigma_t)}{h(\theta^j)}; \quad B = \frac{1}{n}\sum_{j-1}^{n} \frac{q(\theta^j \mid X, Y, \Gamma_t, \Sigma_t)}{h(\theta^j)},$$

and approximate the posterior variance as the ratio C/B, where

$$C = \frac{1}{n}\sum_{j-1}^{n} \frac{(\theta^j - E(\theta^j \mid X, Y, \Gamma_t, \Sigma_t))(\theta^i - E(\theta^j \mid X, Y, \Gamma_t, \Sigma_t))' q(\theta^t \mid X, Y, \Gamma_t, \Sigma_t)}{h(\theta^i)}$$

Von Davier and Sinharay (2004) use a multivariate t distribution with four degrees of freedom as $h(\theta)$, the importance-sampling density. The mean and variance of the importance-sampling density are the same as the mode and curvature of the posterior distribution of θ. Because the posterior mode and posterior curvature (information) are already computed in the MGROUP program, this is a convenient choice. Simulated data and real data analyses indicate that the importance-sampling EM algorithm is a viable alternative to CGROUP for fitting latent regression models.

Multilevel IRT Using Markov Chain Monte Carlo Methods

Fox and Glas (2001, 2003) as well as Johnson and Jenkins (2004) and Johnson (2002) suggested a Bayesian estimation approach that employed the MCMC algorithm (e.g., Gelman et al. 2003; Gilks et al. 1996) to combine the three steps (calibration, population modeling, and variance estimation) of the analytic procedures. Johnson et al. (2004) use a Bayesian hierarchical IRT model similar to the one proposed by Fox and Glas (2001, 2003) that includes the IRT component and the linear regression component of the ILSA model discussed earlier. In addition, their model accommodates clustering of the examinees within schools and PSUs, using a linear mixed effects (LME) model.

The model accommodates clustering of the examinees within schools and PSUs with the help of an LME model of the form

$$\theta_i \mid v_{s(i)}, \Gamma, \mathbf{x}_i, \Sigma \sim N(v_{s(i)} + \Gamma'\mathbf{x}_i, \Sigma),$$

$$v_s \mid \eta_{p(s)}, T \sim N(\eta_{p(s)}, T),$$

$$\eta_{p(s)} \mid \Omega \sim N(0, \Omega),$$

where *s(i)* is the school that examinee *i* attends and *p(s)* is the PSU (school district, state, etc.) to which school *s* belongs. This idea is similar to that proposed by Longford (1995) and Raudenbush et al. (1999).

Appropriate prior distributions are assumed where required. The joint posterior distribution of the item and ability parameters as well as the mean and variance parameters of the regression model are computed. A Gibbs sampler is used to draw from the joint posterior distribution. However, Johnson and Jenkins (2004) deal only with unidimensional examinee proficiency. Johnson (2002) handles data with three-dimensional examinee proficiency but incorporates only three background variables.

Stochastic Approximation

To fit the latent regression model in ILSAs, von Davier and Sinharay (2010) proposed the application of a stochastic approximation approach in the above-mentioned EM algorithm. In this approach, one plugs in a draw from the posterior distribution of the examinee ability variable instead of plugging in the posterior mean of the examinee ability variable in the above-mentioned M step to obtain updated estimates $\Sigma^{(t+1)}$ and $\Gamma^{(t+1)}$. The Metropolis–Hasting algorithm (e.g., Gelman et al. 1995) was employed to produce draws from the posterior distribution of the examinee ability. The iterative updating of the estimates $\Sigma^{(t+1)}$ and $\Gamma^{(t+1)}$ is performed using principles of a decreasing sequence of weights as outlined in Gu and Kong (1998) and Cai (2010). The results presented by von Davier and Sinharay (2010) indicate that stochastic approximation is a method that can be used to avoid some of the technical assumptions underlying current operational methods. Among these is the Laplace approximation used in the current operational approach. In addition, stochastic approximation relies on the draws from the posterior distribution of the statistic of interest. This allows one to generate imputations (plausible values) without the assumption of posterior normality. This is an advantage of the stochastic approximation methods, which may play out favorably in cases where, for example, due to reduced dimensionality of the background model or a limited number of items in the measurement model, the normality assumption in the current operation approach may not be appropriate.

Von Davier and Sinharay (2010) note that the convergence behavior of stochastic methods needs close monitoring. They also note that further research is needed for optimizing the choice of the annealing process to minimize the number of iterations needed, while at the same time ensuring that the process generates a solution that maximizes the likelihood function.

Other Approaches

The YGROUP approach (von Davier and Yu 2003) implements seemingly unrelated regressions (Zellner 1962; Greene 2004) and offers generalized least-squares (GLS) solutions. Li et al. (2009) present a multilevel approach that can be used to incorporate group-level covariates. Xu and von Davier (2008a) applied the general diagnostic model, which is a member of the recently popular series of models referred to as cognitively diagnostic models, to data from NAEP. Xu and von Davier (2008b) suggested the replacement of the current operational model by a multiple-group multidimensional IRT model. Rijmen et al. (2013, see Chapter 24 in this volume) suggested a third-order IRT model to analyze data from ILSAs and applied the model to data from TIMSS.

Outlook

The methodology currently used in ILSAs to estimate population distributions is the focus of ongoing research. Suggestions either to ignore auxiliary information when estimating distributions or to incorporate this information in a different way are discussed in the literature. Research that focuses on extending the model or making the current model less restrictive (Li et al. 2009; Thomas 2002; von Davier and Sinharay 2010) may be viewed as more promising when the goal is to make it more appropriate for increasingly complex samples.

The complexity of the ILSA operational model, as well as its integration into a timely operational analysis and reporting scheme that needs to produce results in a reliable way, makes trials with different model extensions a critical matter. Therefore, an evolution of increasingly flexible models seems preferable to a revolution replacing the current model by some even more complex alternative. At the same time, to study the impact on reporting results of different ways of modeling the data, alternative modeling methods must be explored. In addition, the convergence between IRT and generalized linear and nonlinear mixed models (e.g., De Boeck and Wilson 2004; Rijmen et al. 2003; see also Chapter 24 in this volume by Rijmen et al.) should be taken into account. If possible, desirable features of these general modeling approaches should be incorporated into the operational marginal estimation procedures used in the analysis of ILSA data.

Research that aims at weakening or removing some technical assumptions and approximations seems promising. The increased availability of powerful computers makes some approximations that were necessary in the past obsolete. Among these seemingly minor modifications are methods that aim at producing more general estimation algorithms (e.g., von Davier and Sinharay 2007, 2010). This type of approach is valuable because it maintains continuity, but at the same time allows greater flexibility. It may not have the same appeal as a totally new method, but it has the advantage of gradually introducing generalizations while keeping fallback positions at hand, as well as retaining the option to use well-established solutions for common issues in estimation.

References

Adams, R. J., Wilson, M., and Wu, M. 1997. Multilevel item response models: An approach to errors in variables regression. *Journal of Educational and Behavioral Statistics*, 22(1), 47–76.

Adams, R. J. and Wu, M. L. 2007. The mixed-coefficient multinomial logit model: A generalized form of the Rasch model. In M. von Davier and C.H. Carstensen (Eds.) *Multivariate and Mixture Distribution Rasch Models: Extensions and Applications* (pp. 57–76). New York: Springer Verlag.

Beaton, A. E. 1988. *Expanding the New Design: The NAEP 1985–86 Technical Report*. Princeton, NJ: Educational Testing Service.

Cai, L. 2010. High-dimensional exploratory item factor analysis by a Metropolis–Hastings Robbins–Monro algorithm. *Psychometrika*, 75, 33–57.

Cramer, J. S. 2004. The early origins of the logit model. *Studies in History and Philosophy of Science Part C: Studies in History and Philosophy of Biological and Biomedical Sciences*, 35(4), 613–626.

De Boeck, P. and Wilson, M. (Eds.) 2004. *Explanatory Item Response Models: A Generalized Linear and Nonlinear Approach*. New York: Springer.

Dempster, A. P., Laird, N. M., and Rubin, D. B. 1977. Maximum likelihood from incomplete data via the EM algorithm. *Journal of the Royal Statistical Society, B*, 39, 1–38.

Efron, B. 1982. *The Jackknife, the Bootstrap and Other Resampling Plans*. CBMS-NSF Regional Conference Series in Applied Mathematics, Monograph 38, SIAM, Philadelphia.

Fay, R. E. 1989. Theory and application of replicate weighting for variance calculations. In *Proceedings of the Section on Survey Research Methods, American Statistical Association* (pp. 212–217). Alexandria, VA: American Statistical Association.

Fox, J. P. and Glas, C. A. W. 2001. Bayesian estimation of a multi-level IRT model using Gibbs sampling. *Psychometrika*, 66, 271–288.

Fox, J. P. and Glas, C. A. W. 2003. Bayesian modeling of measurement error in predictor variables. *Psychometrika*, 68, 169–191.

Gelman, A., Carlin, J. B., Stern, H. S., and Rubin, D. B. 2003. *Bayesian Data Analysis*. New York, NY: Chapman & Hall.

Geweke, J. 1989. Bayesian inference in econometric models using Monte Carlo integration. *Econometrica*, 57, 1317–1339.

Gilks, W. R., Richardson, S., and Spiegelhalter, D. J. 1996. *Markov Chain Monte Carlo in Practice*. New York, NY: Chapman & Hall.

Gu, M. G. and Kong, F. H. 1998. A stochastic approximation algorithm with Markov chain Monte-Carlo method for incomplete data estimation problems. *Proceedings of the National Academy of Sciences, USA*, 95, 7270–7274.

Haberman, S. J., von Davier, M., and Lee, Y. 2008. Comparison of multidimensional item response models: Multivariate normal ability distributions versus multivariate polytomous ability distributions. RR-08-45. ETS Research Report.

Heinen, T. 1996. *Latent Class and Discrete Latent Trait Models: Similarities and Differences*. Thousand Oaks, CA: Sage Publications.

Hsieh, C., Xu, X., and von Davier, M. 2009. Variance estimation for NAEP data using a resampling-based approach: An application of cognitive diagnostic models. *Issues and Methodologies in Large Scale Assessments*, Vol. 2. Retrieved 6/11/2012: http://www.ierinstitute.org/fileadmin/Documents/IERI-Monograph/IERI-Monograph-Volume-02-Chapter-07.pdf

Johnson, E. G. 1989. Considerations and techniques for the analysis of NAEP data. *Journal of Education Statistics*, 14, 303–334.

Johnson, E. G. and Rust, K. F. 1992. Population inferences and variance-estimation for NAEP data. *Journal of Educational Statistics*, 17(2), 175–190.

Johnson, M. S. 2002. *A Bayesian Hierarchical Model for Multidimensional Performance Assessments*. Paper presented at the annual meeting of the National Council on Measurement in Education, New Orleans, LA.

Johnson, M. S. and Jenkins. F. 2004. *A Bayesian Hierarchical Model for Large-Scale Educational Surveys: An Application to the International Assessment of Educational Progress*. To appear as an ETS Research Report. Princeton, NJ: Educational Testing Service.

Judkins, D. R. 1990. Fay's method for variance estimation. *Journal of Official Statistics*, 6(3), 223–239.

Li, D., Oranje, A., and Jiang, Y. 2009. On the estimation of hierarchical latent regression models for large-scale assessments. *Journal of Educational and Behavioral Statistics*, 34, 433–463.

Little, R. J. A. and Rubin, D. B. 2002. *Statistical Analysis With Missing Data* (2nd edition). New York: Wiley.

Longford, N. T. 1995. *Model-Based Methods for Analysis of Data From the 1990 NAEP Trial State Assessment*. NCES 95-696. Washington, DC: National Center for Education Statistics.

Lord, F. M. 1980. *Applications of Item Response Theory to Practical Testing Problems*. Hillsdale, NJ: Lawrence Erlbaum Associates.

Lord, F. M. and Novick, M. R. 1968. *Statistical Theories of Mental Test Scores*. Reading, MA: Addison-Wesley.

Mislevy, R. 1984. Estimating latent distributions. *Psychometrika*, 49(3), 359–381.

Mislevy, R. J. 1985. Estimation of latent group effects. *Journal of the American Statistical Association*, 80(392), 993–997.

Mislevy, R. J., Johnson, E. G., and Muraki, E. 1992. Scaling procedures in NAEP. *Journal of Educational Statistics*, 17(2), 131–154.

Mislevy, R. J. and Sheehan, K. M. 1987. Marginal Estimation Procedures. In A. E. Beaton (Ed.), *Implementing the New Design: The NAEP 1983–84 Technical Report* (No. 15-TR-20, pp. 293–360). Princeton, NJ: Educational Testing Service, National Assessment of Educational Progress.

Mislevy, R. J. and Sheehan, K. M. 1992. *How to Equate Tests With Little or No Data*. ETS Research Report RR-92-20-ONR. Princeton, NJ: Educational Testing Service.

Muraki, E. 1992. A generalized partial credit model: Application of an EM algorithm. *Applied Psychological Measurement*, 16(2), 159–177.

Rasch, G. 1960. *Probabilistic Models for Some Intelligence and Attainment Tests*. Chicago, IL: University of Chicago Press.

Raudenbush, S. W., Fotiu, R. P., and Cheong, Y. F. 1999. Synthesizing results from the trial state assessment. *Journal of Educational and Behavioral Statistics*, 24, 413–438.

Rijmen, F., Tuerlinckx, F., De Boeck, P., and Kuppens, P. 2003. A nonlinear mixed model framework for item response theory. *Psychological Methods*, 8, 185–205.

Rogers, A., Tang, C., Lin, M.-J., and Kandathil, M. 2006. *DGROUP (Computer Software)*. Princeton, NJ: Educational Testing Service.

Rubin, D. B. 1987. *Multiple Imputation for Nonresponse in Surveys*. New York, NY: John Wiley & Sons Inc.

Rust, K. F. 1985. Variance estimation for complex estimators in sample surveys. *Journal of Official Statistics*, 1(4), 381–397.

Rust, K. F. and Rao, J. N. K. 1996. Variance estimation for complex surveys using replication techniques. *Statistical Methods in Medical Research*, 5, 283–310.

Thomas, N. 1993. Asymptotic corrections for multivariate posterior moments with factored likelihood functions. *Journal of Computational and Graphical Statistics*, 2, 309–322.

Thomas, N. 2002. The role of secondary covariates when estimating latent trait population distributions. *Psychometrika*, 67(1), 33–48.

Thurstone, L. L. 1925.A method of scaling psychological and educational tests. *Journal of Educational Psychology*, 16, 433–451.

von Davier, M., Gonzalez, E., and Mislevy, R. 2009. What are plausible values and why are they useful? In: *IERI Monograph Series: Issues and Methodologies in Large Scale Assessments*, Vol. 2. Retrieved 9/27/2011: http://www.ierinstitute.org/IERI-Monograph-Volume-02-Chapter-01.pdf

von Davier, M. and Sinharay, S. 2004. *Application of the Stochastic EM Method to Latent Regression Models*. ETS Research Report RR-04-34. Princeton, NJ: Educational Testing Service.

von Davier, M. and Sinharay, S. 2007. An importance sampling EM algorithm for latent regression models. *Journal of Educational and Behavioral Statistics*, 32(3), 233–251.

von Davier, M. and Sinharay, S. 2010. Stochastic approximation for latent regression item response models. *Journal of Educational and Behavioral Statistics*, 35(2), 174–193. http://jeb.sagepub.com/content/35/2/174.abstract

von Davier, M., Sinharay, S., Oranje, A., and Beaton, A. 2006. Statistical Procedures used in the National Assessment of Educational Progress (NAEP): Recent

Developments and Future Directions. In C. R. Rao and S. Sinharay (Eds.), *Handbook of Statistics (Vol. 26): Psychometrics*. Amsterdam: Elsevier.

von Davier, M. and Yon, H. 2004. *A Conditioning Model with Relaxed Assumptions.* Paper presented at the annual meeting of the National Council on Measurement in Education, San Diego, CA.

von Davier, M. and Yu, H. T. 2003. *Recovery of Population Characteristics from Sparse Matrix Samples of Simulated Item Responses.* Paper presented at the annual meeting of the National Council on Measurement in Education, Chicago, IL.

Wang, W., Wilson, M., and Adams, R. J. 1997. Rasch models for multidimensionality between and within items. In M. Wilson and G. Engelhard, (Eds.) *Objective Measurement: Theory into Practice*. Vol. IV. Norwood, NJ: Ablex.

White, H. 1980. A heteroscedasticity-consistent covariance matrix estimator and a direct test for heteroscedasticity. *Econometrica*, 48, 817–838.

Xu, X. and von Davier, M. 2008a. Fitting the structured general diagnostic model to NAEP data. RR-08-27, ETS Research Report.

Xu, X. and von Davier, M. 2008b. Comparing multiple-group multinomial loglinear models for multidimensional skill distributions in the general diagnostic model. RR-08-35, ETS Research Report.

Zellner, A. 1962. An efficient method of estimating seemingly unrelated regressions and tests for aggregation bias. *Journal of the American Statistical Association*, 57(298), 348–368.

8

Imputing Proficiency Data under Planned Missingness in Population Models

Matthias von Davier*
Educational Testing Service

CONTENTS

Introduction

From the perspective of missing data, in most assessments the proficiency of interest is a latent variable that is missing for *all* respondents. We never observe "intelligence" or "reading skill" directly; we only observe how respondents react to certain tasks that are assumed to be indicative of these underlying variables. The responses to these tasks, however, are fallible indicators at best, and only in naïve approaches to measurement are these directly equated to the underlying variable of interest.

* I thank the anonymous reviewers, Irwin Kirsch, Kentaro Yamamoto, and Frank Rijmen for comments on previous versions of this chapter. All remaining errors are, of course, my responsibility.

Because the proficiency variable is missing, an assumption has to be made regarding how observed performance indicators are related to underlying student proficiency. In large-scale educational surveys such as the National Assessment of Educational Progress (NAEP), Trends in Mathematics and Science Study (TIMSS), Progress in International Reading Literacy Study (PIRLS), and the Programme for International Student Assessment (PISA), an item response theory (IRT; Lord and Novick 1968) model—a special case of a latent structure model (Haberman 1974; Haberman et al. 2008)—is used to derive a model-based estimate of this missing proficiency variable for each respondent. Maximum likelihood estimates, or expected a posteriori (EAP) estimates, together with their associated measures of uncertainty—model-based standard errors or posterior variances—are used in IRT models (compare Chapter 7 of this volume) to derive estimates of proficiency.

The statistical procedures used in educational large-scale assessments extend IRT by involving a latent regression of the respondent proficiency variable(s) on a set of predictor variables (Mislevy 1991). Von Davier et al. (2006) describe this approach as well as recent extensions in some detail and show how it is implemented in NAEP. The rationale behind this combined IRT latent regression approach is the use of all available information to impute accurate distributions of student proficiencies given all available information. This approach follows the rationale developed by Rubin (1987, 1996) suggesting that all available information should be used in an imputation model if the data are missing at random (MAR, see the section on missing data below), that is, if the conditional distribution of the variable (and the indicator variable that shows whether a value is missing or not) depends on the values of observed quantities. In this sense, common practice in large-scale educational assessments follows standard recommendations in the treatment of missing data by the inclusion of all available observed variables in the imputation model.

A deviation of the customary approach to include all variables is the use of rotated background questionnaires in which some sets of variables are not presented to some respondents. Fetter (2001) treats the case of imputations in covariates, a case very similar in design to case cohort studies, and the case of incomplete (rotated) background questionnaires. His results show that if mass imputation of covariates is involved, there can be more, less, or an equal amount of bias in the estimates of relationships with other variables.

Also, the techniques to treat missingness in regression predictors so far applied to latent regressions are not entirely satisfactory. Finally, the fact that substantial parts of the students' background data that were not taken by the student are constant may introduce bias for some or all student groups due to effects related to omitted variable bias in regressions (Hanushek and Johnson 1977; Greene 2003). This paper describes the imputation model used in large-scale assessments such as NAEP, PISA, TIMSS, and PIRLS and shows how incomplete background data by design relates to issues discussed in relevant literature on missing data and (mass) imputation. In addition, the need for a careful examination of the goals of large-scale survey assessments is pointed out, and a call for reevaluation of the desire to introduce more planned missingness for the sake of broader coverage of a more diverse set of variables is voiced.

Imputation and Missing Data

In the previous section we argued that the latent variable θ can be viewed as a variable that is missing for all respondents. We introduced the specialized imputation model based on a combination of IRT with a multiple linear regression model. In this section we continue with a more general approach to missing data and introduce an important distinction of missing data mechanisms that can be used to explain the rationale of imputation using all observed data in the larger context.

Missing Data Mechanisms

Rubin (1987) developed a system that today is the gold standard of classifying and describing missing data mechanisms. This classification also has important consequences with respect to the treatment of missingness in statistical analysis. It is important to note that the assumptions made and the categorization of missing data mechanisms operates on the level of indicator variables that show whether a value in a data set is missing, not the variables containing the data of interest. More formally, assume there are $n = 1,\ldots,N$ respondents who are expected to provide data on $Y_1,\ldots,Y_K$ variables, so that y_{nk} denotes the value of variable k obtained for respondent n. Note that we do not make a distinction any longer between item responses $X_1,\ldots,X_I$ and covariates $Z_1,\ldots,Z_L$; we use the same letter Y for all (potentially) observed variables. Rubin (1987, 1996) defines missing data indicator variables $M_1,\ldots,M_K$ with

$$m_{nk} = \begin{cases} 0: & y_{nk} \text{ observed} \\ 1: & y_{nk} \text{ missing} \end{cases}$$

so that each respondent is characterized by two vectors of data, $(y_{n1},\ldots,y_{nK})$ representing the variables of interest, and $(m_{n1},\ldots,m_{nK})$ representing the indicators that show whether each of the variables of interest was observed or is missing for respondent n. Each vector of variables $\boldsymbol{y}_n = (y_{n1},\ldots,y_{nK})$ can be split into a vector $\boldsymbol{y}_{n,\text{obs}}$ of all observed responses of respondent n and a vector $\boldsymbol{y}_{n,\text{miss}}$ of what should have been observed if the data were not missing. Note that we do not know the values in $\boldsymbol{y}_{n,\text{miss}}$. The indicator variables $\boldsymbol{m}_n = (m_{n1},\ldots,m_{nK})$ will then tell us how to recombine the observed vector $\boldsymbol{y}_n = (y_{n1},\ldots,y_{nK})$ from $\boldsymbol{y}_{n,\text{obs}}$ by inserting the appropriate values in those positions where $m_{nk} = 0$.

Then, we can define the following three missing data mechanisms:

MCAR: Y_i is missing completely at random if $P(M_i \mid \boldsymbol{y}_{n,\text{obs}}, \boldsymbol{y}_{n,\text{miss}}) = P(M_i)$.
MAR: Y_i is missing at random if $P(M_i \mid \boldsymbol{y}_{n,\text{obs}}, \boldsymbol{y}_{n,\text{miss}}) = P(M_i \mid \boldsymbol{y}_{n,\text{obs}})$.
NMAR: Y_i is not missing at random if $P(M_i \mid \boldsymbol{y}_{n,\text{obs}}, \boldsymbol{y}_{n,\text{miss}}) \neq P(M_i \mid \boldsymbol{y}_{n,\text{obs}})$.

Note that the missingness condition NMAR states that the occurrence of missing data depends on the missing data $\boldsymbol{y}_{n,\text{miss}}$ itself. Obviously, this is the least desirable situation, which requires more complex modeling in order to find a predictive distribution for Y_i. The three definitions above do not talk about $P(Y_i)$ and potential dependencies between Y_i and the remaining variables in $\boldsymbol{Y}^{(i)}_{n,\text{obs}} = \boldsymbol{Y}_{n,\text{obs}\setminus Y_i}$ and $\boldsymbol{Y}^{(i)}_{n,\text{miss}} = \boldsymbol{Y}_{n,\text{miss}\setminus Y_i}$. This implies that examining

$$P(Y_i \mid \boldsymbol{y}^{(i)}_{n,\text{obs}}),$$

that is, the conditional distribution of Y_i given the observed data as a means to find a predictive distribution. This would be a useful approach when it is determined that missingness is MAR or MCAR. It turns out that many imputation methods indeed use this conditional distribution to derive values for Y_i if it was not observed and missingness is MAR or MCAR. A few examples of these approaches will be given in the next subsection. If missingness is NMAR, as is often the case for ability-related nonresponse in psychological and educational testing, more involved models are required that make use of the missingness indicators $\boldsymbol{m}_n = (m_{n1},\ldots,m_{nK})$ to come up with a predictive distribution for missing variables.

Imputation Methods in a Nutshell

Roughly, imputation describes the generation of (random) replacements of observations that are missing for one reason or another in a data set. Often, imputations are used to provide "complete" data sets in order to allow the use of methodologies that require complete data. Without imputations, many important standard methods that require complete data would be applicable only after list-wise deletion of all cases with missing data, which may introduce bias if the occurrence of missingness is not MCAR.

As an example, if gender is completely observed, that is, it is part of $Y_{.,obs}$, and Y_1 is a variable indicating that the respondent is a smoker, and we have $P(M_1 = 1|\text{male}) > P(M_1 = 1|\text{female})$ (males omit the response to the smoking question more often than women) as well $P(Y_1 = 1|\text{male}) > P(Y_1 = 1|\text{female})$ (males are more likely to smoke), we would obtain a smaller value for $P(Y_1 = 1)$, the proportion of smokers, than expected if we delete cases listwise. That is because we would delete more records of males who are more likely to smoke and would omit the response to the smoking question.

Imputations may help in this case if there are methods that can take these types of relationships into account. However, it may not be enough to know the gender in order to impute missing self-reports on smoking. It may also be useful to know whether the respondents' parents were smokers, or if there was other substance use, as well as knowledge about blood pressure, body-mass index, type of job, or other variables that may have a statistical relationship to smoking. When imputing, it seems, it would be good to use all the information we have, and it appears to be so when looking at standard recommendations put forward, for example, by Little and Rubin (2002). Indeed, Collins et al. (2001) show that what they call the inclusive strategy is greatly preferred (an imputation model that contains as many covariates as possible and available) compared to the restrictive strategy (an imputation model with a limited number of variables, for example, only the core variables in a rotated student questionnaire).

There are many imputation methods, and this chapter discusses only a select number of these methods. Interested readers are referred to Little (1988) as well as Little and Rubin (2002) for a discussion of a range of imputation methods. Among the ones frequently found in the literature are

1. *Hot deck imputation*—A seemingly simple way of coming up with a procedure that "imputes" values for missing observations. Under this approach, a respondent with missing data is simply compared to all other respondents (on the set of observed variables that are not missing), and a "donor" is chosen that is maximally similar to the respondent with missing values. Then, the respondent receives "copies" of the variable values that are missing from this donor.
2. *Multivariate normal distribution*—Another commonly used method for imputation of missing observations due to non-response, which allows generating conditional distributions of variables with missing observations. Obviously, this approach assumes that all observed variables are at least interval level and are (approximately) normally distributed. This approach is therefore not very useful for completion of questionnaire and test data. Instead, methods that allow for modeling multidimensional categorical data are suitable for imputations (e.g., Vermunt et al. 2008).
3. *Multiple imputation by chained equations*—If no multivariate distribution can be found to model all available data, multiple imputation by chained equations is a suitable alternative (e.g., van Buuren and Groothuis-Oudshoorn 2011). This method specifies a multivariate imputation model on a variable-by-variable basis by a set of conditional densities, one for each incomplete variable. After an initial imputation, this method draws imputations by iterating over the

imputation model for the different variables. Typically, a low number of iterations are used. The approach is attractive because it can be implemented by means of imputing first for the variable with the fewest missing values, using only complete variables in the model, then moving to the next variable with the second-lowest number of missing values, and so on, until values for all variables are imputed and a more comprehensive imputation model can be used for subsequent imputations.

Most imputation methods are typically considered for small amounts of missing data based on item-level non-response, and many of the recommendations about the use of only five imputed data sets come from the fact that only small percentages of missingness are considered. If large amounts of missing data are considered, this situation is known as "mass imputation" (e.g., Haslett et al. 2010), and standard methods of imputation may not yield the expected quality of the imputed values (Fetter 2001) in terms of the appropriate recovery of statistical relationships between variables in the database.

Also, as Piesse and Kalton (Westat Research Report 2009) put it: "The dominant concern with imputation is that it may affect the associations between items. In general, the association between an imputed item and another item is attenuated towards zero unless the other item is included as an auxiliary variable in the imputation scheme." As Rubin (1996) puts it: "More explicitly, when Q or U involves some variable X, then leaving X out of the imputation scheme is improper and generally leads to biased estimation and invalid survey inference. For example, if X is correlated with Y but not used to multiply-impute Y, then the multiply-imputed data set will yield estimates of the XY correlation biased towards zero."

This explains why all available observed variables customarily are included in the imputation model applied to produce multiple "completed" data sets. If mass imputation is involved, or if many variables are affected by missing data, this approach can only be followed to some extent, because for many imputations, other observations for the same respondent may be missing as well. In the case of mass imputation, when large sets of variables are jointly missing, the imputation problem becomes circular, and the result may depend on what was chosen to be imputed initially as well as the sequence in which variables were "completed" in the process.

Conditional Independence and Imputation

The categorization of the different missing data mechanisms in the above section used the concept of conditional independence casually without a formal introduction. Because this is a central concept of importance for many of the issues discussed in this chapter, we reiterate the definition and discuss a few implications of conditional independence assumptions more formally.

Let X denote a binary variable that indicates performance on a complex computer simulation task that is intended to measure computer literacy, with $x_n = 1$ indicating that respondent n solves the task, and $x_n = 0$ otherwise. Let Z_* denote a variable that indicates whether at least one of the respondent's parents has a college degree, $z_{n*} = 1$, or not, $z_{n*} = 0$. Let Z_A denote a variable that indicates whether the respondent has his or her own computer at home, $z_{nA} = 1$, or not, $z_{nA} = 0$. Then we can define the following:

The random variables X and Z_A are conditionally independent given Z_* if and only if

$$P(X, Z_A \mid Z_*) = P(X \mid Z_*)P(Z_A \mid Z_*),$$

that is, if conditional independence is given for X and Z_A, we can factorize the joint probabilities into a product of marginal probabilities. This definition leads to the following important corollary

$$P(X \mid Z_*, Z_A) = P(X \mid Z_*),$$

that is, if X and Z_A are conditionally independent given Z_*, then the conditional distribution of X given Z_* and Z_A equals the conditional distribution of X given only Z_*. By symmetry, we have

$$P(Z_A \mid Z_*, X) = P(Z_A \mid Z_*)$$

and we write

$$X \perp Z_A \mid Z_*.$$

This means that if conditional independence of two variables X and Z_A given a third Z_* is established, then we can use only the third variable Z_* and leave the second, Z_A, out of the imputation model for X, and vice versa. Of course, this assumes we keep the conditioning variable (here, only Z_*) in the imputation model, that is, the variable required to establish conditional independence.

Conditional independence can be considered a strong assumption about the relationships among observed variables. Typically, a conditional independence assumption is used in order to define idealized (latent) variables that make all other influences "vanish" or allows us to factorize the likelihood of the joint occurrence of a series of variables by assuming an underlying probabilistic "cause" (Suppes and Zanotti 1995). One of the standard examples where conditional independence (also called local independence in that context) is assumed is in latent structure models, such as IRT (see the Introduction for this chapter as well as Lord and Novick 1968) or latent class models (Lazarsfeld and Henry 1968).

In imputation models, one typically avoids such a strong assumption, mainly out of recognition that wrongly assuming conditional dependence and eliminating variables from the imputation model may introduce bias in estimates of relationships (e.g., Rubin 1996; Collins et al. 2001; Piesse and Kalton 2009). As a consequence, if certain conditional relationships are incorrectly assumed to be nonexistent for some groups of variables, then bivariate or higher-order dependencies between these variables may not be reproduced accurately. Therefore, imputation models typically are maximally inclusive by using all available observed variables for imputation of missing observations whenever possible.

Let us now assume that we have a fourth variable, Z_B, which may indicate that a respondent reports a high ($Z_B = 1$) or low ($Z_B = 0$) level of self-efficiency in terms of computer literacy. Note that conditional independence is not a condition that can be extended to this additional variable in trivial terms. More specifically,

$$X \perp Z_A \mid Z_* \not\Leftrightarrow X \perp Z_A \mid (Z_*, Z_B),$$

that is, if X and Z_A are conditionally independent given Z_*, that does not necessarily imply that X and Z_A are conditionally independent given Z_* and Z_B. As an example, consider Table 8.1, which represents the joint distribution of four binary variables.

It can be seen in Table 8.1 that the sum of all probabilities adds up to 1.0 in either rows or columns of the table. All four conditional tables, when normalized to a sum of 1.0 within the 2 × 2 table, describe the distribution of X and Z_A given the levels of Z_* and Z_B. Table 8.2 shows the conditional tables of X and Z_A given Z_*.

The two tables above show that the conditional distribution of X and Z_A is independent given Z_*. For example, $0.32/0.5 = (0.32/0.5 + 0.08/0.5)^2$, $0.02/0.5 = (0.02/0.5 + 0.08/0.5)^2$, and so on. This means that by collapsing tables across the two levels of Z_B we produced two tables in which conditional independence holds. Finally, Table 8.3 shows that the two variables X and Z_A are not independent when not conditioning on any variable.

TABLE 8.1

Joint Distribution for Which We Do Not Show an Association between and Z_A, but Conditional Independence Holds as in $X \perp Z_A | Z_*$, but $X \perp Z_A | (Z_*, Z_B)$ Does Not Hold

		$Z_* = 0$				$Z_* = 1$		
		0	1			0	1	
$Z_B = 0$	0	0.31	0.01		0	0.01	0.07	
	1	0.01	0.01		1	0.07	0.01	
				0.34				0.16
		0	1			0	1	
$Z_B = 1$	0	0.01	0.07		0	0.01	0.01	
	1	0.07	0.01		1	0.01	0.31	
				0.16				0.34

TABLE 8.2

Conditional Distributions with Conditional Independence for $X \perp Z_A | Z^*$

		$Z_* = 0$				$Z_* = 1$		
		0	1			0	1	
$Z_B = 0 \vee 1$	0	0.32	0.08		0	0.02	0.08	
	1	0.08	0.02		1	0.08	0.32	
				0.5				0.5

The example shows that it is not enough to find a set of variables for which conditional independence holds. *When more predictors are obtained, conditional independence relationships may change.* More specifically, it may be that the distribution $P(X|Z_*,Z_A,Z_B)$ does differ from $P(X|Z_*,Z_B)$, even if it was found that $P(X|Z_*) = P(X|Z_*,Z_A)$. Therefore, if a predictive distribution for imputations is wanted, it appears a good strategy is to include all available variables in order to ensure that the relationship between X and the other variables is maintained when values for X are imputed.

Let us also assume that the additional variable Z_B is assessed together with X and Z_* in only one half of the sample (Sample B), while Z_A is assessed in the other half (Sample A) together with X and Z_*. Let us further assume that conditional independence does NOT hold for X and Z_B given Z_*, but does hold for X and Z_A, given Z_*. We are facing the following situation: We cannot define an imputation model for X given all three variables Z_*, Z_A, Z_B because they are never jointly observed without making additional strong assumptions that allow us to construct a synthetic joint distribution (Little and Rubin 2002).

If we try to devise an imputation model without making additional assumptions, we are bound to use the variables that are used in each of the half samples. If we assess a respondent in Sample B who has a missing response on X, we are well advised to use an imputation model that contains both Z_* and Z_B because conditional independence does not hold. If, however, we have a missing response on X for a respondent from Sample A, the imputation model for this person can only contain Z_*, and optionally, but without consequence, Z_A. Z_B cannot be added because only Z_A was observed, but X and Z_A are conditionally independent given Z_*. This means that when imputing missing X responses in Sample A, we can only use Z_* but an important

TABLE 8.3

Marginal Distribution with a Positive Association between X and Z_A

		$Z_* = 0 \vee 1$		
		0	1	
$Z_B = 0 \vee 1$	0	0.34	0.16	
	1	0.16	0.34	
				1.0

piece of the information is lost because Z_B, of which X is not conditionally independent given Z_*, was not collected in Sample A.

Just to drive the point home: If the background variable Z_B were omitted in some way (not collected, or only collected on a subset of the sample), the joint distribution under a missingness of X of these four variables would not be recovered correctly in an imputation model that contains only Z_* and Z_A. Moreover, an imputation model using the latter two variables would actually suggest that $X \perp Z_A$, and thus only $P(X|Z_*)$, could be used, and thus, information is lost and bias introduced in secondary analyses that would involve imputed values of X and variables that are statistically associated or correlated with Z_A, Z_B.

Imputation of Proficiency in Large-Scale Assessments

In large-scale educational survey assessments, 100% of the student proficiency data is imputed using a specialized imputation model based on statistical procedures that are tailored to incorporate both cognitive response data and student background data (Mislevy 1991; von Davier et al. 2006, 2009; and Chapter 7 in this volume). These procedures are referred to as latent regression models and provide EAP estimates and estimates of posterior variance of proficiency. These estimates are based on a Bayesian approach and thus are using a prior distribution of proficiency, a feature that opens an avenue to the introduction of conditional prior distributions, for example, in the shape of a multiple-group IRT model (Bock and Zimowski 1997; Xu and von Davier 2008).

Latent regression models extend the multiple-group IRT model and provide a different conditional prior distribution for each respondent's proficiency based on a set of predictor variables (Mislevy 1991; von Davier et al. 2006).

Item Response Theory

Let θ denote the latent variable of interest. This may be a variable representing mathematics, reading, or science proficiency. Let $X_1, \ldots, X_I$ denote variables that represent the responses to test items that are assumed to be governed by θ, that is, assume that for each respondent n, the responses to X_i are denoted by x_{ni} and that

$$P_i(x_{ni} = 1 \mid n) = P_{\zeta_i}(x_{ni} = 1 \mid \theta_n)$$

with some item-dependent parameters ζ_i and with θ_n denoting the n-th respondent's proficiency. While most educational assessments use multiparameter IRT models such as the 2PL and 3PL IRT model (Lord and Novick

1968) for binary data, and the generalized partial credit model (Muraki 1992) for polytomous data, PISA has been using the Rasch model (Rasch 1960/80) as the basis (Adams et al. 2007). The probability function defining the IRT models used can be written as

$$P_{\zeta_i}(x_{ni}=1\mid\theta_n)=c_i+(1-c_i)\frac{\exp(\alpha_i(\theta_n-\beta_i))}{1+\exp(\alpha_i(\theta_n-\beta_i))}$$

and $P_{\zeta_i}(x_{ni}=0\mid\theta_n)=1-P_{\zeta_i}(x_{ni}=1\mid\theta_n)$. This model may be used for binary responses $y \in 0,1$ with item parameters $\zeta_i = (\alpha_i, \beta_i, c_i)$, whereas

$$P_{\zeta_i}(X=x_i\mid\theta_n)=\frac{\exp\left(\sum_{z=1}^{x_i}\alpha_i(\theta_n-\beta_{iz}^*)\right)}{1+\sum_{z=1}^{k_i}\exp\left(\sum_{w=1}^{z}\alpha_i(\theta_n-\beta_{iw}^*)\right)}$$

may be assumed for polytomous ordinal responses $y \in \{0,\ldots,k_i\}$ with item parameters $\zeta_i = (\alpha_i,\beta_{i1},\ldots,\beta_{iki})$. Together with the usual assumption of local independence of responses given θ we obtain

$$P_{\zeta}(x_{n1},\ldots,x_{nI}\mid\theta_n)=\prod_{i=1}^{I}P_{\zeta_i}(X=x_{ni}\mid\theta_n).$$

Customarily, it is assumed in IRT models that the ability parameter follows a distribution $f(\theta;\eta)$ where η describes the parameters of the distribution that may be fixed or estimated from the data, depending on whether the latent scale is set and how other parameters such as ζ are constrained in order to remove indeterminacy in scale and location. The distribution $f(\theta;\eta)$ is then used to derive the marginal probability of a response pattern, that is,

$$P_{\zeta,\eta}(x_{n1},\ldots,x_{nI})=\int_{\theta}\left[\prod_{i=1}^{I}P_{\zeta_i}(X=x_{ni}\mid\theta)\right]f(\theta;\eta)d\theta.$$

Maximum likelihood methods are the customary approach to obtain estimates of the parameters η,ζ. For this purpose, the log-likelihood function

$$\ln L(\eta,\xi;X)=\sum_{n=1}^{N}\ln\left(\int_{\theta}\left[\prod_{i=1}^{I}P_{\zeta_i}(X=x_{ni}\mid\theta)\right]f(\theta;\eta)d\theta\right),$$

where X represents the $N\times I$ matrix of item responses from all respondents, is maximized with respect to η and ζ.

The item response data $\boldsymbol{X} = [(x_{11},\ldots,x_{1I}),\ldots,(x_{N1},\ldots,x_{NI})]$ modeled by IRT may contain missing responses by design (MCAR) as well as item-level nonresponses that often are assumed to be ignorable as well. While the missingness by design is introduced by means of matrix sampling of item responses (Mislevy 1991) and can be assumed to be MCAR and thus ignorable, the missingness by means of respondents choosing to not provide an answer to one or more items can be assumed to be MAR at best, but there is strong indication in real data that nonresponse is informative, that is, NMAR (e.g., Rose et al. 2010).

Usually the number of omitted responses is small, but for larger amounts of informative (NMAR) missing data, model-based approaches to handle missingness have been devised. For these approaches, we refer interested readers to Moustaki and Knott (2000), Glas and Pimentel (2008), as well as Rose et al. (2010). We will not discuss this particular topic of non-ignorable missing item response data further because it is not within the focus of this chapter.

Latent Regression

In addition to the observed item response variables X_i, latent regression models assume there are additional variables collected for each respondent. Let these variables be grouped into collections of variables $Z_* = (Z_{*1},\ldots,Z_{*I^*})$ as well as $Z_A = (Z_{A1},\ldots,Z_{AI^A})$, $Z_B = (Z_{B1},\ldots,Z_{BI^B})$, and $Z_C = (Z_{C1},\ldots,Z_{CI^C})$, or more groups of covariates. One of these variable groups may represent measures of home background and socioeconomic status (say Z_A) while another group (say Z_B) may represent student attitudes and motivational variables, and so on.

It is common practice in most national and international large-scale assessments to collect observations on all variable groups $\boldsymbol{Z} = (Z_*, Z_A, Z_B, Z_C)$ for all respondents who provide responses on the test items $X_1,\ldots,X_I$. This practice is motivated by the use of these covariates in the latent regression that we will define formally below. Using these groups of covariates, the conditional distribution of the latent trait variable θ is defined as

$$P(\theta \mid z_{n^*}, z_{nA}, z_{nB}, z_{nC}) = f(\theta; \boldsymbol{z}_n)$$

with $\boldsymbol{z}_n = (z_{n^*}, z_{nA}, z_{nB}, z_{nC})$ denoting the realizations of $\boldsymbol{Z} = (Z_*, Z_A, Z_B, Z_C)$ for respondent n. If all the covariates are continuous, one may use

$$\mu_\theta(z_{n^*}, z_{nA}, z_{nB}, z_{nC}) = \sum_{w \in \{*,A,B,C\}} \sum_{i=1}^{I^w} \gamma_{iw} z_{nwi} + \gamma_0,$$

that is, a linear predictor based on the covariates with regression parameters $\Gamma = (\gamma_{*1},\ldots,\gamma_{CI_w},\gamma_0)$, and a common conditional variance

$$\Sigma = V(\theta \mid Z_{*}, Z_A, Z_B, Z_C)$$

for which an estimate can be obtained using ordinary least-squares (OLS) regression methods within an EM algorithm (Mislevy 1991; von Davier et al. 2006). To obtain estimators $\hat{\Sigma}$ and $\hat{\Gamma}$, the estimation needs to take into account that θ is unobserved (latent) and therefore the regression does not have access to the values of the dependent variable. Instead, the conditional density of θ given z_n by

$$f(\theta; z_n, \Gamma, \Sigma) = \phi(\theta; \Gamma, \Sigma)$$

is used in an iterative scheme to estimate Γ and Σ (Mislevy 1991; von Davier et al. 2006).

Imputing Plausible Values from the Posterior Distribution

While the marginal distribution of responses in ordinary IRT does not depend on any other observed variables, the marginal distribution of a response pattern in latent regressions does depend on the observed covariates z_n. We obtain

$$P(x_{n1}, .., x_{nI} \mid z_n) = \int_{\theta} \left[\prod_{i=1}^{I} P_{\zeta_i}(X = x_{ni} \mid \theta) \right] f(\theta; z_n, \Gamma, \Sigma) d\theta_{\theta}.$$

Then, the predictive distribution of θ given item responses and covariates $x_1 \ldots x_I$, z_n is

$$p(\theta \mid x_{n1} \ldots x_{nI}, z_n) = \frac{\left[\prod_{i=1}^{I} p(x_i \mid \theta) \right] f(\theta; z_n)}{\int_{\theta^*} \left[\prod_{i=1}^{I} p(x_i \mid \theta^*) \right] f(\theta; z_n)}$$

It is important to understand that the above expression is an imputation model for θ given $x_1, \ldots, x_I$, z_n in that it allows us to derive an expected value and a variance of the proficiency variable given the item responses $x_1, \ldots, x_I$ and the covariates $z_1, \ldots, z_D$. Even more important to understand is that the relationship between θ and $x_1, \ldots, x_I$ is model-based and prescribed to follow a strong monotone relationship given by an IRT model (Hemker et al. 1996), while the relationship between θ and $z_1, \ldots, z_D$ is estimated using a linear regression as described in the previous section.

The latent regression IRT model is employed to derive imputations for public use files that can be used by secondary analysts. More specifically, each proficiency variable is represented in the public use data by a set of *M* imputations, in this context called plausible values (PVs), for each respondent in the data. These PVs depend on the item responses (proficiency indicators) as well as covariates such as student self-reports on activities, attitudes and interests, socioeconomic indicators, and other variables of interest for reporting. The rationale is to include all covariates that are available and may be of analytic interest in order to reduce bias in secondary analysis (Mislevy 1991; von Davier et al. 2009). Following the rationale put forward by Rubin (1987, 1996) and Mislevy (1991), all available background and item response data are used in the imputation in order to minimize bias in the estimates of statistical quantities involving these imputations.

Proficiency Imputation under Missingness in Background Data

In the previous section we learned that the inclusion of as many relevant variables as possible into the imputation model is superior to a strategy that includes only a minimum set of variables (Rubin 1996; Collins et al. 2001). This means that if a large amount of variables have a substantive proportion of missing data, it is to be expected that some bias in the estimation of relationships between the partially observed and other variables will be introduced.

This reasoning put forward by Rubin (1987, 1996) and others led to the convention to use as many variables as possible that will provide data about the student in the population modeling in all PISA (and other survey assessments) cycles, with the exception of PISA 2012. Therefore, use of rotated background questionnaires appears not to follow longstanding advice to use all available information on students in the population model to derive multiple imputations of student proficiency (Mislevy 1991; Rubin 1996; von Davier et al. 2006). We will illustrate two versions of planned missingness in background data that deviate from the imputation rationale put forward by Rubin (1987, 1996; Little and Rubin 2002) and others in Tables 8.4 and 8.5.

TABLE 8.4

File Matching Case of Missing Data

Rotation	X_1	X_2	X_3	θ	Z_*	Z_A	Z_B	Z_C
I	X	X			X	X		
II		X	X		X		X	
III	X		X		X			X
IV	X	X			X		X	
VI		X	X		X			X
VII	X		X		X	X		
...	...	...	...	...	...	...	...	...
??	X		X		X		X	

TABLE 8.5

Incomplete Background Data Collection Using Rotation

Rotation	X_1	X_2	X_3	θ	Z_*	Z_A	Z_B	Z_C
I	X		X		X	X	X	
II		X	X		X		X	X
III	X	X			X	X		X
IV	X	X			X		X	X
VI		X	X		X	X		X
VII	X		X		X	X	X	
...	...	...	...	...	...	...	...	...
??	X		X		X		X	X

In these tables, an empty cell denotes missing observations and a X denotes observed data. Note that all cells below θ are empty because the proficiency is missing for all respondents and has to be imputed for 100% of all cases. There is also planned missingness in the observed item responses $X_1, \ldots, X_I$, of which we will assume for simplicity, but without limiting generality, that these observed variables are collected in three blocks X_1, X_2, X_3 that are combined into three booklets I, II, III. The IRT model assumption does provide a conditional independence assumption that allows us to model the joint distribution of the observed responses given ability θ even though they are not jointly observed. By means of the IRT model assumption of an underlying *probabilistic cause* (Suppes and Zanotti 1995)—which we refer to as latent variable θ in IRT—we are able to derive the joint conditional distribution even if not all observed variables were collected jointly on all respondents (see, e.g., Zermelo 1929).

It is important to note that the same reasoning does not apply with respect to the sets of covariates Z_*, Z_A, Z_B, Z_C unless another set of model assumption is made and all variables collected in these covariates are assumed to be related to one or more underlying latent variables. While this may be a reasonable assumption for some subsets of the covariates, it does not appear trivial to develop a comprehensive latent variable model that is defensible and contains explanatory relationships between a set of latent variables and all covariates in Z_*, Z_A, Z_B, Z_C.

If some significant part of the background information is not collected on a large portion of the students, as is true in rotated designs discussed and implemented so far, this violates a basic imputation rationale (Rubin 1987, 1996) used in the majority of large-scale assessments such as PISA, TIMSS, PIRLS, and NAEP. Also, as Little and Rubin (2002, p. 7) point out in their introduction to missing data problems, if there are two (or more) sets of variables that are never jointly observed, associations between these two sets of variables cannot be estimated accurately. Little and Rubin (2002, p. 7) call this missing data design the *file matching* case, and the data collection design that leads to this file matching can be illustrated as in Table 8.4. We will use the same notation as in previous sections, but will now look at the case where different patterns of missingness are found in the covariates Z_*, Z_A, Z_B, Z_C. Table 8.4 shows a balanced block design for the item responses X_1, X_2, X_3 but a file matching data collection design for Z_*, Z_A, Z_B, Z_C.

In the case depicted in Table 8.4, only the covariates in Z_* are collected in all cases, but the variables collected in Z_A are never jointly observed with variables in Z_B or Z_C. Given this setup, an imputation model for Z_A would need to assume that Z_A given Z_* is conditionally independent of Z_B, Z_C, that is, an imputation model would produce a conditional distribution $P(Z_A|Z_*)$, which involves neither Z_B nor Z_C even though it may likely be that these are not conditionally independent given Z_*. As Little and Rubin (2002, p. 7) point out, the partial associations between the rotated, never jointly observed, variable sets are unknown, and that "in practice, the analysis of data with this pattern typically makes the strong assumption that these partial associations are zero." While this case would probably not be considered for obvious reasons stated above, it is important to note that the missingness introduced by this design requires a strong independence assumption as pointed out by Little and Rubin (2002). A less extreme case that could be described as a relaxed file matching case is illustrated in Table 8.5.

In the example depicted in Table 8.5, an imputation model for Z_A could use the conditional distribution $P(Z_A|Z_*, Z_C)$, assuming conditional independence of Z_A and Z_B given Z_*, Z_C. Alternatively, one could use the conditional distribution $P(Z_A|Z_*, Z_B)$, assuming conditional independence from Z_C, to impute values in Z_A. This would lead to two competing imputation models that may produce different sets of predictive distributions for the same set of "true" but unobserved values of Z_*, Z_B, Z_C. If all variables would have been jointly observed for some subsample, or (strong) model assumptions are used to come up with a joint distribution, a predictive distribution $P(Z_A|Z_*, Z_B, Z_C)$ that involves all other variable groups except Z_A could be used instead. Without this, students with identical background profiles, but from different background questionnaire rotation groups, may receive different predicted imputed values based on the rotation that was received.

The design in Table 8.5, while less sparse than the case depicted in Table 8.4, will still need additional (strong) model assumptions, or sequential (chained) methods in order to derive an imputation model that involves all covariates. Note that chained methods have to start with only observed variables, and usually choose the variables that are complete for all cases (here Z_*). Thus, they would be penalized by a design like the one in the tables because this would essentially amount to a mass imputation problem that has been shown to potentially lead to unpredictable bias in estimates involving imputations (Fetter 2001).

Assumptions Needed for Imputation with Incomplete Background Data

The underlying issue to address is whether there is evidence that the different rotation schemes in background questionnaires may introduce bias due to omitted variables in the latent regression models used for population modeling in large-scale assessment. Given the above discussions of customary imputation methods that tend to include all available data as well as the description of the latent regression model that provides imputation of proficiency for 100% of all respondents, there are some central questions to be answered before a decision about using incomplete questionnaires.

1. When imputing proficiency values for θ, what are the implications of using incomplete questionnaire data? If there are questionnaire and test booklet rotations such as I, II, III, IV, ... as illustrated in the table, there are different options for the use of these data as they are introduced into the latent regression model. We list a range of theoretically possible approaches to imputing θ and mention some of the assumptions and consequences on sample size and uniqueness of the imputation models involved:
   a. One option would be to use a model that incorporates only observed covariates that are available for all respondents: $P(\theta \mid X_2, X_3, Z_*; \zeta, \Gamma, \Sigma)$ or $P(\theta \mid X_1, X_3, Z_*; \zeta, \Gamma, \Sigma)$ would be the three possible combinations of background variables Z_* and blocks of item responses X_1, X_2, X_3 in the example design introduced above. Note that these could be estimated in one comprehensive latent regression model because the IRT model can handle MCAR and MAR observations in item response data X_1, X_2, X_3. In this case we would make the strong (Little and Rubin 2002) and untestable assumption that the posterior distribution of θ does not depend on Z_A or Z_B or Z_C or any combination of these, that is, θ and any combination of the omitted covariate blocks are indeed conditionally independent given one of the three combinations of X variables and Z_*. The assumption is untestable because the four sets of covariates are not observed jointly.
   b. It could be considered using different latent regression imputation models for different subpopulations: All combinations of cognitive items X_1, X_2, X_3 may be combined in the IRT part, while three different latent regressions would be estimated that include only those respondents who were assessed with either the jointly occurring Z_*, Z_A, Z_B; Z_*, Z_A, Z_C; or Z_*, Z_B, Z_C background rotations. The imputation models would be $P(\theta \mid X_1, X_2, X_3, Z_*, Z_A Z_B; \zeta, \Gamma^{(C)}, \Sigma^{(C)})$, $P(\theta \mid X_1, X_2, X_3, Z_*, Z_A Z_C; \zeta, \Gamma^{(B)}, \Sigma^{(B)})$, or $P(\theta \mid X_1, X_2, X_3, Z_*, Z_B Z_C; \zeta, \Gamma^{(A)}, \Sigma^{(A)})$, each of which is defined by the inclusion of all available background data, that is, each is missing only one of the rotated sets of background variables. This separate estimation of the imputation model parameters $\Gamma^{(C)}$, $\Sigma^{(C)}$ (those without parameters for Z_C), as well as $\Gamma^{(B)}$, $\Sigma^{(B)}$ (those without Z_B and $\Gamma^{(A)}$, $\Sigma^{(A)}$

(without Z_A) would cut the sample size for the latent regression models by a factor of 3, and would produce three latent regression estimates, one for each of the three subsamples. Again, certain conditional independence assumptions would be necessary, without being testable, for the same reason as above.

c. Another option would be a joint estimation using an imputation model that includes all cognitive variables and all sets of background variables when observed, and a placeholder if missing by design. Instead of separately estimating an imputation model for each of the three rotations, one joint model will be defined. For rotations I, placeholder values (mode if the variables are categorical, mean if the variables are continuous) are defined and set to be constant for all respondents who have missing data. Additionally, an indicator variable will be defined that shows which variables are missing and were replaced by modal or mean values. Let us denote this constant vector Z_C^{miss} if values in Z_C are missing by design, and Z_B^{miss} and Z_A^{miss} likewise. Now the data matrix that is used to estimate $P(\theta \mid X_1, X_2, X_3, Z_*, Z_A Z_B, Z_C; \zeta, \Gamma, \Sigma)$ is defined by replacing missing values in $Z_A Z_B$, Z_C by the *constant* vectors $Z_A^{\text{miss}} Z_B^{\text{miss}}, Z_C^{\text{miss}}$ and the imputation model parameters $\Gamma^{(\text{miss})}, \Sigma^{(\text{miss})}$ would be estimated. Note that it is a well-known fact that in linear regressions, this approach does not lead to consistent estimates and will underestimate the correlations of the background variables and the dependent variable θ (e.g., Jones 1996). More specifically, we will have $E\left(\Gamma^{(\text{miss})}\right) \neq \Gamma$ and $E\left(\Sigma^{(\text{miss})}\right) \neq \Sigma$, that is, the replacement of data missing by design will introduce data that are constant for large parts of the sample, which in turn will tend to bias the relationship between proficiency variable θ and predictors Z_A, Z_B, Z_C downward. Two different simulation studies (Rutkowski 2011; PISA TAG(1204)6b) have shown that this is a result that can also be seen in latent regressions. If in addition there are domains to be imputed without associated cognitive data, this effect is likely to be even more pronounced. The basic underlying problem is that data that were vectors "completed" using the constant vectors $Z_A^{\text{miss}}, Z_B^{\text{miss}}, Z_C^{\text{miss}}$ do not represent the statistical dependencies in the data accurately, and models estimated based on this replacement will underestimate the relationships in the data.

d. Finally, using different but overlapping latent regression models could be considered: Different combinations of cognitive items X_1, X_2, X_3 may be present in the IRT part, while the latent regressions would be based on two-thirds of the sample: All respondents who took either Z_*, Z_A, Z_B or Z_*, Z_A, Z_C would be modeled by a regression model that predicts θ by $P(\theta \mid X_1, X_2, X_3, Z_*, Z_A; \zeta, \Gamma^{(B,C)}, \Sigma^{(B,C)})$, and another two-thirds of the sample would be used in a regression that allows prediction using $P(\theta \mid X_1, X_2, X_3, Z_*, Z_B; \zeta, \Gamma^{(A,C)}, \Sigma^{(A,C)})$ and $P(\theta \mid X_1, X_2, X_3, Z_*, Z_C; \zeta, \Gamma^{(A,B)}, \Sigma^{(A,B)})$. In this case, each respondent would obtain two (potentially different) predictive distributions. A respondent who took Z_*, Z_B, Z_C would be assigned imputations based on the two different imputation models $P(\theta \mid X_1, X_2, X_3, Z_*, Z_B; \zeta, \Gamma^{(A,C)}, \Sigma^{(A,C)})$ and $P(\theta \mid X_1, X_2, X_3, Z_*, Z_C; \zeta, \Gamma^{(A,B)}, \Sigma^{(A,B)})$. Each of these imputation models is using two-thirds of the sample and would assume that conditional independence holds for θ and two out of the four sets of BQ covariates.

2. Prior to imputing θ, a procedure could be devised that allows completing the data for the covariates Z_*, Z_A, Z_B, Z_C. The model size and dependency on model assumptions (e.g., normality of the latent trait) will require some decisions to make the imputation problem tractable. For example,

   a. A model that allows us to specify a joint distribution $P(Z_*, Z_A, Z_B, Z_C \mid \omega)$ given certain distributional assumptions could be employed. This would ignore the cognitive responses X_1, X_2, X_3 and thus imply certain strong assumptions regarding the conditional independence of the imputed Z sets and X_1, X_2, X_3 given the other background variable (Z) sets. The question is whether this can be justified without the use of the cognitive data, which are typically significantly correlated with many of the background variables assessed in the Z sets of variables.

   b. Alternatively, a model that contains the cognitive data could be used as well, for example, in the form of a joint distribution $P(X_1, X_2, X_3, Z_*, Z_A, Z_B, Z_C \mid \omega, \rho)$. This is a model of quite impressive proportions. Many large-scale assessments use item pools of 120 items and more. PISA uses about 170–200 items in three dimensions—reading, science, and mathematics—so that $\theta = (\theta_R, \theta_M, \theta_S)$. In addition, the data are quite sparse, with only a fraction of the X variables observed, because sparse balanced incomplete block designs are employed that use 13 or more instead of only three test forms in our example. Once a decision has been made on which approach to take, it is possible to develop an imputation model for each of the missing variables. As an example, if variable set Z_A is missing, the imputation model

$$P(Z_A \mid Z_*, Z_B, Z_C, X_1, X_2, X_3; \omega, \rho)$$

   could be used to generate multiple draws for each respondent who was not given Z_A. The same rationale applies to the imputation of other sets of variables Z_B, Z_C, Z_*. While there is not a question about the ability to mechanically execute such a procedure, the question remains whether bias can be avoided in this type of mass imputation (Fetter 2001; Haslett et al. 2010).

The main question that emerges is which conditional independence assumptions can be justified, if any. More specifically, can the generation of data for public use files using imputation models be justified to be carried out using a predictive distribution that involves only a subset of the background variables? From the enumeration of cases above, it should be clear that it is not only the choice of variables that is limited when using incomplete rotated background questionnaires, but it is also the sample sizes available for the different imputation models that are limited, and the generality of the imputed values may be limited due to the noninclusion of potentially important predictors.

We have so far not discussed the case where certain background characteristics are collected only in conjunction with certain cognitive domains. As an example, assume that there are two cognitive domains θ_1 and θ_2 and each is assessed by presenting different sets of tasks, say X_{11}, X_{12} for θ_1 and X_{21}, X_{22} for θ_2. In addition, assume that there are specific covariates Z_1 and Z_2, each of which will be assessed only when the task sets for θ_1 and θ_2 are tested. In this case, Z_1, X_{11}, X_{12} and Z_2, X_{21}, X_{22} are jointly observed or missing. This means that the imputation model for θ_1 and θ_2, assuming both will be imputed for all respondents, would be based on a set of predictors that includes task responses and covariates for one, but not the other, skill variable. If only Z_1, X_{11}, X_{12} are assessed and θ_1 as well as θ_2 are both imputed, then we will base our imputation model on the assumption that θ_2 and Z_2, X_{21}, X_{22} are conditionally independent given only Z_1, X_{11}, X_{12}. More concretely, if θ_1 represents reading proficiency and θ_2 is math proficiency, we would end up imputing math skills having only indicators of reading skills X_{11}, X_{12} and reading interest and self-reported reading practices Z_1 without either math skill indicators X_{21}, X_{22} or self-reported interest or practices in math.

This may seem a somewhat far-fetched case, but it is one that illustrates the extent to which assumptions are being made about what can be left out of the imputation model if important context variables are collected in a rotated data collection design.

Implications

The following condensed list of implications can be compiled from the discussion of the literature on missing data and imputation, as well as the discussion of the imputation model necessary for reporting proficiency data by means of PVs:

1. We have shown above that conditional independence is not a safe bet if it was established on a limited set of variables. More specifically, if X and Z_A are conditionally independent given Z_*, that is

$$P(X \mid Z_*) = P(X, Z_A, Z_*)$$

this does not imply that X and Z_A are conditionally independent given Z_*, Z_B. That is, it may be that

$$P(X \mid Z_*, Z_B) \neq P(X \mid Z_A, Z_*, Z_B) \quad \text{while } P(X \mid Z_*) = P(X, Z_A, Z_*).$$

This suggests that Z_A should NOT be left out of an imputation model if, by adding more predictors, the conditional independence established using a limited set of predictors will be violated. This is the basis for the commonly used imputation rationale that all available variables should be included in the predictive model used to impute missing observations in X.

2. In addition we have shown that whether or not a rotated questionnaire design is used, the imputation of proficiency variables θ is necessary for 100% of the sample. The proficiency variables θ are latent variables, since they are not directly observed. In all large-scale assessments, these variables are derived by means of a complex imputation model (called conditioning model or latent regression model) that contains an item response theory model and a latent regression. In that sense, even without rotations of the background questionnaire, the model used in large-scale assessments depends on accurately collected covariates for a successful imputation of proficiency values. When rotated background questionnaire sections are introduced, this approach is based on an even sparser foundation of observed variables, and thus becomes more vulnerable for potential bias in the estimates.
3. In PISA, the sparseness is further increased by the use of imputations of proficiency variables for which no item responses have been collected. In this assessment, students may take only a subset of the observed item response sets X_1, X_2, X_3 where X_i is the observation base for θ_i. Each student may only take booklets from two (say X_1, X_2) out of the three sets of X variables, and for the missing X_3 variable it is assumed that θ_3 can be imputed by assuming conditional independence of θ_3 and X_3 given X_1, X_2 and the background variables Z_A, Z_*, Z_{cB}. If background variables are rotated, the set of the Z variables is also incomplete, making the observed data for the imputation more sparse.
4. Following Little and Rubin (2002) we have shown that a case where nonoverlapping rotations are used requires strong assumptions

with respect to conditional independence of two variable sets given a third. These strong assumptions cannot be tested unless all variables are jointly observed (e.g., in a separate sample), so they have to be made without empirical evidence if only rotated questionnaires are used. The rotation design that uses K rotation blocks and administers $K–M$ of these two for each student also makes conditional independence assumptions when attempting to build an imputation model. More specifically, without a complete data matrix, the M blocks left out have to be assumed to be conditionally independent given the observed responses X and the $K–M$ blocks of background data.

5. A design that rotates test questions is already applied in the cognitive part. The X_1, X_2, X_3 variable sets are not completely observed and observations are collected in blocks of test items that are combined in a way that maximizes the association with the proficiency of interest across blocks. That is, each of the blocks within X_1 is targeted to assess the same underlying variable θ_1. Put another way, the blocks of test questions defined based on X_1 are designed so that each maximizes the correlation with the proficiency of interest, θ_1. The same holds for blocks based on variable sets X_2, X_3 that are designed to maximize statistical association with θ_2, θ_3. This follows the same rationale as pointed out in the discussion of conditional independence and is compatible with the common assumption made in IRT models. The idea is basically the same as the one put forward in Rässler et al. (2002): Define rotated sections of variables that minimize correlations within sections, while maximizing correlations across sections. Rässler et al. produce variable sets in which essentially each is a replicate of the other by ensuring that each variable Q gets a *sister variable* Q' in another rotated block that is highly associated with Q. Along the same lines, Graham et al. (1996, 2006) report that distributing response variables that belong to the scale across rotated sections "was clearly better from a statistical standpoint." While Graham (2006) amends this statement by saying that "the strategy that is better in a statistical sense turns out to be a big headache in a logistical sense: For every analysis, one must use the best FIML or multiple imputation procedures," he concludes that a recommendation to avoid this headache by keeping scales intact would "change as the missing data analysis procedures advance to the point of being able to handle large numbers of variables." In our view, this stage is reached by advances in computational statistics, latent variable modeling, and imputation methods. A split of variables belonging to the same scale across rotation sections would be possible in background questionnaire rotations as well, but it would require not relying on a file matching (or the relaxed versions with

multiple sets per respondent) design where different sets of variables end up in different rotations. Quite the contrary, the design suggestion put forward by Rässler et al. (2002) would discourage such an approach and would rather distribute indicators of the same underlying construct into different rotations, just as it is customary in the design of the cognitive part of the assessment since 2000 in PISA, 1995 in TIMSS, and 1983 or earlier in NAEP. These types of rotated designs can be handled by general latent variable models (e.g., Moustaki 2003; Rijmen et al. 2003; von Davier 2005, 2009).

Based on these implications it appears that rotations that violate the imputation rationale or are versions of the file-matching missing data design are not suitable data collection designs for the necessary imputation of proficiency variables. A design that balances the rotated sections in a way that each section contains indicators of the same underlying variables seems more in line with the need to minimize the use of un-testable conditional independence assumptions since it would, rather than kicking variables out of the imputation model, retain all variables by means of measurement of these variables with somewhat different sets of indicators in each rotation.

Summary

We hope to have provided enough evidence that a discussion is needed on how much we are willing to base our inferences on untested conditional independence assumptions. What are the unintended consequences if a faulty assumption was made with respect to conditional independence that leads to ignoring certain background variables in 50% or more of the imputed θ values? Recall that 100% of all proficiency values are imputed, typically with the use of item responses (cognitive indicators) X. Note, however, that these indicators sometimes may not be available as illustrated in the above example. In PISA, for example, some of the core domains of reading, mathematics, and science are not observed for all students, that is, some respondents may receive booklets that assess mathematics and science, but not reading. If imputations on all three domains are desired, the imputation of reading proficiency would be based on math and science item responses, and on the background data, of which both sets may be incomplete. This puts an even heavier burden on the imputation model (whichever is chosen) when some of the proficiencies are not only indirectly observed, but not observed at all and rather imputed by proxy through other correlated assessment domains and incomplete background data.

We have shown in this paper that the imputation of proficiencies in order to generate proficiency distributions in large-scale assessments relies heavily on the availability of reliable item response data and background questionnaire data that are suitable for inclusion in a prediction model. Proficiency estimation in large-scale educational assessments use a complex imputation model based on an extension of IRT methods by means of a latent regression model, in which typically as many variables as possible are used for imputation in order to avoid potentially faulty conditional independence assumptions.

It was also shown by example that incomplete representation of the available data in the prediction model may implicitly make the false assumption of conditional independence, while conditional dependency may emerge once more when conditioning variables are included in the imputation model. If some variables are indeed conditionally independent of the proficiencies of interest, these could be left out of the imputation model, given the remainder of the background data. However, if that is the case and some of the background variables are indeed conditionally independent of the proficiency of interest, why should they be collected at all in an assessment that is focused on reporting distributions of proficiency and the relationships of proficiencies with policy-relevant background variables?

In addition, various options of defining an imputation model under incomplete background data and their underlying conditional independence assumptions were discussed. It was explained how either sample size would be decreased when using certain models, or how different imputation models with potentially conflicting predictions and assumptions would be obtained if respondents were pooled to augment and background data that were jointly observed in samples used to define the prediction model for imputation.

Also, prior research (ACER PISA document TAG(1204)6b) has shown that correlations between imputed values and background variables are reduced when incomplete background data are used. Jakubowski (2011) has presented results that show increased uncertainty under imputation with incomplete background data. Note that none of these reports provide strong evidence in favor of a general recommendation with respect to rotated designs. Both papers show the implications of a rotated design in a simulation based on real data from only a single cycle of PISA, and both studies show that associations between outcome variables and background variables are reduced under the rotated design, and/or that the uncertainty of comparisons is increased when introducing rotated designs.

At the very least, rotation schemes should use a similar design based on assessing constructs with incomplete representation of observables, as is already in place in the cognitive assessment (for a questionnaire design procedure that follows many of the design principles applied in cognitive parts of large-scale assessments, see Rässler et al. 2002). Leaving constructs completely out of the design in some rotations, as in the relaxed file matching design, should be avoided because strong and often untestable assumptions have to be made (Little and Rubin 2002), and different rotations would then include the same constructs, but be assessed with different subsets of observed response data.

Finally, some results obtained in studies on mass imputation for survey data indicate that, depending on the imputation model and the selection of variables, correlations of variables that involve imputations can be biased upward, downward, or almost unaffected (Fetter 2001). This leaves the question of whether any decision rule can be devised with respect to which imputation model to use and what variables to include in cases where substantial amounts of missingness are present due to incomplete collection designs.

Because the imputations from large-scale assessments enter the public use files as one of the main data products in the form of PVs required for online reporting tools and secondary analysis, we believe that a careful and diligent reevaluation of the use of incomplete background data collections is in order. The complexity of the models used to generate these data products, and the sparseness of the item response data collected, warrants more thorough examination of the risks involved when using incomplete background data for mass imputation of proficiency data.

References

Adams, R. J., Wu, M. L., and Carstensen, C. H. 2007. Application of multivariate Rasch models in international large scale educational assessment. In M. von Davier and C. H. Carstensen (Eds.) *Multivariate and Mixture Distribution Rasch Models: Extensions and Applications* (pp. 271–280). New York, NY: Springer-Verlag.

Bock, R. D. and Zimowski, M. F. 1997. Multiple group IRT. In W. J. van der Linden and R. K. Hambleton (Eds.), *Handbook of Modern Item Response Theory*. New York, NY: Springer-Verlag, 433–448.

Collins, L. M., Schafer, J. L., and Kam, C. M. 2001. A comparison of inclusive and restrictive strategies in modern missing data procedures. *Psychological Methods*, 6, 330–351.

Fetter, M. 2001. Mass imputation of agricultural economic data missing by design: A simulation study of two regression based techniques. *Federal Conference on Survey Methodology*. http://www.fcsm.gov/01papers/Fetter.pdf, downloaded 07/23/2012.

Glas, C. A. W. and Pimentel, J. 2008. Modeling nonignorable missing data in speeded tests. *Educational and Psychological Measurement*, 68, 907–922.

Graham, J. W., Hofer, S. M., and MacKinnon, D. P. 1996. Maximizing the usefulness of data obtained with planned missing value patterns: An application of maximum likelihood procedures. *Multivariate Behavioral Research*, 31, 197–218.

Graham, J. W., Taylor, B. J., Olchowski, A. E., and Cumsille, P. E. 2006. Planned missing data designs in psychological research. *Psychological Methods*, 11(4), 323–343.

Greene, W. H. 2003. *Econometric Analysis*. 5th ed. New Jersey: Prentice-Hall.

Haberman, S. J. 1974. *The Analysis of Frequency Data*. Chicago: University of Chicago Press.

Haberman, S. J., von Davier, M., and Lee, Y. 2008. Comparison of multidimensional item response models: Multivariate normal ability distributions versus multivariate polytomous ability distributions. RR-08-45. ETS Research Report.

Hanushek, Eric A. and John E. Jackson. 1977. *Statistical Methods for Social Scientists*. New York: Academic Press, Inc.

Haslett, S. J., Jones, G., Noble, A. D., and Ballas, D. 2010. More for less? Comparing small-area estimation, spatial microsimulation, and mass imputation. Paper presented at JSM 2010.

Hemker, B. T., Sijtsma, K., Molenaar, I. W., and Junker, B. W. 1996. Polytomous IRT models and monotone likelihood ratio of the total score. *Psychometrika*, 61, 679–693.

Jakubowski, M. 2011. Implications of the student background questionnaire rotation on secondary analysis of PISA data. Unpublished manuscript.

Jones, M. P. 1996. Indicator and stratification methods for missing explanatory variables in multiple linear regression. *J Am Stat Assn*. 1996; 91:222–230.

Lazarsfeld, P. F. and Henry, N. W. 1968. *Latent Structure Analysis*. Boston: Houghton Mifflin.

Little, R. J. A. 1988. Missing-data adjustments in large surveys. *Journal of Business & Economic Statistics*, 63, 287–296.

Little, R. J. A. and Rubin, D. B. 2002. *Statistical Analysis With Missing Data* (2nd edition). New York, NY: Wiley.

Lord, F. M. and Novick, M. R. 1968. *Statistical Theories of Mental Test Scores*. Reading, MA: Addison-Wesley.

Mislevy, R. J. 1991. Randomization-based inference about latent variables from complex samples. *Psychometrika*, 56, 177–196.

Moustaki, I. and Knott, M. 2000. Weighting for item non-response in attitude scales using latent variable models with covariates. *Journal of the Royal Statistical Society, Series A*, 163(3), 445–459.

Moustaki, I. 2003. A general class of latent variable models for ordinal manifest variables with covariate effects on the manifest and latent variables. *British Journal of Mathematical and Statistical Psychology*, 56(2), 337–357.

Muraki, E. 1992. A generalized partial credit model: Application of an EM algorithm. *Applied Psychological Measurement*, 16, 159–176.

Piesse, A. and Kalton, G. 2009. A Strategy for Handling Missing Data in the Longitudinal Study of Young People in England (LSYPE), Westat Research Report No. DCSF-RW086.

PISA TAG(1204)6b. 2012. On the use of rotated context questionnaires in conjunction with multilevel item response models.

Rässler, S., Koller, F., and Mäenpää, C. 2002. A Split Questionnaire Survey Design applied to German Media and Consumer Surveys. *Proceedings of the International Conference on Improving Surveys, ICIS 2002*, Copenhagen.

Rijmen, F., Tuerlinckx, F., De Boeck, P., and Kuppens, P. 2003. A nonlinear mixed model framework for item response theory. *Psychological Methods*, 8, 185–205.

Rose, N., von Davier, M., and Xu, X. 2010. *Modeling Non-Ignorable Missing Data with IRT*. ETS-RR-10-10. Princeton: ETS Research Report Series.

Rubin, D. B. 1987. *Multiple Imputation for Nonresponse in Surveys*. New York, NY: John Wiley & Sons.

Rubin, D. B. 1996. Multiple imputation after 18+ years. *Journal of the American Statistical Association*, 91, 473–489.

Rutkowski, L. 2011. The impact of missing background data on subpopulation estimation. *Journal of Educational Measurement*, 48(3), 293–312.

Suppes, P. and Zanotti, M. 1995. *Foundations of Probability with Applications: Selected Papers*, pp. 1974–1995. Cambridge: Cambridge University Press.

van Buuren, S. and Groothuis-Oudshoorn, K. 2011. MICE: Multivariate Imputation by Chained equations in R. *Journal of Statistical Software*, 45(3), 1–67.

Vermunt, J. K., Van Ginkel, J. R., Van der Ark, L. A., and Sijtsma, K. 2008. Multiple imputation of categorical data using latent class analysis. *Sociological Methodology*, 33, 369–297.

von Davier, M. 2005. A general diagnostic model applied to language testing data (ETS Research Report No. RR-05-16). Princeton, NJ: ETS.

von Davier, M. 2009. Mixture distribution item response theory, latent class analysis, and diagnostic mixture models. In: S. Embretson (Ed.), *Measuring Psychological Constructs: Advances in Model-Based Approaches*. ISBN: 978-1-4338-0691-9. pp. 11–34. Washington, DC: APA Press.

von Davier, M., Sinharay, S., Oranje, A., and Beaton, A. 2006 Statistical procedures used in the National Assessment of Educational Progress (NAEP): Recent developments and future directions. In C. R. Rao and S. Sinharay (Eds.), *Handbook of Statistics (vol. 26): Psychometrics*. Amsterdam: Elsevier.

von Davier, M., Gonzalez, E., and Mislevy, R. 2009. What are plausible values and why are they useful? In M. von Davier and D. Hastedt (Eds.), *IERI Monograph Series: Issues and Methodologies in Large Scale Assessments (Vol. 2)*. IEA-ETS Research Institute.

Xu, X. and von Davier, M. 2008. Comparing multiple-group multinomial loglinear models for multidimensional skill distributions in the general diagnostic model. RR-08-35, ETS Research Report.

Zermelo, E. 1929. Die Berechnung der Turnier-Ergebnisse als ein Maximumproblem der Wahrscheinlichkeitsrechnung. (The calculation of tournament results as a maximum problem of probability calculus: in German). *Mathematische Zeitschrift*, 29, 436–460.

9

Population Model Size, Bias, and Variance in Educational Survey Assessments

Andreas Oranje
Educational Testing Service

Lei Ye
Educational Testing Service

CONTENTS

Introduction

Standard methodologies in typical testing applications involve calculating a score for each assessed individual with respect to some domain of interest. In situations where the assessment booklets administered to each test-taker contain large numbers of items and each test form is constructed to be psychometrically parallel, standard analysis approaches work reasonably well. Similar to most educational surveys (e.g., Programme for International Student Assessment [PISA], Trends in International Mathematics and Science Study [TIMSS], and Progress in International Reading Literacy Study [PIRLS]), the National Assessment of Educational Progress (NAEP), a U.S. large-scale educational survey assessment conducted in Grades 4, 8, and 12 in subjects such as reading, mathematics, and science, has a key design characteristic that requires a nonstandard methodology. Each student answers only a systematic, small portion of the cognitive item pool, ensuring that all students combined provide an approximately equal number of responses to each item in the item pool and, under the most common design, that any two items appear together in at least one form. As a result, students cannot be compared to each other directly and reliable individual proficiency estimates cannot be obtained. However, consistent estimates of proficiency (e.g., averages) and the dispersion of proficiency (e.g., variances) in various reporting groups of interest (Mazzeo et al. 2006) can be estimated by employing a latent regression model (Mislevy 1984, 1985).

The basic model for the univariate case (i.e., a single-proficiency variable) can be depicted by the following regression equation

$$\theta = x'\gamma + \varepsilon$$

where θ is proficiency, which is not directly observed (i.e., latent), x is a vector representing the population model and contains several variables specified under 1.2, γ is a vector of regression weights, and ε is a residual term, which is assumed to be distributed normal with a mean equal to 0 and variance equal to σ_W^2. The W subscript denotes (average) within-student residual variance. A between-student variance component can be defined as $\sigma_B^2 = \sigma_T^2 - \sigma_W^2$, denoting the total proficiency variance as σ_T^2. As mentioned, proficiency is not observed directly, but is assumed to be defined through an item response model. Particularly, a combination of the three-parameter logistic (3PL) model and the generalized partial credit model (GPCM; e.g., van der Linden and Hambleton, 1996) is used for multiple choice and constructed response items, respectively. Under the 3PL model, the probability of answering a dichotomous item correctly for given proficiency is defined as

$$P(y_{ij} = 1 | \theta_i) = 1 - P(y_{ij} = 0 | \theta_i) = c_j + (1 - c_j)\frac{e^{Da_j(\theta_i - b_j)}}{1 + e^{Da_j(\theta_i - b_j)}}$$

where y_{ij} is the item response for student i to item j assuming values 1 for correct and 0 for incorrect. Also, a number of IRT item parameters are defined, including pseudo-guessing (c_j), discrimination (a_j), difficulty (b_j), and a constant scaling parameter (D), which, in combination and across items, will be referred to as β. We will not further define the GPCM. The probability of answering any two items correctly is the product of the probability of answering each item correctly (i.e., local item independence). In addition, test-takers are considered independently observed. This can be expressed as an item response likelihood

$$P(\mathbf{Y} | \boldsymbol{\theta}) = \prod_i P(\mathbf{Y} = \mathbf{y}_i | \theta_i) = \prod_i \prod_j P(Y = y_{ij} | \theta_i)$$

Combining both the latent regression ("the population model") and the item response likelihood ("the measurement model"), the conditional probability of observing a particular response pattern, given IRT item parameters, regression coefficients, residual variance, and proficiency, can be defined as

$$P\left(\mathbf{Y}|\,\boldsymbol{\theta},\beta,X,\gamma,\sigma_W^2\right)=P\left(\mathbf{Y}|\,\boldsymbol{\theta},\beta\right)\phi\left(\boldsymbol{\theta}|\,X,\gamma,\sigma_W^2\right)$$

where ϕ denotes a normal distribution with mean $X'\gamma$ and variance σ_W^2. The log of this probability can be marginalized and derivatives with respect to the parameters of interest can be developed for estimation. These equations can be easily extended to the multivariate case.

NAEP's population model, specifying student groups of interest, has grown from a small model with relatively few variables (5–10) to a very large model with many variables (more than 200) and dozens of interactions between key reporting variables. The premise of these large models is that unbiased group estimates can only be obtained for a particular variable if that variable is included in the population model (Mislevy and Sheehan 1987; Mislevy et al. 1992b) and that the program is responsible for publishing results for a large number of groups. Biased group estimates that stem from (partially) omitting group indicators from the model are generally referred to as secondary biases.

The *best* approach to specifying and estimating population models continues to be the subject of debate and can be characterized by two divergent views about what such a model should look like. At one end is a single comprehensive model that encompasses the full range of variables in which users are likely to be interested (Mislevy 1984, 1985; Mislevy et al. 1992a) under the philosophy that results for all those variables are readily available. At the other end are one-variable-at-a-time models (Cohen and Jiang 1999, 2002), under the philosophy that users themselves will only generate those models in which they are interested. The first approach has the advantage that a single consistent set of results is provided that caters to a large audience (Mazzeo et al. 2006) and the disadvantage is that secondary biases may occur for variables and interactions not included in the model (Thomas 2002). Reversely, the second approach is nominally not subject to secondary biases, but the inconsistency of results across models might be quite dramatic.

Previous applied research in NAEP has yielded several findings that are relevant to this debate:

- Overall NAEP proficiency distributions appear to be predominantly skewed (left), and a larger model, being a mixture of many normal distributions, is more apt to accommodating such a skew (Moran and Dresher 2007). There appeared to be an optimal ratio of observations (i.e., students) to variables of about 11 for the data used in that study, where optimal is defined as the minimum bias the results under those models carry.
- NAEP data are hierarchical by nature (students within schools) and a relatively large model can serve as a fixed-effect hierarchical model to represent hierarchies accurately and, thereby, improving the accuracy

of variance estimates. Alternatively, a model with parameters that explicitly define those hierarchical relations (Li and Oranje 2006, 2007) can be used, but will be more complicated to estimate.

In summary, there are good reasons to use a larger model as long as it is not too large, at which point the cost of an inflated variance may outweigh the benefit of unbiased results. However, what the model should look like and how large is "large" is not exactly clear. Mazzeo et al. (1992) studied secondary biases for two model-selection and dimensionality-reduction approaches. In both cases, principal components were computed from the background variables. However, under the first approach, simply the minimum number of components was taken that explains a certain percentage of variance, while under the second, a *logit* transformation of an observed percent-correct score of cognitive items was computed and an R^2 criterion was used to select the number of components. The results were similar across the two methods.

Assume, at least for the foreseeable future, that NAEP continues to be committed to providing a single, consistent set of results based on a single large model that can be used by secondary analysts to verify the reported results and investigate a variety of policy questions. The question is then how the public (i.e., both consumers of reports and publicly accessible analysis tools and secondary analysts) is best served in terms of their ability to acquire accurate results and to make effective inferences. This is foremost a policy issue: which variables are most important and to what extent can we trade accuracy with variability, and vice versa? For example, suppose the true average for males in a particular state is 231. Would we rather have an estimate of 233 and a standard error of 2.2, or an estimate of 232 with a standard error of 3.9? In both cases, the true average is within the estimated confidence bounds. The first estimate is more biased, but also provides more power for comparisons. The second estimate is less biased, but has far wider confidence bounds, which means that comparisons with other groups are more likely to be nonsignificant. This example is depicted in Figure 9.1.

This example captures exactly the trade-off of model size that will be at the core of the studies described in this chapter. The studies presented in the ensuing sections address the key questions of how and to what extent the NAEP population models should be reduced. The "how" is not only a question of what variable selection and/or dimensionality-reduction technique should be applied, but also what the consequences of a smaller model are and how adverse consequences should be mitigated. "To what extent" is not only a question of how much reduction should occur, but also how this affects the reporting goals of the program.

NAEP collects many types of variables, including

- Demographics, for example, gender, race/ethnicity, and parental education
- General student factors, for example, number of books in the home

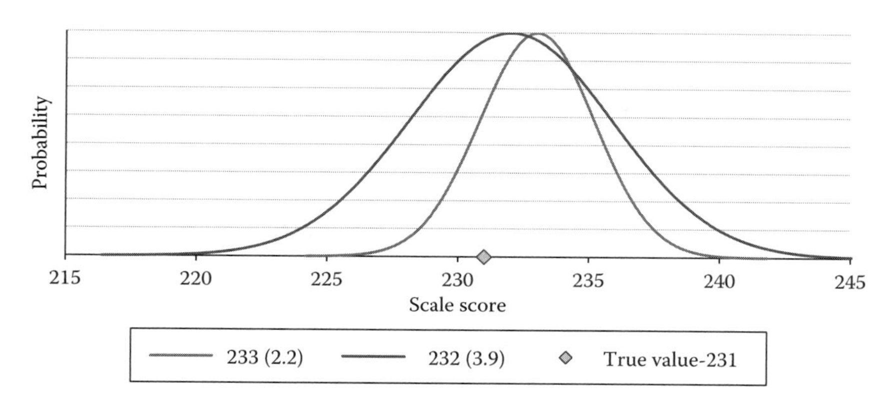

FIGURE 9.1
Sampling distributions for two estimators.

- Subject-specific student factors, for example, number of nonfiction books read for pleasure
- School factors, for example, public or private, and location
- Teacher factors, for example, professional training

In total, several hundred of these variables are collected from school records and through several questionnaires administered to participating students, their teachers, and their school officials. These variables are used to define student groups for whom results are disaggregated in reports and on the web (e.g., www.nces.ed.gov/nationsreportcard/naepdata). In addition, this online data analysis tool allows the user to make three-way comparisons (e.g., the difference between males and females who are living in an urban area and who do not receive free or reduced-price lunch). To minimize secondary biases, all these variables are included in the marginal estimation model in some form, as well as two-way interactions between major reporting variables. Figure 9.2 provides a schematic of the process from collected variables to variables that are used for marginal estimation.

The following steps are taken in the current operational process:

1. All the variables from the various sources are dummy-coded. Furthermore, dummy-code two-way interactions are created between the main reporting variables (e.g., gender by race/ethnicity and race/ethnicity by parental education). This results in a very large matrix (up to 2000 variables for each student) with many multicollinearities.
2. Two procedures are applied to reduce the dimensionality:
   a. First, multicollinearities between variables are removed by use of the SWEEP algorithm (Goodnight 1978), based on a correlation matrix of the (dummy-coded) variables and interactions. Those

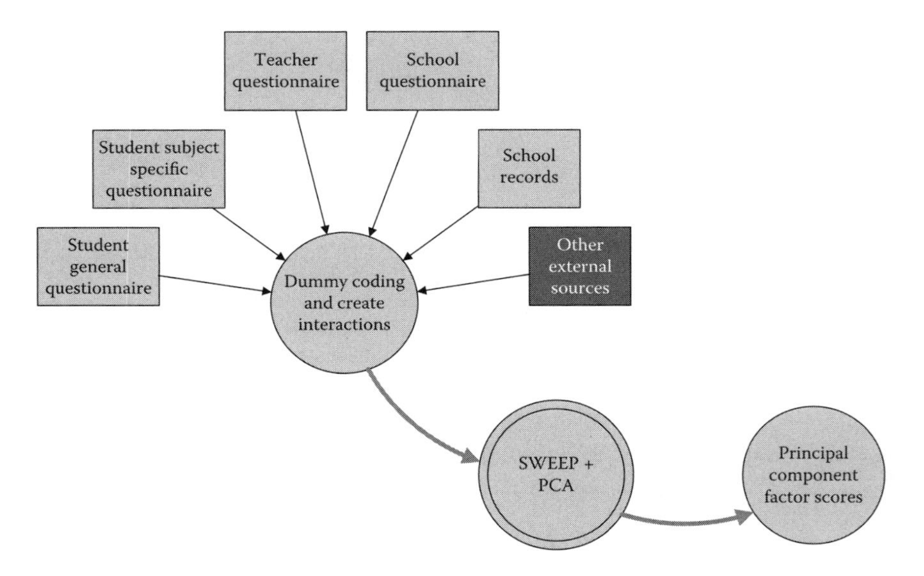

FIGURE 9.2
Schematic of current operational procedures.

contrasts that do not have a substantial residual variance relative to all other group defining variables are removed.

b. Second, principal components are extracted based on the remaining contrasts. The smallest set of components that explains 90% of the variance is retained for further analysis. Subsequently, principal component scores are derived and used as independent variables in the marginal estimation model.

Implicit to this strategy is the notion that there is nothing that guarantees that any of the student grouping variables is fully represented in the population model. In practice, representation is verified for all contrasts. Nevertheless, secondary biases cannot be entirely avoided and should be expected, at least to some extent. This goes back to the debate between tailored single-variable and comprehensive one-size-fits-all models, where the former model could not be prone to secondary biases and the latter could be, depending on the level of dimension reduction that is required. In the past, a practical reason for dimensionality reduction has been computational tractability, which is at present not a major concern, as substantial computing power is easily available. However, even near-multicollinearities have proven to be problematic in the estimation of NAEP's latent regression model. In other words, while an algorithm such as the SWEEP algorithm and a pivot criterion of .001 seems reasonably capable of detecting multicollinearities in most applications, for the latent regression, this does not seem sufficiently robust. The additional dimension reduction via principal components becomes necessary to estimate group effects.

The number and type of students, the number and type of background variables, and the number and type of cognitive items each affect the estimation of student group effects. While the focus of this research is primarily on background variables, it is important to view the findings in a larger context. Necessarily, sample size and the items are factors or conditions in this research.

The interaction of these three factors is quite complex, but by keeping two of the factors constant, one can hypothesize what is likely to happen when one factor is manipulated. Besides producing correct results for student groups (i.e., low bias), we are in this context mostly interested in

1. The total variance (σ_T^2) of the overall proficiency distribution: in addition to average estimates, the program also reports on achievement levels, hence, the importance of a correct distribution of proficiency for student groups. The total variance can be disaggregated as follows:
   a. Between-student variance reflected by the population model effect parameters (σ_B^2).
   b. Between- and within-student variance not reflected by the population model (σ_W^2), also referred to as the residual variance.
2. The average squared regression error (ε^2): this is an important component of the measurement standard error, reflecting how well the population model explains the differences in individual student proficiencies.
3. The average student posterior variance (σ_i^2): this is the other important component of the measurement standard error and reflects the confidence of student proficiency estimates given available cognitive information about that student and the group(s) to which that student belongs.

It should be noted that we simplify "sample size" considerably by using "effective sample size," which can be interpreted as the comparable simple-random-sample sample size that is available after between-unit dependencies are accounted for. NAEP is a survey conducted in a hierarchical sample and, therefore, variability of estimates is not linearly related to sample size. A recent evaluation of several standard error estimators in the NAEP context can be found in Oranje et al. (2009). Also note that the residual variance (σ_W^2) and the average regression error (ε^2) are, in fact, not independent statistics and presenting them as such here is another simplification.

There are at least two related compensatory processes at work: one at the group distribution level and one at the student level.

1. As the population model increases, more between-student variance is explained by the model and less is residual. As a result, the

between- and within-variance components *trade* variance, but the overall standard deviation of the proficiency distribution remains the same. Similarly, as the number of students increases, more complex between-student effects exist and, therefore, the model will be less able to explain these. As a result, the within variance increases and the between variance decreases.

2. As the number of cognitive items per student increases, the posterior distribution for a student's proficiency becomes tighter. However, the prior model will be given less weight in predicting student group effects and, therefore, the average regression error will increase. The posterior variance and the regression error *trade* variance, but the overall combined error does not change.

The above makes one important assumption, which is that the estimation of these statistics is unbiased. This is not necessarily the case in practice. For example, with few cognitive items, the item response likelihood *may* be so erratic that neither of the statistics can be accurately determined and results from the model cannot be trusted. Similarly, a highly saturated population model (where every participating student becomes essentially a group by him- or herself) leads to model indeterminacy. What exactly constitutes a saturated model is debatable and depends not only on the number of students (relative to the number of model parameters), but also the number of unique cognitive item response patterns.

The goals of this study are to investigate how alternative model selection choices within the NAEP marginal estimation methodology affect bias and variance estimates of student group means. The results are intended to provide information about the current state of affairs in NAEP model selection and guidance on the extent to which alternative approaches may better serve the reporting goals of the program.

Method

Under the first task, alternative methods to select a model have been examined within the current operational framework of model selection. Currently, a principal component analysis (PCA) is conducted based on Pearson correlations between dummy-coded background variables. The variations on the current methodology that will be evaluated are to use main reporting variables directly and to perform a conditional PCA on the rest of the variables. The goal is to make sure that the variance of key reporting variables is fully accounted for. A second variation is to use a covariance instead of a correlation matrix to perform the PCA. The goal here is to make sure that highly variable variables are proportionally better represented.

The general topic of model size is of course a well-known statistical issue, and analytic results from past research may prove helpful in this context. Several consistency and asymptotic normality results are available (Haberman 1977a, 1977b) for exponential response models commonly utilized in NAEP. For the general linear model, results from Portnoy (1984, 1988) indicate that $q^2/N \rightarrow 0$ is required for asymptotic normality of $\hat{\gamma}$, where q is the number of parameters and $\hat{\gamma}$ the vector of regression weights. Relating this to NAEP is not straightforward because the dependent variable is latent, where at different stages different parameters are considered known. Therefore, it is not necessarily clear how large q is. However, the important perspective that this result offers is that the penalty incurred by adding variables is quadratic relative to sample size.

Manipulations, Data, and Evaluation

The following manipulations were applied:

- Similar to operational procedures, extract principal components from a correlation matrix and use those as the population model of the latent regression. Take, respectively, the minimum set of principal components that explain 90%, 70%, and 50% of the variance, respectively. We will call these **R90**, **R70**, and **R50**.
- Force in main reporting variables (e.g., gender, race/ethnicity, parental education) as is, compute a residualized correlation matrix with respect to those main reporting variables, and extract principal components on that residualized matrix. Take, respectively, the minimum set of principal components that explain 90%, 70%, and 50% of the variance. We will call these **M90**, **M70**, and **M50**. The rationale behind this approach is simply that the main reporting variables are the most important and those should be included with certainty.
- Extract principal components from a covariance matrix and use those as the population model of the latent regression. Take, respectively, 90% and 50%. We will call these **S90** and **S50**. The rationale behind this approach, relative to a correlation-based approach, is that groups that carry more variability should be relatively better represented in the principal components.

NAEP data from the 2007 mathematics and reading Grade 4 and 8 and the 2005 Grade 12 assessments were used in this study. A number of jurisdictions across grades were selected, representing geographically diverse, large and small states. Table 9.1 provides a listing of the states, grades, and subjects that were used in this study. The jurisdictions in Grade 4 have at least 3000 participating students, while the jurisdictions in Grade 8 have at least 2500.

TABLE 9.1

Data Sets

Grade	Year	Subject	Jurisdictions
4	2007	Reading	California, North Carolina, Illinois, Delaware, Louisiana, Connecticut, Washington, DC
		Mathematics	Texas, South Carolina, Ohio, Nevada, Oregon, Massachusetts, Delaware
8	2007	Reading	Texas, South Carolina, Ohio, Nevada, Oregon, Massachusetts, Delaware
		Mathematics	California, North Carolina, Illinois, Delaware, Louisiana, Connecticut, Washington, DC
12	2005	Reading	National
		Mathematics	National

Some jurisdictions have larger samples because of two reasons: (a) results for several major urban districts (e.g., Boston, Massachusetts) were provided in 2007 based on an additional sample of about 1000–1500 for each district, and (b) large states were oversampled (i.e., California and Texas) at twice the usual rate, in addition to large urban districts that may have participated in those states as well. Note that in 2007, only national samples were assessed in Grade 12, which have a sample size of around 10,000 students each.

In Grade 4 reading, two subscales are recognized, while three subscales are recognized in Grades 8 and 12. In mathematics, five subscales are recognized in Grades 4 and 8, and four in Grade 12. Subsequently, a multivariate version of the model described above is employed. The multivariate integrations in the expectation step are carried out via a Laplace approximation, which is described in detail in Thomas (1993).

While real datasets were used for this study, the evaluation is most useful if there is some notion of truth involved, which usually constitutes some form of simulation. Yet, the study is meant to be useful to the NAEP practice, and given the very complex population at hand, simulation quickly reaches a limit in terms of generalizability. Therefore, NAEP data was used, but responses to cognitive items were simulated. The proficiency results from the NAEP operational analyses could have been taken as "truth" in this simulation. However, since those results are in part a product of a specific dimension reduction (PCA based on a correlation matrix and by retaining 90% of the variance), and because it was important not to set up the simulation in a way that could favor one of the conditions, the following procedure was followed to simulate the data and to *estimate* truth:

1. A relatively noncollinear set of contrasts was obtained by reducing the complete set of available variables and operationally specified interactions, by the SWEEP algorithm and a pivot criterion of .01 (10 times the operational criterion). As mentioned in the "Introduction"

section, the usual operational criterion (of .001, based on rules of thumb derived from past practice) is not sufficient to remove multicollinearities to the extent that the operational marginal maximum likelihood functions. Therefore, a substantially more stringent criterion was used. A number of trial-and-error runs were conducted to arrive at this criterion.

2. The latent regression model was estimated assuming operational item parameters known.
3. The proficiency results from the previous step were taken as "truth" (i.e., using the first plausible value, see Mislevy (1993) for details on plausible values) and item responses were generated. To mimic the real data as well as possible, only those responses that students provided were simulated, while omit codes were carried forward as is from the original data.
4. Each of the eight manipulations described above were applied, and student group effects were estimated. Student group effects were also estimated for two comparison models:
   a. The complete model with all contrasts (after SWEEP) for use as the baseline model in comparisons. This will be called the B or baseline model.
   b. A relatively small model with only the contrasts (after SWEEP) of the variables that are used to make comparisons (see Table 9.3). This will be called the F or focal group model.

These steps were followed once for each of the 30 jurisdictions. The proficiency estimates that are the basis for the third step of course could have been used as baseline model. However, because only one iteration is carried out at this point, a comparison model that was consistent with this one instance of the model was preferred. For one jurisdiction, multiple iterations were carried out to provide an initial check on the usefulness of a single iteration.

As is clear from step 4 above, two comparison models were used: **B** and **F**. While the **B** model, in theory, has no biases for all noncollinear variables, since all those variables were entered into the model, one of the key motivations of this study was concern about the size of the model. Therefore, an alternative comparison model was used that only includes those noncollinear contrasts that describe the groups in Table 9.3, which were specifically used for the evaluation in this study.

For the first step, a pivot criterion of .01 was used. However, for a number jurisdictions, this was not sufficient to obtain a noncollinear set of contrasts that yielded an estimable model in the second step. Therefore, the criterion was increased to .02 and .03 for some jurisdictions. As can be expected, these were generally the smallest states in terms of sample size yields. Table 9.2 provides the details on which criterion was used for each jurisdiction. Table 9.2 also shows the ratio between the number of contrasts in the

TABLE 9.2

Jurisdiction, Ratio, and Pivot Criterion Used

	Reading			Mathematics		
Grade	**Jurisdiction**	**Ratio[a]**	**Pivot Criterion Used**	**Jurisdiction**	**Ratio[a]**	**Pivot Criterion Used**
4	CA	1.80	0.01	DE	1.58	0.02
	CT	1.95	0.01	MA	1.85	0.01
	DE	1.95	0.01	NV	1.87	0.01
	DC	1.70	0.03	OH	1.97	0.01
	IL	1.86	0.01	OR	1.85	0.01
	LS	2.00	0.01	SC	1.55	0.02
	NC	1.68	0.01	TX	1.94	0.01
8	DE	1.86	0.01	CA	1.71	0.01
	MA	1.81	0.01	CT	1.28	0.03
	NV	1.45	0.02	DE	1.24	0.03
	OH	1.52	0.02	DC	1.49	0.02
	OR	1.43	0.02	IL	1.24	0.03
	SC	1.56	0.02	LS	1.34	0.03
	TX	1.81	0.01	NC	1.30	0.02
12	National	1.38	0.01	National	1.36	0.01

[a] Ratio of the number of contrasts used to apply the various manipulations to (i.e., the baseline model) and the number of principal components used in the operational models.

baseline model relative to the number of principal components that were used operationally. The fact that (a) a more stringent criterion was used and (b) the criterion varied across jurisdictions has some implications for the generalizability of the results to operational situations. This will be further discussed in the results.

Central to the evaluation is the bias-variance trade-off. Bias will be computed with respect to the baseline model across a number of different students groups, including both main reporting variables, interactions, and variables collected through questionnaires. Table 9.3 provides the variables that were used. Note that not all the groups identified by these variables are present in the baseline model due to the application of the SWEEP criterion, and that this varies across jurisdictions. The evaluation is still based on all groups.

The following statistics are used in the comparison. The squared bias for a particular jurisdiction and grade is computed as

$$\text{Bias}^2 = \frac{1}{G}\sum_{g=1}^{G}\left(\hat{\mu}_g - \mu_g\right)^2 \tag{9.1}$$

for G groups, and where $\hat{\mu}_g$ is the estimate of the mean proficiency for group g based on the posterior distribution, and μ_g is the mean from the baseline

TABLE 9.3

Groups of Variables Used for Evaluation

Variable Group	Number of Groups Identified by These Variables	Variables (or Interactions)
Main variables	19	Total, gender, race/ethnicity, type of location, school lunch, student disability status, English language learner status
Interactions	48	Gender by race/ethnicity, race/ethnicity by student disability, race/ethnicity by English language learner, and type of location by school lunch
Background variables	39–50 (depending on subject and grade)	Varies by grade: three student background questions, two teacher background questions, and two questions from the school questionnaire

model. The variance for each subgroup in each jurisdiction and grade is computed as the sum of the sampling variance, estimated via a jackknife repeated replications (JRR) approach and measurement variance, estimated via a between-imputations variance approach. Both are described in Allen et al. (2001). A more general description of the JRR can be found in Wolter (1985). The JRR approach is taken because the NAEP sample is a complex sample (students are nested within schools, which are nested within jurisdictions). A discussion of the NAEP samples and variance estimators is well beyond the scope of this paper, and for a concise treatment, Rust and Johnson (1992) and Mislevy (1991) should be consulted. A recent evaluation of several sampling variance estimators in NAEP can be found in Oranje et al. (2008). Across groups, variance is estimated as

$$\text{Variance} = \frac{1}{G}\sum_{g=1}^{G} V\left(\hat{\mu}_g\right) \tag{9.2}$$

where $V(\hat{\mu}_g)$ is the variance associated with the mean proficiency estimate for group g. In combination, the mean squared error (MSE) can be computed as

$$\text{MSE} = \text{bias}^2 + \text{variance} \tag{9.3}$$

The bias, variance, and MSE will be computed both across all groups and for each of the three specified sets of groups separately. They will also be considered at the jurisdiction level and averaged across all jurisdictions. An indicator that will be used to qualify some of the results is the total sample-to-number-of-variables ratio, in an attempt to find some relation to the findings of Dresher (2006), suggesting that about 11 students per variable seemed to yield the least-biased results.

Results

Two types of results will be presented. First, the data and manipulations will be evaluated in terms of the extent they accomplish model reduction and how that compares to typical operational situations. The second type of results is simply the MSE-based comparisons.

This study is intended to be operationally relevant and, therefore, it is important to verify that, despite the increased pivot criterion, there are still enough contrasts in each dataset to reduce the model to something operational-like, at least for the conditions **R90** and **M90**, which are most similar to an operational model. As mentioned before, Table 9.2 shows the ratio of the number of initial contrasts (after SWEEP has been applied) and the number of principal components used operationally for each of the jurisdictions. Ideally, this ratio should be well above 1, which is the case for most jurisdictions, especially when a pivot of .01 is applied. A ratio close to or below 1 would indicate that the initial model is already about the size of or smaller than an operational model and that the results might compare poorly to operational situations.

Next, some of the estimation properties, such as convergence, residual variances, and subscale correlations were inspected, verifying that a larger model with more parameters takes more cycles to converge, has smaller residual variances, and, based on previous experience, has a larger conditional correlation (i.e., off-diagonal elements of the standardized residual variance matrix). However, the marginal correlations (i.e., the correlations between two subscales taking the model into account) would be expected to be largely constant irrespective of model size. In other words, increasing the model size (i.e., reducing the sample size per variable) is essentially a matter of trading within-student variance for between-student variance, without affecting the total proficiency variance or covariances between subscales of proficiency. Table 9.4 shows for each of the manipulations and—averaged over all the jurisdictions, the average number of students per population model variable (i.e., the parsimony of the model)—the average number of cycles, average residual variance, and average conditional and marginal correlation. The last two statistics are also averaged across pairs of subscales and all correlations are averaged via a Fisher transformation. All of these statistics are ordered relative to the number of students per variable (column 2).

It appears that convergence and both the marginal and conditional correlations behave as expected relative to model size, although the drop-off in marginal correlation for the baseline model is surprising. The pattern of residual variances relative to model size is relatively straightforward except for the focal model. It should be noted that because the main models **M**, **R**, and **S** are not nested, there is no guarantee that an ordered relationship is found.

TABLE 9.4

Convergence, Residual Variance, and Correlations

Manipulation	Average Sample Size per Variable	Number of Cycles	Average Residual Variance	Average Conditional Correlation	Average Marginal Correlation
Baseline	10.1	658	0.261	0.946	0.673
M90	18.0	426	0.399	0.836	0.793
R90	21.2	350	0.430	0.806	0.792
S90	24.4	313	0.468	0.779	0.779
M70	30.0	247	0.493	0.759	0.771
R70	40.6	196	0.532	0.729	0.783
M50	49.4	176	0.564	0.721	0.787
Focal	81.4	150	0.717	0.737	0.788
R50	86.5	145	0.616	0.709	0.767
S50	112.0	139	0.661	0.717	0.761

Finally, the effectiveness of the manipulations in terms of model reduction was inspected and the average of percentage model reduction across jurisdictions is provided in Table 9.5 alongside the operational results. The operational results are based on similar procedures as **R90** except that the pivot criterion is only .001, resulting in a much larger initial set of variables. Nevertheless, Table 9.5 shows that a relatively wide range of model sizes has been obtained and that the larger models are comparable in size to the average operational model. Note that both removing multicollinearities and extracting principal components (and taking a subset of those) are two procedures that accomplish something similar, which is to obtain a smaller set of variables that represent the initial set sufficiently. Therefore, starting with a very large set of contrasts, where most of those will not contribute much unique information and, therefore, a small proportion of components is retained,

TABLE 9.5

Average Percent Reduction of Dimensions across Jurisdictions for Each Condition

Condition	Initial Number of Variables	Final Number of Variables	Percent Reduction
Operational	827	269	67.5
R50	447	52	88.4
R70	447	111	75.2
R90	447	212	52.6
S50	447	40	91.0
S90	447	184	58.8
M50	447	91	79.6
M70	447	150	66.4
M90	447	250	44.1

is not much different from starting with a small, relatively noncollinear set, where a large proportion of components is retained.

For the main results, we will focus on the MSE. Figure 9.3 provides the MSE averaged across jurisdictions for each of the manipulations relative to the baseline and focal models. The MSEs are quite similar across the manipulations and appear to be well aligned with model size. Overall, **M90** provides the best bias-variance trade-off, having the lowest, albeit marginally, overall MSE. It is important to note that the bias in the **M** models is not zero for the main reporting variables, which is against expectation. There are two reasons for this. The first is that results for both the baseline model and each of the manipulations are based on estimates that are obtained through imputation (Mislevy 1991). Therefore, some random variation is inherent to this process. Second, some main reporting groups were not included in the model due to a relatively stringent SWEEP criterion, but are part of the set of main reporting variables in the comparison. While any comparison based on less than 62 students, following NAEP reporting standards, was removed, there are still a number of generally smaller groups that did not survive SWEEP and are subject to the most variability across imputations. A logical follow-up analysis could be to eliminate those groups from the comparison as well.

There are two confounded issues that play a role here. The first issue is that the **M** models are generally somewhat larger, which might in turn indicate that bias, more so than variance, drives the MSE among the conditions in this study. The second issue is that the **M** models by design do better for main variables, and that those variables are an important component in the evaluation. Therefore, disaggregating results by variable type is useful, as is done in

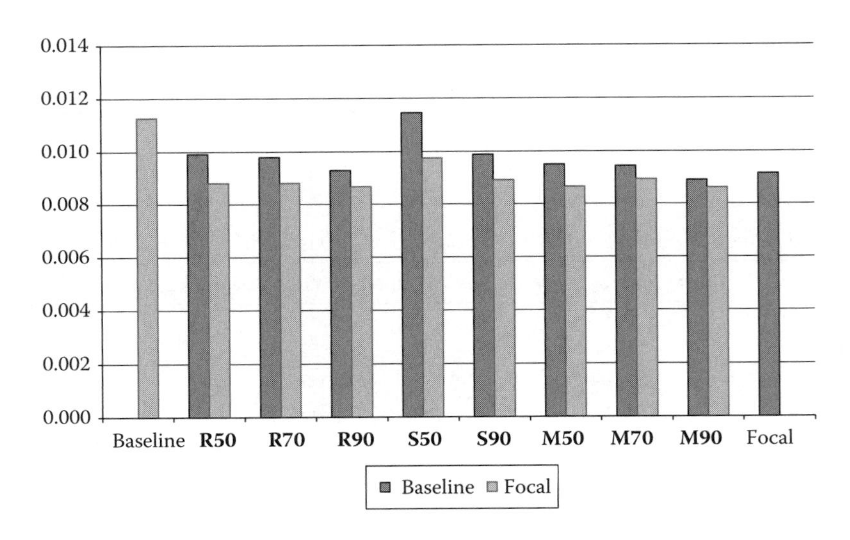

FIGURE 9.3
Average mean squared error relative to the baseline and focal models.

TABLE 9.6

MSE Disaggregated by Variable Type

	Main		Background		Interactions	
Condition	**Baseline**	**Focal**	**Baseline**	**Focal**	**Baseline**	**Focal**
R50	0.0078	0.0088	0.0118	0.0126	0.0105	0.0115
R70	0.0070	0.0067	0.0115	0.0107	0.0098	0.0092
R90	0.0064	0.0060	0.0101	0.0105	0.0089	0.0086
S50	0.0098	0.0058	0.0124	0.0095	0.0135	0.0081
S90	0.0075	0.0079	0.0113	0.0109	0.0102	0.0112
M50	0.0069	0.0064	0.0120	0.0104	0.0093	0.0089
M70	0.0065	0.0061	0.0110	0.0110	0.0088	0.0084
M90	0.0060	0.0060	0.0101	0.0104	0.0087	0.0084

Table 9.6. Naturally, for the main reporting variables, the **M** models give the lowest MSE, but this is not necessarily true for the questionnaire variables. In fact, a relatively inconsistent pattern is found where **M70** has a higher MSE than **M50** and **M90** relative to the focal model, although the differences appear to be very minor. The MSEs based on the interactions mostly favor the **M** models, which might also be an indicator of to what extent these interactions have any added value above the main effects. The **S** models seem to be least favorable. Based on this data, one could conclude that if main reporting variables are the most important, these **M** models would be the most appropriate.

While the differences show several interesting patterns, it is important to realize that the MSEs in general are relatively similar across conditions. One important reason for this could be the fact that the variance in NAEP contains two parts: measurement and sampling. Furthermore, sampling variance accounts for approximately 85–90% of the variance. While a larger model, in theory, would be better able to describe the hierarchical nature of the data and, thus, would affect the sampling variance in some way, this can probably be considered a minor factor. Therefore, MSEs were computed that only reflect the imputation variance (i.e., measurement variance). Figure 9.4 is the equivalent of Figure 2.3, with the MSE based on only imputation variance. While overall patterns do not change, the differences are decidedly larger and, relative to the focal model, differences between **M90** and **R90** are very small.

Results disaggregated by variable type can be found in Table 9.7. Patterns are relatively consistent across the variable types. The **M90** model continues to yield the lowest MSE. It is interesting to note that for the main variables, the **M** models show an increase in MSE relative to the focal model as the model gets larger. Given that both the **M** models and the focal model explicitly define the main variables of interest in the model, the only difference between these conditions is the size of the model other than the main variables. Hence, it is to be expected that the MSE increases as the model gets larger because all standard errors increase as, essentially, noise is added to all estimates. Yet, the

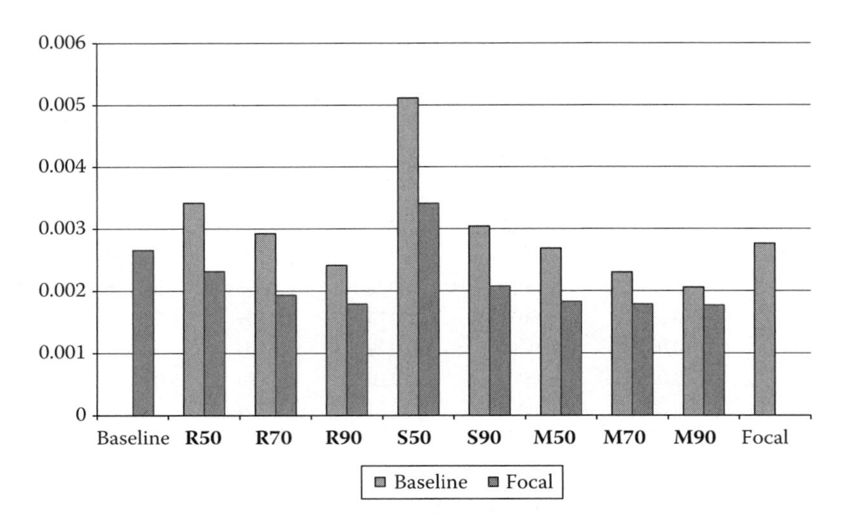

FIGURE 9.4
Average mean squared error relative to the baseline and focal models using only imputation variance to compute the MSE.

amount of random variation that follows from the imputation model should not be underestimated either. Furthermore, the size of the imputation variance relative to the squared bias is quite small, suggesting that Figure 9.4 and Table 9.7 are driven by the bias and not much else.

To shed further light on the relative contributions of the sampling and imputation variance (σ_i^2), Table 9.8 provides the percentage imputation variance relative to the total error variance. This percentage is relatively stable, suggesting that the size of the model does not affect either source of variance differentially, except for exceptionally large (baseline) or small (focal) models.

TABLE 9.7
Average Mean Squared Error Disaggregated by Variable Type Relative to the Baseline and Focal Models Using Only Imputation Variance to Compute the MSE

	Main Variables		Background Variables		Interactions	
	Baseline	**Focal**	**Baseline**	**Focal**	**Baseline**	**Focal**
R50	0.00265	0.00161	0.00312	0.00202	0.00434	0.00304
R70	0.00201	0.00117	0.00276	0.00175	0.00373	0.00258
R90	0.00159	0.00104	0.00230	0.00168	0.00307	0.00234
S50	0.00473	0.00296	0.00369	0.00220	0.00747	0.00518
S90	0.00244	0.00139	0.00268	0.00180	0.00401	0.00273
M50	0.00166	0.00098	0.00282	0.00183	0.00319	0.00230
M70	0.00145	0.00100	0.00243	0.00178	0.00267	0.00224
M90	0.00128	0.00113	0.00204	0.00162	0.00256	0.00236

TABLE 9.8
Percentage Imputation Variance across Jurisdictions

	All Variables (%)	Main Variables (%)	Background Variables (%)	Interactions (%)
Baseline	9.3	9.5	7.4	12.3
R50	11.6	10.3	8.9	16.3
R70	11.7	10.9	9.1	16.4
R90	11.4	10.2	9.2	15.1
S50	12.0	11.0	9.0	17.1
S90	11.1	10.7	8.0	16.4
M50	11.6	11.6	8.2	17.2
M70	10.8	10.0	8.4	14.9
M90	11.2	11.5	8.2	16.0
Focal	13.5	12.2	10.3	18.4

The combined error variance disaggregated by variable type is shown in Figure 9.5, confirming a relative insensitivity to model size. This is somewhat counterintuitive relative to theoretical results, but does confirm the practical experience that there are many sources of variation in NAEP and that model size is only a small factor. Alternatively, it is of course possible that some of the theoretical trade-offs described in the introduction do not hold true for some cases.

For brevity's sake, a detailed jurisdiction level has been omitted. Generally, comparable patterns hold for the individual jurisdictions although there are some counterintuitive patterns where, for example, the **R70** model has

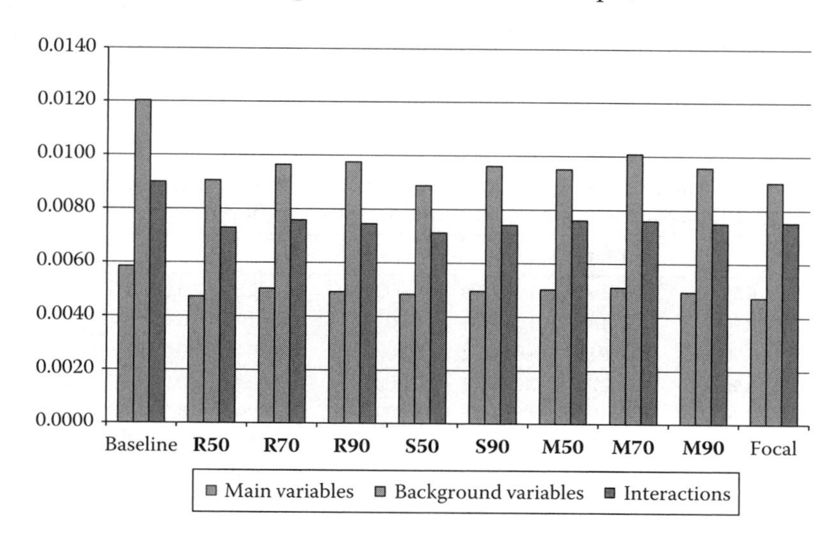

FIGURE 9.5
Average combined variance across jurisdictions disaggregated by variable type.

TABLE 9.9

Squared Biases for Male, Black, and Asian Students in Grade 4 Reading in North Carolina for the Initial Results and Five Replications

		Initial	1	2	3	4	5
Male	**R50**	0.000119	0.000063	0.000064	0.000070	0.000030	0.000119
	R70	0.000591	0.000451	0.000501	0.000476	0.000341	0.000591
	R90	0.000136	0.000111	0.000195	0.000119	0.000055	0.000136
Black	**R50**	0.001866	0.001864	0.001287	0.001426	0.002257	0.001866
	R70	0.003249	0.003408	0.002298	0.002744	0.003613	0.003249
	R90	0.002030	0.001957	0.001506	0.001297	0.001883	0.002030
Asian	**R50**	0.000161	0.000240	0.000056	0.000948	0.000493	0.000161
	R70	0.000530	0.002669	0.002639	0.000582	0.000986	0.000530
	R90	0.000056	0.000575	0.000221	0.000006	0.000084	0.000056

a higher MSE than either the **R50** or the **R90** model, probably attributable to the issues stated above (random noise, inclusion of sampling variance). Residual variance patterns hold up uniformly in the sense that larger models have smaller residual variances.

To find out to what extent reliance on a single replication is problematic and responsible for the counterintuitive results in the **R** models, five additional replications were carried out for Grade 4 reading in North Carolina and for the **R** models. For reference, the results for Grade 4 North Carolina showed a similar counterintuitive pattern where from **R50** to **R90** the MSE is 0.0086, 0.0090, and 0.0085. In addition, while the overall variance was relatively stable, the overall squared bias had a peculiar pattern as well, being 0.0016, 0.0020, and 0.0012. Table 9.9 shows the squared bias for five replications for male, black, and Asian students, and Table 9.10 shows the

TABLE 9.10

Imputation Variance for Male, Black, and Asian Students in Grade 4 Reading in North Carolina for the Initial Results and Five Replications

		Initial	1	2	3	4	5
Male	**R50**	0.000189	0.000199	0.000192	0.000224	0.000205	0.000215
	R70	0.000185	0.000181	0.000173	0.000194	0.000194	0.000195
	R90	0.000049	0.000041	0.000051	0.000048	0.000056	0.000057
Black	**R50**	0.000270	0.000261	0.000264	0.000239	0.000255	0.000295
	R70	0.000191	0.000175	0.000195	0.000209	0.000208	0.000203
	R90	0.000058	0.000067	0.000070	0.000088	0.000069	0.000068
Asian	**R50**	0.002760	0.003012	0.002975	0.003149	0.003284	0.003185
	R70	0.002527	0.002702	0.002416	0.002060	0.002504	0.002077
	R90	0.003331	0.003690	0.003704	0.003787	0.003786	0.004149

imputation variance. Black students make up about 26% of the population in North Carolina and Asian students about 2%. The results are very consistent across replications, indicating that it was justified to use a single replication. Note that the imputation variance is decreasing instead of increasing as the model increases in size for these student groups. This may be related to the fact that in a larger model each of these groups are better represented (reducing the imputation variance, σ_i^2) and that a larger model better represents hierarchical relations (reducing the residual variance σ_W^2 for these groups substantially). Further investigation of this issue is obviously warranted, as we were not able to find a satisfactory explanation.

Finally, the overall standard deviation of the proficiency distribution is inspected. An important finding of the Moran and Dresher (2007) work was that under particularly sparse conditions, the overall standard deviation was inflated. In theory, the standard deviation should not be affected by the size of the model, as a change in model size represents a trade-off of fixed and random-effect variance. In other words, a larger model means that more variance is explained by the model (between-student) and less resides in the residual (within-student) variance. Figure 9.6 shows that the results here are similar to Moran et al. in the sense that there seems to be a relationship between the size of the model and the standard deviation. The larger the model, or the smaller the sample-to-variable ratio, the larger the standard deviation of the proficiency distribution. It is not clear why this relationship exists.

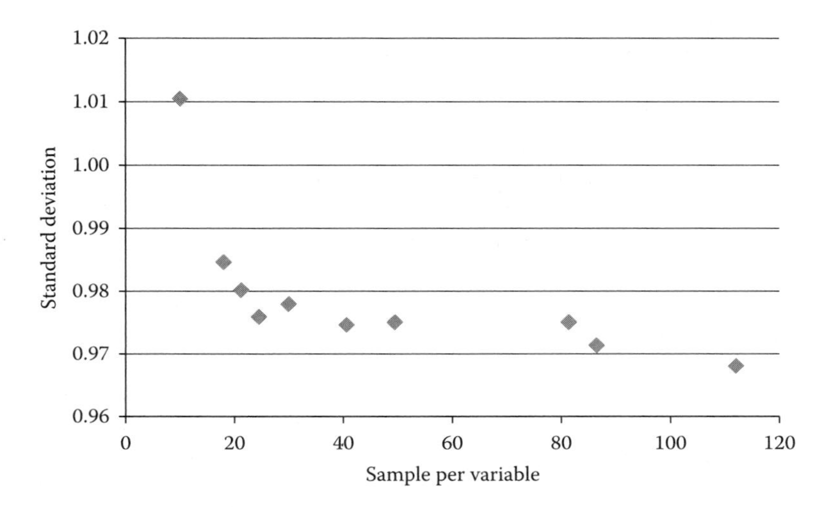

FIGURE 9.6
Average overall standard deviation across jurisdictions for each condition as indicated by the ratio of sample and the number of variables.

Discussion and Conclusion

In this study, three approaches to model reduction have been studied in the context of latent regression models for large-scale educational survey assessments, specifically the NAEP. The essential analysis challenge is that students only receive a relatively small portion of the instrument and that common methods for student proficiency estimation do not apply. Instead, marginal estimation methodologies have been developed, through which cognitive responses are aggregated across student groups using an IRT-based latent regression model. The independent variables used for these models are often several hundred student group indicators obtained through school records and student, teacher, and school questionnaires that are being administered concurrently with the cognitive items. A model with several hundred student group indicators becomes quickly intractable and provides several computational challenges, mostly due to multicollinearities in the data. In addition, choices need to be made as to whether a large model that generally insures against biases is worth the increase in variance estimates. Once the multicollinearities are removed, this becomes the key question and is the main interest of this study.

The three approaches used in this study included the current operational approach and two alternatives. Under the current operational approach, principal components are extracted from the correlation matrix based on all the available covariates (and some interactions between covariates). Subsequently, the smallest set of components that explains 90% of the variance is retained and factor scores are computed for each student. These factor scores become the covariates in the latent regression. In this study, lower percentages were also used. These types of models are denoted by **R**. An alternative method used in this study was to base the extraction of principal components on a covariance matrix instead of a correlation matrix. These types of models are denoted by **S**. The final method used in this study was to use a number of covariates that are considered critical for reporting "as is," thereby ensuring that they are fully represented in the model. Subsequently, a residual correlation matrix was computed and principal components were extracted. The direct covariates together with the factor scores were used as covariates in the latent regression. These types of models are denoted by **M**.

A simulation study was developed in which several NAEP datasets from a number of jurisdictions across Grades 4, 8, and 12 and across reading and mathematics were used. The data were used directly with only cognitive responses simulated to follow a specific model to maintain the characteristics of the real data as closely as possible. The results were provided relative to a baseline model with all noncollinear covariates included, as the squared bias, variance, and MSE. A second comparison model was also used, which incorporated only a specific set of variables and interactions (about 30) that were used for the evaluation.

Based on all variables, the **M** model where 90% of the variance of principal components for noncritical covariates is retained provided the best bias-variance trade-off across all models. This is under the assumption that (the squared) bias and variance are equally important. However, even a significantly smaller model such as the **M** with 50% retained has an MSE that appears only trivially larger. Nonetheless, with the range of models at hand, a minimum MSE that is typical in a bias-variance trade-off situation was not found. So far, the reduction in bias by increasing the model appears to still pay off. Do note that the number of points is small, which makes it at best unlikely to find a minimum.

It appears that, if bias would be considered a more important driver, the **M** models seem to be doing best, including among questionnaire variables that are not part of the critical variables that are used "as is" in the model. This comes of course, as can be expected, at the cost of a larger variance. The **S** models appeared to be generally unfavorable. Proportional representation of covariates by variability seemed to increase bias relative to the variance and compared to the standard representation.

Generally, the intuition holds within each of the methods that the larger the model, the smaller the bias and the larger the variance. However, we did encounter some exceptions, notably for the **R** models, which display a larger MSE when 70% is retained than when either 50% or 90% is retained. We looked into the number of replications and disaggregated by sampling and imputation variance, and found that the pattern seems mostly driven by the sampling variance and was consistent across replications. A larger model that possibly better reflects the hierarchical structure will also better transcend this structure (i.e., between-school variability) into the imputations as opposed to attributing this to residual variance. Therefore, a larger model could be better positioned to provide a reliable sampling variance estimate. It is not clear that this should always be higher or lower (trading against residual variance) and the patterns in this study certainly do not provide evidence in a particular direction.

In summary, the results indicated that the **M** models, where critical covariates are directly included in the model, provide promising results. Yet, it should be noted that differences found in this study were very small and that the influence of the population model is relatively modest, at least with respect to the composite.

There are some important limitations and follow-up work to note. An obvious follow-up to this study would be to collect some additional data points in terms of the percentage of variance explained in each of the models. This might reveal specific patterns and also might reveal where a minimum for the MSE might be for some of the jurisdictions. In addition, a more detailed investigation of the imputation process and how various sources of variance are reflected would be useful. Finally, looking at subscales, especially those with relatively few items, may prove useful, although the general high correlation between subscales may not reveal vastly different results from the composite.

In terms of unexpected results that are candidates for further study, the patterns of the **R** models need further attention. In addition, the practice of taking a certain percentage variance explained within the principal component structure leaves the analyst relatively naive to the object of inference, namely, proficiency. More direct approaches could include the use of proficiency proxies to select variables or to use the cognitive response patterns and use the model that best predicts those. Finally, there are many methods that could be investigated that are substantially different from the current model-reduction paradigm, such as graphical modeling, cluster analysis, and so on. An investigation in those areas would be an important follow-up.

Acknowledgments

This work was funded by the National Center of Education Statistics (NCES) within the U.S. Department of Education under contract ED-07-R-0007 for the National Assessment of Educational Progress. The findings and opinions expressed in this chapter are solely those of the authors and do not represent NCES, Educational Testing Service (ETS), or any of their affiliates. This work benefited tremendously from conversations with and reviews by John Mazzeo, John Donoghue, Rebecca Zwick, Mathew Johnson, and the National Assessment of Educational Progress Design and Analysis Committee.

References

Allen, N., Donoghue, J., and Schoeps, T. 2001. *NAEP 1998 Technical Documentation.* Washington, DC: National Center of Education Statistics, Institute of Education Statistics.

Cohen, J. and Jiang, T. 1999. Comparison of partially measured latent traits across nominal subgroups. *Journal of the American Statistical Association.* 94, 1035–1044.

Cohen, J. and Jiang, T. 2002. *Direct Estimation of Statistics for the National Assessment of Educational Progress (NAEP).* Unpublished paper, Washington, DC: American Institutes for Research.

Dresher, A. 2006. Results from NAEP Marginal Estimation Research. Paper presented at the 2006 annual meeting of the National Council for Measurement in Education, San Diego, CA.

Goodnight, J.H. 1978. *The Sweep Operator: Its Importance in Statistical Computing.* SAS Technical Report No. R-106. Cary, NC: SAS Institute Inc.

Haberman, S.J. 1977a. Maximum likelihood estimates in exponential response models. *The Annals of Statistics*, 5(5), 815–841.

Haberman, S.J. 1977b. Log-linear models and frequency tables with small expected cell counts. *The Annals of Statistics*, 5(6), 1148–1169.

Li, D. and Oranje, A. 2006. *On the Estimation of Hierarchical Latent Linear Models for Large Scale Assessments.* (Research Report 06–37). Princeton, NJ: Educational Testing Service.

Li, D. and Oranje, A. 2007. *Estimation of Standard Errors of Regression Effects in Latent Regression Models* (Research Report 07–09). Princeton, NJ: Educational Testing Service.

Mazzeo, J., Donoghue, J.R., Li, D., and Johnson, M.S. 2006. *Marginal Estimation in NAEP: Current Operational Procedures and AM.* Paper prepared for National Center of Education Statistics.

Mazzeo, J., Johnson, E., Bowker, D., and Fai Fong, Y. 1992. *The Use of Collateral Information in Proficiency Estimation for the Trial State Assessment.* Paper presented at the annual meeting of the American Educational Research Association, Orlando, FL.

Mislevy, R.J. 1984. Estimating latent distributions. *Psychometrika*, 49(3), 359–381.

Mislevy, R.J. 1985. Estimation of latent group effects. *Journal of the American Statistical Association*, 80(392), 993–997.

Mislevy, R.J. 1991. Randomization-based inference about latent variables from complex samples. *Psychometrika*, 56(2), 177–196.

Mislevy, R.J. 1993. Should "multiple imputations" be treated as "multiple indicators"? *Psychometrika*, 58(1), 79–85.

Mislevy, R.J. and Sheehan, K.M. 1987. Marginal estimation procedures. In A.E. Beaton (Ed.). *Implementing the New Design: The NAEP 83-84 Technical Report* (Report No: 15-TR-20), 293–360. Princeton, NJ: Educational Testing Service.

Mislevy, R.J., Beaton, A. E., Kaplan, B., and Sheehan, K. M. 1992a. Estimating population characteristics from sparse matrix samples of item responses. *Journal of Educational Measurement*, 29, 133–161.

Mislevy, R.J., Johnson, E.G., and Muraki, E. 1992b. Scaling procedures in NAEP. *Journal of Educational Statistics*, 17, 131–154.

Moran, R. and Dresher, A. 2007. *Results from NAEP Marginal Estimation Research on Multivariate Scales.* Paper presented at the annual meeting of the National Council for Measurement in Education, Chicago, IL.

Oranje, A., Li, D., and Kandathil, M. 2008. *On the Role of Background Variables in Large-Scale Survey Assessment Analysis.* Paper presented at the 2008 annual meeting of the National Council on Measurement in Education, New York, NY.

Oranje, A., Li, D., and Kandathil, M. 2009. *Evaluation of Methods to Compute Complex Sample Standard Errors in Latent Regression Models (RR-09-49).* Princeton, NJ: Educational Testing Service.

Portnoy, S. 1984. Asymptotic behavior of *M*-estimators of p regression parameters when p^2/n is large. I. Consistency. *The Annals of Statistics*, 12(4), 1298–1309.

Portnoy, S. 1988. Asymptotic behavior of likelihood methods for exponential families when the number of parameters tends to infinity. *The Annals of Statistics*, 16(1), 356–366.

Rust, K. and Johnson, E. 1992. Sampling and weighting in the national assessment. *Journal of Education Statistics*, 17(2), 131–154.

Thomas, N. 1993. Asymptotic corrections for multivariate posterior moments with factorized likelihood functions. *Journal of Computational and Graphical Statistics*, 2, 309–322.

Thomas, N. 2002. The role of secondary covariates when estimating latent trait population distributions. *Psychometrika*, 67, 33–48.
Van der Linden, W.J. and Hambleton, R.K. 1996. *Handbook of Modern Item Response Theory.* New York, NY: Springer-Verlag.
Wolter, K. 1985. *Introduction to Variance Estimation.* New York, NY: Springer-Verlag.

10

Linking Scales in International Large-Scale Assessments

John Mazzeo
Educational Testing Service

Matthias von Davier
Educational Testing Service

CONTENTS

Introduction

One principal way that results of international large-scale survey assessments (ILSAs) can be of value to policy makers and researchers is in providing information on changes over time in the levels of the academic or workforce skills and competencies that ILSAs are designed to measure. Attention can be focused on these trends within each participating country, rates and direction of change can be compared across countries, and the presence of statistical relationships among trend outcomes and a variety of input and policy variables can be investigated. However, for ILSAs to be used for such purposes, the instrumentation, administration, and analysis of these assessments must be carefully controlled to ensure the results produced by a sequence of assessments can be reported on a scale that maintains a stable, comparable meaning over time. We use the term *linking scales* to refer to the process of accomplishing this task.

In this chapter, we discuss selected survey design and psychometric analysis considerations pertinent to linking scales in ILSAs. As the designs and methods used in ILSAs are, in some respects, modeled after those used in the U.S. National Assessment of Educational Progress (NAEP), we will discuss NAEP methodologies as well.

This chapter has four sections. In the section "Some General Design Considerations," we present some general principles that need to be considered in the design of assessment instruments and the ongoing conduct of the assessments in order to facilitate the maintenance of comparable scales. We then explain in the section "Some Typical Data Collection Designs" some of the specific designs and administration procedures that have been used in ILSAs—the Programme for International Student Assessment (PISA), Trends in International Mathematics and Science Study (TIMSS), and Progress in International Reading Literacy Study (PIRLS)—to provide the necessary data structures that make it possible to link scales over time. In the section "Analysis Procedures: How Scale Linking Is Carried Out in PISA, TIMSS, and PIRLS," we describe the psychometric analysis procedures applied to the data to produce linked scales. (This material assumes familiarity with Chapter 7 of the current volume.) Finally, in the section "Current Challenges and Future Directions," we discuss various ongoing challenges to linking scales that are still being addressed with current ILSAs and point to future challenges that will emerge as the extent and nature of ILSAs evolve in coming decades.

Some General Design Considerations

Almost all assessment programs—whether they are testing programs in the United States designed to provide test scores to individual test takers such as the SAT, or large-scale assessments such as NAEP, PISA, TIMSS, and PIRLS—face the challenge of providing results over time that are comparable. Testing programs that provide individual test takers with scores that will be used for high-stakes decisions such as college admission must offer different editions of the test (i.e., different collections of test items) at various times throughout the year, and over years. Different editions, or test forms (a commonly used term for these different editions), need to change so students testing later in time are not advantaged by having advance knowledge of the particular items. Such advance knowledge would raise questions of fairness as well as validity. However, if the different editions contain different collections of items, how can the testing program ensure that scores on different editions mean the same thing? Do some editions contain easier items than others, and if so, should a score on a harder edition of the test be given more credit than a score of equal value on an easier edition? Perhaps just as importantly, are the same skills and proficiencies being measured? Does one edition of a mathematics test emphasize geometry and factual and formula recall, while another emphasizes algebra and problem solving?

ILSAs face similar challenges concerning the comparability of results from different "editions" of the assessment.* The need to publicly release some portion of their item pools—to help users of results better understand what the assessments are measuring or simply allow for the continual evolution and relevance of the material—necessitates the use of different item pools from one assessment to the next. As discussed in Chapters 4 and 7 of this volume, ILSAs make use of large pools of items for each assessment that are typically configured into multiple assessment booklets. The different editions of an ILSA, then, refer to the different collections of assessment booklets that are administered from one assessment to the next. Does the difficulty of these collections of test booklets change between assessments and, if so, how can the assessment be analyzed and reported in a manner that permits results to be validly compared? How can it be ensured that the same skills and proficiencies are being measured from one assessment to the next?

Assuring the same mix of skills from one assessment to the next is largely about assessment construction practices. Test blueprints or item-pool specification documentation are commonly used to control the content and item types included in the different editions of tests in an attempt to ensure that the same kinds of skills and proficiencies are being measured. *Scale linking*, the main focus of this chapter, is about determining the relative difficulty of different additions and adjusting the analysis of results for any unintended differences in difficulty. For a successful scale linking that results in comparable scores over time, an appropriate control of the makeup of the item pool, an appropriate data collection design, and appropriate analysis procedures are all required and equally critical. Analysis procedures (the topic of the

---

* ILSAs face challenges as well with respect to the comparability of results across languages and cultural settings.

section "Analysis Procedures: How Scale Linking Is Carried Out in PISA, TIMSS, and PIRLS") alone cannot solve linking challenges if the assessment instruments have not been constructed to measure comparable mixes of content and skills and the data have not been collected in an appropriate fashion to support valid inferences.

The data collection principle that underlies scale linking appears simple. Each new edition of the assessment is made up of items unique to that edition along with items that have been administered as part of previous editions of the test (i.e., linking items). When analyzed appropriately—in particular, using the statistical models and methods described in Chapter 7—the data from the common portion of the test provide the means of scoring the current assessment in a way so its results are expressed on scales comparable to those used for earlier editions.

In order for the models and methods to produce valid results, however, it is essential that the psychometric properties associated with each repeated item (e.g., item difficulty level, discrimination, or susceptibility) are the same in each of the assessment cycles in which it appears. As we will discuss in some detail in the remainder of this section, care must be taken that the repeated assessment material is appropriately representative of the content of the full assessment and is presented and scored in a similar fashion across assessments, and that other aspects of the test administration are sufficiently standardized. Insufficient content representatives and/or changes in context—the way material is presented or other aspects of the administration procedures—can compromise the ability to carry out valid scale linking.

On the Content Representativeness of Linking Items

If the assumptions of item response theory (IRT; see, e.g., van der Linden and Hambleton 1997, pp. 1–25) models that are used to analyze survey assessments were correct (i.e., the educational constructs defined by the item pools used in the assessments were truly unidimensional), then the size and makeup of the set of the linking items would matter little. The rank order of countries in any given assessment cycle or the changes in average scores for a given country would depend little, if at all, on which subset of the total item pool was selected for use in subsequent assessments. In fact, it is generally acknowledged that educational constructs are multifaceted and the unidimensional summaries typically used to analyze such surveys capture, at best, the overall average performance across the different facets of the construct (Mislevy 1990).

Consequently, ILSAs can be expected to show variability in trend results if one looks within the overall collection of items used to measure trend. Countries differ in terms of curricula emphases and instructional practices, both at a given point in time as well as over time. Add to that the additional challenges associated with translation and ensuring comparable scoring for constructed-response items, and it would be surprising *not* to find considerable item-by-country or item-by-administration-cycle interactions. In our view, such interactions are not an indication of unstable trend measurement *per se*.

However, successful scale linking and stable trend measurement does require that country-by-country trend results (e.g., changes in country averages over time) be relatively invariant over multiple *collections of linking items*, each of which was considered *a priori*, an appropriate measure of the intended construct. When dissected, each such collection might show variability in country-level results for subsets of items. However, as long as country-level results based on each of the full set of item collections are stable across collections, we would argue that stable trend measurement vis-à-vis the overall construct has been achieved. Such considerations highlight the importance of making the set of repeated items sufficiently large and representative of the full content domain framework to ensure that the assessment results serve what we see as their intended purpose—to report reliably on the overall trend with respect to the full construct defined by that assessment's framework.

On the Importance of Context

The importance of controlling context is well illustrated by NAEP's experience in the 1980s with its reading anomaly (Beaton and Zwick 1990). Educational Testing Service (ETS) had conducted its first NAEP assessments in reading and writing in 1984, introducing IRT scaling and the use of marginal estimation and plausible values—methodologies discussed in Chapter 7—and reexpressing results from NAEP assessments back to 1971 on these newly derived IRT scales. NAEP reading assessments were planned for 1986 along with assessments in mathematics and science and the introduction of IRT scaling methods to these content domains. The 1986 NAEP reading assessment results were to be reported on the scales established in 1984 and, to this end, a number of test items from 1984 were also part of the 1986 instrument.

Several changes, however, were made to the design of the NAEP reading assessment, in particular to the booklet design, between 1984 and 1986:

1. Passages/items from 1984 were repeated in 1986 but not as part of intact clusters. Therefore, both local context and item positions within cluster were different.
2. Other content domains within the booklet were different. In 1984, reading clusters were paired with writing, while in 1986, reading was paired with mathematics, science, and, for students at age 17, computer competence, history, and literature.
3. Timing of clusters was different—1–2 minutes longer in 1986 than in 1984, and the number of items per block increased in 1986, given the extra time that was allocated.

4. Booklet formats differed with respect to ink color, line length of reading passages, and response format (i.e., circle the correct option in 1984 versus fill in the oval in 1986).

The original analysis of the 1986 reading data showed anomalous results. Average scores dropped precipitously for two of the three age groups assessed, and the magnitude of the change over this 2-year period far exceeded anything observed from the inception of the NAEP assessment program in 1971.

The 1986 reading results were not published at that time. Instead, an experiment was planned and conducted in 1988 to disentangle the aggregate effect of the reading assessment design changes made between 1984 and 1986. Two randomly equivalent samples at each age level were selected. One sample was assessed with the identical instruments and procedures used in 1984. The second group was assessed with the 1986 instruments and procedures. Data from both samples were separately analyzed and compared to each other, with the differences providing an estimate of the impact of the collective set of changes described above. Results differed by age group, ranging from 2–8 NAEP scale score points. In effect-size terms, this amounted to changes between .05 and .22. The larger differences (.16 and .22 in effect-size terms for age 13 and age 17, respectively) exceeded in magnitude any of the reported changes in NAEP reading results between 1971 and 1984.

As a result of this experience, the NAEP program has adopted a conservative stance with respect to keeping context as consistent as possible from one assessment cycle to the next. When changes to test designs that can impact context are considered in NAEP, they are accompanied by bridge studies modeled after the reading anomaly experiment that are designed to estimate and appropriately adjust results for the potential impact of said changes. In the remainder of this section, we discuss some of the design features of ILSAs, with an eye toward examining how those design features need to be managed from one assessment cycle to the next to ensure appropriate data for the linking of scales.*

Considerations in Maintaining a Common Context

Use of Individual Linking Items versus Intact Linking Blocks

As noted above, linking scales across assessments requires that material from prior assessment cycles be repeated in the current assessments. More specifically, it requires repeating individual test items, or small collections of test items based on a common passage. These linking items can be assembled into blocks and booklets along with the other material intended for the current assessment. However, care needs to be exercised in constructing

* For a general discussion of ILSA design, see Chapter 7 of this volume.

the blocks and booklets for the current assessment to avoid introducing changes in context—such as position within a separately timed section, cueing, fatigue effects, or differential speededness* associated with the material the linking items are paired with—that could have an unwanted impact on its psychometric functioning. Changes in the psychometric functioning of a small number of these items can often be detected and accounted for by adjustments to analysis procedures. However, if the number of such items is large, it may be difficult to disentangle the cumulative effect of such methodologically induced changes from the changes in proficiency that one is trying to measure. If context changes cannot be avoided, data collection designs like those used to investigate the NAEP reading anomaly described above may be required to either demonstrate that the changes have not had an impact, or to provide the data that would allow the impact of such changes to be adjusted for.

Several test assembly strategies can be adopted to mitigate some of the potential for context effects. For example, care is usually taken to avoid introducing large changes in the within-block serial position of linking items from one assessment to the next. Similarly, factors that impact block speededness, such as the number and mix of item types in each separately timed section, are often held as constant as possible from one assessment to the next. Assembled blocks and booklets can be carefully reviewed to double check for unwanted cueing of correct responses.

One commonly used approach to mitigate the potential for context effects is to repeat only entire intact blocks of test items from previous assessments. The use of intact linking blocks maintains stricter control over at least the local context in which an item appears. When these intact blocks correspond to separately timed testing sessions, particularly tight control can be maintained over context effect. Each linking item appears in the identical serial position within the block, and each item is paired with the same test items from assessment to assessment so that any practice, fatigue, or cueing effects present should remain consistent. Moreover, assuming that timing conventions are kept consistent, the introduction of differential speededness can be effectively controlled. Adherents of such an approach argue that this greater control over position and context makes the assumptions of consistent psychometric functioning for test items that undergird the analysis procedures used for linking scales more plausible. Practical experience in programs that use such an approach, such as NAEP, suggest this approach works reasonably well.

---

* We use the term *speededness* here in a general sense to refer to the relative adequacy of the time allotted to complete a collection of items. For our purposes, we define a speeded test as one in which student performance would improve were the student given additional time to complete the test. While most survey assessments are intended to be largely unspeeded, they are likely speeded to at least some degree. Different test booklets within an assessment can display differing degrees of speededness, depending on the number and nature of the items they include.

Controlling Block Position within Booklet

The test booklets that comprise the content of ILSAs are typically organized in terms of some number of blocks of items, with each block of items appearing in multiple test booklets within a given assessment year. Sometimes, though not always, a block will correspond to a separately timed section within the total time period allowed. For example, as described in greater detail in a subsequent section, each TIMSS test booklet is composed of four blocks of items, which are further organized for administrative purposes into two separately timed parts, each of which presents two blocks of items. As a result, in TIMSS as well as other ILSAs, a particular block of items can, in principle, appear in a number of serial positions within a test booklet and often does appear in different serial positions in the different test booklets within a given assessment year.

When using IRT to analyze ILSA data, an assumption is usually made—at least at the outset—that the serial position of a block within the test booklet does not affect the psychometric functioning of the items contained in that block. This assumption is often, at best, an approximation. Ample evidence exists from most ILSAs to suggest that such assumptions are false to varying degrees (see, e.g., Adams and Wu 2002, pp. 282–85; Gonzalez et al. 2004, p. 264 and Appendix D). A number of strategies have been used—test design and administration strategies as well as, in some cases, data-analytic strategies—to attempt to either mitigate the potential impacts of these effects or at least partially mitigate to adjust for them statistically.

NAEP has relied largely on test design and administrative strategies to mitigate the potential impacts of block position. For most NAEP content domains, an attempt is made to balance block position across test booklets within each assessment so that, in any given assessment cycle, each block typically appears an equal number of times in each position. Moreover, within each NAEP test administration session, test booklets are distributed in such a way that as many different booklets as possible are administered and assigned to test takers according to a random process. The design and administration procedures are intended to work together to ensure there is little or no association between school and test-taker characteristics—the characteristics that define the groups for which NAEP results are reported—and the test booklet assigned. In this way, potential biases that might arise due to unaccounted-for block position effects are minimized. PISA, on the other hand, has used explicit, post hoc statistical adjustments to the results of initial IRT analyses to attempt to correct for booklet effects* (Adams and Carstensen 2002, pp. 157–262).

* ILSAs are typically designed to result in the assignment of each test booklet to randomly equivalent samples of test takers. If the assumptions about equivalence of item functioning across booklets are tenable, IRT analyses should produce scaled score results for the subsets of test takers assigned each test booklet that are equivalent within the tolerance expected due to test-taker sample and item selection (sampling and measurement variance). We use the term *booklet effects* to refer to those situations in which IRT analyses do not produce the intended equivalence of results between the booklets.

Given the potential for position effects, it is typically wise to make sure that when intact blocks of linking items are used, block positions are controlled carefully from one administration to the next. One way to control block positions is to allow blocks to appear in multiple positions while ensuring that the relative frequency of block positions is maintained across assessment cycles. If, in prior assessments, linking blocks appeared an equal number of times in each possible position, then similar design constraints should probably be imposed on subsequent assessments. If, in contrast, linking blocks appeared with unequal frequency, or perhaps in a single position in prior assessments, similar conventions should probably be imposed in subsequent assessments to avoid introducing changes in item functioning due to changes in position that might invalidate trend analyses.

There may be instances where practical considerations prevent current assessment designs from imposing identical constraints on block positions in the design phase. Similarly, intentional, or in some cases, unintentional, changes in booklet distribution or test administration procedures may result in differences across administrations in the relative frequency of the linking blocks in each of the possible positions. In such cases, it may be wise to restrict the *linking analyses* to a subset of the data (as described in greater detail below). Such a restriction may help to control, after the fact, any impact such differences might have on the validity of the linking analyses. Alternatively, post hoc adjustments like those used in PISA might be considered.

Controlling for Speededness

Generally speaking, the sponsors and producers of ILSAs work hard to create blocks and test booklets that the vast majority of test takers can complete within the time period allotted. Moreover, the intention of test developers is that typical test takers have sufficient time to display what they know and are able to do. Despite the best efforts of test sponsors, producers, and developers, some ILSA test takers will experience a degree of speededness because of individual differences in proficiency and familiarity with effective test-taking strategies, as well as practical constraints. Many test takers may simply run out of time and never reach items at the end of separately timed sections. Occasionally, even items that appear earlier in the test may be skipped by large numbers of test takers, perhaps in part because they decided that insufficient time was available to complete the test. In some cases, data from particular test items may be dropped from the overall analysis because of too few test-taker responses.

The potential presence of speededness has an important implication in the measurement of trend through the use of linking items. Overt changes to blocks—such as adding more items or increasing the proportion of constructed-response items compared to multiple-choice items—could have an impact on the psychometric characteristics of the linking items the blocks contain and thereby compromise the ability to effectively link scales and report valid trend results. Similarly, actions such as changes in time limits for blocks and booklets can materially affect the functioning of linking items and thereby affect the measurement of trends. In general, best practice with respect to linking scales is to keep as much of the block and booklet design and test administration procedures constant over time as is practical.

Multiple-Content Domain versus Single-Content Domain Designs

In any given testing cycle, almost all ILSAs conduct assessments in more than one content domain. For example, each of the five cycles of PISA carried out to date has included assessments in reading, mathematics, and science literacy. Each of the five cycles of TIMSS carried out to date has included assessments of both mathematics and science. In NAEP, each assessment cycle typically includes assessments in multiple content domains (e.g., reading, mathematics, and science in 2009; civics, geography, and U.S. history in 2010).

One of the first design features that must be addressed in such assessments is how to distribute the assessment material from the multiple content domains among the test takers. There are three obvious strategies: (1) present each student with material from all domains; (2) present each student with material from multiple, but not necessarily all, domains; and (3) present each student with material from only one content domain. We refer to the first two strategies as multiple-content domain (MCD) designs and to the latter as a single-content domain (SCD) design. The different design options offer different benefits and challenges—from an overall efficiency perspective, as well as from the point of view of what needs to be controlled from assessment cycle to assessment cycle to best ensure comparable trend results.

MCD designs offer obvious efficiency advantages over SCDs in that overall they tend to require smaller numbers of test takers. Assuming a three-content domain assessment and equal target sample sizes per domain (denoted here as n), an SCD design will require three times the sample size as required by an MCD. More specifically, to obtain the $3n$ observations needed for such an assessment requires $3n$ test takers under an SCD design but only n test takers under an MCD design, where each test taker provides three observations. Because obtaining the required number of test takers often presents considerable financial and logistical challenges, such efficiency advantages make MCD designs quite attractive. An additional advantage of such designs is that they provide data for examining the statistical relationships between performances in the different content domains and exploring how these relationships might be mediated by other variables collected as part of the survey. Such analyses are obviously not possible given the data that SCD assessments provide. With these advantages, it is perhaps not surprising that MCD designs have been the approach of choice for both TIMSS and PISA.

MCD designs do, however, present some challenges. Because test takers respond to items in multiple content domains in MCD designs, they typically require that each test taker be tested for a longer period of time than do SCD designs. Requiring test takers to test for relatively long periods can present challenges to obtaining adequate survey participation. As ILSAs provide little direct benefit to the participating schools or test takers, overly long testing times may negatively impact the willingness of schools and test takers to participate. Moreover, the potential impact of longer testing times on test-taker fatigue and level of effort could jeopardize the validity of the assessment results. For these reasons, the second variant of the MCD design described above, in which each test taker is tested in a subset of the possible different content domains, is a compromise that is less burdensome for test takers.

An additional challenge involved with MCD designs is the maintenance of a constant assessment context over time. One aspect of context that may be important to control in MCDs is the nature and amount of the other content domain with which a particular focal content domain is paired. For example, the difficulty of a given block of mathematics items may be different depending on whether the mathematics block is paired with reading or science blocks within test booklets. Perhaps more subtly, the difficulty of a given block of mathematics items may change depending on whether the block is paired with a relatively small or large amount of material from another content domain. Such differences in difficulty could arise due to differential fatigue, subtle cuing or practice effects, or subtle differences in speededness. The specific design and analysis procedures used in ILSAs to attempt to control such context effects are discussed further in the next section.

Comparability of Constructed-Response Scoring over Time

When choosing sets of linking items, it is important that they are representative of the full range of item types that make up the entire assessment item pool. Therefore, linking items in ILSAs are typically made up of both multiple-choice and constructed-response items. Multiple-choice items can be automatically scored as correct or incorrect, while some or all of the items that allow for an open (or constructed) response typically require ratings by human scorers. These scorers evaluate the response and classify it in one of several response categories. These response categories are often also dichotomous, so that correct versus incorrect is distinguished, or partially correct responses are considered by allowing for a score with multiple, ordered categories that reflect the level of correctness. Constructed-response items used in ILSAs may allow for three or even four ordered levels of correctness in which the highest level indicates a complete and correct answer, and lower levels indicate partially correct responses.

For constructed-response items to be used effectively for linking, it is imperative that scorers are trained so constructed-response scoring remains comparable across assessment cycles. Changes over time in how scoring rubrics are applied will change the psychometric properties of constructed-response items and, in some cases, render them useless for linking. This leads to the question of how to test whether the scoring process adhered to the same rules across scorers and assessment cycles. One way is to embed responses from previous assessments in the scoring process and have them rescored by scorers working on current assessments to check whether old responses are scored in the same way as responses from the current cycle. Such rescoring activities are most effective if they can be carried out and the results analyzed and used on an ongoing basis to monitor and correct for any unwanted changes (i.e., greater leniency or stringency) in applying scoring rubrics. Scorers can be given feedback about their performance compared to other scorers as they go along, retraining or scorers can be conducted if necessary, and rescoring of subsets of the responses may be carried out to keep the scoring process on track. If all else fails, information from these scoring studies can inform the IRT analyses after the fact to properly adjust for the absence of comparable scoring.

Some Typical Data Collection Designs

All data collection designs presented in this subsection share features that show their ancestry. TIMSS was initiated at a time when NAEP would be the obvious model for designing an international educational survey, while the different designers of PISA and PIRLS could use the examples provided by NAEP and TIMSS. At a very high level of abstraction, all the assessments exemplified in this section start with a decision about how much task material, in the aggregate, is necessary and practical to develop and present during a particular assessment cycle to cover the content domains adequately. The next decision is how long to test any individual student—how long can students be expected to focus on the subject-matter material, how much time will be needed to obtain information about demographic background and educational experiences, and how much testing time in total are participating schools likely to tolerate. Typically, no more testing time than the duration of one, or at most two, instruction periods (of about 60 minutes) can be used unless an extended break between sessions can be accommodated.

In order to increase the ability to present items in multiple block positions, the subject-matter testing periods are further broken down into testing slots (say, e.g., 20- or 30-minute slots), and blocks of tasks are assembled by content domain that are expected to require the same amount of testing time. Each task typically appears only in one block, and multiple blocks are combined in systematic ways into a number of booklets following combination rules that ensure that block positions are balanced in certain ways. While PISA and TIMSS may combine blocks from two or more different content domains into a booklet, PIRLS combines blocks from two subdomains of reading. How this plays out in practice will be demonstrated by discussing recent designs from PISA, TIMSS, and PIRLS.

PISA

A distinguishing feature of the PISA design is that assessment domains rotate from one assessment cycle to the next in terms of being treated as a major or minor domain. In any given cycle, one content domain is major and the other two are minor. For PISA 2009 (OECD 2012), reading was the major domain, and science and mathematics were the minor domains. In 2009, there were 13 clusters of assessment material—three clusters used for each of the two minor domains and seven clusters used for reading. This represented a total of 350 minutes of assessment material.

Each PISA booklet contains four of the available clusters of content-matter items and represents 2 hours of student testing time. In 2009, the PISA standard design called for 13 booklets. These 13 booklets made up the standard design (the layout is given in Table 10.1).* The 13 standard booklets in PISA

TABLE 10.1

PISA Assessment Design for the 2009 Cycle with Reading as Major Domain

Booklet ID	Cluster			
1	**M1**	**R1**	R3	M3
2	**R1**	*S1*	R4	R7
3	*S1*	R3	**M2**	S3
4	R3	R4	*S2*	**R2**
5	R4	**M2**	R5	M1
6	R5	R6	R7	R3
7	R6	M3	S3	R4
8	**R2**	**M1**	*S1*	R6
9	**M2**	*S2*	R6	**R1**
10	*S2*	R5	M3	*S1*
11	M3	R7	**R2**	**M2**
12	R7	S3	**M1**	*S2*
13	S3	**R2**	**R1**	R5

Note: The reading clusters R3 and R4 are present in two variants, the standard version A and an easier version B.

* In 2009, an additional two blocks of easier reading items were also included in the PISA assessment. By exchanging these clusters with two of the standard reading clusters (R3 and R4 in Table 10.1), seven additional booklets were produced, resulting in a somewhat easier 13-booklet alternative collection that was available and administered in lieu of the standard collection at the discretion of the participating countries.

for 2009 follow a Youden squares design where each block appears once in each position of the booklet, and each block appears with every other block in the design. The resulting booklets contain one or more clusters of reading items, 9 of the 13 booklets contain one or more mathematics clusters, and 9 of the 13 contain one or more science clusters. Six of the booklets include at least one cluster of material from all three content domains, six booklets include material from two of the three content domains, and one booklet contains only reading clusters.

As noted earlier, the linking of results across years requires repeating items from earlier assessment cycles. For the 2009 PISA assessment, two of the seven reading clusters (28 of 131 items) were identical to those included in the 2006 PISA assessment. For the mathematics assessment, the three clusters used were a subset of the mathematics clusters used in 2006. For these blocks, which are shown in bold within Table 10.1, the local context was fairly tightly controlled across administration cycles. For science, the situation is different. Two blocks, shown in italics, consist of material presented in 2006 when science was the major domain. However, this material was newly reassembled into clusters and appears with different material in 2009 and, in some cases, in clusters of different length, compared to 2006.

Table 10.2 presents an alternative perspective on the PISA design. The rows in the table correspond to the 13 clusters of items used in the standard booklets in 2009. The columns indicate the serial position within each test booklet to which the clusters could be assigned, and the entries in the body of the table indicate the booklet number in which a particular cluster was assigned to a particular serial position. For example, mathematics cluster one (M1) appeared in the first position in Booklet 1 (B1), the second position in Booklet 8 (B8), the third position in Booklet 12 (B12), and the fourth position in Booklet 5 (B5).

Table 10.2 highlights that each block appears once in each of the four positions. The balancing of positions for each of the item clusters attempts to control for position because the results of earlier PISA assessments have demonstrated an effect of the cluster position on the way respondents answer the items in a cluster. It is well known that fatigue may reduce performance on tests (Nunnally 1967; Thorndike 1914); therefore, students may not perform as well on test items given in later block positions compared to those in early block positions. For PISA 2006, Le (2009) has shown that item difficulty increases with block position, particularly for open response items, but also for closed (multiple-choice) response formats. Mazzeo and von Davier (2008) reported similar findings for PISA and to a lesser extent for NAEP. This means that when the same item parameters are assumed for an item appearing in different positions, this is only an approximation of reality because items appear to become somewhat more difficult with presentation in later cluster positions.

In addition to effects of cluster position, the current PISA rotation between minor and major domains, and the change in context for the common items

TABLE 10.2

Booklets as a Function of Cluster Positions in PISA 2009

	Cluster Position			
Cluster	**1**	**2**	**3**	**4**
M1	B1	B8	B12	B5
M2	B9	B5	B3	B11
M3	B11	B7	B10	B1
S1	B3	B2	B8	B10
S2	B10	B9	B4	B12
S3	B13	B12	B7	B3
R1	B2	B1	B13	B9
R2	B8	B13	B11	B4
R3	B4	B3	B1	B6
R4	B5	B4	B2	B7
R5	B6	B10	B5	B13
R6	B7	B6	B9	B8
R7	B12	B11	B6	B2

that this design feature induces, have been identified as potential sources of distortion in trend results over time (Carstensen 2012; Gebhardt and Adams 2007; Urbach 2012). It seems that the reduction of the item set used in minor domains may distort trend results for some countries more than others. A further systematic study of different design options revealed evidence that the rotation between major and minor domains may indeed lead to more instability compared to a balanced assessment of all three domains in each cycle (see Chapter 11 in this volume). One potential explanation is that the effects of ubiquitous country-by-item interactions—the variation of statistical properties of items across countries—may be exacerbated by the change in coverage between minor and major domains. The actual size of the impact of the minor/major design rotation on estimates of average performance is, of course, difficult to assess. We refer the interested reader to the articles cited above for details on the assumptions under which the current results were obtained.

TIMSS

The TIMSS assessment design is based on the combination of material from the domains of science and mathematics. The 2011 TIMSS assessment contained a total of 28 clusters (referred to as blocks in TIMSS) of subject-matter material, 14 blocks each for mathematics and science. TIMSS assessment sessions are divided into two separately timed parts separated by a break. Each TIMSS assessment booklet contains four blocks of subject-matter items—two per part.

Table 10.3 shows the TIMSS assessment design for the 2011 cycle that was used for both the fourth- and eighth-grade assessment. It is the same design as used in 2007. Each TIMSS booklet consists of a separately timed mathematics and science part. The testing parts are 36 minutes long in the fourth-grade assessment, and 45 minutes long in the eighth-grade assessment. Two blocks are administered within each of these parts.

In half the books, mathematics is presented first, while in the other half, science is the first subject presented. Each block appears in 2 of the 14 booklets. Unlike PISA, block position is not completely balanced. However, some degree of balance is maintained at the block level by ensuring that each block appears once as the first block within a part and once as the second block. More specifically, a block of items will appear once in position 1 (i.e., the first block in part 1) and once in position 4 (i.e., the second block in part 2), or once in position 2 (i.e., the second block in part 1) and once in position 3 (i.e., the first block in part 2). The rationale appears to be that potential block positions are considered to be reset by the break between the two testing parts. That is, positions 1 and 3 are considered equivalent (as both are the first after starting or restarting the testing session), as are positions 2 and 4 in the TIMSS design.

TIMSS repeats intact blocks from previous assessments to provide the necessary data for scale linking. In the 2011 design, a total of 16 blocks (eight for each of the two content domains) were common to 2007. Context was

TABLE 10.3

TIMSS 2011 Design for Grades 4 and 8

Student Achievement Booklet	Assessment Blocks			
	Part 1		Part 2	
Booklet 1	M01	M02	S01	S02
Booklet 2	S02	S03	M02	M03
Booklet 3	M03	M04	S03	S04
Booklet 4	S04	S05	M04	M05
Booklet 5	M05	M06	S05	S06
Booklet 6	S06	S07	M06	M07
Booklet 7	M07	M08	S07	S08
Booklet 8	S08	S09	M08	M09
Booklet 9	M09	M10	S09	S10
Booklet 10	S10	S11	M10	M11
Booklet 11	M11	M12	S11	S12
Booklet 12	S12	S13	M12	M13
Booklet 13	M13	M14	S13	S14
Booklet 14	S14	S01	M14	M01

Note: Parts 1 and 2 are separated by a break and consist of 36 minutes of testing for fourth grade and 45 minutes for eighth grade.

carefully controlled, as these blocks had been presented in an essentially identical booklet design in 2007.

PIRLS

PIRLS is an international assessment that focuses on reading literacy in the elementary school grades. In 2006, 40 countries participated in PIRLS with a total sample size of 215,137. In 2006, the assessment contained 10 item blocks. Each item block is preceded by a text that students have to read to answer the items. The number of items within an item block varies from 11 to 14 (Martin et al. 2007), with a total of 126 items. Passages and their associated item sets are classified into one of two purposes-of-reading domains—reading for information (I) and reading for literary experience (L). Table 10.4 shows the booklet design used in PIRLS 2006. A total of 13 booklets were used. Each booklet contains two blocks. Six booklets consist of two blocks from the same reading domain, while the remainder contains one block from each of the domains. Eight of the 10 blocks (L1 to L4, and I1 to I4) appear in three booklets. Position is only partially balanced, as each of these blocks appears twice in the first position and once in the second position (or vice versa). The final two blocks (L5 and I5) appear only in the Reader booklet.

The Reader booklet is produced in a magazine-like style and in color, while the other booklets are printed in black and white. The Reader is assigned to students at three times the rate of the other booklets, providing a more robust sample of students for the Reader (one-fifth of the overall sample, instead of one-thirteenth, as it would appear from looking at the design). Rotating the Reader at three times the rate of the rest of the booklets also

TABLE 10.4

PIRLS 2006 Booklet Design

Booklet	Part 1	Part 2
1	**L1**	**L2**
2	**L2**	L3
3	L3	L4
4	L4	**I1**
5	**I1**	**I2**
6	I2	I3
7	I3	I4
8	I4	**L1**
9	**L1**	**I1**
10	**I2**	**L2**
11	L3	I3
12	I4	L4
Reader	L5	I5

ensures that each item is presented to approximately the same number of students throughout the design. The lack of linkage between the Reader and the other booklets means that the PIRLS design relies on the assumption that randomization of the booklet assignment will ensure that the results of the Reader can be compared to the results of the other 12 booklets (Mullis et al. 2006, p. 41). The other 12 booklets, however, are designed in such a way that each block appears in each block position at least once.

To allow for linking of scales across years, 4 of the 10 blocks included in 2006 (two from each of the reading domains) were identical to blocks that had been part of the 2001 PIRLS assessment. These blocks are indicated in bold in Table 10.4. The design in 2001 was similar to that of 2006 in that each block appeared in three booklets—twice in first position and once in the second position, or vice versa (Gonzalez 2003). Thus, the use of intact blocks and a similar design results in a similar context across years for the repeated material.

A Closing Comment on Designs

It is worth noting that all assessment designs face challenges by the limitations of testing time, the need to incorporate new subdomains or emphasize one domain over the other, or the need to test a lot of students with one approach, which is less authentic than envisioned while maintaining a subsample that receives more authentic testing material. These challenges led PISA to adopt a rotation of minor and major domains and PIRLS to use an unlinked, but more authentic booklet. These decisions, in turn, have an effect on the strength of the linking of blocks across booklets and of booklets over time. All current assessments—and most likely all assessments for the foreseeable future—make some concessions and adapt the design to enable the incorporation of desired features. References throughout the section "Some General Design Considerations" provide examples of methodological research that address the effects of incorporating design features that deviate from an ideal, completely balanced, linking design. Operational analytic procedures are informed by these studies and are reviewed on a regular basis to ensure that best practice is adopted. The next section discusses current procedures used in scaling and linking procedures.

Analysis Procedures: How Scale Linking Is Carried Out in PISA, TIMSS, and PIRLS

As noted in Chapter 7, similar IRT-based analysis procedures are used by the major ILSAs, both in general and specifically in regards to scale linking. In most respects, the procedures used in TIMSS and PIRLS are modeled after those used with NAEP—the first large-scale assessment to employ

IRT methods. NAEP, along with TIMSS (since the 1995* cycle; Yamamoto and Kulick 2000) and PIRLS, makes use of the two- and three-parameter logistic models (2PL and 3PL) and the generalized partial-credit models. PISA makes use of binary and partial-credit Rasch models (Adams and Wu 2002) and uses scale-linking methods more common to that family of models. All the major ILSAs carry out their analyses in four distinct stages—scaling, conditioning, generation of plausible values, and final scale transformation. As the analysis procedures required for establishing links in TIMSS and PIRLS involve all four stages, we provide a brief review here but refer the reader to Chapter 7 for a more extensive discussion.

In Stage 1, IRT item parameters are estimated along with the parameters of a simplified population proficiency distribution—typically a normal distribution. As discussed in more detail below, in the base year of an assessment, the scaling models assume a single population. In subsequent years, parameters for multiple distinct proficiency distributions are estimated. For TIMSS and PIRLS, this scaling phase has been accomplished with a version of SSI's PARSCALE (Muraki and Bock 1997), while in PISA, the scaling work has been conducted in ConQuest (Wu et al. 1997). In most cases, an overall scale is estimated for reporting each content domain (e.g., mathematics and science in TIMSS), as well as separate subscales within a content domain (e.g., in mathematics, algebra, and geometry).

The scaling step is typically carried out on a specially constructed scaling sample that is made of equal-sized randomly selected samples from each of the participating countries. In this way, data from each of the participating countries contribute equally to the determination of the item parameters used to score the assessment. As discussed in the sections "PISA," "TIMSS," and "PIRLS," owing to the booklet designs used in ILSAs, any given item is administered to only a subset of the test-taker sample in each jurisdiction. However, because the scaling stage involves aggregating across jurisdictions and—as noted below in the case of TIMSS and PIRLS—data are pooled from adjacent assessment cycles, per-item sample sizes for this scaling step are quite large (several thousand or more).

In the second stage—*conditioning*—the first-stage item parameters are treated as fixed and known, and used, along with test-taker data—to estimate more complex population models for each jurisdiction (typically countries in ILSAs) for which results will be reported. The second-stage population models (sometimes referred to as *conditioning models*) typically take the form of univariate or multivariate linear regression models, with IRT proficiency as the latent (i.e., unobserved) dependent variable and demographic and other student, teacher, or school characteristics collected as part of the assessment as the observed independent variables (Mislevy 1991).

* The 1995 TIMSS assessment results were originally analyzed and reported using the Rasch and partial credit models. The data from this assessment was later rescaled using 2PL, 3PL, and the generalized partial credit model and revised results were issued.

Thus, in these more complex models, IRT proficiency is modeled as normally distributed, conditional on background variables, with a common variance. These second-stage models are what generate the reported statistics that make up the publications and secondary-use data files associated with these assessments. However, for a number of practical reasons, this generation takes place through two subsequent steps.

In the third stage, the second-stage regression model estimates, test-taker responses, and background variables are used to produce a set of plausible values for each test taker—the details of which are described in Chapter 7. The plausible values are random draws from the estimated posterior distribution of student proficiency, given the item responses, background variables, and the regression model estimates produced in Stage 2. Second-stage estimates are produced separately for each of the jurisdictions participating in the assessment using a common set of item parameters, ensuring that the results of these separate runs are on a common scale.

In the fourth and final stage of the analysis, these plausible values are linearly transformed to the metric used by each of the assessments for the reporting of results. In the base year of the assessment, this transformation is typically established arbitrarily to provide a convenient reporting metric. In subsequent assessment cycles, this transformation is derived so current assessment results are expressed in terms of the same metric established in the base year. The distribution of these transformed plausible values—more specifically, various functions thereof, such as the means, variances, percentiles, and percentages above fixed benchmarks—constitute the reported results of ILSAs.

Within an overall IRT analysis framework, a number of strategies exist for linking the scales from multiple years of assessments on the same scale (see, e.g., Donoghue and Mazzeo 1992; Hedges and Vevea 1997; Petersen et al. 1989; Yamamoto and Mazzeo 1992). To date, TIMSS and PIRLS have used a variant of the concurrent calibration approach applied in NAEP. Specifically, scales are linked by conducting an ongoing series of concurrent calibrations, using the data from successive pairs of adjacent assessments in the time series (e.g., year 1 and year 2; year 2 and year 3; and year 3 and year 4) that share common items (see Jenkins et al. 1999, for an example from NAEP). Donoghue and Mazzeo (1992) and Hedges and Vevea (1997) compared the concurrent calibration used for NAEP to some of the alternatives. Both studies suggest the procedure works well in maintaining the comparability of results across assessment years.

For a given pair of assessment years, the first step in the linking process occurs at Stage 1—the IRT scaling step. Data from the current assessment are pooled with data from the assessment immediately prior, and a two-group concurrent scaling is carried out. Only countries that participate in both assessments are used in this concurrent calibration. For example, for the 2007 TIMSS assessment, data from the countries common to both the 2003 (prior) and 2007 (current) assessments were pooled and concurrently scaled, with the two groups defined by the prior and current assessment (Foy et al. 2008, pp. 235–236).

In the concurrent scaling approach, the parameter estimates for the items that appeared in both assessments—the trend items—are obtained subject to the constraint that the same set of item parameters characterize the psychometric functioning of the items in both years. In other words, for the trend items, data from both the prior and current year are used to obtain a single set of trend-item parameters. Assuming common item-characteristic curves for items that appear in different assessment years may or may not be consistent with the data, and if the data are not consistent with common item characteristic curves (ICCs), the validity of various comparisons across the different assessment cycles can be affected. Assessment programs that use this concurrent scaling approach routinely check whether these assumptions are appropriate and take corrective action (typically estimating separate ICCs for the current and prior year) when necessary.

PISA assessments have taken a common-item approach consistent with the Rasch model, where possible. The data from each assessment cycle are scaled separately at Stage 1. That is, unlike the approach used in TIMSS and PIRLS, in PISA, the item parameters estimates in a given cycle are obtained using only the data from that cycle. For example, the item parameter estimates used to produce the 2009 PISA results were estimated with data from the 2009 assessment only (OECD 2012, pp. 197–198). No explicit constraints regarding equality of item parameters across cycles are imposed on the estimates for the trend items included in the current assessment. The degree to which these current estimates for the trend items align with estimates from earlier cycles in a manner predicted by the Rasch model is also checked and appropriate corrective action (not including this item in determining the linking of the current scales to the reporting metric) taken where necessary.

In the Stage 1 scaling phase for TIMSS and PIRLS, the proficiency distributions for the scaling samples* for each year are assumed to be normal, but separate means and variances are simultaneously estimated with the item parameters. The linear indeterminacy of IRT scales can be resolved any number of ways. In TIMSS and PIRLS, this indeterminacy is resolved by constraining the combined scaling data set (prior and current assessment) to have a proficiency mean of 0 and combined proficiency variance of 1. In PISA, normality for the current year's scaling sample is assumed and the indeterminacy is resolved by constraining the mean of the item difficulty parameter estimates to a constant for items included in the current assessment.

In Stages 2 and 3, all three assessments, TIMSS, PIRLS, and PISA, estimate jurisdiction-specific *conditioning* models and produce plausible values based on the Stage 1 item parameters. However, these plausible values are on a

* As noted earlier, the scaling samples used in ILSAs are aggregations made up of approximately equal-sized samples from a collection of the participating countries that meet certain qualifying conditions.

provisional metric. For PISA, this metric can, in principle, be used to compare country results within the current assessment. For PIRLS and TIMSS, the constraint of common item parameters for the trend items establishes a common provisional metric on which the resulting sets of plausible values for the current and prior assessments can, in principle, be directly compared. However, owing to the linear indeterminacy associated in general with IRT latent variable scales (Yen and Fitzpatrick 2006, p. 123), these provisional metrics are not identical to the reporting scales established in the base year. Further work in Stage 4 is required to link these provisional metrics to the official reporting scales.

There are two general approaches to obtaining the final transformation to reporting scales—one based on item-parameter estimates and one based on population proficiency distributions. Consistent with its use of Rasch models, PISA has tended to rely largely on item-parameter-based methods, whereas TIMSS and PIRLS have tended to follow the procedures used in NAEP and rely on proficiency distribution-based methods.

The procedures used in PISA have varied to some extent with changes in assessment designs. Therefore, it is probably best to illustrate the general approach by discussing a particular assessment administration—in this case, the 2009 assessment (OECD 2012, pp. 229–230). For the mathematics and science assessments—the minor domains in 2009—a "Mean-Mean" common-item equating approach (Yen and Fitzpatrick 2006, p. 134) was used. As noted above, the 2009 scaling stage produced, by definition, an average item difficulty value of zero in the Rasch model logit metric for the common items. As the common items are a subset of the items included in 2006, no such restriction existed for the average item difficulty in the 2006 scaling step. The difference between 0 and the 2006 logit-metric average item difficulty for the common items was determined, and this difference was used to adjust the 2009 plausible values to the 2006 logit-metric scale. These adjusted plausible values were, in turn, linearly adjusted to the reporting metric, using the same transformation that was used to convert the 2006 results to the PISA reporting metric.

For the major domain of reading in 2009, a two-step procedure that made use of both common-item and common-population approaches was used.* Step 1 was analogous to the common-item adjustment used for the minor domains. A scaling of the standard books was carried out using the full set of items. For the trend items, the difference in average item difficulty estimates between the 2009 and 2006 scaling was determined. This difference made up one component of the trend adjustment. Step 2 was a common-population

* The procedure described above is discussed in some detail in Chapter 12 of the PISA 2009 Technical Report. The reasoning behind the two-step procedure is not entirely clear to the current authors. It appears that those responsible for the analysis were concerned that framework and assessment content revisions for 2009 Reading introduced the possibility that trend results would be different when scaling with trend items only or with trend and new items in the major domain. The two-step procedure was intended to address any main effect associated with this concern.

adjustment. A second scaling using the same international scaling data set was carried out. This second scaling used only the 2009/2006 trend items and was carried out under the constraint that the mean difficulty of these items was zero. The difference in mean proficiency for this OECD-only data set under the two scalings made up the second component of the trend adjustment. As with the minor domains, the 2009 PISA plausible values were adjusted by the two components and then linearly adjusted to the reporting metric using the same transformations used in prior PISA assessments.

Proficiency-distribution-based approaches, like those used in PIRLS and TIMSS, capitalize on the fact that for the jurisdictions that participated in the prior assessment, two sets of plausible values can be obtained. One set—used to report prior assessment results—is based on the Stage 1 item-parameter estimates from the prior administration. A transformation of this first set to the reporting scales already exists. A second set of plausible values can be obtained based on the Stage 1 item-parameter estimates produced for the current assessment. The transformation to the reporting scales for the current assessment can then be obtained by following these steps:

1. Determine the mean and variance for the proficiency distribution using prior-assessment plausible values.
2. Transform the mean and variance to determine the mean and variance of the proficiency distributions using current-assessment plausible values.
3. Determine the linear equation that sets the means and variances in Step 2 equal to those in Step 1.
4. Concatenate the transformation in Step 3 with the transformation used to express prior-year assessment results on the reporting scales.

Current Challenges and Future Directions

Comparability of Scale Scores across Countries

No statistical model fits real data perfectly. This insight led Box and Draper (1987) to the well-known adage that all models are wrong, while some models are useful. Whether tests are linked by means of adjustment methods such as test equating (Kolen and Brennan 2004) or by methods that define a common measurement model for the two forms to establish ability variables that are on the same scale, statistical and model error will always contribute to the level of uncertainty about the resulting linkage. In ILSAs, there are additional sources of concern because these assessments are conducted in multiple languages and put on the same reporting scale. This, in itself, is an example of a complex linkage of statistical models over multiple populations. Criticisms have been raised that appear to imply that the scales established across countries do not hold the promise of comparability (Ercikan and Koh 2005).

While the issues raised in these studies should not be taken lightly, the analyses conducted on them rely on comparably small country-level samples and are therefore prone to rely on estimates that are associated with substantially larger uncertainty than the international estimates that are the target of criticism.

Consider, for example, the PISA assessment. Each item cluster appears in about one-third of the booklets. Given that PISA targets a sample size of 4500 students per country, this practice yields about 1500 responses per cluster (and per item because each item is a part of exactly one cluster-assessment cycle). Moreover, PISA (like TIMSS and PIRLS) is a school-based survey, in which samples of about 150 schools per country are selected, followed by a sampling of students within these schools. This clustered sampling leads to a further reduction of the effective sample size because observations with schools tend to be statistically dependent (see Chapter 6 by Rust in this volume). Typical sizes of this reduction from observed to effective sample size (design effects) are in the range of 2–6 (but more extreme values are possible) in these types of surveys, so the effective sample size per item ranges from 250 to 750.

Given these considerations, country-level effective sample sizes per item may be insufficient to allow reliable estimation of IRT models. This means that the apparent low correlations between item-parameter estimates reported in these studies, which show that country-level scales are not comparable, may be due in large part to the unreliable country-specific item-parameter estimates. Gonzalez (2008) has shown that even when using country-specific item parameters to estimate proficiency results, the effects on cross-country comparisons are minimal. A study by Barth et al. (2009) showed that even after allowing each country to eliminate items that are not covered in its national curriculum, the scale scores for international comparisons are virtually unchanged.

Instead of relying on country-level item parameters, alternative approaches that combine item-based fit diagnostics (Molenaar 1983; Rost and von Davier 1994) with generalized concurrent calibration-based linking methods (von Davier and von Davier 2007) can be developed (Oliveri and von Davier 2011).

Maintaining Comparable Construct Meaning for Scale Scores over Time

While comparing results across countries within an assessment cycle is based on the same set of items used in that cycle, linking to report trends over time has the added challenge that only a part of the assessment (typically a subset of item clusters) is common between two assessment cycles. Assessment domains as described in assessment frameworks are subject to sometimes subtle and sometimes more obvious changes (Stacey 2012). Experts review the assessment domains and frameworks to ensure they (still) reflect current research and practice in the content domain of interest. Thus, the newly developed items, unique to the current assessment, may involve somewhat different skills and competencies than the unique items from prior assessment years and, thereby, change the substantive meaning of the scales derived through the use of IRT methods. Owing to the addition of item blocks based on a revised framework in the renewal process, there is confounding of assessment cycle (cohort) changes and changes in scale meaning due to the new item material. Adams (2009) referred to these issues when, for the first PISA Research Conference, he titled his invited address, "Trends: Are They an Outrageous Fortune or a Sea of Troubles?"

Linking Computer-Delivered and Paper-and-Pencil Assessments

The successful completion of a task given to respondents is determined not only by the ability of the respondent and task features but also by the lack of nuisance factors such as noise level and distractions. In addition, there are systematic factors that may influence the probability of a successful task completion. One of these factors is the mode of delivery of the task: When moving from a paper-and-pencil assessment to a computer-delivered assessment, factors such as familiarity with information technology may have an impact on the likelihood of successful task completion. Even if great care was taken to make the task look the same on the computer as on paper, the mode of input is different and remains different even if tablets with dynamic capturing software for written responses are used. Computers may make it harder to solve a task for some test takers who, for example, are less familiar with technology than others. However, computers do increase the ease with which respondents can produce text, at least for those who are used to typing on a keyboard rather than writing with a pencil.

To examine and quantify the effects of moving assessments to the computer, researchers conduct mode-effect studies, in which the same assessment is given in both computer and paper-and-pencil formats. These studies in themselves often are linking studies because respondents cannot be given the same assessment twice, and, typically, switching between modes in the middle of the testing session is avoided to eliminate potential effects. Therefore, mode-effect studies usually use an assessment design that administers the computer and paper-and-pencil assessments to randomly equivalent samples.

To determine what can be done if mode effects are found, a distinction has to be made between the level and type of mode effects detected in the study. In the ideal case, there is no mode effect, and items can be assumed to function the same way across modes of delivery. Somewhat more realistic is a finding that detects mode effects in some but not all items; for instance, some items administered on the computer might seem more difficult while others might seem easier. In this case, methods similar to the generalized IRT linking procedures (von Davier and von Davier 2007) can be applied to link the assessments across modes of delivery.

If it appears that some respondents find it challenging to perform tasks on the computer, but the assessment appears to work across modes for the majority of respondents, a hybrid design can route less technologically familiar respondents to the paper-and-pencil version. While these methods have been shown to work well in terms of accounting for and controlling the impact of mode effects on comparability (Beller 2012; Kirsch et al. 2012), a single study on moving an assessment from a paper-and-pencil format to a computer-delivered format cannot account for what will happen in the future, when the assessment is given on more advanced computers or on different types of devices. Approaches for handling this ongoing evolution of delivery platforms will be discussed in the next section.

The above discussion pertains to situations in which the same kinds of items are delivered in both paper-and-pencil and computer-delivered format. An additional challenge associated with computer delivery of assessments is that the delivery technology may allow the introduction of novel assessment tasks that change the nature of the domain being assessed. For example, the introduction of computer-delivered tasks that require students to conduct simulated experiments may enable more direct measurement of scientific inquiry skills than was feasible with paper-and-pencil testing. In such instances, considerations like those discussed in the section "Maintaining Comparable Construct Meaning for Scale Scores over Time" are pertinent as well.

Fast-Paced Technological Development and Linking Computer-Delivered Assessments

Moore's law (Moore 1965), while recently disputed, has quite accurately predicted for more than four decades that computing power doubles every 2 years. Today's smartphones have more computing power than the personal computers of 10 years ago. The impact of this technological development on content delivered by computer, phones, tablets, and so on is profound in general; it has led to more dynamic content (i.e., animation and video), which in turn leads test takers to expect a certain look and feel from technology-delivered content.

Moore's law also implies that computer-delivered items from 2012 will most likely look outdated to users in 2021. This leads to the question of how to maintain comparable scales and linking assessments against the backdrop of fast-paced technological advances. While some may argue that mathematics skills and knowledge do not change, others might argue that many tasks, including higher mathematics, become easier with more powerful computers. Do we still need to test student abilities to perform tasks that are essentially made obsolete by computers? Is a science inquiry item that requires the student to produce a graph using mouse clicks and keystrokes comparable to one that uses voice recognition and only requires students to say aloud which variables to graph?

The question is whether surface characteristics that change quickly due to technological advances will lead to changes in the requirements of underlying skills and knowledge. If students increasingly use technology in everyday activities, and if these technologies become easier to apply to everyday problems over time, then the traditional concept of linking assessments over time by means of tasks that stay the same and look the same becomes unsuitable. Consequently, psychometric and content experts will need to develop an evidence base as to whether a trend scale can still be maintained over several cycles in this situation. One reason for this concern is that at least some proportion of test takers will consider linking items from previous cycles to be outdated and the activities needed to solve these items will become increasingly unfamiliar, thus introducing construct-irrelevant variance.

Another factor is the more diverse set of countries participating in these assessments, in which the introduction of these technological developments takes place at different points in time. While some countries are moving toward using tablets to access cloud computing instead of using paper-based textbooks, other countries still have not yet introduced networked technology into everyday instruction. The inevitable move to new modes of assessment, most likely delivered by technology, poses challenges in how to account for this diversity across the 80 or so countries that will likely participate in the next round of ILSAs.

In summary, the challenges of linking ILSAs will likely grow with the increasing use of fast-developing information technology in educational and everyday contexts. Linking approaches and psychometric models for complex linkages exist, but only future data collections can show how well the assumptions behind these linkages work when building analytical models for such a diverse sample of participating jurisdictions. While current applications that compare computer-delivered to paper-and-pencil assessments have shown that several challenges have to be met by psychometric methodologies used to link ILSAs, the main aim is to ensure that these assessments are indeed comparable across participating countries and over time. Future changes to the assessment frameworks, technological advances in testing and in schools, and participating countries' increasing appetite for more information about their students will produce the need for further advances in the methodologies used in these assessments and a different understanding of what it means to say that results are comparable over time.

References

Adams, R. 2009. Trends: Are they an outrageous fortune or a sea of troubles? Invited address at the *1st PISA Research Conference*, Kiel, Germany.

Adams, R. and Carstensen, C. 2002. Scaling outcomes. In *PISA 2000 Technical Report*, ed. R. J. Adams and M. Wu. Paris: OECD.

Adams, R. J. and Wu, M. 2002. *PISA 2000 Technical Report*. Paris: OECD.

Barth, J., Rutkowski, L., Neuschmidt, O., and Gonzalez, E. 2009. Curriculum coverage and scale correlation on TIMSS 2003. *IERI Monograph Series*, 2, 85–112.

Beaton, A. E. and Zwick, R. 1990. *Disentangling the NAEP 1985–86 Reading Anomaly*. (NAEP Report No. 17-TR-21). Princeton, NJ: Educational Testing Service.

Beller, M. 2012. Technologies in large scale assessments: New directions, challenges, and opportunities. In *The Role of International Large Scale Assessments: Perspectives from Technology, Economy, and Educational Research*, ed. M. von Davier, E. Gonzalez, I. Kirsch, and K. Yamamoto. New York, NY: Springer.

Box, G. E. P. and Draper, N. R. 1987. *Empirical Model Building and Response Surfaces*. New York, NY: John Wiley & Sons.

Carstensen, C. H. 2012. Linking PISA competencies over three cycles—Results from Germany? In *Research in the Context of the Programme for International Student Assessment*, ed. M. Prenzel, M. Kobarg, K. Schöps, and S. Rönnebeck. New York, NY: Springer.

Donoghue, J. R. and Mazzeo, J. 1992. *Comparing IRT-Based Equating Procedures for Trend Measurement in a Complex Test Design*. Paper presented at the annual meeting of the National Council on Measurement in Education, San Francisco.

Ercikan, K. and Koh, K. 2005. Examining the construct comparability of the English and French versions of TIMSS. *International Journal of Testing*, 5(1), 23–35.

Foy, P., Galia, J., and Li, I. 2008. Scaling the data from the TIMSS 2007 mathematics and science assessments. In *TIMSS 2007 Technical Report*, ed. J. F. Olson, M. O. Martin, and I. V. S. Mullis. Chestnut Hill, MA: TIMSS & PIRLS International Study Center, Boston College.

Gebhardt, E. and Adams, R. J. 2007. The influence of equating methodology on reported trends in PISA. *Journal of Applied Measurement*, 8(3), 305–322.

Gonzalez, E. 2003. Scaling the PIRLS reading assessment data. In *PIRLS 2001 Technical Report*, ed. M. O. Martin, I. V. S. Mullis, and A. M. Kennedy: 151–168. Chestnut Hill, MA: TIMSS & PIRLS International Study Center, Boston College.

Gonzalez, E. 2008. *Interpreting DIF in the Context of IEA's International Assessments*. Paper presented at the annual meeting of the National Council on Measurement in Education, New York.

Gonzalez, E., Galia, J., and Li, I. 2004. Scaling methods and procedures for the TIMSS 2003 mathematics and science scales. In *TIMSS 2003 Technical Report*, ed. M. O. Martin, I. V. S. Mullis, and S. J. Chrostowski. Chestnut Hill, MA: TIMSS & PIRLS International Study Center, Boston College.

Hedges, L. V. and Vevea, J. L. 1997. *A Study of Equating in NAEP*. Palo Alto, CA: American Institutes for Research.

Jenkins, F., Chang, H.-H., and Kulick, E. 1999. Data analysis for the mathematics assessment. In *The NAEP 1996 Technical Report*, ed. N. L. Allen, J. E. Carlson, and C. A. Zelenak: 255–289. Washington, DC: National Center for Education Statistics.

Kirsch, I., Lennon, M. L., von Davier, M., Gonzalez, E., and Yamamoto, K. 2012. On the growing importance of international large scale assessments. In *The Role of International Large Scale Assessments: Perspectives from Technology, Economy, and Educational Research*, ed. M. von Davier, E. Gonzalez, I. Kirsch, and K. Yamamoto. New York, NY: Springer.

Kolen, M. J. and Brennan, R. L. 2004. *Test Equating, Scaling, and Linking*. New York, NY: Springer.

Le, L. T. 2009. Effects of item positions on their difficulty and discrimination—A study in PISA science data across test language and countries. In *New Trends in Psychometrics*, ed. K. Shigemasu, A. Okada, T. Imaizumi, and T. Hoshino. Tokyo: Universal.

Martin, M. O., Mullis, I. V. S., and Kennedy, A. M. 2007. *PIRLS 2006 Technical Report.* Chestnut Hill, MA: TIMSS & PIRLS International Study Center, Boston College.

Mazzeo, J. and von Davier, M. 2008. *Review of the Programme for International Student Assessment (PISA) Test Design: Recommendations for Fostering Stability in Assessment Results.* http://www.oecd.org/dataoecd/44/49/41731967.pdf (accessed December 12, 2008).

Mislevy, R. 1990. Scaling procedures. In *Focusing the New Design: The NAEP 1988 Technical Report*, ed. E. G. Johnson and R. Zwick. (Report No. 19-TR-20). Princeton, NJ: ETS.

Mislevy, R. J. 1991. Randomization-based inference about latent variables from complex surveys. *Psychometrika*, 56, 177–196.

Molenaar, I. W. 1983. Some improved diagnostics for failure in the Rasch model. *Psychometrics*, 48, 49–75.

Moore, G. E. 1965. Cramming more components onto integrated circuits. *Electronics Magazine*, 38(65), 1–4. ftp://download.intel.com/museum/Moores_Law/Articles-Press_Releases/Gordon_Moore_1965_Article.pdf (accessed November 11, 2006).

Mullis, I., Kennedy, A., Martin, M., and Sainsbury, M. 2006. *PIRLS 2006 Assessment Framework and Specifications*, 2nd ed., Newton, MA: TIMSS & PIRLS International Study Center, Boston College. http://timssandpirls.bc.edu/PDF/P06Framework.pdf (accessed April 30, 2012).

Muraki, E. and Bock, R. D. 1997. *PARSCALE: IRT Item Analysis and Test Scoring for Rating Scale Data.* Chicago, IL: Scientific Software.

Nunnally, J. 1967. *Psychometric Theory.* New York, NY: McGraw-Hill.

OECD. 2012. *PISA 2009 Technical Report.* http://www.oecd.org/document/19/0,3746,en_2649_35845621_48577747_1_1_1_1,00.html (accessed April 2, 2012).

Oliveri, M. E. and von Davier, M. 2011. Investigation of model fit and score scale comparability in international assessments. *Psychological Test and Assessment Modeling*, 53(3), 315–333. http://www.psychologie-aktuell.com/fileadmin/download/ptam/3-2011_20110927/04_Oliveri.pdf (accessed September 29, 2011).

Rost, J. and von Davier, M. 1994. A conditional item fit index for Rasch models. *Applied Psychological Measurement*, 18, 171–182.

Petersen, N. S., Kolen, M. J., and Hoover, H. D. 1989. Scaling, norming, and equating. In *Educational Measurement*, ed. R. L. Linn: 221–222. New York, NY: Macmillan.

Stacey, K. 2012. *The International Assessment of Mathematical Literacy: PISA 2012 Framework and Items.* Paper presented at the 12th International Congress on Mathematical Education, COEX, Seoul, Korea.

Thorndike, E. L. 1914. *Educational Psychology. Mental Work and Fatigue and Individual Differences and Their Causes*, Vol. III. New York, NY: Teachers College, Columbia University.

Urbach, D. 2012. An investigation of Australian OECD PISA trend results. In *Research in the Context of the Programme for International Student Assessment*, ed. M. Prenzel, M. Kobarg, K. Schöps, and S. Rönnebeck. New York, NY: Springer.

van der Linden, W. J. and Hambleton, R. K. 1997. Item response theory: Brief history, common models, and extensions. In *Handbook of Modern Item Response Theory*, ed. W. J. van der Linden and R. K. Hambleton. New York, NY: Springer-Verlag.

von Davier, M. and von Davier, A. 2007. A unified approach to IRT scale linkage and scale transformations. *Methodology*, 3(3), 115–124.

Wu, M. L., Adams, R. J., and Wilson, M. R. 1997. ConQuest; Multi-aspect test software [computer program]. Camberwell, Victoria, Australia: Australian Council for Education Research.

Yamamoto, K. and Kulick, E. 2000. Scaling methodology and procedures for the TIMSS mathematics and science scales. In *TIMSS 1999 Technical Report: Description of the Methods and Procedures Used in IEA's Repeat of the Third International Mathematics and Science Study at the Eighth Grade*, ed. M. O. Martin, K. D. Gregory, and S. E. Stemler: 237–263. Chestnut Hill, MA: TIMSS & PIRLS International Study Center, Boston College.

Yamamoto, K. and Mazzeo, J. 1992. Item response theory scale linking in NAEP. *Journal of Educational Statistics*, 17, 155–173.

Yen, W. M. and Fitzpatrick, A. R. 2006. Item response theory. In *Educational Measurement (Fourth Edition)*, ed. R. L. Brennan. Westport, CT: American Council on Education and Praeger Publishing.

11

Design Considerations for the Program for International Student Assessment

Jonathan P. Weeks
Educational Testing Service

Matthias von Davier
Educational Testing Service

Kentaro Yamamoto
Educational Testing Service

CONTENTS

The Programme for International Student Assessment (PISA) is a collaborative effort commissioned by the Organization for Economic Co-operation and Development (OECD) to provide international comparisons of student performance in reading, math, and science. These comparisons include snapshots of achievement in a given year in addition to trend information. The assessment is administered every 3 years and has a unique design; every time PISA is administered, one of the three content areas is treated as a major domain while the other two are treated as minor domains. This major/minor domain design is implemented by reducing the number of items for each of the minor domains relative to the full set of items that would have been used for a major domain; there is a reduction in items of around 40–75%, depending on the content domain. This reduction in the number of items has the potential to introduce bias into estimates of country means. The question of interest for this chapter is how the PISA design can be modified in a way that maintains an emphasis on the major/minor domain perspective while minimizing bias and potential issues in the linking process, which are likely to affect trend results.

PISA items are administered in a wide range of countries; hence, there is a strong potential for item-by-country interactions. That is, items may not function identically across all participating countries.* In current practice, international item parameters are estimated with the Rasch model using a representative sample of item responses from all of the participating OECD countries. This is done so that country results can be reported on a common scale, yet these international parameters may be biased relative to national (country-specific) item parameters that would have been obtained if they were estimated for each country separately (Gebhardt and Adams 2007). As such, the use of international item parameters may result in biased estimates of country means. By administering a large sample of items, there is an implicit assumption that the mean bias across items for each country is relatively small; yet, this may not be the case for smaller subsets of items (e.g., items associated with minor domains). This can be problematic in a single year, but the issue is likely to be exacerbated in trend results due to changes in the number of items associated with each content area as part of the move from major to minor domains and back again. Previous examinations of PISA trends (Carstensen 2013; Gebhardt and Adams 2007; Urbach 2013) address the issue of bias due to item-by-country interactions; however, none of these studies explicitly addresses the impact of the major/minor domain design.

PISA Design

For each of the three content domains in PISA there are around 120 items as part of the major domain item pool. These items are distributed across seven item blocks (in PISA, these are called clusters) in each content area.† When a given area is the major domain, all seven clusters are administered. When reading is a minor domain only two clusters are administered, and when

* While these interactions are monitored over the course of analyses, the interactions by country are not eliminated (they are documented). Items are only eliminated in rare cases when the variability of item difficulties across countries is very large.

† In the first administration of PISA in 2000, when reading was the major domain, there were nine reading clusters; in 2003, when math was the major domain, there were two additional problem-solving clusters.

math and/or science are minor domains, two to four clusters are administered. The combined set of major- and minor-domain item clusters are compiled into a set of booklets comprising four clusters each; they are distributed across booklets in a balanced incomplete block (BIB) design. The advantage of the BIB design is that it allows for more expansive content coverage without overburdening the examinees (e.g., Rutkowski, Gonzalez, Joncas, and von Davier 2010; von Davier, Gonzalez, and Mislevy 2009); however, the PISA design does not take full advantage of this framework. All seven clusters in each content area could be included readily as part of the BIB setup, yet the major/minor domain design was adopted instead.* A complete design would include 21 booklets as opposed to the 13 booklets used under the current design. One possible explanation for this decision is resource allocation; it may be financially and/or administratively prohibitive to include the additional booklets, particularly for a paper-based assessment. On the other hand, the design may have been chosen intentionally in order to emphasize certain content areas in different years.

Figure 11.1 illustrates the representation of the PISA content area item clusters from 2000 to 2009. The change in content representation associated with the major/minor domain design is obvious, yet it is also important to notice that the representation of the minor domain content areas is, in general, inconsistent over time. For instance, in the years where science is a minor domain (2000, 2003, and 2009) there are four, two, and three administered item clusters, respectively. This change in the number of item clusters from one administration to the next can present challenges when linking the scales.

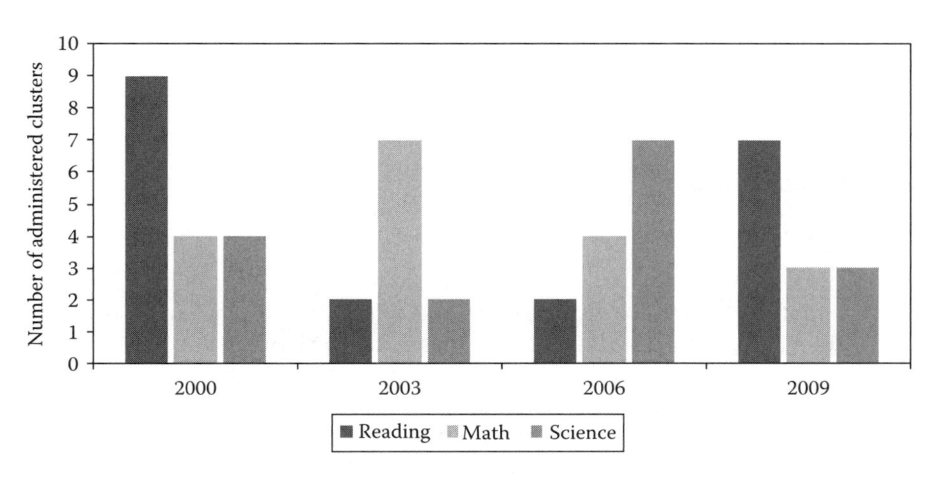

FIGURE 11.1
PISA content representation.

* There is no justification for the adoption of the major/minor domain design in any published reports.

There are a number of alternative designs that could have been adopted for PISA; however, we limit our examination to those designs that keep the spirit of the major/minor distinction intact. Given the current design, it is not immediately clear how the major/minor domain emphasis can be incorporated without reducing the number of items for the minor domains (or, conversely, increasing the items for the major domain). A less obvious choice might be increasing/reducing the number of examinees taking items in each domain. Our approach to this problem coincides more with the latter perspective by focusing on the sampling of examinees as opposed to the sampling of item clusters. In short, by maintaining the total set of item clusters in each content area and oversampling the number of examinees taking items in a given content area (to establish a major domain), the accuracy of performance estimates can be maintained or increased for each content domain. By administering all item clusters there should be less potential for bias due to item-by-country interactions, although the trade-off is likely to be less precision in the performance estimates for the minor domains. Our goal is to examine the trade-off between precision and bias under the current PISA design, an alternate design that allows for more flexibility in the specification of content emphasis (our proposed design), and a completely balanced design where all domains are "major" and all of the item clusters are equally represented.

Before proceeding with this comparison it is important to address the impact of systematic versus random error and why item-by-country interactions under the current PISA design are particularly problematic. This discussion is intended to serve as a rationale for our proposed design. Further, it is important to consider previous empirical examinations of PISA results to illustrate operational issues associated with the current design and to provide an evidentiary comparison to our simulation results.

Sources of Error

All large-scale assessments, including PISA, are subject to two types of errors: (1) systematic errors that introduce bias and affect the accuracy of results, and (2) random errors that do not introduce bias but reduce precision, thus increasing uncertainty in the results. Important determinants of these errors—apart from the overall sample size and the reliability of the test—are the test design, sampling methods, operational procedures, and the psychometric modeling and data analyses. Adjustments can be made to these facets to reduce random and systematic errors, yet given limited resources there is often a trade-off between reducing one source of error at the expense of another. An increase in random errors will reduce the ability to detect small group differences in average proficiency as well as the ability to detect larger average proficiency differences between small groups. An increase in systematic errors, on the other hand, not only reduces the ability to detect group differences, but may lead to the attribution of false differences (i.e., differences that are considered significant when the true differences are zero). As such, a reduction in systematic errors is generally preferred over a reduction in random errors.

Under the current design, PISA emphasizes the major domain at the cost of the quantity of data collected for the minor domains, primarily by reducing content representation while keeping the number of responses for the retained minor domain items at a level comparable to that of the major domain items.* One underlying premise of the current design is that the sample size for the minor domain items is sufficient for estimating item parameters. However, the reduction in the number of items from about 120 for a major domain to about 30 for minor domains brings with it the potential for systematic errors due to item-by-country interactions that are more pronounced in smaller item samples (Monseur, Sibberns, and Hastedt 2008). This type of error is likely to be realized in one of two ways: statistical bias (i.e., differential item functioning across countries) or construct bias (i.e., unintended bias due to poor representation of the construct—relative to construct representation in the major domain). For this study we focus on the problem of statistical bias.

Previous Studies of PISA Trends

There are several studies that illustrate the problem of item-by-country interactions related to the move from major to minor domains in each assessment cycle, particularly with respect to trend estimation. Gebhardt and Adams (2007) examined the influence of the choice of equating method on PISA trends and found that there can be significant effects associated with changes from the major to minor domain item sets for some countries. In particular, they found that when item-by-country interactions are present in the linking items, trend results can look notably different depending on whether the linking is conducted based on international item parameters versus country-specific item parameters. In the former case, trend interpretations may be misleading, while in the latter case, comparability between country trends is likely to decrease.

In two studies that share similar research questions, Urbach (2013) and Carstensen (2013) examined PISA trend results in Australia and Germany, respectively. Urbach compared the proficiency distributions between adjacent assessment years (2000–2003 and 2003–2006) in each content area for scales based on country-specific versus international item parameter estimates. He indicates that the magnitude of increases/declines in performance at certain percentile ranges (reported in the media) based on international

* A sufficient sample size per item is maintained in each administration.

parameter estimates would have been reported with a somewhat different emphasis if national parameter estimates had been used to report trends. In short, he showed that item-by-country interactions appear to have a noticeable effect on the results of PISA. This is consistent with the findings from Gebhardt and Adams (2007).

Carstensen discusses some of the limitations of trend analyses based on data from PISA and the changes between rotating major and minor domains across assessment cycles. He suggests that a more careful analysis of item-by-time-point interaction, a concept analogous to the item-by-country interaction that looks at changes of item difficulty over cycles, may lead to insights on the issue. His study concludes with the development of a model that utilizes only the minor domain Reading clusters as well as the common items of the minor domains in Mathematics and Science to determine a country-specific measure of trend. While a design like this has the potential to provide more stable trend estimates, it does not address the issue of item-by-country interactions.

To address the issue of item sampling in trend estimation, Monseur et al. (2008) examined trend results based on different subsets of items (through the elimination of item clusters). Using the Progress in International Reading Literacy Survey (PIRLS) as an example, they found that variance in the estimated item parameters, when estimated concurrently across countries, was primarily accounted for by item sampling (i.e., the number of items) and item-by-country interactions. Only a small portion of the variability was explained by item-by-time point-by-country interactions (i.e., changes in country-specific bias over time). These findings suggest that potential issues in PISA trend results are likely to be exacerbated under the major/minor design.

While none of the above studies directly quantifies the change in reported country mean scores that can be expected for a certain level of model misfit introduced by item-by-country interactions, all of the examples provide evidence of a lack of consistency of results when moving from major to minor domain sets.

Methods

Large-scale international assessments are typically analyzed by estimating the item parameters for each content area using item response data from all countries. While there are procedures in place to select those items that show the least variation in item difficulties across countries, it cannot be expected that the sample item statistics (as well as the expected value of item parameters) of all items agree perfectly across countries. Apart from model error, the estimated international item parameters will differ somewhat from the estimates that would be obtained in each country individually. This means that for each item cluster—composed of a limited subset of items—an estimate of the cluster-by-country interaction can be calculated based on country-specific versus international differences in difficulty in the items used in a given cluster. If these cluster differences exist (and it is most likely they will in a set of 60–70 countries that require translated versions of the source items developed in English or French), it is conceivable that cluster differences vary somewhat across the seven-item clusters used in PISA. More specifically, the country-specific cluster differences for minor domains will depend heavily on the item clusters that have been chosen to become the minor domain assessment.

To address the issue of item cluster-by-country interactions we conducted a simulation by varying the booklet design (i.e., the sampling of items) and the amount of bias associated with item clusters for nine simulated countries. The goal of the simulation is twofold: to examine (1) the extent to which bias is present in country means when different booklet designs are used, and (2) the associated standard errors around the estimated parameters under the various designs. These two goals are intended to address the trade-off between bias and random error.

As a first step, we created a booklet design that incorporated all of the content area clusters; that is, a design where all of the items in all content areas would be administered (see Table 11.1). We denote these clusters as {R1, R2, . . ., R7, M1, M2, . . ., M7, S1, S2, . . ., S7} for Reading (R), Math (M), and Science (S), respectively. The booklets were created using a BIB design similar to PISA. Conceptually, the booklet design for PISA can be treated as a weighted subset of the complete booklet design. The rightmost column in Table 11.1 illustrates (approximately) the number of examinees, given a total of 1000 examinees, who would take each booklet under the 2006 PISA design where Science was the major domain. It is important to notice that there are a number of booklets that would not be administered (i.e., booklets with weights equal to zero). This clearly characterizes the major/minor domain setup. As an alternative to the PISA weights, a different set of weights could be chosen so that each of the 21 booklets from the complete design would be included—at least to some extent. In the simplest case, equal weights could be chosen for all booklets. This is characterized in Table 11.1 by the column for the balanced design.

However, to maintain the major/minor distinction without reducing the number of administered items for the minor domains, weights could be chosen so that a greater proportion of booklets comprised predominantly of Science clusters would be included, relative to the other booklets.* Using this framework we established three design conditions: (1) a *balanced item design* where an equal representation of all booklets in the complete design are included, (2) a *trend design* (our proposed design) where different weights are chosen for the booklets in the complete design to correspond to major

* A linear programming approach similar to that presented in van der Linden (1998, 2004) could be used to identify an optimal set of weights; however, our intent is simply to show how potential bias is affected by different, plausible, weights.

TABLE 11.1

Complete Booklet Design

					# of Booklets Taken per 1000 Examinees		
Booklet	**Section 1**	**Section 2**	**Section 3**	**Section 4**	**Balanced**	**Trend**	**PISA**
1	S1	R1	R2	R3	48	20	0
2	S2	R2	R6	R4	48	20	0
3	S3	S1	S6	R7	48	110	77
4[a]	S4	S2	S1	M1	48	110	77
5[a]	S5	S3	S2	M2	48	110	77
6	S6	S7	S3	S1	48	130	77
7	S7	S5	S4	S2	48	130	77
8[b]	R3	S6	S5	M3	48	100	192
9[b]	R6	S4	S7	M4	48	100	192
10	M6	M5	R4	S3	48	20	77
11	M4	M7	R5	S4	48	20	77
12	R4	R5	M5	S5	48	20	0
13	R7	R6	M6	S6	48	20	0
14[a]	M3	M1	M2	S7	48	20	77
15	R1	R4	R7	R5	48	10	0
16	R2	R7	R3	R6	48	10	0
17	M7	M3	M1	M5	48	10	0
18	M2	M4	M7	M6	48	10	0
19	R5	R3	R1	M7	48	10	0
20	M1	M2	M3	R1	48	10	0
21	M5	M6	M4	R2	48	10	0

[a] Booklets 4, 5, and 14 include two math and two science clusters in the actual PISA design (see Table 11.2).

[b] There are five variations of the {R,S,S,M} booklet in the actual PISA design. In this illustration, booklets 8 and 9 are weighted 2.5 times each to arrive at the ≈192 examines per booklet.

and minor domains, and (3) the current *PISA design*. The booklet design presented in Table 11.1 is used for the balanced item and trend designs. A restructured set of booklets (see Table 11.2) is used for the PISA design to maintain consistency with how the tests are administered in practice.

Approximately 60% of the items in the PISA design correspond to the major domain and 20% of the items correspond to each of the minor domains. For our proposed Trend design we chose weights* so that 64% of the total clusters—used as a proxy for item representation—were associated with the major domain (Science) and around 18% of the clusters were associated with

* The following weights were used: $w = 0.01$ for booklets 15–21; $w = 0.02$ for booklets 1, 2, 10–14; $w = 0.10$ for booklets 8–9; $w = 0.11$ for booklets 3–5; $w = 0.13$ for booklets 6–7.

TABLE 11.2

PISA Booklet Design[a]

Booklet	Section 1	Section 2	Section 3	Section 4
1	S1	S2	S4	S7
2	S2	S3	M3	R3
3	S3	S4	M4	M1
4	S4	M3	S5	M2
5	S5	S6	S7	S3
6	S6	R6	R3	S4
7	S7	R3	M2	M4
8	M1	M2	S2	S6
9	M2	S1	S3	R6
10	M3	M4	S6	S1
11	M4	S5	R6	S2
12	R3	M1	S1	S3
13	R6	S7	M1	M3

[a] The 2006 booklet design with Science as the major domain was used as the reference for the PISA design. See the OECD technical report for more information: http://www.oecd.org/dataoecd/0/47/42025182.pdf

each of the minor domains. Based on these weights, all of the Science clusters, across booklets, were similarly represented; that is, a similar proportion of examinees were "administered" each cluster (see Table 11.3). Three of the Reading forms (R3, R6, and R7) were predominantly represented (compared to the two clusters in the PISA design where Reading was a minor domain), and four of the Math clusters (M1–M4) were predominantly represented.

In addition to the three booklet designs, we considered two-item cluster-by-country bias conditions. For the first condition, values of 0 or 0.1 logits were added to or subtracted from the item parameters for items associated with each of the seven clusters in each content area. This value was chosen since it is slightly higher than the standard error of the means in practice. The pattern of additions/subtractions differed across countries; however, the

TABLE 11.3

Proportional Representation of Each Cluster: Trend Design

	Item Cluster							
	1	2	3	4	5	6	7	Total
Reading	0.013	0.015	**0.035**	0.018	0.015	**0.038**	**0.038**	0.170
Math	**0.038**	**0.038**	**0.035**	**0.035**	0.015	0.015	0.013	0.188
Science	**0.093**	**0.093**	**0.093**	**0.090**	**0.090**	**0.090**	**0.095**	0.643

Note: Bold values identify the item clusters that are predominantly represented for each content area under the trend design.

mean bias for each country across all seven clusters (in each content area) was equal to zero. Similarly, the mean bias within clusters, across countries, was equal to zero. As such, we should expect the estimated item parameters for each cluster and the country means—when all seven clusters are administered—to be unbiased. On the other hand, when only a subset of the clusters are administered, as in the PISA design, we should still expect the item parameters to be unbiased, but we should observe a predetermined amount of bias for each country, in each content area. In Reading, two countries had a mean bias of −0.2 logits, two countries had a mean bias of 0.2 logits, and the remaining five countries had a mean bias equal to zero. In Math, two countries had a mean bias of −0.1 logits, two countries had a mean bias of 0.1 logits, and the remaining five countries had a mean bias equal to zero. Since all seven clusters were presented in Science, the mean bias for all countries was equal to zero. We refer to this as the *balanced bias* (BB) condition.

For the second condition, the item difficulties for each cluster for each country were adjusted by a random value drawn from a normal distribution with a mean of zero and a standard deviation of 0.2 logits.* These additions/subtractions were centered across all countries for each cluster so that, as in the BB condition, the estimated item parameters should be unbiased. However, no constraints were placed on the mean bias across clusters for each country in each content area. This is intended to serve as a more realistic condition. We refer to this as the *non-balanced bias* (NBB) condition.

Hypotheses

Below is a summary of our hypothesized expectations prior to conducting the simulation.

1. In all of the simulation conditions the estimated item parameters across countries in each content area should be unbiased.
2. In the BB condition, the country means should be unbiased for all booklet designs in Science and for the balanced item design in both Reading and Math. We expect very little bias in the country means for the Trend and PISA designs, although slightly more bias for the PISA design. We expect to see the greatest impact in Reading. There should still be some bias in the country means for both the Trend and PISA designs, but the bias should be substantially larger under the PISA design.
3. For the NBB condition, we expect the bias in the estimated country means for all booklet designs in Science, and for the Balanced Item design in both Reading and Math to be essentially the same as

* If magnitudes in logits are treated as a proxy for effect sizes, in most instances the effect of the bias for a given cluster should be small to moderate, with a relatively small mean bias across clusters.

the simulated bias. For the Trend design in both Reading and Math we expect there to be very little difference between the simulated and estimated bias. For the PISA design, we expect there to be larger differences between simulated and estimated bias in Math relative to the Trend design, and significantly larger differences in Reading relative to the Trend design.

4. Given that more examinees are "administered" the subset of PISA items, we expect the standard errors for the country means to be smaller in Reading and Math relative to the standard errors under the Trend design. We do not expect much difference in the standard errors for the country means in Science across conditions.

Simulation Setup and Procedures

The following section describes the specific steps used to conduct the simulation.

Step 1: Examinee abilities for each of the three content areas were generated for 5000 examinees in nine countries. These values were drawn from a multivariate normal distribution with means equal to zero, standard deviations equal to one, and correlations (between all content areas) equal to (a realistic value for these kinds of assessments of) 0.8 for each country.

Step 2: Starting with the reported PISA item parameters from the years in which each content area was the major domain, cluster-by-country item parameters were established using the two bias conditions.

Step 3: Item responses were generated for all examinees for all items based on the Rasch model.

Step 4: Examinees in each country were assigned booklets for each of the three booklet designs, and item responses for nonadministered clusters—for each examinee—were coded as missing.

Step 5: Item parameters and country means were estimated concurrently for each content area in ConQuest (Wu, Adams, Wilson, and Haldane 2007) using the Rasch model and a multi-group framework where the countries were treated as separate populations.

Step 6: Steps 3–5 were repeated 20 times.

Step 7: Unadjusted (i.e., original) PISA item parameters were compared to the estimated item parameters in each content area for each booklet design and bias condition using the root mean square difference (RMSD) and mean absolute difference (MAD) statistics.

Step 8: True (bias-adjusted) country means were compared to the posterior means of the estimated country means using the RMSD and MAD statistics.

Results

The RMSD, MAD, and mean standard errors (based on the standard deviation of the estimates) for the item difficulties and country means are presented in Table 11.4. The results are generally consistent with what we expressed in our hypotheses. There was very little difference between the true and estimated item difficulties across all conditions. This is consistent with our expectation of unbiased item parameters. It should also be noted that in both the BB and NBB conditions the mean standard errors were notably larger for the Trend design in Reading and Math and slightly smaller in Science relative to the balanced item and PISA design. Since the difficulty estimates in Reading and Math under the Trend design are based on responses from a smaller sample of examinees, one should expect less precision in these estimates. Conversely, since the Science items were oversampled relative to the balanced item and PISA design, one should expect increased precision in these estimates.

The small differences between the true and estimated item parameters are useful for illustrating that the simulation performed as expected, yet the more interesting results are found in the differences between the

TABLE 11.4

Simulation Results

	Country Bias	Item Design	Item Difficulty			Country Means		
			Reading	Math	Science	Reading	Math	Science
Root mean square difference	Balanced	Balanced	0.013	0.022	0.018	0.017	0.009	0.018
		Trend	0.019	0.027	0.008	0.072	0.048	0.012
		PISA	0.012	0.019	0.009	0.132	0.074	0.010
	Non-balanced	Balanced	0.024	0.017	0.023	0.030	0.014	0.024
		Trend	0.023	0.024	0.010	0.087	0.036	0.015
		PISA	0.022	0.017	0.010	0.222	0.059	0.011
Mean absolute difference	Balanced	Balanced	0.010	0.015	0.015	0.012	0.007	0.014
		Trend	0.015	0.022	0.007	0.059	0.036	0.009
		PISA	0.010	0.014	0.007	0.096	0.057	0.008
	Non-balanced	Balanced	0.020	0.014	0.021	0.024	0.013	0.021
		Trend	0.017	0.018	0.008	0.074	0.032	0.013
		PISA	0.021	0.014	0.009	0.193	0.054	0.009
Mean standard error	Balanced	Balanced	0.024	0.026	0.025	0.010	0.014	0.009
		Trend	0.042	0.039	0.018	0.016	0.019	0.007
		PISA	0.022	0.023	0.020	0.016	0.014	0.008
	Non-balanced	Balanced	0.027	0.028	0.025	0.011	0.011	0.011
		Trend	0.039	0.040	0.018	0.014	0.018	0.007
		PISA	0.022	0.022	0.020	0.014	0.012	0.008

true country means (adjusted for bias)[*] and the estimated country means. It is important to note that our basis for comparison is the true country means ± the mean simulated bias (introduced through the simulated item cluster-by-country cluster interactions) across all forms (see Figures 11.2 and 11.3 for the BB and NBB conditions, respectively).[†] To be clear, our goal in this simulation was not to obtain unbiased estimates of country means—we

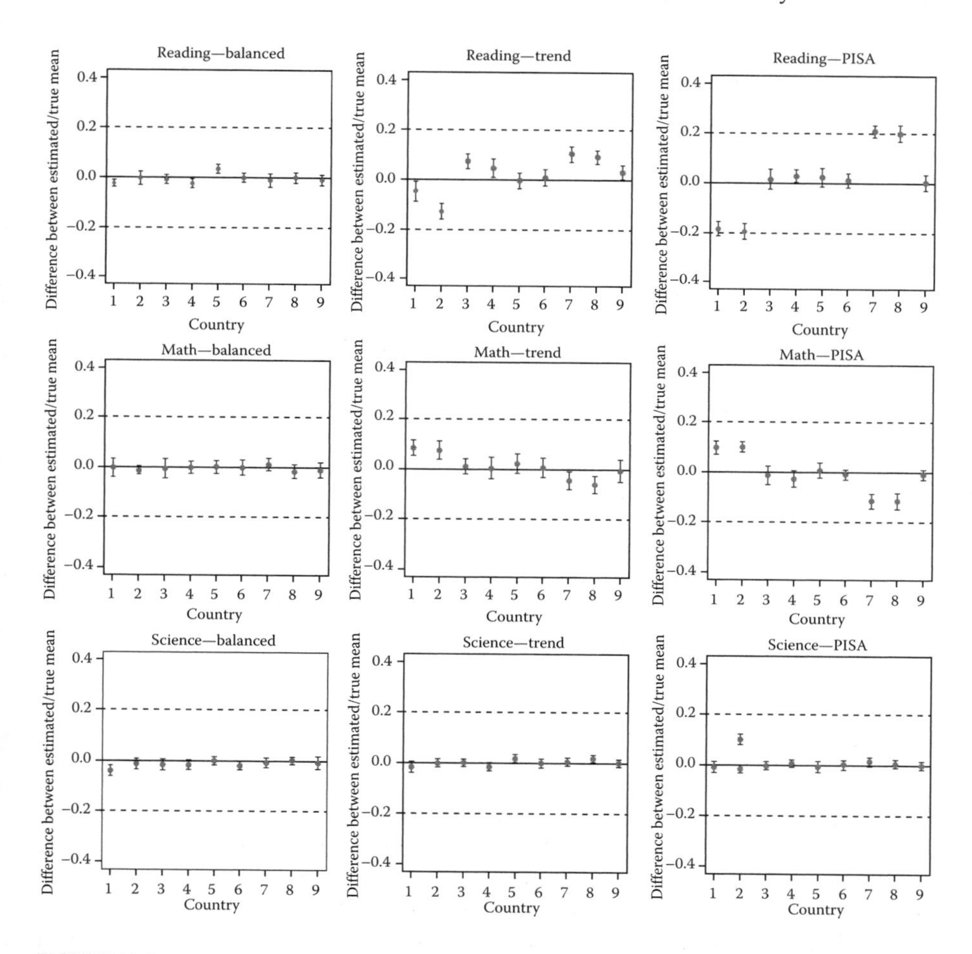

FIGURE 11.2
Difference between true and estimated country means, balanced bias condition.

[*] We hereafter refer to the bias-adjusted true means as simply true means.

[*] The error bars in Figures 11.2 and 11.3 represent the empirical standard deviation of the estimated country means. We used 20 replications in this simulation to ensure that the bias results are evaluated based on a comparison of expected bias to a range of observed biases under multiple replications. This way, differences between simulated and estimated mean bias can be evaluated against the sampling variance of the estimator under consideration.

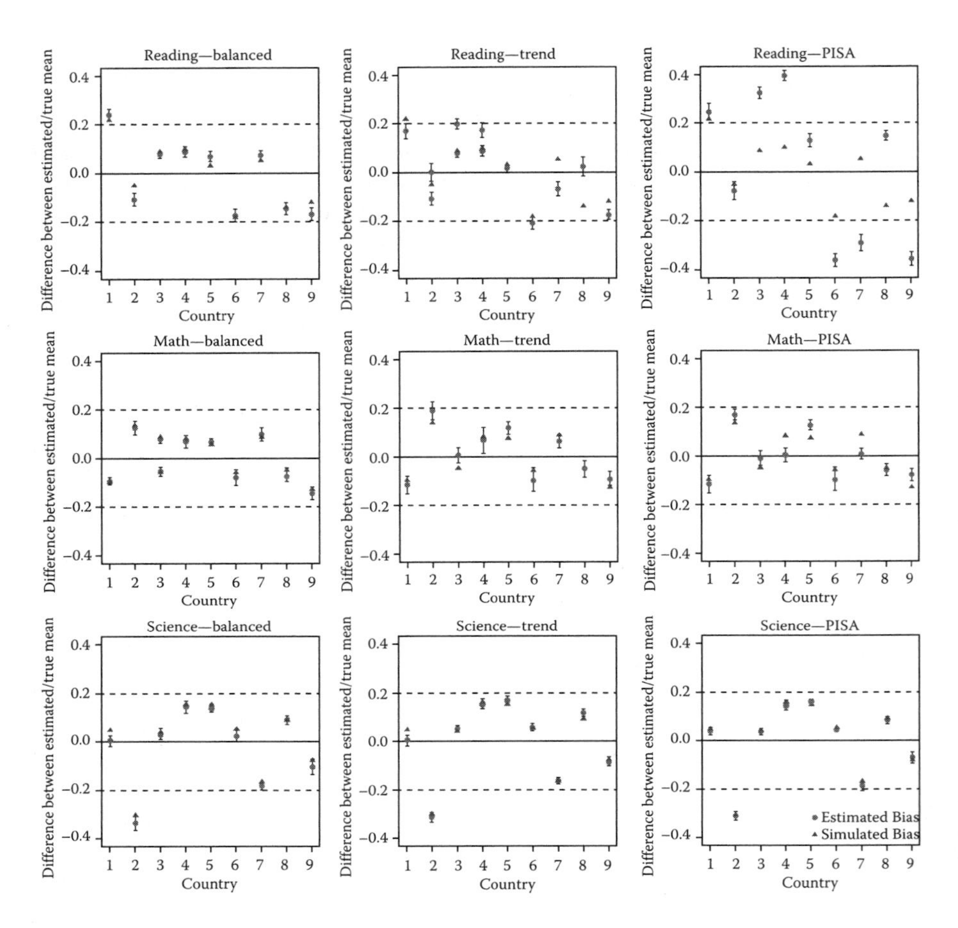

FIGURE 11.3
Difference between true and estimated country means, nonbalanced bias condition.

indeed acknowledge in one of the simulation conditions that country means could be biased if item-by-country interactions are not completely balanced (NBB) for each country across items. Rather, the focus was on determining the relative increase in bias under different designs. To illustrate the impact of the different designs, we examine the results for each content area; in turn, differences between the BB and NBB conditions are addressed within each content area.

Reading

Since Reading is the minor domain with the smallest representation in the PISA design, it is not surprising that the greatest impact of design choice is observed in this content area. In the Balanced Item design all seven item clusters are administered; hence, in both the BB and NBB conditions, the true country means were recovered very well. The RMSD and MAD were smaller for this design relative to the other two designs and the mean standard error was also somewhat smaller. For the Trend design, although all of the Reading items were included, estimates of the country means still exhibited some bias relative to the true means. In the BB condition, the observed bias was near zero or close to −0.1 or 0.1 logits (the bias associated with individual clusters); however, for the PISA design under the BB condition, the full magnitude of the simulated bias (based on the two administered clusters) was observed.

For the NBB condition, the results are more pronounced. In general, the bias in the estimated mean was larger than the simulated bias for both the Trend and PISA designs. For the Trend design, the MAD was 0.07 logits while for the PISA design the MAD was 0.19 logits. The results for the PISA design are not unexpected, although they seem surprisingly large. The difference between the simulated and estimated bias illustrates the potential impact of the minor domain item reduction. In some cases, the observed bias is substantially higher than the expected (simulated/true) bias; in other instances, the direction of the bias changes altogether (e.g., countries 7 and 8). Under the Trend design, the bias for countries 7 and 8 changes direction as well, although the magnitude of the change is much smaller. In summary, the Balanced Item design performed better (i.e., resulted in less random error and bias) than both the Trend and PISA design, although the Trend design performed notably better than the PISA design.

Mathematics

The results in Mathematics were similar to those for Reading, with the key difference being the magnitude of estimated biases. Again, the RMSD, MAD, and mean standard error were smallest under the Balanced Item design for both the BB and NBB conditions. The MAD was around 0.3 logits for the Trend design and around 0.5 logits for the PISA design, although the mean standard error was somewhat higher for the Trend design. These results suggest that the Trend design should still be preferred to the PISA design due to the decrease in potential bias, even though there is a small reduction in precision.

Science

In Reading and Math, the Balanced Item design served primarily as a reference for the Trend and PISA designs under the expectation that there is less bias and greater precision; however, in the case where Science is the major domain, the Balanced Item design performed worse than the other two designs. This was a bit surprising, though not unreasonable, since the Science items were administered to a larger sample of examinees for both the Trend and PISA designs compared to the Balanced Design. Based on the booklet weights the Science forms were oversampled in the Trend design relative to the Balanced Item design. For the PISA design, fewer booklets were administered, so more examinees responded to Science item clusters. For both the BB and NBB conditions, the differences between the true and estimated country means were very small. This suggests that, in practice, no additional bias is likely to be observed for the major domain.

Discussion

The results from the simulation illustrate the potential for bias related to item-by-country interactions under the current PISA design. Although we only examined bias for a single administration, the implications of these biases should carry over into comparisons of administrations as part of reporting trend results. Since we did not explicitly examine the impact of the design on trend results (or the associated linking), we recommend that this be addressed in future research to ensure that there are no unanticipated additional issues. Prior empirical examinations of PISA trend results by Gebhardt and Adams (2007) as well as Urbach (2013) and Carstensen (2013) showed that differences similar to those from our simulation are indeed found in actual PISA data for item sets representing major versus minor domains. We believe that our simulation study, while only varying one source of error—country-specific bias in the difficulty of item clusters—is a consistent and plausible explanation of the results presented by these authors. The existence of item-by-country interactions cannot be denied, but we were able to show that the effect of these interactions can be notably reduced by holding the set of item clusters administered in each cycle constant and adjusting the sampling of examinees taking each booklet.

We showed that both the Balanced Item and Trend designs should be preferred to the current PISA design if the goal is to reduce potential bias. However, given that the Balanced Item design is essentially a variation of the Trend design where the weights for all booklets are equal, our proposed design should not be thought of as a reformulation of the major/minor domain emphasis, but as a flexible design that allows varying the relative weight of each domain. For example, under the current PISA design (or even with a Balanced Item design) all countries must conform to the specified structure. When Reading is the major domain, all countries are forced to treat it as such. On the other hand, with our proposed design it would be possible for each country to choose more than one major domain as an emphasis in a given administration. The key trade-off would be some reduction in precision for less-emphasized domains, but an increase in accuracy for the domain(s) of interest.

Finally, when considering the potential impact of this design for the future it is important to consider the administration format. We examined the item design in the context of a paper-and-pencil test administration, yet with advances in technology there is likely to be an increase in computer-based administrations. If the current PISA design were to be maintained, there would be little advantage—from a psychometric perspective—in moving to an electronic platform, since the booklet setup would likely remain unchanged. The flexibility of the proposed design, on the other hand, is ideally suited for this type of administration, particularly given that weights can be determined independently for each content area without having to consider the composition of clusters within booklets.

References

Carstensen, C. H. 2013. Linking PISA competencies over three cycles—Results from Germany. In: M. Prenzel, M. Kobarg, K. Schöps, and S. Rönnebeck (Eds.) *Research on PISA: Research Outcomes of the PISA Research Conference 2009*. New York, NY: Springer.

Gebhardt, E. and Adams, R. J. 2007. The influence of equating methodology on reporting trends in PISA. *Journal of Applied Measurement, 8*(3), 305–322.

Monseur, C., Sibberns, H., and Hastedt, D. 2008. Linking errors in trend estimation for international surveys in education. *IERI Monograph Series: Issues and Methodologies in Large-Scale Assessments, 1,* 113–122.

Rutkowski, L., Gonzalez, E., Joncas, M., and von Davier, M. 2010. International large-scale assessment data: Issues in secondary analysis and reporting. *Educational Researcher, 39,* 142–151.

Urbach, D. 2013. An investigation of Australian OECD PISA trend results. In: M. Prenzel, M. Kobarg, K. Schöps, and S. Rönnebeck (Eds.) *Research on PISA: Research Outcomes of the PISA Research Conference 2009*. New York, NY: Springer.

van der Linden, W. J. 1998. Optimal assembly of psychological and educational tests. *Applied Psychological Measurement, 22,* 195–211.

van der Linden, W. J. 2005. *Linear Models for Optimal Test Design*. New York: Springer-Verlag.

von Davier, M., Gonzalez, E., and Mislevy, R. 2009. What are plausible values and why are they useful? *IERI Monograph Series: Issues and Methodologies in Large Scale Assessments, 2,* 9–36. http://www.ierinstitute.org/fileadmin/Documents/IERI_Monograph/IERI_Monograph_Volume_02_Chapter_01.pdf

Wu, M. L., Adams, R. J., Wilson, M. R., and Haldane, S. 2007. *ACER ConQuest 2.0: Generalised Item Response Modeling Software* [computer program manual]. Camberwell, Australia: ACER Press.

12

Innovative Questionnaire Assessment Methods to Increase Cross-Country Comparability

Patrick C. Kyllonen
ETS—Research and Development

Jonas P. Bertling
ETS—Research and Development

CONTENTS

The primary purpose of international large-scale assessments has been to compare student achievement levels across educational systems, but a secondary purpose, growing in importance, has been to compare student responses on background questionnaires. For example, at a recent international meeting, a representative of one country's educational system announced to the audience that he had "good news and bad news" to report. The good news was that his country was "once again, one of the top countries in the world in student achievement." The bad news was that his country "was among the worst in students' attitudes towards school." This statement, essentially implying parity between achievement scores and background factors, reflects the seriousness with which educational leaders are now treating data obtained from background questionnaires. On the other hand, one wonders whether such data, based almost exclusively on self-assessments using simple rating scales (often called Likert or Likert-type scales), are of sufficient quality to permit this kind of inference concerning national comparisons. Do students from this country truly believe their school climate is worse than the school climates of other countries? Or does the finding simply reflect a difference in response style, the general tendency to respond to rating scales in a particular way, without regard to the construct being measured? Or is it a mixture of both construct and response style differences?

Although there is literature on the importance of response styles in international comparative research (e.g., Buckley 2009), there are many cross-national comparisons found in the literature on variables such as personality and attitudes that are taken at face value despite being based solely on responses to rating scales. For example, research on cultural dimensions (Hofstede 2001; House et al. 2004) relies on individuals' responses largely to rating scale items to draw inferences on national differences. Research on personality similarly has investigated the relationship between national mean personality scores (based on self-ratings) and other nation-level variables such as perceptions of national character (Terracciano et al. 2005), and Global Competitiveness and Human Development indices (Bartram 2013). Sometimes, findings seem to be curious, particularly with those involving simple rating scales, such as a strong negative relationship at the country level between conscientiousness and the World Economic Forum's Global Competiveness Index (McRae 2002). Although a substantive explanation for such a relationship might be possible, it is also possible that the data obtained from simple rating scales are not useful for comparing nations.

Attitude–Achievement Anomaly

Related to this odd country-level finding is a peculiar but consistent finding in international large-scale assessment research. Specifically, there are a number of questionnaire scales, such as attitude scales, that have been shown to be positively correlated with achievement within a country at the individual level, but negatively correlated with achievement at the country level. For example, in the Programme for International Student Assessment (PISA) 2003, the mean within-country correlation (across 37 countries) between the five-item mathematics self-concept scale (example item: "I learn mathematics quickly—'strongly agree,' 'agree,' 'disagree,' 'strongly disagree'"), and mathematics achievement was $r = .40$. At the same time, the correlation between country-mean mathematics self-concept and country-mean mathematics achievement ($N = 37$ countries) was $r = -.20$. That is, lower achievement-scoring countries tended to have higher self-concepts on average, and the better performing countries had lower self-concepts, on average. This negative county-level relationship* was not limited to mathematics self-concept but was shown for a number of scales, such as mathematics interest, attitudes toward school, and others. In these cases, positive correlations with achievement within a country were found simultaneously with negative correlations with achievement at the country level. Of course there is no reason why aggregate (ecological) correlations must follow individual-level correlations—well-known discrepancies include positive correlations between Protestantism and suicide in Europe, and immigration and literacy in the United States (both true at the aggregate but not individual level; Freedman 1999). Failure to recognize the possibility of a different correlation at the individual and aggregate level is known as the *ecological fallacy*. Still, although reference group effects have been suggested to account for some of this discrepancy (Van de Gaer et al. 2012), it is not obvious why there should be a different direction of relationship between background questionnaire scales and achievement at the individual and country levels, and at the very least the anomaly invites explorations for why this might happen.

Problems with Rating Scales

There are several hints in the literature that simple rating scales, at least for some constructs, are problematic for making cross-country comparisons. One is found in a cross-cultural study of personality, which compared personality questionnaires based on rating scales (the *NEO Personality Inventory* and the *Big Five Inventory*) with a forced-choice format (the *Occupational Personality Questionnaire*), where respondents were asked to choose the statement (of either 3 or 4 presented) that was "most like them" or "least like them." One might expect that personality scales such as conscientiousness, which indicates proclivity to achieve, be dependable, and be organized, would correlate with national indicators such as the World Economic Forum's Global Competitiveness Index and the United Nations Human Development Index. The former measures the competitive performance of economies around the world based on 12 variables, including health and primary education, higher education and training, technological readiness, infrastructure, and innovation. The latter measures development through indicators of life expectancy, educational attainment, and income. There are documented positive correlations between *conscientiousness* and many of these variables at the individual level (e.g., Bogg and Roberts 2013).

* This has been called a "paradoxical relationship" (Van de Gaer et al. 2012); or, in the context of race-ethnicity studies within the United States, the "attitude–achievement paradox" (Mickelson 1990). In this chapter, we sometimes use the term *anomaly* rather than *paradox* to avoid implying that there might be anything inherently contradictory about individual and ecological correlations having different signs.

The forced-choice format, compared to the Likert items, yielded higher correlations with these two country-level measures (Bartram 2013, Table 11). Another indication of rating scale problems is the finding that there are systematic national differences in how respondents from different countries rate common individuals that are described in vignettes (Mõttus et al. 2012). In this study, respondents rated 30 hypothetical people who were described in short vignettes. There were clear differences between respondents' tendency to use or not use the rating scale endpoints (i.e., extreme responses), and furthermore, this tendency was correlated with self-reported *conscientiousness*. Correcting responses for extreme responding resulted in higher correlations with *conscientiousness*.

A third pertinent finding is that not all ratings are problematic. In the PISA 2003 data, for example, *Mathematics Self-Concept* and *Mathematics Interest* showed the anomalous pattern of positive average-within-country correlations and negative country-level correlations, while other scales, particularly *Mathematics Self-Efficacy*, did not. This latter scale consists of concrete items such as "I can solve an equation like $2(x + 3) = (x + 3)(x - 3)$, 'strongly agree,' 'agree,' 'disagree,' 'strongly disagree.'" This item is highly correlated with achievement both at the average-within-country individual level and at the between-country level. Other items that do not show the anomalous pattern are *socioeconomic status* items, such as a count of the number of books in the home, which correlates with achievement at both the individual and country level and *Mathematics Anxiety* items. These ask respondents whether when preparing for mathematics tests and assignments they "worry that it will be difficult," "get very tense," "get very nervous," "feel helpless," or "worry that I will get poor grades." Arguably, all of these items—self-efficacy, socioeconomic status, anxiety—diverse as they are, differ from the items that do show the anomalous pattern in that either they use a concrete response scale (e.g., books in the home) or the prompt itself is sufficiently behavioral that the anchors (agree–disagree) have a clear meaning.

Innovative Item Types as Alternatives to Likert Scales

When there are differences between countries in a background variable, such as interests or attitudes, those differences might reflect true differences between countries in the construct, or they might reflect construct-irrelevant differences in response style. It could be that students from some of the higher-achieving countries really do have worse attitudes toward school, less interest in mathematics, or lower levels of conscientiousness than students from countries with lower levels of achievement, or it could be that there are national differences in response style that manifest themselves on the particular rating scales commonly used in background questionnaires. A way to address this issue would be to experiment with alternatives to simple rating scales and evaluate whether the country-level differences are affected.

We had an opportunity to do just this in the PISA 2012 Field Trial. We experimented with a number of different methods, including anchoring vignettes, forced-choice comparisons, situational-judgment-tests, behavioral items, and signal-detection methods for measuring topic familiarity. In the interest of conserving space, we focus on three of these methods, described here separately.

Anchoring Vignettes

The anchoring vignettes technique (King et al. 2004) is a method to increase the validity of rating-scale responses by rescaling them based on responses to a set of common vignettes that serve as anchors. The method has been used in social, political science, and other surveys, but had not been used in international large-scale survey assessments (ILSAs) prior to the PISA 2012 Field Trial (see Buckley, 2008, for an educational application). The method requires respondents to rate a vignette or vignettes describing persons or situations on a categorical rating scale, then to rate one's own situation (or person) on that same rating scale. Any number, k, of vignettes could be used, but we used $k = 3$, designed roughly to reflect a range of the attribute being measured (i.e., low, medium, and high exemplars). One's self-rating then is rescaled using the vignette ratings as anchors. There are various statistical methods to accomplish the rescaling (e.g., Wand et al. 2011), but we found that a fairly simple nonmetric method (King and Wand 2007) worked well with PISA data. The nonmetric method rescales responses to the $2k + 1$ points at, above, below, or between the k vignette ratings, so 3 vignettes yields a 7-point rating scale (there are various ways to deal with vignette rating ties and reversals).

For PISA 2012, we created anchors for the *Teacher Support* scale, which consists of four items: My teacher "...lets students know they need to work hard," "... provides extra help when needed," "...helps students with their learning," and "...gives students the opportunity to express opinions," which are rated on the standard 4-point rating scale ("strongly agree" to "strongly disagree"). Prior to rating these four items, students rated three vignettes on the same 4-point scale (vignettes were authored by Richard Roberts).

An American English version of the vignettes follows (vignettes were adapted and translated appropriately following PISA protocols):

1. Ms Anderson assigns mathematics homework every other day. She always gets the answers back to students before examinations. Ms Anderson is concerned about her students' learning.
2. Mr Crawford assigns mathematics homework once a week. He always gets the answers back to students before examinations. Mr Crawford is concerned about his students' learning.
3. Ms Dalton assigns mathematics homework once a week. She never gets the answers back to students before examinations. Ms Dalton is concerned about her students' learning.

Based on students' ratings of (a) Ms Anderson, (b) Mr Crawford, and (c) Ms Dalton, the students' responses to the four *Teacher Support* items were rescaled.

The results were clear and striking. First, for the 63 countries participating ($N \approx 200$–1000 per country), the mean within-country correlation between the uncorrected *Teacher Support* scale score and achievement was $r = .03$. With the anchoring-adjusted correction, that correlation rose to $r = .13$. More dramatically, the unadjusted country-level correlation (with $k = 63$) was $r = -.45$, an example of the anomaly. However, with the anchoring adjustment, that correlation rose to $r = +.29$. In other words, the anchoring adjustment seemingly resolved the anomaly, aligning the within- and between-country correlations. This finding is evidence that response styles—whether due to cultural, linguistic, or other differences—vary greatly at the national level, and that they can be mitigated with an anchoring adjustment. Summarizing the argument thus far, anchoring-vignette adjustment of rating-scale responses increases the correlation between a background scale and achievement, both within a country, and more dramatically, at the country level.

If response styles are fairly general individual-difference factors, then it might be possible to adjust scales other than the *Teacher Support* scale using the *Teacher Support* anchors. Such an adjustment could be justified if individuals tended to use response styles, such as extreme or acquiescence, consistently, regardless of the construct. We applied this logic to rescale students' responses to two other scales, *Student–Teacher Relations*, and *Mathematics Interest*, using their responses to the *Teacher Support* scale anchors. Interestingly, the adjustment worked just as well for these scales as it did for the target scale. For *Student–Teacher Relations*, the unadjusted within-country correlation with achievement was $r = .05$, rising to $r = .14$ with adjustment based on the *Teacher Support* anchors. For *Mathematics Interest* the comparable correlations were $r = .17$ and .22. The results for the between-country adjustment were dramatic. The unadjusted between-country correlations with achievement were $r = -.41$ for *Student–Teacher Relations* and $r = -.50$ for *Mathematics Interest*, changing to $r = +.40$ and $r = +.14$ for the two scales, respectively, with the anchoring adjustment. Again, the anchoring adjustment, even when using anchors from a different scale, seems to have eliminated the anomaly.

Forced Choice

A straightforward alternative to a rating-scale format is the preference or forced-choice item. We created these for the PISA 2012 Field Trial for several *Learning Strategy* scales, *Control, Elaboration,* and *Memorization*. The method for creating items was fairly simple. It involved taking three rating-scale items from the PISA 2003 questionnaire, one each from *Control, Elaboration,* and *Memorization,* and asking students which learning strategy method they preferred. We also administered the simple rating-scale items by themselves. The results were somewhat similar to the results from the anchoring vignettes. The forced-choice versions of the questions increased slightly the within-country correlations with achievement, but raised the between-country correlations with achievement dramatically. For example, for the *Control Strategy,* the country-level correlation with achievement was $r = -.47$ for the simple Likert scale version, but $r = +.60$ for the forced-choice version.

Signal Detection Correction for Topic Familiarity

There was interest in measuring opportunity to learn (OTL) in the PISA 2012 Field Trial. Textbook and curriculum surveys were deemed impractical, and so several methods were attempted for this purpose involving presenting students with example mathematics items, or topic concepts, and asking them about their prior exposure and familiarity with these items or concepts. For example, students were presented with a list of mathematics concepts, such as *logarithm, rational number,* and *divisor,* and asked how familiar they were with those concepts on a six-point scale (from "never heard of it" to "quite familiar with it"). Mixed in with the concepts were several foils (e.g., *subjunctive scaling*). Two scores were computed from students' responses. One was a simple mean of their familiarity scores on the 6-point scale, and the other took that mean and subtracted from it the mean familiarity score of the foil concepts. (On average, students indicated a 3.6 familiarity rating for the mathematics concepts and a 2.3 familiarity rating for the foils.) The average within-country correlations between the familiarity ratings and achievement were $r = .45$ and $r = .44$ for the unadjusted and adjusted ratings, respectively; that is, they were essentially the same. However, as with the other adjustment methods, the differences were striking at the between-country level. The country-level correlation between the unadjusted topic familiarity score and achievement was $r = .17$, and with adjustment that correlation increased to $r = .58$. Also, the foil score from the topic familiarity test could be used to correct OTL measures based on familiarity with various mathematics items, and with PISA items, indicating again that a general response style factor may have contributed to response patterns on related measures.

Summary

A common finding in ILSAs has been a difference in magnitudes and even signs of correlations between background variables and achievement at the individual versus school and country levels. Background variables such as attitudes or interests have been shown to correlate positively with achievement within a country, while simultaneously correlating negatively at the country level. This pattern could reflect true differences between countries—students from high-performing countries might really have poorer attitudes toward school on average—but if alternative measurement methods change the correlations, then that suggests that response-style effects might obscure true construct relationships. Here we showed that several alternatives to simple rating scales, including anchoring vignettes, forced-choice methods, and signal-detection-type corrections all had dramatic effects on country-level correlations, providing evidence supporting the response-style-effect hypothesis. All three methods were shown to align between- and within-country correlations resolving the attitude–achievement anomaly. Furthermore, because response styles are fairly general, it may be possible to adjust a diverse set of Likert-type responses using a small number of relatively narrow correction techniques and still observe benefits in increased validity. These findings are important for increasing the quality of data obtainable from ILSA questionnaires, and may be even more generally useful in addressing cross-cultural (Bartram 2013) and even cross-subgroup (Mickelson 1990) comparability issues.

Acknowledgments

We wish to thank the OECD Programme for International Student Assessment (PISA), the Australian Council for Educational Research (ACER), and ETS R&D initiative funding for the Center for Academic and Workforce Readiness and Success (CAWRS) for support for this chapter. We also thank Richard Roberts, Bobby Naemi, Jeremy Burrus, Ted Blew and his data analysis staff, and the PISA 2012 Questionnaire Expert Group (Eckhard Klieme, chair) for contributions to the project, and Frank Rijmen, Matthias von Davier, and Lale Khorramdel for their reviews of earlier drafts.

References

Bartram, D. 2013. Scalar equivalence of OPQ32: Big five profiles of 31 countries. *Journal of Cross-Cultural Psychology*, 44, 61–83.

Bogg, T. and Roberts, B. 2013. The case for conscientiousness: Evidence and implications for a personality trait marker of health and longevity. *Annals of Behavioral Medicine*, 45, 278–288.

Buckley, J. 2008. *Survey Context Effects in Anchoring Vignettes*. Working Paper 822. (retrieved from http://polmeth.wustl.edu/mediaDetail.php?docId=822 on 1 March 2013. Washington University in St. Louis.

Buckley, J. 2009. Cross-National Response Styles in International Educational Assessments: Evidence from PISA 2006. https://edsurveys.rti.org/PISA/documents/Buckley_PISAresponsestyle.pdf

Freedman, D.A. 1999. Ecological inference and the ecological fallacy, In N. J. Smelser and P. B. Baltes (Eds.), *International Encyclopedia of the Social & Behavioral Sciences* 6: 4027–4030. Oxford: Elsevier.

Hofstede, G. 2001. *Culture's Consequences: Comparing Values, Behaviors, Institutions, and Organizations across Nations* (2nd ed.). London: Sage.

House, R. J., Hanges, P. J., Javidan, M., Dorfman, P. W., and Gupta, V. (Eds.) 2004. *Culture, Leadership, and Organizations: The GLOBE Study of 62 Societies*. Thousand Oaks, CA: Sage Publications.

King, G. and Wand, J. 2007. Comparing incomparable survey responses: New tools for anchoring vignettes. *Political Analysis*, 15, 46–66.

King, G., Murray, C.J.L., Salomon, J. A., and Tandon, A. 2004. Enhancing the validity and cross-cultural comparability of measurement in survey research. *American Political Science Review*, 98, 191–207.

McRae, R.R. 2002. NEO-PI-R data from 36 cultures: Further intercultural comparisons. In R. R. McCrae and J. Allik (Eds.), *The Five-Factor Model of Personality across Cultures* (pp. 105–125). New York: Kluwer Academic/Plenum.

Mickelson, R.A. 1990. The attitude-achievement paradox among black adolescents. *Sociology of Education*, 63, 44–61.

Mõttus, R., Allik, J., Realo, A., Rossier, J., Zecca, G., Ah-Kion, J. et al. 2012. The effect of response style on self-reported conscientiousness across 20 countries. *Personality and Social Psychology Bulletin*, 38, 1423–1436.

Terracciano, A., Abdel-Khalek, A.M., Adam, N.A., Adamovova, L., Ahn, C.-k., Ahn, H.-n. et al. 2005. National character does not reflect mean personality trait levels in 49 cultures. *Science*, 310, 96–100.

Van de Gaer, E., Grisay, A., Schulz, W., and Gebhardt, E. 2012. The reference group effect: An explanation of the paradoxical relationship between academic achievement and self-confidence across countries. *Journal of Cross-Cultural Psychology*, 43(8), 1205–1228.

Wand, J., King, G., and Lau, O. 2011. Anchors: Software for anchoring vignettes data. *Journal of Statistical Software*, 42, 1–25.

13

Relationship between Computer Use and Educational Achievement

Martin Senkbeil
Leibniz Institute for Science and Mathematics Education at the University of Kiel

Jörg Wittwer
University of Göttingen

CONTENTS

Introduction

International large-scale assessments (ILSAs) such as the Programme for International Student Assessment (PISA), Trends in Mathematics and Science Study (TIMSS), and Progress in International Reading Literacy Study (PIRLS) are carried out not only to describe educational achievement but also to examine factors that influence educational achievement. These factors are called *productivity factors* by Walberg (2006). The aim of studying productivity factors in educational research is to advance the understanding of the phenomenon *educational achievement* by elucidating the question of how productivity factors produce this phenomenon. In educational research, an answer to this question usually results in a scientific explanation. The quality of such a scientific explanation, however, might greatly vary depending, among others, on the research designs used to investigate productivity factors and on the theoretical underpinnings of how productivity factors bring about their effects on educational achievement.

In this chapter, we address methodological and conceptual issues in studying productivity factors and discuss their role for the development of a scientific explanation on productivity factors in educational achievement. To do so, we specifically focus on the relationship between computer use at home and educational achievement. Since the rise of the Internet, computers have become an indispensable part of a student's daily life. Therefore, the question arises as to whether computer use at home is an important out-of-school activity that influences educational achievement. First, we present empirical findings on the relationship between computer use at home and educational achievement. Second, we address methodological and conceptual issues in the study of computer use at home and educational achievement. Third, as an illustration, we provide an example of how to acknowledge the methodological and conceptual issues when investigating computer use at home and educational achievement in large-scale assessments.

Computer Use and Educational Achievement: Empirical Findings

One of the productivity factors in Walberg's (2006) model that influences educational achievement is the use of out-of-school time. Today, nearly every student in industrialized countries possesses a computer and has access to the Internet (OECD 2011). Therefore, computer use at home is an important out-of-school activity. Accordingly, educational research has recently begun to examine the effects of computer use at home on educational achievement. For example, Attewell and Battle (1999) investigated the role of having a computer at home for reading and mathematical achievement. The results showed that having a computer at home was positively correlated with reading and mathematical achievement. This was true even after controlling for the influence of social background. Similarly, Papanastasiou et al. (2003) found a positive relationship between frequency of computer use at home and level of science achievement while controlling for the influence of social background. In other words, students who more frequently used a computer at home had a higher level of science achievement than students who less frequently used a computer at home. Also, the OECD (2006) showed that mathematical achievement was positively related with the availability of a computer at home and the frequency of using it when controlling for the influence of social background. Finally, Luu and Freeman (2011) found a positive relationship between computer experience and science achievement while controlling for the influence of student-related variables (e.g., social background) and school-related variables (e.g., school size). Thus, more experienced students had a higher level of science achievement.

Taken together, the results of the studies suggest that various aspects of computer use at home (e.g., frequency of computer use at home, computer experience) promote educational achievement. This seems to be true for such domains as reading, mathematics, and science. Thus, a scientific explanation for why the level of educational achievement is high would claim, for example, that a high frequency of computer use at home or a high computer experience produces a high level of educational achievement.

Methodological and Conceptual Issues in Studying the Relationship between Computer Use at Home and Educational Achievement in Large-Scale Assessments

In the following, we will address some methodological and conceptual issues in studying the relationship between computer use at home and educational achievement in large-scale assessments. In doing so, we shed light on potential difficulties associated with explaining and interpreting the effects of computer use at home on educational achievement. A general review of methodological and conceptual issues in studying productivity factors in educational achievement is beyond the scope of this chapter.

Methodological Issues I: Observational Research Designs

In large-scale assessments of educational achievement, observational research designs such as ex-post-facto research designs are often used to examine the effects of productivity factors on educational achievement. These designs refer to productivity factors and educational achievement that have already occurred (e.g., Nachtigall et al. 2008). Thus, in contrast to experimental designs, observational research designs do not allow for the possibility of manipulating productivity factors in order to study their effects on educational achievement. Accordingly, establishing cause-and-effect relationships between productivity factors and educational achievement in observational research designs requires caution (e.g., Schneider et al. 2007).

When examining the effects of computer use at home on educational achievement, the computer use at home is regarded as a productivity factor and, thus, as a cause that brings about the effect on educational achievement. However, in ex-post-facto designs, particularly, in cross-sectional studies, there is no asymmetry in the relationship between productivity factors and educational achievement. Therefore, a productivity factor could be the cause for educational achievement. However, the other way around could also be true. That is, educational achievement could be the cause for a productivity factor. Thus, when studying computer use at home and educational achievement in large-scale assessments, it cannot be ruled out that educational achievement is the cause that brings about the effect on computer use at home.

Methodological Issues II: Endogeneity Bias

Walberg's (2006) model postulates a number of productivity factors to influence educational achievement. Thus, when examining the effects of possible productivity factors on educational achievement, it is important that the statistical analysis takes into account not only the productivity factor under investigation but also the productivity factors that have already been shown to influence educational achievement (e.g., social background, cognitive abilities). Otherwise, a so-called *endogeneity bias* is likely to occur. In the context of productivity factors, endogeneity refers to a (causal) relationship between two or more productivity factors. When a productivity factor under investigation is correlated both with educational achievement and other productivity factors, but the other productivity factors are omitted from the statistical analysis, then the effects of the productivity factor under investigation on educational achievement are biased (cf. Schneider et al. 2007).

Hence, when examining the effects of computer use at home on educational achievement, it is necessary to include other productivity factors. Otherwise, the estimated effects of computer use at home, such as the frequency of computer use at home, on educational achievement are biased to the extent to which computer use at home is correlated with these other productivity factors (e.g., Stone and Rose 2011). As a result, it is possible that the effects of computer use at home on educational achievement are overestimated or underestimated. From the perspective of scientific explanation, a good scientific explanation includes information not only about causes but also about the statistical relevance of the causes (e.g., Salmon 1997). In fact, the reported correlations between computer use at home and educational achievement (e.g., Attewell and Battle 1999; Luu and Freeman 2011;

OECD 2006; Papanastasiou et al. 2003) demonstrate the statistical relevance of computer use at home for educational achievement (e.g., Salmon 1997). In addition, the reported studies address the endogeneity bias by including productivity factors other than computer use at home in the statistical analysis. This, however, raises the general question as to which productivity factors to include in the statistical analysis at all. Even though educational research has already cataloged a number of productivity factors that influence educational achievement (e.g., cognitive abilities), there is always the risk to obtain biased effects due to omitted variables. Accordingly, a scientific explanation with computer use at home as a cause for a high level of educational achievement might be in general true. However, the explanation might be flawed with respect to the statistical and, thus, explanatory relevance of computer use at home for educational achievement.

Conceptual Issues I: Proximity to Educational Achievement

Educational research shows that productivity factors vary along a dimension of proximity to educational achievement. *Proximal* productivity factors, such as the instructional material used in classroom teaching in schools, are closest to the learning experiences of students. Thus, they exert more influence on educational achievement than do *distal* productivity factors, such as a student's community, that are rather removed from the learning experiences of students (e.g., Ellet and Teddlie 2003).

In this context, computer use at home can be conceptualized as a distal productivity factor because it is not necessarily targeted at learning in a purposeful way. Therefore, it can be expected that computer use at home produces a smaller effect on educational achievement than do more proximal factors. This conclusion is important not only for a better theoretical understanding of the interplay of (proximal and distal) productivity factors with respect to educational achievement, but also for informing practitioners about which educational practices are most effective in promoting educational achievement. Thus, when thinking about how to increase educational achievement, promoting computer use at home should not be in the first place. From an explanatory perspective, a good scientific explanation for why the level of educational achievement is high would include all productivity factors that bring about an effect on educational achievement (see also the section "Methodological Issues II: Endogeneity Bias").

Conceptual Issues II: Psychological Mechanisms

Many approaches to scientific explanation in the philosophy of science assume causation to be fundamental to scientific explanation. However, which events count as causes of a phenomenon has been largely disputed in literature on scientific explanation (see also Schneider et al. 2007). Apart from identifying causes for a phenomenon, there are accounts that require a scientific explanation to specify the (causal) mechanisms underlying the relationship between cause and effect (e.g., Salmon 1997). For example, the bacterial theory of ulcers not only identifies *Helicobacter pylori* as the cause for duodenal ulcers but also describes the mechanisms by which the bacterium produces the disease (see, e.g., Thagard 1998).

When examining the effects of productivity factors on educational achievement at an individual level, educational achievement can be conceptualized as a psychological phenomenon. Accordingly, when studying computer use at home as a cause for a high level of educational achievement, it is necessary to specify the psychological mechanisms by which computer use at home promotes educational achievement. A scientific explanation that only mentions, for example, the frequency of using a computer at home as cause for a high level of educational achievement is, intuitively, not very satisfactory precisely because it fails to specify the operating mechanisms.

Example of Investigating the Relationship between Computer Use at Home and Educational Achievement in Large-Scale Assessments

In the following, we will provide a simple example of examining the effects of computer use at home on educational achievement, still keeping in mind the previously mentioned methodological and conceptual issues. In our example, we use data from students who took part in the PISA 2006 study in Germany. As the educational domain, we look at the students' mathematical achievement.

First, the PISA 2006 is a cross-sectional study. Therefore, it is not possible to address the difficulties associated with interpreting cause-and-effect relationships in nonexperimental research designs. Second, in order to minimize the influence of an endogeneity bias, we include not only variables related to computer use at home but also other productivity factors such as social background, cognitive abilities, immigration background, and gender in the statistical analysis. Third, we conceptualize computer use at home as a distal productivity factor in educational achievement. Thus, we expect computer use at home to produce, if at all, just a small effect on educational achievement. Fourth, we theoretically elucidate the psychological mechanisms by which computer use at home might produce an effect on educational achievement. To do so, we focus on the cognitive effects of computer use at home. The debate about the effectiveness of media use in school in the early 1980s (e.g., Clark 1994) has shown that computers have no influence on educational achievement per se. Rather, effects of computer use on educational achievement depend on how a computer is used (e.g., Lei and Zhao 2007).

In our example, we follow Jonassen (2004) and assume that students differ in the extent to which they use a computer as a cognitive tool. Using a computer as a cognitive tool engages students in problem-solving activities. For example, students might be required to use problem-solving strategies to solve tasks such as using spreadsheets or searching the Internet for information. For example, when searching for information, problem-solving skills are needed to find information sources, extract and organize relevant information from every source, and, finally, integrate the information with each other. Similarly, working with quantitative information when using spreadsheets engages analytical reasoning because it is necessary to consider the implications of conditions and options (Jonassen et al. 1998). Accordingly, it can be expected that using a computer at home as a cognitive tool promotes educational achievement.

Participants

PISA assesses the educational achievement of 15-year-old students in the domains of mathematics, science, and reading. In Germany, a total of 4981 students (49% female, 51% male) participated in the 2006 PISA cycle. Of the 4981 students, 293 students with learning disabilities were not able to answer the PISA questionnaires. Therefore, these students were excluded from further analyses.

Instruments

Mathematical achievement. A total of 48 items were used to assess mathematical achievement. The achievement scores were scaled in such a way that 500 points represent the average score across all students who participated in PISA 2006. The standard deviation is 100 points (for more details, see OECD 2009).

Index of economic, social, and cultural status (ESCS). The ESCS index is derived from three variables related to a student's family background: highest level of parental education (in number of years of education), highest parental occupation, and number of home possessions (for details, see OECD 2007).

Immigration background. Students were asked to indicate whether one or both of their parents were not born in Germany.

Cognitive abilities. Cognitive abilities were assessed by using the subtest *Figurale Analogien* of the German test *Kognitiver Fähigkeitstest* (Heller and Perleth 2000). The subtest, consisting of 25 items, required students to identify analogies among different figures.

Computer use at home. In order to assess a student's computer use at home, we used three measures. In the international ICT familiarity questionnaire, students were asked to indicate their (1) experience with using a computer (0 = *less than five years*, 1 = *more than five years*) and to indicate the (2) frequency with which they used a computer at home (0 = *less than almost every day*, 1 = *almost every day*). In the national ICT familiarity questionnaire, the students were asked to indicate (3) the extent to which they used a computer at home as a cognitive tool on a 4-point rating scale ranging from 1 (= *disagree*) to 4 (= *agree*). The use of the computer at home as a cognitive tool was assessed by using 5 items (e.g., "When confronted with difficult tasks, for example, when searching for information, I use the Internet to get information").

Statistical Analysis

For the statistical analyses, we used the software *WesVar 4.2* that is capable of analyzing complex survey data containing plausible values. We included the categorical variables (e.g., gender, immigration background, computer experience, and frequency of computer use at home) as dummy variables and the continuous variables as *z*-standardized variables in the statistical analyses. Missing data concerning the indices and scales were imputed by applying the EM algorithm using the program *NORM 2.03*. Missing data concerning the immigration background were imputed by using a missing indicator. To present the results of the statistical analyses, we use the unstandardized regression coefficient (b) and the standardized regression coefficient (β).

Results

Model I

We computed three regression models to statistically analyze the effects of computer use at home on mathematical achievement. In the first regression model, we included the productivity factors *computer experience* and *frequency of computer use at home* as predictors while we controlled for the effects of *social background*. This procedure is similar to the one used in some of the reported studies (e.g., OECD 2006). The results show that computer experience and frequency of computer use at home are significantly and positively related with mathematical achievement (Table 13.1). In total, 18% of the variance in mathematical achievement is explained by the computer variables and the social background.

Model II

In the second regression model, we included *cognitive abilities, immigration background*, and *gender* as additional productivity factors. The results show that cognitive abilities have the strongest significant effect on mathematical achievement. The effect of computer experience is significant but the magnitude of this effect is lower than in the first regression model. In addition, a significant effect of frequency of computer use at home is no longer observable in the second regression model. Instead, immigration background and

TABLE 13.1

Relationship between Mathematical Achievement and Computer Use at Home While Accounting for Other Productivity Factors

	Model 1			Model 2			Model 3		
	b	S.E. (b)	*β*	*b*	S.E. (b)	*β*	*b*	S.E. (b)	*β*
Constant	500.95**	3.91		489.47**	3.69		490.69**	3.71	
Cognitive abilities				5.64**	0.19	0.66	5.62**	0.20	0.66
ESCS	36.27**	2.06	0.40	15.88**	1.49	0.17	15.50**	1.48	0.17
No immigration background				15.94**	3.17	0.08	16.36**	3.13	0.08
Immigration background unknown				−7.40	5.07	−0.02	−7.10	4.88	−0.02
Gender (male)				19.28**	2.37	0.11	20.14**	2.34	0.12
Computer experience (more than 5 years)[a]	11.19**	2.88	0.07	4.56*	2.09	0.03	3.80	2.11	0.02
Computer use at home (almost daily)[a]	11.15**	3.14	0.06	2.03	2.55	0.01	−0.13	2.66	0.00
Cognitive tool							4.25**	1.19	0.05
R^2	0.18			0.60			0.60		

[a] Reference group: Less than 5 years computer experience and less than almost daily computer use at home.
$^*p < .05$; $^{**}p < .01$.

gender produce a stronger significant effect on mathematical achievement than do computer experience and frequency of computer use at home. In total, the variance in mathematical achievement explained by all productivity factors increases from 18% in the first regression model to 60% in the second regression model.

Model III

In the third regression model, we included the *computer use at home as cognitive tool* as an additional productivity factor. The results show that computer use at home as a cognitive tool is significantly and positively correlated with mathematical achievement. Interestingly, computer use at home as a cognitive tool produces a stronger significant effect on mathematical achievement than does computer experience. Computer experience even loses its significant effect on mathematical achievement. In total, the variance in mathematical achievement explained by all productivity factors remains the same.

Intercorrelations among Productivity Factors

Table 13.2 displays the intercorrelations among the productivity factors. Computer experience and frequency of computer use at home are

TABLE 13.2

Intercorrelations among Productivity Factors

	2	3	4	5	6	7	8
1. Cognitive abilities	0.32**	0.19**	−0.09**	−0.04	0.11**	0.11**	−0.03
2. ESCS	1	0.24**	−0.07**	0.05*	0.19**	0.18**	0.05**
3. No immigration background		1	−0.41**	0.01	0.07**	0.02	−0.02
4. Immigration background unknown			1	−0.01	−0.02	−0.03	0.00
5. Gender (male)				1	0.14**	0.17**	0.02
6. Computer experience (more than 5 years)					1	0.21**	0.13**
7. Computer use at home (almost daily)						1	0.33**
8. Cognitive tool							1

*$p < .05$; **$p < .01$.

substantially correlated with other productivity factors such as social background. This result suggests that an endogeneity bias is likely to occur when productivity factors other than computer use at home are not included in the statistical analysis. In addition, computer use at home as a cognitive tool is also correlated with computer experience and frequency of computer use at home. This result indicates that the psychological mechanism of using a computer at home as a cognitive tool is in fact underlying, at least partly, computer experience and the frequency of computer use at home.

Discussion

First, the results of the first regression model are in line with empirical findings reported in prior research. They suggest that computer experience and frequency of computer use at home are positively related with educational achievement. Second, however, as suggested by results of the second regression model, the significant correlations of computer experience and frequency of computer use at home with mathematical achievement disappear or decrease when additional productivity factors such as cognitive abilities are included in the model. Hence, the correlations observed in the first regression model are in part spurious and, thus, not the result of causal effects of computer use at home on mathematical achievement. Third, the results of the bivariate correlations among the productivity factors show that computer experience and frequency of computer use at home are correlated with other productivity factors such as social background. Therefore, the omission of other productivity factors in the analysis of computer use at home would result in an endogeneity bias and, thus, in an overestimation of the effects of computer use at home on educational achievement (e.g., Wittwer and Senkbeil 2008). Fourth, the results of the third regression model show that the operationalization of a psychological mechanism of computer use at home, that is, the use as a cognitive tool, produces a stronger effect on mathematical achievement than does a computer variable, such as the frequency of computer use at home.

In our example of investigating the relationship between computer use at home and educational achievement using data from the PISA 2006 study, we were not able to provide a solution to the problem with using nonexperimental research designs (but see, e.g., Stone and Rose 2011). Therefore, in our analysis, it cannot be ruled out that, for example, students with a higher level of educational achievement more often use a computer at home as a cognitive tool because they like to do so. To rule out this possibility, it is necessary to conduct experimental research that tests specific assumptions about cause-and-effect relationships. However, the results of large-scale assessments such as those illustrated in our example can empirically inform the design of such experiments (Schneider et al. 2007).

Furthermore, it is noteworthy that the use of additional control variables is often not sufficient to solve the problem of endogeneity bias. This is because it is impossible to rule out that some omitted variables may bias the true relation (see also the section "Contextual Indicators in Adult Literacy Studies: The Case of PIAAC" of this chapter). To address this problem, there are several methods of analysis that control for the endogeneity bias and that make causal inferences with observational data such as large-scale datasets. These methods, including fixed effects models, instrumental variables, propensity score matching, and regression discontinuity designs, can be used to approximate randomized controlled experiments (see Schneider et al. 2007, or Stone and Rose 2011, for an overview).

References

Attewell, P. and Battle, J. 1999. Home computers and school performance. *Information Society*, 15, 1–10.

Clark, R. E. 1994. Media will never influence learning. *Educational Technology Research and Development*, 42, 21–29.

Ellett, C. D. and Teddlie, C. 2003. Teacher evaluation, teacher effectiveness and school effectiveness: Perspectives from the USA. *Journal of Personnel Evaluation in Education* 17, 101–128.

Heller, K. A. and Perleth, C. 2000. *Kognitiver Fähigkeitstest für 4.-12. Klassen, Revision (KFT 4-12 + R) [Cognitive ability test for 4th to 12th grades, Revision (KFT 4-12 + R)]*. Göttingen: Hogrefe.

Jonassen, D. H. 2004. *Learning to Solve Problems: An Instructional Design Guide.* San Francisco: Pfeiifer/Jossey-Baas.

Jonassen, D. H., Carr, C., and Yueh, H.-P. 1998. Computers as mindtools for engaging learners in critical thinking. *Tech-Trends* 43, 24–32.

Lei, J. and Zhao, Y. 2007. Technology uses and student achievement: A longitudinal study. *Computers & Education* 49, 284–296.

Luu, K. and Freeman, J. G. 2011. An analysis of the relationship between information and communication technology (ICT) and scientific literacy in Canada and Australia. *Computers & Education* 56, 1072–1082.

Nachtigall, C., Kröhne, U., Enders, U., and Steyer, R. 2008. Causal effects and fair comparison: Considering the influence of context variables on student competencies. In *Assessment of Competencies in Educational Contexts.* ed. J. Hartig, E. Klieme, and D. Leutner, 315–335. Cambridge, MA: Hogrefe.

OECD. 2006. *Are Students Ready for a Technology-Rich World? What PISA Studies Tell Us.* Paris: OECD.

OECD. 2007. *PISA 2006. Science Competencies for Tomorrow's World. Volume 1: Analysis.* Paris: OECD.

OECD. 2009. *PISA 2006. Technical Report.* Paris: OECD.

OECD. 2011. *PISA 2009 Results: Students on Line. Digital Technologies and Performance (Volume VI).* Paris: OECD.

Papanastasiou, E., Zembylas, M., and Vrasidas, C. 2003. When computer use is associated with negative science achievement. *Journal of Science Education and Technology* 12, 325–332.

Salmon, W. C. 1997. Causality and explanation: A reply to two critiques. *Philosophy of Science* 64, 461–477.

Schneider, B., Carnoy, M., Kilpatrick, J., Schmidt, W. H., and Shavelson, R. J. 2007. *Estimating Causal Effects. Using Experimental and Observational Designs.* Washington, DC: AERA.

Stone, S. I. and R. A. Rose. 2011. Social work research and endogeneity bias. *Journal of the Society for Social Work and Research* 2, 54–75.

Thagard, P. 1998. Explaining disease: Correlations, causes, and mechanisms. *Minds and Machines* 8, 61–78.

Walberg, H. J. 2006. Improving educational productivity. An assessment of extant research. In *The Scientific Basis of Educational Productivity,* ed. R. F. Subotnik, and H. J. Walberg, 103–160. Greenwich, CT: Information Age Publishing.

Wittwer, J. and Senkbeil, M. 2008. Is students' computer use at home related to their mathematical performance at school? *Computers & Education* 50, 1558–1571.

14

Context Questionnaire Scales in TIMSS and PIRLS 2011

Michael O. Martin
TIMSS and PIRLS International Study Center

Ina V. S. Mullis
TIMSS and PIRLS International Study Center

Alka Arora
TIMSS and PIRLS International Study Center

Corinna Preuschoff
Jones and Bartlett Learning

CONTENTS

Introduction

Since their inception, the International Association for the Evaluation of Educational Achievement (IEA) Trends in International Mathematics and Science Study (TIMSS) and Progress in International Reading Literacy Study

(PIRLS) international assessments have provided extensive information about home supports and school environments for teaching and learning.* In 2011, the trend cycles of TIMSS (every 4 years) and PIRLS (every 5 years) came together, producing a synergy that led to advancements in the quality of the background data collected in both studies, and in particular to updates in the way context questionnaire data were scaled and reported.

Because both TIMSS and PIRLS assess students in the fourth grade (TIMSS also assesses students in the eighth grade), the alignment of the two projects provided the opportunity for countries to assess the same fourth-grade students in reading, mathematics, and science in conjunction with the extensive background data collected by these assessments. Given the need for a concerted approach meeting the needs of both projects, much of the questionnaire development work was accomplished during joint meetings of the TIMSS 2011 Questionnaire Item Review Committee and the PIRLS 2011 Questionnaire Development Group. The National Research Coordinators for TIMSS and those for PIRLS had multiple opportunities for reviewing the full array of questionnaires and making modifications.

The goals for updating the TIMSS and PIRLS questionnaires for 2011 include

- Updating the TIMSS and PIRLS contextual frameworks and questionnaires to reflect current policies and practices for teaching and learning
- Enhancing the validity of TIMSS and PIRLS 2011 background data by better aligning contextual frameworks and background questionnaires
- Maximizing the opportunity for comparisons across the three subject areas of reading, mathematics, and science
- Including scales that measure effective contexts for learning and meet the criteria for rigorous measurement
- Retaining questions for reporting trends on background contexts for learning
- Minimizing the response burden

The goals for updating background questionnaires in TIMSS and PIRLS 2011 reflect the desire to advance the questionnaires conceptually and empirically with each assessment cycle. Background questionnaire development began with updating the TIMSS contextual framework (Mullis et al. 2009b) and the PIRLS contextual framework (Mullis et al. 2009a) to reflect recently published research literature about effective educational policies and practices. The questionnaires were then updated in alignment with the

* TIMSS has reported trends in students' mathematics and science achievement at fourth and eighth grades every 4 years, beginning with TIMSS 1995. Similarly, PIRLS has reported trends in fourth grade students' reading comprehension every 5 years, beginning with PIRLS 2001.

frameworks so that they measured important aspects of effective learning environments.

As a 2011 initiative, renewed emphasis was placed on developing reliable scales that would provide valid measurement of effective home, school, and classroom environments for learning. Because of the need to keep response burden to a minimum, attention was restricted to constructs expected to have a positive relationship with student achievement in reading, mathematics, or science in the TIMSS or PIRLS data. The questionnaire development effort included adding questionnaire items to strengthen existing measures, such as the student self-confidence in learning scales for reading, mathematics, and science, and the index of school discipline and attendance problems. However, considerable effort also was invested in developing new scales to measure important aspects of effective classrooms, such as student engagement, teachers' efforts to stimulate student engagement, and teachers' confidence in teaching mathematics and science to their class.

As a further 2011 initiative, questionnaire items were chosen so that the response data from students, teachers, principals, and parents could be scaled using the 1-parameter item response theory (IRT; Rasch) partial-credit measurement model. Sufficient items were included for each scale so that the Rasch model could be reliably applied. Many questionnaire items had four response categories (e.g., "agree a lot," "agree a little," "disagree a little," and "disagree a lot"); a scale composed of such items required 6–7 of them to meet the minimum requirements of Rasch scaling. Scales composed of items with three response categories required more items.* Generally, one or two extra items were developed for each scale to allow for attrition after field testing.

Field Testing the Context Questionnaire Scales

Developing the TIMSS and PIRLS assessment instruments included conducting a full-scale field test in each participating country. The field test was designed to provide data on the measurement properties of the achievement and context questionnaire items and scales, and serve as a "dress rehearsal" operationally for the assessment. More than 60,000 students from 48 countries participated in the field test for TIMSS and PIRLS in the fourth grade, and an additional 60,000 in the TIMSS eighth-grade field test.

Using the data from the field test, the measurement properties of each of the items intended to form a background scale were analyzed to ensure

* As a guideline, Suen's (1990) formula was used to determine the minimum requirements for Rasch scaling: (number of items) × (number of response categories – 1) ≥ 20.

that items would be suitable for scaling with the 1-parameter IRT (Rasch) measurement model following the 2011 data collection. This involved checking the assumption of an underlying unidimensional construct and estimating the reliability of the resulting scale. As valid indicators of effective environments for learning, the background scales should be related to student achievement. Thus, the relationship between the item sets and student achievement in reading, mathematics, and science was also investigated.

Checking for Unidimensionality

Although there is no absolute criterion for a set of items to constitute a unidimensional scale, a scale typically may be considered "sufficiently unidimensional" if a single underlying construct is the dominant influence on the item responses (Embretson and Yang 2006; Reckase 1979). In factor-analytic terms, this implies the existence of a single large factor accounting for most of the covariance among the items.

In the TIMSS and PIRLS field test, the dimensionality of the items for each scale was evaluated using principal components analysis. Field test data from all participants were combined for these analyses, with all countries contributing equally regardless of their sample size. To evaluate the credibility of a unidimensional underlying construct, an analysis based on the first principal component was conducted. If there was evidence of more than one component, the item set was revised to better conform to a single dimension. That is, items not loading on the first component were flagged for elimination from the scales. Consistent with Comrey and Lee's (1992) rule of thumb, items with factor loadings (correlations between each item and the overall scale) less than 0.3 were eliminated from a scale unless they were considered to have crucial conceptual importance for measuring the construct.

The item sets for most scales were found to be sufficiently unidimensional, although in a few instances, items were identified that did not contribute to the measurement of the construct. Such items were eliminated.

Estimating Reliability (Internal Consistency)

To provide an indicator of the reliability of the proposed scales based on the field test data, Cronbach's Alpha, a measure of internal consistency, was computed for each scale. A scale was considered sufficiently reliable if Cronbach's Alpha was at least 0.7 (Nunally and Bernstein 1994). Most of the proposed scales had Cronbach's Alpha coefficients for the field test that exceeded 0.7. A few individual items, however, were identified that did not increase the reliability of the scale and thus were considered for elimination.

Evaluating the Relationship between the Proposed Scales and Student Achievement

For indicators of effective learning environments, a positive relationship with student achievement is an important aspect of validity. To examine the relationship between the proposed scales and student achievement for the field test data, a preliminary score was constructed for each scale. This was done by assigning a numerical value to each item response category and summing across the items in the scale. For example, if the response options for a set of five questions were *disagree a lot* (coded 1), *disagree a little* (2), *agree a little* (3), and *agree a lot* (4), the maximum score was 20 and the minimum score was 5. The responses were coded so that a high score indicated a supportive learning environment and a lower score a less supportive learning environment.

Although questionnaire responses were provided by students, parents, teachers, and school principals, depending on who completed the questionnaire, the information pertained to the learning environment of the student, so the field test analysis was conducted with the student as the unit of analysis, consistent with the TIMSS and PIRLS reporting approach. That is, parents', teachers', and school principals' responses were attached to student records so that scales describing learning environments could be related to student achievement results.

For the field test analysis, the score distribution of each scale was divided into three categories: above the 75th percentile (top 25%), above the 25th percentile but not above the 75th percentile (middle 50%), and below the 25th percentile (bottom 25%). Average achievement in reading, mathematics, and science was computed for the students in each scale category. Those students in the most supportive learning environments (top 25%) were expected to have higher achievement than those in the middle category, and those in the middle category higher than those in the bottom category. Most of the proposed scales had such a positive relationship with achievement within and across countries and thus were valid indicators of effective learning environments. A few scales, however, lacked the expected relationship with student achievement and were eliminated.

Reviewing and Finalizing the TIMSS and PIRLS 2011 Background Questionnaires

During the summer of 2010, the results of the field test analysis of the questionnaire items were reviewed by the TIMSS Questionnaire Item Review Committee, the PIRLS Questionnaire Development Group, the TIMSS National Research Coordinators, and the PIRLS National Research

TABLE 14.1

Number of Context Questionnaire Scales for the TIMSS and PIRLS 2011 Data Collection

	Grade 4	Grade 8
	Number of Scales	
Home questionnaire	8	—
Student questionnaire	11	9
School questionnaire	8	6
Teacher questionnaire	10	18[a]

[a] 9 scales were based on mathematics teachers' responses and 9 scales were based on science teachers' responses.

Coordinators. Most of the field test scale items were found to have good measurement properties and were retained for the main data collection. To minimize response burden, items that did not make a contribution to construct measurement or lacked a relationship with student achievement were not included in the final questionnaires. The field test analysis ensured that the proposed contextual scales were likely to provide countries with valid and reliable indicators of effective environments for learning with the least possible response burden.

The field test analysis resulted in the adoption of items for the final background questionnaires for 37 background scales for TIMSS and PIRLS in the fourth grade and 33 for TIMSS in the eighth grade, as summarized in Table 14.1.

Conducting the TIMSS and PIRLS 2011 Assessments

The TIMSS and PIRLS data collection took place at the end of 2010 in countries with school systems on the southern hemisphere time line and in early 2011 in northern hemisphere countries. TIMSS participants included 63 counties and 14 benchmarking entities (regional jurisdictions of countries, such as states or provinces). Countries and benchmarking participants could elect to take part in the fourth-grade assessment, the eighth-grade assessment, or both: 52 countries and seven benchmarking entities participated in the fourth-grade assessment, and 45 countries and 14 benchmarking entities participated in the eighth-grade assessment. Several of the countries, where fourth- and eighth-grade students were expected to find the TIMSS assessments too difficult, administered the fourth- and eighth-grade assessments to their sixth- and ninth-grade students, respectively. In total, more than 600,000 students participated in TIMSS 2011. The TIMSS 2011 mathematics results are reported in Mullis et al. (2012a) and the science results in Martin et al. (2012).

The PIRLS 2011 fourth-grade assessment of reading comprehension included 49 countries and nine benchmarking entities. Four countries assessed their sixth-grade students, and three countries participated in prePIRLS, a less difficult version of PIRLS inaugurated in 2011 to be a steppingstone to PIRLS. More than 300,000 students participated in PIRLS or prePIRLS, with most of these students also participating in TIMSS 2011 at the fourth grade. The results are reported in Mullis et al. (2012b).

Reporting Context Questionnaire Scales in TIMSS and PIRLS 2011

As an example illustrating the TIMSS and PIRLS approach to reporting context questionnaire data, Figure 14.1 presents the PIRLS 2011 *Students Confident in Reading* scale. As the name suggests, this scale seeks to measure how confident students feel about their ability to read, in terms of their level of agreement with seven statements about their reading. For each of the seven statements, students were asked to indicate the degree of their agreement: agree a lot, agree a little, disagree a little, or disagree a lot. Using IRT partial-credit scaling, student responses were placed on a scale constructed so that the mean scale score across all PIRLS countries was 10 and the standard deviation was 2. Statements expressing negative sentiment were reverse

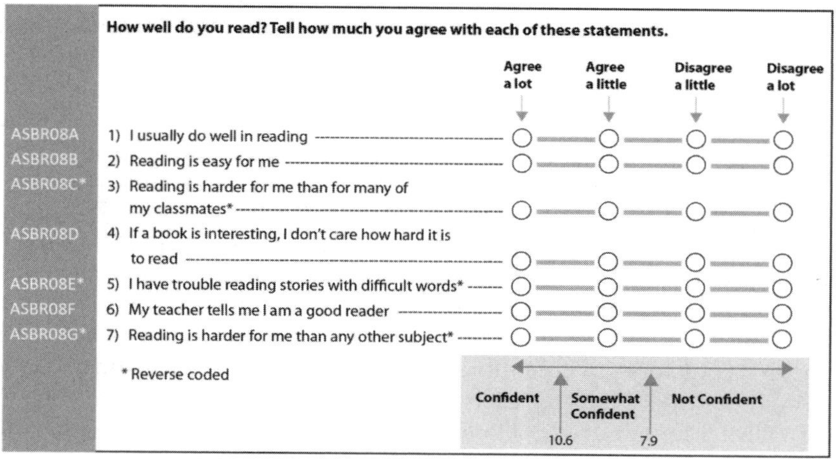

How well do you read? Tell how much you agree with each of these statements.

		Agree a lot	Agree a little	Disagree a little	Disagree a lot
ASBR08A	1) I usually do well in reading	○	○	○	○
ASBR08B	2) Reading is easy for me	○	○	○	○
ASBR08C*	3) Reading is harder for me than for many of my classmates*	○	○	○	○
ASBR08D	4) If a book is interesting, I don't care how hard it is to read	○	○	○	○
ASBR08E*	5) I have trouble reading stories with difficult words*	○	○	○	○
ASBR08F	6) My teacher tells me I am a good reader	○	○	○	○
ASBR08G*	7) Reading is harder for me than any other subject*	○	○	○	○

* Reverse coded

Confident | 10.6 | Somewhat Confident | 7.9 | Not Confident

FIGURE 14.1
Questionnaire items for the PIRLS 2011 *Students Confident in Reading* scale.

coded during the scaling (statements 3, 5, and 7). Students **confident** in their reading had a scale score greater than or equal to 10.6, the point on the scale corresponding to agreeing a lot on average with four of the seven statements, and a little with three of the statements. Students **not confident** in their reading had a score no higher than 7.9, the point on the scale corresponding to disagreeing a little with four of the statements on average, and agreeing a little with three of them.

Scaling Procedure

Partial-credit IRT scaling is based on a statistical model that relates the probability that a person will choose a particular response to an item to that person's location on the construct underlying the scale. In the *Students Confident in Reading* example scale, the underlying construct is confidence in reading, and students who agree in general with the seven statements are assumed to be more confident in their reading ability, and students who disagree with the statements are assumed to be less confident.

The partial-credit model is shown below

$$P_{x_i}(\theta_n) = \frac{e^{\sum_{k=0}^{x}(\theta_n - \delta_i + \tau_{ij})}}{\sum_{h=0}^{m} e^{\sum_{k=0}^{h}(\theta_n - \delta_i + \tau_{ik})}}; \quad x_i = 0,1,\ldots,m_i$$

where $P_{x_i}(\theta_n)$ denotes the probability that person n with location θ_n on the latent construct would choose response level x to item i out of the m_i possible response levels for the item. The item parameter δ_i gives the location of the item on the latent construct and τ_{ij} notes step parameters for the response levels. For each scale, the scaling procedure involves first estimating the δ_i and τ_{ij} item parameters, and then using the model with these parameters to estimate θ_n of the score on the latent construct, for each on the n respondents. Depending on the scale, respondents may be students, parents, teachers, or school principals.

The TIMSS and PIRLS 2011 context questionnaire scaling was conducted using the ConQuest 2.0 software (Wu et al. 2007).

In preparation for the context questionnaire scaling effort, the TIMSS and PIRLS International Study Center developed a system of production programs that could effectively calibrate the items on each scale using ConQuest and produce scale scores for each scale respondent. Each assessment population (TIMSS fourth grade, TIMSS eighth grade, and PIRLS fourth grade) consisted of approximately 300,000 students, as well as their parents, teachers, and school principals. The estimation of the item parameters, a procedure also known as item calibration, was conducted on the combined data

TABLE 14.2

Item Parameters for the PIRLS 2011 *Students Confident in Reading* Scale

Item	delta	tau_1	tau_2	tau_3
ASBR08A	−1.38588	−0.06832	−0.67050	0.73882
ASBR08B	−1.41524	−0.12326	−0.41288	0.53614
ASBR08C[a]	−0.70497	−0.15123	0.25172	−0.10049
ASBR08D	−1.01512	0.44367	−0.39057	−0.05310
ASBR08E[a]	−0.13353	−0.54074	0.47905	0.06169
ASBR08F	−0.85116	−0.29990	−0.45939	0.75929
ASBR08G[a]	−0.81967	0.16729	0.25788	−0.42517

[a] Reverse coded.

from all countries, with each country contributing equally to the calibration. This was achieved by weighting each country's student data to sum to 500. Table 14.2 shows the international item parameters for the *Students Confident in Reading* scale. For each item, the delta parameter δ_i shows the estimated overall location of the item on the scale, and the tau parameters τ_{ij} show the location of the steps, expressed as deviations from delta.*

Once the calibration was complete and international item parameters had been estimated, individual scores for each respondent (students, teachers, principals, or parents) were generated using weighted maximum likelihood estimation (Warm 1989). All cases with valid responses to at least two items on the scale were included in the calibration and scoring processes.

The scale scores produced by the weighted likelihood estimation are in the logit metric and range from approximately −5 to +5. To convert to a more convenient reporting metric, a linear transformation was applied to the international distribution of logit scores for each scale, so that the resulting distribution across all countries had a mean of 10 and a standard deviation of 2. Table 14.3 presents the scale transformation constants applied to the international distribution of logit scores for the *Students Confident in Reading* scale to transform them to the (10,2) reporting metric.

On the TIMSS and PIRLS achievement scales in mathematics, science, and reading, the low, intermediate, high, and advanced international benchmarks of achievement are specific reference points on the scale that can be used to monitor progress in student achievement. Using a scale anchoring procedure

* Although typically the values of the item step parameters estimated from the data are in the same order as their response categories, this is not the case with the tau parameters for the *Students Confident in Reading* scale. However, as described in Adams et al. (2012), this does not imply that the data do not fit the scaling model, but rather reflect the distribution of respondents across the response categories of the items. As with many of the TIMSS and PIRLS context questionnaire scales, students were very positive in their responses to the items on this scale, with more than half the students reporting that they "agree a lot" to almost all of the items. This does not prevent the scale from effectively summarizing the item responses, but does result in the disordered tau parameters.

TABLE 14.3

Scale Transformation Constants for the PIRLS 2011 *Students Confident in Reading* Scale

Scale Transformation Constants	
A = 9.96677	Transformed scale score = 9.96677 + 2.18490 · logit scale score
B = 2.18490	

(Mullis 2012), student performance at each benchmark is described in terms of the mathematics, science, or reading (depending on the subject) that students reaching that benchmark know and can do. The percentage of students reaching each of these international benchmarks can serve as a profile of student achievement in a country.

To provide an analogous approach to reporting the context questionnaire scales, a method was developed to divide each scale into high, middle, and low regions and provide a content-referenced interpretation for these regions. The interpretation is content referenced to the extent that the boundaries of the regions were defined in terms of identifiable combinations of response categories. The particular response combinations that defined the regions, boundaries, or cutpoints, were based on a judgment of what constituted a high or low region on each individual scale. For example, based on a consideration of the questions making up the *Students Confident in Reading* scale, it was determined that in order to be in the high region of the scale and labeled "confident," a student would have to agree a lot, on average, with at least four of the seven statements and agree a little to the other three. Similarly, it was determined that a student who, on average, at most agreed a little with four of the statements and disagreed with the other three would be labeled "not confident."

The scale region cutpoints were quantified by assigning a numeric value to each response category, such that each respondent's responses to the scale's questions could be expressed as a "raw score." Assigning 0 to "disagree a lot," 1 to "disagree a little," 2 to "agree a little," and 3 to "agree a lot" results in raw scores on the *Students Confident in Reading* scale ranging from 0 (disagree a lot with all seven statements) to 21 (agree a lot with all seven). A student who agreed a lot with four statements and agreed a little with the other three would have a raw score of 18 ($4 \times 3 + 3 \times 2$). Following this approach, a student with a raw score of 18 or more would be in the "confident" region of the scale. Similarly, agreeing a little with three statements and disagreeing a little with four statements would result in a raw score of 10 ($3 \times 2 + 4 \times 1$), so that a student with a raw score less than or equal to 10 would be in the "not confident" region.

A property of a Rasch scale is that each raw score has a unique scale score associated with it. Table 14.4 presents a raw score–scale score equivalence table for the *Students Confident in Reading* scale. From this table, it can be seen that a raw score of 10 corresponds to a scale score of 7.9 (after rounding), and

TABLE 14.4

Equivalence Table of the Raw Score and Transformed Scale Scores for the PIRLS 2011 *Students Confident in Reading* Scale

Raw Score	Transformed Scale Score	Cutpoint	Raw Score	Transformed Scale Score	Cutpoint
0	1.97908		11	8.09145	
1	3.87424		12	8.37394	
2	4.75552		13	8.67056	
3	5.36024		14	8.98289	
4	5.83733		15	9.32033	
5	6.23940		16	9.69670	
6	6.59960		17	10.12500	
7	6.92758		18	10.64147	10.6
8	7.23353		19	11.29688	
9	7.52532		20	12.25876	
10	7.80940	7.9	21	14.35923	

a raw score of 18 corresponds to a scale score of 10.6. These scale scores were the cutpoints used to divide the scale into the three regions.

Validating the TIMSS and PIRLS 2011 Context Questionnaire Scales

As evidence that the context questionnaire scales provide comparable measurement across countries, reliability coefficients were computed for each scale for every country and benchmarking participant, and a principal components analysis of the scale items was conducted. Table 14.5 presents the results of this analysis for the *Students Confident in Reading* scale. The Cronbach's Alpha reliability coefficients generally were at an acceptable level, with most above 0.6 or 0.7, although in a few countries, the value was below 0.5. The exhibit also shows the percentage of variance among the scale items accounted for by the first principal component in each country. In most cases, this was acceptably high, indicating that the items could be adequately represented by a single scale. The factor loadings of each questionnaire item from the principal components analysis are positive and substantial, indicating a strong correlation between each item and the scale in every country.

As indicators of effective environments for learning, a positive relationship with achievement is an important aspect of validity for the TIMSS and PIRLS context questionnaire scales. For the *Students Confident in Reading* scale, Table 14.6 presents the Pearson correlation with reading achievement in PIRLS 2011

TABLE 14.5

Cronbach Alpha Reliability Coefficients and Principal Component Analysis of the Items in the PIRLS 2011 *Students Confident in Reading* Scale

Country Name	Cronbach Alpha Reliability Coefficient	Percent of Variance Explained	Factor Loadings for Each Item						
			ASBR08A	ASBR08B	ASBR08C[a]	ASBR08D	ASBR08E[a]	ASBR08F	ASBR08G[a]
Australia	0.72	40	0.73	0.78	0.71	0.46	0.61	0.34	0.69
Austria	0.75	43	0.78	0.81	0.68	0.39	0.51	0.63	0.71
Azerbaijan	0.58	29	0.39	0.53	0.69	0.28	0.61	0.49	0.68
Belgium (French)	0.66	35	0.71	0.71	0.64	0.32	0.55	0.47	0.66
Bulgaria	0.78	46	0.77	0.83	0.70	0.48	0.51	0.68	0.71
Canada	0.71	40	0.72	0.77	0.73	0.34	0.60	0.39	0.73
Chinese Taipei	0.72	39	0.79	0.81	0.56	0.60	0.27	0.62	0.57
Colombia	0.48	26	0.55	0.57	0.62	0.24	0.45	0.42	0.61
Croatia	0.75	44	0.79	0.78	0.72	0.33	0.44	0.74	0.69
Czech Republic	0.77	44	0.73	0.82	0.72	0.38	0.56	0.64	0.71
Denmark	0.73	41	0.76	0.79	0.75	0.38	0.69	0.41	0.59
England	0.73	42	0.74	0.80	0.75	0.38	0.66	0.26	0.73
Finland	0.69	40	0.76	0.75	0.71	0.24	0.58	0.46	0.75
France	0.69	38	0.73	0.71	0.71	0.29	0.55	0.53	0.66
Georgia	0.63	35	0.69	0.70	0.63	0.20	0.51	0.63	0.61
Germany	0.76	44	0.78	0.80	0.74	0.33	0.54	0.64	0.70
Hong Kong SAR	0.69	36	0.77	0.75	0.56	0.60	0.41	0.48	0.57
Hungary	0.77	46	0.76	0.81	0.73	0.31	0.52	0.76	0.71
Indonesia	0.53	27	0.21	0.28	0.82	0.08	0.65	0.24	0.77
Iran, Islamic Republic of	0.54	29	0.56	0.64	0.65	0.17	0.52	0.48	0.62
Ireland	0.71	40	0.73	0.78	0.72	0.37	0.62	0.34	0.71

Israel	0.66	35	0.52	0.70	0.69	0.38	0.55	0.57	0.66
Italy	0.66	35	0.69	0.73	0.64	0.27	0.54	0.61	0.58
Lithuania	0.74	43	0.80	0.82	0.70	0.25	0.49	0.67	0.66
Malta	0.71	39	0.68	0.76	0.67	0.39	0.53	0.58	0.66
Morocco	0.38	24	0.04	0.61	0.62	0.25	0.32	0.59	0.62
Netherlands	0.78	46	0.77	0.82	0.76	0.24	0.61	0.60	0.77
New Zealand	0.67	36	0.69	0.74	0.65	0.39	0.53	0.44	0.64
Northern Ireland	0.71	39	0.71	0.75	0.70	0.39	0.64	0.31	0.70
Norway	0.67	37	0.73	0.75	0.64	0.37	0.59	0.42	0.62
Oman	0.54	29	0.60	0.69	0.59	0.20	0.39	0.58	0.57
Poland	0.76	44	0.76	0.79	0.71	0.23	0.66	0.63	0.70
Portugal	0.73	41	0.77	0.77	0.62	0.37	0.58	0.70	0.54
Qatar	0.52	30	0.34	0.56	0.75	−0.07	0.61	0.41	0.74
Romania	0.74	42	0.76	0.78	0.65	0.40	0.52	0.69	0.66
Russian Federation	0.64	40	0.76	0.71	0.71	−0.06	0.48	0.68	0.71
Saudi Arabia	0.58	29	0.38	0.58	0.68	0.34	0.59	0.48	0.65
Singapore	0.69	37	0.72	0.77	0.65	0.48	0.51	0.39	0.65
Slovak Republic	0.77	45	0.76	0.81	0.67	0.33	0.63	0.66	0.70
Slovenia	0.77	45	0.77	0.80	0.67	0.29	0.68	0.70	0.67
Spain	0.61	32	0.70	0.69	0.59	0.43	0.27	0.62	0.55
Sweden	0.74	42	0.74	0.80	0.71	0.36	0.63	0.44	0.75
Trinidad and Tobago	0.68	37	0.69	0.74	0.66	0.35	0.50	0.57	0.65
United Arab Emirates	0.57	30	0.50	0.65	0.67	0.24	0.54	0.43	0.66
United States	0.71	39	0.72	0.77	0.73	0.34	0.55	0.41	0.72

continued

TABLE 14.5 (continued)

Cronbach Alpha Reliability Coefficients and Principal Component Analysis of the Items in the PIRLS 2011 *Students Confident in Reading* Scale

Country Name	Cronbach Alpha Reliability Coefficient	Percent of Variance Explained	Factor Loadings for Each Item						
			ASBR08A	ASBR08B	ASBR08C[a]	ASBR08D	ASBR08E[a]	ASBR08F	ASBR08G[a]
Sixth Grade Participants									
Botswana	0.53	28	0.53	0.59	0.73	0.28	0.34	0.40	0.68
Honduras	0.47	26	–0.02	0.03	0.79	–0.20	0.70	–0.21	0.79
Kuwait	0.59	30	0.15	0.59	0.71	0.35	0.63	0.43	0.73
Morocco	0.40	26	–0.19	0.58	0.69	0.22	0.45	0.50	0.69
Benchmarking Participants									
Alberta, Canada	0.71	40	0.71	0.78	0.75	0.33	0.63	0.31	0.73
Ontario, Canada	0.71	39	0.73	0.76	0.71	0.33	0.60	0.39	0.72
Quebec, Canada	0.71	40	0.72	0.78	0.74	0.30	0.58	0.41	0.72
Maltese-Malta	0.73	40	0.75	0.79	0.63	0.50	0.51	0.64	0.55
Eng/Afr (5)-RSA	0.60	30	0.50	0.59	0.67	0.36	0.58	0.38	0.67
Andalusia, Spain	0.59	31	0.71	0.69	0.55	0.43	0.25	0.62	0.53
Abu Dhabi, UAE	0.56	29	0.48	0.65	0.65	0.25	0.52	0.47	0.64
Dubai, UAE	0.61	32	0.59	0.66	0.69	0.26	0.57	0.41	0.67
Florida, USA	0.72	41	0.74	0.79	0.75	0.34	0.52	0.42	0.75
Pre-PIRLS Countries									
Colombia	0.45	24	0.57	0.60	0.47	0.36	0.19	0.65	0.41
South Africa	0.48	26	0.54	0.56	0.63	0.23	0.45	0.42	0.62
Botswana	0.45	29	0.53	0.53	–0.50	0.52	–0.55	0.56	–0.55

[a] Reverse coded.

TABLE 14.6

Relationship between the *Students Confident in Reading* Scale and PIRLS 2011 Reading Achievement

Country	Pearson's Correlation with Reading Achievement		Variance in Reading Achievement Accounted for by Difference between Regions of the Scale (η^2)
	(r)	(r^2)	
Australia	0.45	0.21	0.20
Austria	0.36	0.13	0.12
Azerbaijan	0.27	0.07	0.07
Belgium (French)	0.40	0.16	0.15
Bulgaria	0.41	0.17	0.17
Canada	0.40	0.16	0.14
Chinese Taipei	0.34	0.12	0.10
Colombia	0.34	0.12	0.08
Croatia	0.37	0.14	0.13
Czech Republic	0.38	0.15	0.15
Denmark	0.44	0.20	0.18
England	0.42	0.18	0.16
Finland	0.37	0.14	0.14
France	0.40	0.16	0.14
Georgia	0.35	0.12	0.10
Germany	0.39	0.15	0.14
Hong Kong SAR	0.34	0.12	0.10
Hungary	0.49	0.24	0.22
Indonesia	0.29	0.08	0.08
Iran, Islamic Republic of	0.33	0.11	0.11
Ireland	0.37	0.14	0.13
Israel	0.41	0.17	0.16
Italy	0.30	0.09	0.08
Lithuania	0.43	0.19	0.17
Malta	0.46	0.21	0.21
Morocco	0.30	0.09	0.08
Netherlands	0.31	0.10	0.09
New Zealand	0.44	0.19	0.17
Northern Ireland	0.37	0.14	0.13
Norway	0.35	0.12	0.13
Oman	0.41	0.17	0.16
Poland	0.45	0.20	0.22
Portugal	0.40	0.16	0.15
Qatar	0.49	0.24	0.20

continued

TABLE 14.6 (continued)

Relationship between the *Students Confident in Reading* Scale and PIRLS 2011 Reading Achievement

Country	Pearson's Correlation with Reading Achievement		Variance in Reading Achievement Accounted for by Difference between Regions of the Scale (η^2)
	(r)	(r^2)	
Romania	0.47	0.22	0.22
Russian Federation	0.37	0.14	0.13
Saudi Arabia	0.43	0.19	0.18
Singapore	0.38	0.15	0.14
Slovak Republic	0.41	0.17	0.15
Slovenia	0.43	0.19	0.18
Spain	0.35	0.12	0.11
Sweden	0.39	0.15	0.14
Trinidad and Tobago	0.50	0.25	0.23
United Arab Emirates	0.42	0.18	0.16
United States	0.40	0.16	0.15
International Median	0.40	0.16	0.14
Sixth Grade Participants			
Botswana	0.49	0.24	0.21
Honduras	0.32	0.10	0.07
Kuwait	0.39	0.15	0.13
Morocco	0.31	0.10	0.08
Benchmarking Participants			
Alberta, Canada	0.40	0.16	0.16
Ontario, Canada	0.39	0.15	0.14
Quebec, Canada	0.40	0.16	0.13
Maltese-Malta	0.36	0.13	0.12
Eng/Afr (5)-RSA	0.44	0.20	0.17
Andalusia, Spain	0.34	0.12	0.11
Abu Dhabi, UAE	0.45	0.20	0.18
Dubai, UAE	0.39	0.16	0.15
Florida, USA	0.43	0.19	0.16
Colombia	0.44	0.19	0.17
South Africa	0.36	0.13	0.09
Botswana	0.43	0.19	0.16
International Median	0.43	0.19	0.16

for each country, together with the proportion of variance in reading achievement attributable to the *confident* scale (r-square). These figures show a moderate positive relationship in every country. Also shown is the proportion of variance in reading achievement attributable to differences between the regions of the *confident* scale. This is very similar to the proportion of variance explained by the scale as a whole, indicating that dividing the scale into regions retains most of the relationship between the scale and achievement.

The item parameter estimates and item and scale statistics presented above for the *Students Confident in Reading* scale are examples of the statistics available for each of the TIMSS and PIRLS 2011 context questionnaire scales. The complete list of scales and the assessment populations with which they were used may be found at http://timssandpirls.bc.edu/methods/t-context-q-scales.html.

References

Adams, R. J., Wu, M. L., and Wilson, M. 2012. The Rasch rating model and the disordered threshold controversy. *Educational and Psychological Measurement*, 72(4), 547–573. Retrieved from http://epm.sagepub.com/content/72/4/547

Comrey, A. L. and Lee, H. B. 1992. *A First Course in Factor Analysis.* Hillsdale, NJ: Lawrence & Erlbaum Associates.

Embretson, S. and Yang, X. 2006. Item response theory. In J. L. Green, G. Camilli, and P. B. Elmore (Eds.), *Handbook of Complementary Methods in Education Research.* Hillsdale, NJ: Lawrence & Erlbaum Associates.

Martin, M. O., Mullis, I. V. S., Foy, P., and Stanco, G. M. 2012. *TIMSS 2011 International Results in Science.* Chestnut Hill, MA: TIMSS & PIRLS International Study Center, Boston College.

Mullis, I. V. S. 2012. Using scale anchoring to interpret the TIMSS and PIRLS 2011 achievement scales. In M. O. Martin and I. V. S. Mullis (Eds.), *Methods and Procedures in TIMSS and PIRLS 2011.* Chestnut Hill, MA: TIMSS & PIRLS International Study Center, Boston College. Retrieved from http://timssandpirls.bc.edu/methods/t-achievement-scales.html

Mullis, I. V. S., Martin, M. O., Kennedy, A. M., Trong, K. L., and Sainsbury, M. 2009a. *PIRLS 2011 Assessment Framework.* Chestnut Hill, MA: TIMSS & PIRLS International Study Center, Boston College.

Mullis, I. V. S., Martin, M. O., Ruddock, G. J., O'Sullivan, C. Y., and Preuschoff, C. 2009b. *TIMSS 2011 Assessment Framework.* Chestnut Hill, MA: TIMSS & PIRLS International Study Center, Boston College.

Mullis, I. V. S., Martin, M. O., Foy, P., and Arora, A. 2012a. *TIMSS 2011 International Results in Mathematics.* Chestnut Hill, MA: TIMSS & PIRLS International Study Center, Boston College.

Mullis, I. V. S., Martin, M. O., Foy, P., and Drucker, K. T. 2012b. *PIRLS 2011 International Results in Reading.* Chestnut Hill, MA: TIMSS & PIRLS International Study Center, Boston College.

Nunally, J. C. and Bernstein, I. H. 1994. *Psychometric Theory* (3rd ed.). New York: McGraw-Hill.

Reckase, M. D. 1979. Unifactor latent trait models applied to multifactor tests: Results and implications. *Journal of Educational Statistics*, 4, 207–230.

Warm, T. A. 1989. Weighted likelihood estimation of ability in item response theory. *Psychometrika*, 54(3), 427–450.

Wu, M. L., Adams, R. J., Wilson, M. R., and Haldane, S. 2007. *Conquest 2.0 [computer software]*. Camberwell, Australia: Australian Council for Educational Research.

15

Motivation and Engagement in Science around the Globe: Testing Measurement Invariance with Multigroup Structural Equation Models across 57 Countries Using PISA 2006

Benjamin Nagengast
University of Tübingen

Herbert W. Marsh
University of Western Sydney
University of Oxford
King Saud University

CONTENTS

International large-scale assessment studies offer unprecedented opportunities for studying achievement and achievement-related processes from a cross-cultural perspective. In this chapter, we focus on the cross-cultural study of motivational processes using large-scale assessment data (see Brunner et al. 2009; Chiu and Klassen 2010; Marsh and Hau 2004; Nagengast and Marsh 2012; Nagengast et al. 2011; Seaton et al. 2009, for recent applied examples). While the data quality of large-scale assessment studies such as the Program for International Student Assessment (PISA) or the Trends in International Maths and Science Study (TIMSS) is typically very high due to the rigorous testing protocols and quality assurance measures, we focus on the issue of comparability of scales across countries that cannot be taken as a given (van de Vijver and Leung 2000). Cross-cultural comparisons require that the meaning of constructs remains invariant over the studied countries, so that comparisons are meaningful (Marsh and Grayson 1994). Multigroup confirmatory factor analyses (MG-CFA) and structural equation modeling (MG-SEM) have emerged as standard tools for testing measurement invariance of scales in cross-cultural research (e.g., Lee et al. 2011; Marsh et al. 2006) and are increasingly used to analyze data from international large-scale surveys (e.g., Marsh et al. 2006; OECD 2010). Depending on the parameters that are invariant across countries, different cross-country comparisons are meaningful. In this chapter, we introduce a 13-model taxonomy of measurement invariance models and the possible inferences based on these models (Marsh et al. 2009; Widaman and Reise 1997). We present an exemplary analysis of motivation and engagement measures in the student background questionnaire of PISA 2006. The subject focus is on motivation and engagement in science, an area with important policy implications.

Methodological Focus: Measurement Invariance

Importance of Measurement Invariance

Cross-cultural comparisons of constructs rest on the assumption that the scales have the same meaning in the considered cultures and that there are no cultural biases and differences in the meaning of items across countries (Lee et al. 2011; Marsh et al. 2006). Even though the importance of establishing measurement invariance in cross-cultural research has been well established in the methodological literature (e.g., Chen 2008; Little 1997; Lee, et al. 2011; Marsh et al. 2006), it has not permeated applied research to an acceptable degree. Chen (2008) reported that in 48 cross-cultural studies published in the *Journal of Personality and Social Psychology*, a leading journal in its field and one of the premier journals in psychology, from 1985 to 2005 less than 17% tested measurement invariance. When the author carefully inspected published findings, he also found evidence for violations of invariance of factor loadings, a violation of a crucial assumption, discussed subsequently.

With regard to its potential for cross-cultural comparisons, PISA stands out among other international large-scale surveys. The quality control measures for scale construction and translation of items are particularly high, constructs and their operationalization are selected based on recommendations of international expert groups, and all scales are tested in extensive field trials. In addition, particular care of selecting representative samples and of obtaining high-quality data is taken (see OECD 2009). From the perspective of educational researchers interested in processes related to achievement and engagement, it is especially laudable that the high level of quality control is not only applied to the achievement tests that are the basis of the high-profile league tables of countries that PISA produces, but also to the background questionnaires that assess motivational constructs. However, careful scale construction alone is not sufficient to guarantee construct equivalence across countries. After data have been collected, further steps of statistical validation are necessary to ensure that the items consistently assess the measured constructs in all countries and that the scale properties are not affected by cultural differences. Typically, these properties are established by testing measurement invariance (Meredith 1993; Vandenberg and Lance 2000; Widaman and Reise 1997) using MG-CFA and MG-SEM.

Tests of measurement invariance seek to assure that the items used to measure the constructs of interest are related to the latent variables in the same way across countries. Typically, different degrees of invariance are considered that allow different types of comparisons across countries. MG-SEM is often the method of choice for assessing measurement invariance; hence, we introduce this modeling framework and use it to define the models of the 13-model taxonomy introduced by Marsh et al. (2009) (see also Widaman and Reise, 1997).

MG-CFA Model

In its most general form, the MG-CFA model for p indicators, m latent variables, and g groups is defined by the following equations (Sörbom 1974):

$$x^{(g)} = \boldsymbol{\tau}_x^{(g)} + \boldsymbol{\Lambda}_x^{(g)}\boldsymbol{\xi}^{(g)} + \boldsymbol{\varepsilon}^{(g)}. \quad (15.1)$$

$$E(\boldsymbol{\xi}^{(g)}) = \boldsymbol{\nu}^{(g)} \quad (15.2)$$

$$Var(\boldsymbol{\xi}^{(g)}) = \boldsymbol{\Phi}^{(g)} \quad (15.3)$$

$$Var(\boldsymbol{\varepsilon}^{(g)}) = \boldsymbol{\Theta}_\varepsilon^{(g)} \quad (15.4)$$

In the MG-CFA, the p-dimensional group-specific response vectors $\boldsymbol{x}^{(g)}$ that are typically of a high dimensionality (e.g., corresponding to items in an instrument, indicators of the latent factors) for each of g groups are explained by an underlying set of latent variables $\boldsymbol{\xi}^{(g)}$ of lower dimensionality, an m-dimensional vector. The p x m-dimensional factor loading matrix $\boldsymbol{\Lambda}_x^{(g)}$ specifies the relations of the latent variables and the indicators. The p-dimensional vector $\boldsymbol{\tau}_x^{(g)}$ contains the group-specific intercepts, one for each indicator, and the p-dimensional vector $\boldsymbol{\varepsilon}^{(g)}$ contains the residuals with p x p-dimensional variance–covariance matrix $\boldsymbol{\Theta}_\varepsilon^{(g)}$ that is typically assumed to be a diagonal matrix, implying that residuals associated with each indicator are uncorrelated. The m-dimensional mean vector of the latent variables is given by $\boldsymbol{\nu}^{(g)}$, the m x m-dimensional variance–covariance matrix of relations among the multiple latent factors by $\boldsymbol{\Phi}^{(g)}$. Both the latent variables $\boldsymbol{\xi}^{(g)}$ and the residuals $\boldsymbol{\varepsilon}^{(g)}$ are assumed to be normally distributed. The superscripts (g) indicate that the corresponding vectors and matrices can vary across the multiple groups. The model implies group-specific p x p-dimensional variance–covariance matrices $\Sigma_{xx}^{(g)}$ and p-dimensional mean vectors $\boldsymbol{\mu}_x^{(g)}$ or the observed variables

$$\boldsymbol{\mu}_x^{(g)} = \boldsymbol{\tau}_x^{(g)} + \boldsymbol{\Lambda}_x^{(g)}\boldsymbol{\nu}^{(g)}, \tag{15.5}$$

$$\Sigma_{xx}^{(g)} = \boldsymbol{\Lambda}_x^{(g)}\boldsymbol{\Phi}^{(g)}\boldsymbol{\Lambda}_x^{(g)} + \boldsymbol{\Theta}_\varepsilon^{(g)}. \tag{15.6}$$

The discrepancy between the model implied and the observed mean vectors and covariance matrices constitute the basis for global tests of model fit (see, e.g., Kaplan 2009 for an introduction into model estimation and testing of structural equation models).

As specified in Equations 15.1 through 15.4, the MG-CFA model would not be identified without further constraints. In the present context, we identified the model by choosing a reference indicator for each latent variable and by fixing that indicator's loading to a value of one. To identify the mean structure, the means of the latent variables were fixed to a value of zero in all groups. The latter restriction can be lifted after invariance of the item intercepts is established (see below). In all models with completely invariant item intercepts across countries, only the factor means in the reference group were constrained to zero for identification.

Tests of Measurement Invariance Using MG-CFA

Typically, tests of measurement invariance proceed in the following way (see Marsh et al. 2009; Meredith 1993; Vandenberg and Lance 2000; Widaman and Reise 1997): In the first step, the loading matrices $\boldsymbol{\Lambda}_y^{(g)}$ are restricted to have the same pattern of fixed and freed elements across the groups. This *configural invariance model* (Widaman and Reise 1997) tests whether the factorial structure is the same in all considered countries. In this model, none of the estimated parameters are constrained to be invariant over groups (except for those constrained to fixed values—typically 0 or 1—used to identify the factor structure in each group). If this model does not fit the data, there are fundamental cross-country differences in the dimensionality of the assessed constructs and cross-country comparisons on common scales are fraught with difficulty (see Marsh and Grayson 1994, for discussion of a hierarchy of invariances and what interpretations might be justified without at least partial configural invariance). The configural invariance model also serves as a reference model against which the fit of more restrictive invariance models can be compared. Setting parameters to be invariant across countries will impose further constraints on the mean vector and variance–covariance matrices of the indicators and lead to a decrease of model fit. This creates a taxonomy of partially nested models that can be compared using conventional methods of assessing fit in structural equation models.

The second model to be tested is the *metric invariance model* (Vandenberg and Lance 2000; also referred to as the *weak measurement invariance model*; Meredith 1993). In this model, the factor loading matrices are set to be invariant across the countries, that is, $\mathbf{\Lambda}_y^{(g)}=\mathbf{\Lambda}_y$. When metric invariance holds, the indicators are related to the latent variables in the same way in all countries. Differences in the latent variables get translated into differences in the indicators in a similar way across the countries. Metric invariance is the precondition for meaningful comparisons of the variance–covariance matrices of the latent variables $\mathbf{\Phi}^{(g)}$ across the countries, as they are defined by similar measurement models (Marsh et al. 2006; Meredith 1993; Widaman and Reise 1997).

After establishing metric invariance, there is no universal agreement on what restrictions to test next. Marsh et al. (2009) presented a 13-model taxonomy of measurement invariance that systematically incorporates all combinations of invariance tests. The configural invariance model and the metric invariance model are included as M1 and M2 in this taxonomy. All further models are built on the metric invariance model and further restrict parameters to be invariant across countries. M3 restricts the residual variance–covariance matrix to be invariant across the countries, that is, $\mathbf{\Theta}_\varepsilon^{(g)} = \mathbf{\Theta}_\varepsilon$. M4 restricts the variance–covariance matrix of the latent variables to be invariant, that is, $\mathbf{\Phi}^{(g)} = \mathbf{\Phi}$. The combination of these two restrictions in M6 implies that the reliability of the items and scales does not vary across countries. A variation of M4 does not restrict the complete variance–covariance matrix, but only the correlation matrix of the latent variables to be invariant across countries (M4a). In this model, there are no direct restrictions on the variances of the latent variables, but only on the off-diagonal elements of the variance–covariance matrix. Testing the invariance of the correlations rather than the covariances across countries might be called for when differences in dispersion across countries are not of primary interest. M4 poses additional constraints on the variance–covariance matrix and is nested in M4a.

Another important model that is conventionally included in tests of measurement invariance is the *strong measurement invariance* (Meredith 1993) or *scalar invariance model* (M5; Vandenberg and Lance 2000). Here, the item intercepts are set to be invariant across the countries, that is, $\boldsymbol{\tau}_x^{(g)} = \boldsymbol{\tau}_x$. Strong measurement invariance is the precondition for comparing latent factor means across countries (Marsh et al. 2009; Meredith 1993; Widaman and Reise 1997). If this restriction holds, there are no systematic differences in the average item responses between countries that are not due to differences in the mean level of latent variables.

M7 that combines invariant factor loadings, invariant item intercepts, and invariant residual variances is known as the *strict measurement invariance model* (Meredith 1993). If M7 holds, the manifest scale scores (and not just latent variables) can be meaningfully compared across countries (Marsh et al. 2009; Meredith 1993; Widaman and Reise 1997). Finally, models M8 to M13 further combine various restrictions, including invariance of the latent factor means $\boldsymbol{\nu}^{(g)}$. Using the flexible MG-SEM framework, it is also possible to include covariates (such as gender, achievement, or socioeconomic status) in the model. Doing so allows testing of the relationship between the latent variables and other predictors. This can be done by estimating bivariate correlations between covariates and the latent variables. Alternatively, it is also possible to model the latent variables as dependent variables in a multiple regression using the covariates as predictor variables. The latter type of model is known as multiple-indicator-multiple-causes (MIMIC) model (Jöreskog and Goldberger 1975). Implementing these models in an MG-SEM framework allows testing the invariance of relations between latent variables and the covariates across the countries. Again, the meaningful interpretation of MG models with covariates requires that metric invariance (i.e., M2) holds.

Substantive Focus: Motivation and Engagement in Science

The number of students pursuing careers in science and related fields (e.g., engineering) is declining worldwide (OECD 2007). While schools focus primarily on academic achievement, the fundamental problem is one of motivation to engage in a science-related career path (Nagengast and Marsh 2012). Hence, a crucial first step in reversing this trend requires a precise understanding of the key motivational determinants of career choices and their interplay, as well as the underlying motives to students' disengagement in science subjects (Nagengast and Marsh 2012). Developing strong and well-validated measures of motivation and engagement in science is an important first step in this direction.

Among the development of cross-culturally valid and comprehensive instruments for assessing motivational processes, Marsh et al.'s (2006; see also Artelt 2005; Artelt et al. 2003; Peschar et al. 1999) *Student Approaches to Learning* (SAL), included in the background questionnaire for PISA 2000, ranks as one of the most impressive accomplishments. In its final version, this questionnaire consisted of 14 scales assessing self-regulated learning strategies, self-beliefs, motivation, and learning preferences. Based on a large cross-national pilot study, the final scales were selected from a larger set of 29 constructs following a careful assessment of their psychometric properties by two panels of substantive and statistical experts (see Peschar et al. 1999). In a large cross-national validation study based on the main study of PISA 2000, Marsh et al. (2006) showed that the SAL items loaded on the theoretical scales and showed good cross-cultural measurement properties. MG-CFA models showed that factor loadings, residual variances of items, the factor variance–covariance matrix, and the relations to four criterion variables (socioeconomic status, gender, reading, and math achievement) were reasonably invariant across a sample of 25 mostly Western, OECD, and developed countries. Marsh et al. (2006) avoided testing the invariance of item and scale means across countries, arguing that such comparisons would be difficult for motivational constructs as they would be subject to very different national frame-of-reference and social desirability tendencies. However, these differences should not affect the loading structure and relations of constructs.

While PISA 2000 focused on students' achievement in reading, PISA 2006, the focus of the present investigation, placed emphasis on achievement in science. Again, a large range of motivational constructs were included in the background questionnaire, all with a specific focus on motivational processes in science. Although it was not explicitly used to inform the selection of constructs in PISA 2006, the motivational constructs match very well with self-concept theory (Marsh 2006) and the expectancy-value theory of achievement motivation (Eccles 1983; Wigfield and Eccles 2000). There were two scales that assessed student self-beliefs about their ability in science: *academic self-concept in science*, which was based on questions about students' self-perceptions of their ability in science, and *science self-efficacy*, which assessed students' beliefs about their capabilities of solving specific science-related tasks. Three scales assessed different value domains (Eccles 1983; Wigfield and Eccles 2000) that students attach to science: The two scales, *personal value of science* and *instrumental value of science*, can be classified as assessing the utility value of science focusing on the usefulness of science in their personal life and for reaching their personal goals. The intrinsic value dimension was covered by one scale, *enjoyment of science*. The *future-oriented science motivation* scale assessed whether students intended to pursue a career in a science-related field and can be used as a proxy measure for career choice. The *extracurricular activities in science* scale assessed the frequency with which students were engaging in science-related activities outside of school, for example, attending a science club or watching science-related TV programs. Although the scales were subject to the rigorous quality control measures implemented for PISA 2006 (see OECD 2009) and have been used in a number of substantive studies (e.g., Nagengast and Marsh 2011, 2012; Nagengast et al. 2011), there apparently has been no comprehensive test of measurement invariance for all constructs and all countries at the same time.

Present Investigation

Here, we study measurement invariance of the motivational and engagement constructs included in the background questionnaire of PISA 2006. On the one hand, these analyses serve as a demonstration for cross-cultural studies of psychosocial scales included in the background questionnaires of international large-scale assessment studies. On the other hand, they fill an important gap with respect to the validation of the specific scales used in PISA 2006. Tests of measurement invariance across the participating nations are not reported for the background questionnaires in the technical report (OECD 2009) accompanying the international PISA 2006 database. However, as outlined above, tests of measurement invariance are the minimal precondition for meaningful cross-cultural comparisons. Given the importance of science skills for developed and developing nations, gaining further insights into the structure of science-related motivational constructs is an important research topic in its own right.

Method

Sample

We used the full international sample of PISA 2006 providing responses from 398,750 students from 57 countries (30 OECD countries, 27 partner countries). The samples were collected using a complex two-stage sampling design and were, after employing the appropriate survey weights, representative for the national populations of 15-year-old students (OECD 2009).

Variables

Overall, we used 44 items from the background questionnaire that measured motivation and engagement in science. All items were positively worded. Most of them were scored on a 4-point Likert scale ranging from 1 ("strongly agree") to 4 ("strongly disagree"). Exceptions were the scales *science self-efficacy* (answer categories ranging from "do easily" to "could not do it" on a 4-point Likert scale) and engagement in *extracurricular activities* (answer categories ranging from "very often" to "hardly ever" on a 4-point Likert

TABLE 15.1

Reliabilities of the Engagement Scales Obtained in the Calibration Sample of PISA 2006 and the Present Analyses

Scale	Number of Items	Median α OECD Countries	Median α Non-OECD Countries	Scale Reliabilities Total Sample
Enjoyment	5	0.92	0.87	0.896
Instrumental motivation	5	0.92	0.88	0.905
Future-oriented motivation	4	0.92	0.90	0.908
Self-efficacy	8	0.83	0.79	0.811
Self-concept	5	0.92	0.87	0.902
General value	5	0.75	0.69	0.727
Personal value	5	0.80	0.76	0.794
Extracurricular activities	6	0.78	0.76	0.784

Note: Values for the median values of coefficient α were obtained from OECD (2009) and have been computed based on the calibration samples ($N = 500$, for detail see OECD 2009). Scale reliabilities were obtained from the total sample analysis based on a confirmatory factor analysis with eight factors.

scale). For the analyses reported here, all answers were reverse coded, so that higher scores indicated higher values of the underlying constructs. Further information on the wording of items and scale properties based on the calibration samples of PISA 2006 (based on 500 students per country) is given in OECD (2009). The median internal consistencies across countries as measured by coefficient α based on these analyses are presented in Table 15.1.

Missing Data

The proportion of missing data for the items was small, ranging from 1.3% to 8.5%. The indicators of the two scales that appeared at the end of the background questionnaire (*instrumental motivation in science* and *science self-concept*) had higher missing rates than the indicators for the other scales. We used multiple imputation (Rubin 1987; Schafer 1997) to deal with the missing data. Using a large imputation model, 10 imputed datasets were created using MCMC imputation in SPSS (IBM 2011). The decision to use multiple imputation rather than a full information maximum likelihood (Enders 2010) was based on the need to include a large number of auxiliary variables in the imputation model and the need to use plausible values for the achievement test that require an analytical strategy akin to an analysis of multiple imputed datasets. However, best practice with respect to dealing with missing information in hierarchically nested data is still evolving (Graham 2009) and full information maximum likelihood estimation with auxiliary variables is a useful alternative.

Data Analysis

The data were analyzed using Mplus 6.1 (Muthén and Muthén 1998–2010). Owing to the large number of parameters to be estimated that were beyond the capacity of a high-end 64-bit workstation when raw data was analyzed, the MG-SEM were based on weighted variance–covariance matrices and mean vectors that were calculated using the package *survey* (Lumley 2004, 2011) in R (R Development Core Team 2011). All analyses used the final student weights to obtain population representative results. Using the Mplus default, weights were standardized so that they added up to the total sample size for each country. Separate analyses were run for each of the imputed data sets. The results were combined according to Rubin's (1987) rules to obtain parameter estimates, standard errors, and model fit statistics.

Modeling Strategy

We first tested the fit of the confirmatory factor analysis model in the total sample, not taking the 57 countries into account. Based on the fitted model in the total sample, we calculated the scale reliabilities (Raykov 2009; Raykov and Shrout 2002) that are a more defensible reliability estimate than Cronbach's α that is typically reported (Revelle and Zinbarg 2009; Sijtsma 2009). After establishing that the initial structure held, we then implemented the 13-model taxonomy of invariance models (Marsh et al. 2009) described above. Based on the best fitting invariance model, we explored the relations of the motivational constructs to the covariates gender, science achievement, and socioeconomic status. For this purpose, we investigated the bivariate correlations and multiple regressions of the latent variables on the covariates using a MIMIC model.

Assessing Model Fit

A combination of test statistics, model fit indices, and inspection of model parameters was used to assess the fit and the appropriateness of the structural equation models. Owing to the large sample size, the power to detect very small deviations from the hypothesized model structure with the likelihood ratio test was very large. In fact, not a single one of the specified models passed the χ^2 test of exact model fit. Model fit indices like the root mean squared error of approximation (RMSEA; Browne and Cudeck 1993; Steiger and Lind; 1980), the comparative fit index (CFI; Bentler 1990), and the Tucker–Lewis index (TLI; Bentler and Bonett 1980) do not depend on the total sample size and offer alternative ways of assessing the fit of structural equation models (Marsh et al. 1996, 1988, 2004). Typically, models are considered to show a good fit to the data if the RMSEA is below 0.06, and if TLI and CFI are above 0.95. A model is said to show acceptable fit if the RMSEA is below

0.08 and TLI and CFI are above 0.9 (Chen et al. 2008; Hu and Bentler 1999; Marsh et al. 2004). However, these values only serve as rough guidelines for assessing model fit, not as golden rules (Marsh et al. 2004), as many seemingly arbitrary features of the data (e.g., unique variances, Heene et al. 2011) can influence the behavior of these indices.

The comparison of nested models in the taxonomy of measurement invariance models suffers from similar problems. In the large sample used in the present investigation, the likelihood ratio test that compares the fit of two nested models has a very high power to detect even small differences in model fit. In fact, again, all likelihood ratio tests were statistically significant at $\alpha = 0.001$. In order to assess the decrease of model fit, we turned to an evaluation of change in model fit indices. Using the suggestions of Cheung and Rensvold (2002) and Chen (2007), we considered a decrease in model fit to be practically insignificant, if the RMSEA dropped by less than 0.015 and if CFI dropped by less than 0.01 for the more restrictive model (i.e., the model with more invariant parameters across the 57 countries). Again, we did not treat these suggested cut-off values as golden rules (Marsh et al. 2004), but compared the model fit indices along the 13 models of the invariance taxonomy. As we were testing restrictions on the unique variances and the indicator means, we used the more restrictive (but appropriately nested) null model, following Widaman and Thompson (2003) to calculate the incremental fit indices TLI and CFI.

Results

Total Sample Analysis

We began with a confirmatory factor analysis of the eight motivation and engagement scales in the total sample of 397,850 students. For this analysis only, the indicators were centered around their country mean to remove the impact of response tendencies varying between countries and to more closely mimic the MG analysis. A CFA model with eight correlated factors (shown in Figure 15.1) fit the data well (RMSEA = 0.020, TLI = 0.961, CFI = 0.964). The scale reliabilities based on the factor loadings of the structural equation model for the eight scales based on this model in the total sample are presented in Table 15.1. They were close to the median internal consistencies in the calibration samples that were reported in OECD (2009, see Table 15.1). In this total group model, potential country effects on the structure were not considered, so we next turned to MG-SEM models in which we tested the invariance of measurement models across the 57 countries.

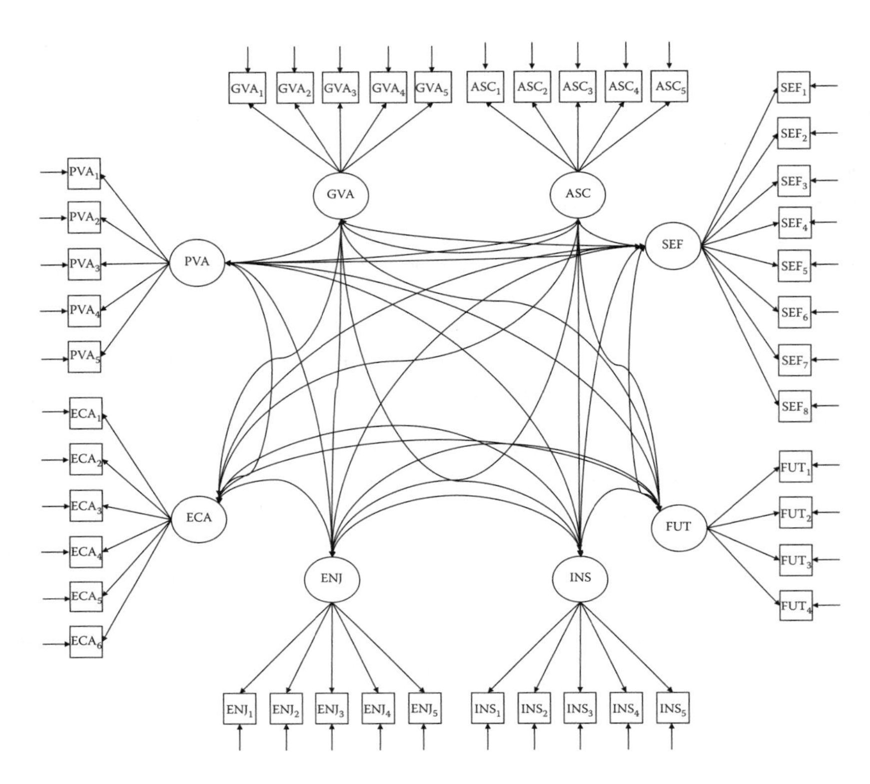

FIGURE 15.1
Path diagram of the eight-factor confirmatory factor analysis model. ENJ: enjoyment; INS: instrumental motivation; FUT: future-oriented motivation; SEF: self-efficacy; ASC: academic self-concept; GVA: general value; PVA: personal value; ECA: extracurricular activities.

Taxonomy of Invariance Models

After establishing the fit of the model in the total sample, we proceeded with testing the invariance of the established measurement structure across the 57 countries using the 13-model taxonomy introduced by Marsh et al. (2009). Table 15.2 gives an overview of parameters that were set invariant across the countries and the resulting model fit indices. The configural invariance model (M1) in which the same measurement model was specified in each of the 57 countries, but no model-estimated parameters were set invariant across the samples fit the data well (RMSEA = 0.036, TLI = 0.949, CFI = 0.957). This model provides a baseline against which the fit of the more restrictive invariance models can be compared.

In the next step, we tested weak measurement invariance, setting the factor loadings of the 45 items on the eight factors invariant across the countries (M2). This model led to a small decrease in model fit indices

TABLE 15.2

Model Fit of the 13 Models of the Invariance Model Taxonomy

	Invariance Constraints	χ^2	df	parms	RMSEA	TLI	CFI
Models without Covariates							
M1	None	497,607	49,818	9120	0.036	0.949	0.957
M2	FL	566,031	51,834	7104	0.038	0.944	0.951
M3	FL, Unq	962,642	54,298	4640	0.049	0.905	0.913
M4	FL, FVCV	673,920	53,850	5088	0.041	0.935	0.940
M4a	FL, Fcor	608,055	53,402	5536	0.039	0.941	0.947
M5	FL, INT	998,397	53,850	5088	0.050	0.901	0.909
M6	FL, Unq, FVCV	1,084,645	56,314	2624	0.051	0.897	0.901
M7	FL, Unq, INT	1,414,041	56,314	2624	0.059	0.864	0.870
M8	FL, FVCV, INT	1,107,076	55,866	3072	0.052	0.894	0.899
M9	FL, Unq, FVCV, INT	1,537,164	58,330	608	0.060	0.857	0.858
M10	FL, INT, FMn	1,182,089	54,298	4640	0.054	0.883	0.892
M11	FL, Unq, INT, FMn	1,597,427	56,762	2176	0.062	0.847	0.852
M12	FL, FVCV, INT, FMn	1,334,225	56,314	2624	0.057	0.872	0.877
M13	FL, UNQ, FVCV, INT, FMn	1,768,433	58,778	160	0.064	0.836	0.836
Models with Covariates							
M4_pred	FL, FVCV	781,314	60,006	6969	0.041	0.927	0.934
M4_pred_a	FL, FVCV, COV(PRED,TARGET)	813,699	61,350	5625	0.042	0.925	0.931
M4_pred_b	FL, FVCV, COV(PRED,TARGET), FVCV(PRED)	846,934	61,686	5289	0.043	0.922	0.928

Note: FL: factor loadings, UNQ: uniquenesses, FVCV: factor variance–covariance matrices, Fcor: factor correlation matrices, INT: item intercepts, FMn: factor means, COV(PRED, TARGET): covariances between covariates and constructs, χ^2: χ^2 test statistic, df: degrees of freedom, parms: number of free parameters, RMSEA: root mean square error of approximation, TLI: Tucker–Lewis index, CFI: confirmatory fit index.

(ΔRMSEA = 0.002, ΔTLI = 0.005, ΔCFI = 0.006) compared to the configural invariance model (i.e., they were smaller than the recommended cut-offs of ΔRMSEA = 0.015, ΔTLI = 0.01, ΔCFI = 0.01 discussed earlier). The absolute values of the model fit indices (RMSEA = 0.038, TLI = 0.944, CFI = 0.957) were also acceptable. This implied that weak measurement invariance held with respect to the eight motivational constructs and the 57 countries participating in PISA 2006. Hence, an important precondition for comparing the interrelations between the motivational constructs across countries was fulfilled.

Next, we tested the invariance of the variance–covariance matrix (M4) and the invariance of the factor correlation matrix (M4a) across the 57 countries. The decrease in RMSEA and TLI compared to M2, which control for model parsimony, were below the prespecified cut-off values for

M4 (ΔRMSEA = 0.003, ΔTLI = 0.009), whereas the difference for CFI slightly exceeded it (ΔCFI = 0.011). For tests of the invariance of the factor correlation matrix (M4a), the tests yielded a more positive picture with all decrements in the fit indices (as compared to M2) below the prespecified cut-off values (ΔRMSEA = 0.001, ΔTLI = 0.003, ΔCFI = 0.004). The difference between models M4 and M4a is whether the variances of the latent variables are restricted to be invariant across countries (in M4 they are restricted, in M4a they can vary). As M4 is nested within M4a, the difference in fit indices can be meaningfully interpreted: Overall, the fit indices did not indicate a large decrease in model fit between M4a and M4 (ΔRMSEA = 0.002, ΔTLI = 0.006, ΔCFI = 0.007), although these decreases were somewhat larger than the ones between M2 and M4a. This finding indicated that setting the variances of the latent variables invariant across countries introduced a larger amount of misfit than restricting only the correlations. While the motivational constructs seemed to differ slightly with respect to their variances, their bivariate linear relations remained reasonably invariant across countries. It should be noted though that the overall fit was acceptable for M4.

Apart from studying relations between constructs in different cultures, cross-cultural researchers are often interested in comparing countries with respect to the mean level of constructs. An important precondition for the comparability of factor means (and, in fact, also of manifest mean comparisons) across countries is strong measurement invariance. Under this condition, both item intercepts and factor loadings have to be invariant across countries. Hence, we next tested whether the item intercepts were invariant across the 57 countries (M5). In comparison to M2, in which only the factor loadings were invariant, the fit indices in M5 indicated a considerable decrease in model fit that was above the prespecified cut-off values (ΔRMSEA = 0.012, ΔTLI = 0.043, ΔCFI = 0.042). Also, with respect to an overall evaluation of model fit, M5 did not provide an acceptable fit to the data: Only the RMSEA was still within the region of acceptable fit (RMSEA = 0.050), whereas the incremental fit indices were not completely satisfactory (TLI = 0.901, CFI = 0.909). This finding indicated that there was no support for invariance of the item intercepts and hence no support for comparing the mean values of the latent motivational constructs across the 57 countries. It would be possible to pursue a strategy of partial invariance (Byrne et al. 1989) to try to locate which items were culturally sensitive, leading to uniform differential item functioning. However, given the problematic nature of cross-country comparisons of item and scale means discussed by Marsh et al. (2006), we did not follow this approach here.

The test of the invariance of the unique item variances across the 57 countries that would be one of the preconditions for comparing manifest scores across the countries showed that strict measurement invariance was not defensible. Both the comparisons of M3 to M2 (ΔRMSEA = 0.011, ΔTLI = 0.039, ΔCFI = 0.038) and of M7 to M5 (ΔRMSEA = 0.009, ΔTLI = 0.037, ΔCFI = 0.039) yielded fit decrements that exceeded the prespecified cut-off values. Also, the global fit indices of M3 and M7 were not satisfactory. Hence, strict measurement invariance was rejected, indicating that countries should not be compared on manifest scale scores of the motivational constructs.

After establishing that Model M4 with invariant factor loadings and invariant variance–covariance matrices across the 57 countries was the most parsimonious model of the 13-model taxonomy that was still defensible in terms of model fit, we proceeded with inspecting the resulting relations between the eight constructs (see Table 15.3 for the resulting correlation matrix). The average correlation was $r = .512$, indicating that the eight motivational constructs were all highly correlated. The largest correlations emerged between personal value of science and general value of science ($r = .771$) and between personal value of science and enjoyment of science ($r = .700$). Personal value of science was also highly correlated with instrumental motivation of science ($r = .634$) and future-oriented motivation in science ($r = .626$). Other large correlations emerged between enjoyment of science and future-oriented science motivation ($r = .626$) and enjoyment of science and engagement in extracurricular science activities ($r = .617$). While academic self-concept in science had above-average correlations with enjoyment of science ($r = .575$), instrumental motivation in science ($r = .533$), personal value of science ($r = .531$), and future-oriented motivation in science ($r = .522$), science self-efficacy consistently had below-average correlations with the other seven constructs. Similarly, general value of science only exhibited the aforementioned large correlation with personal value of science—all other correlations were comparatively small. Personal value of science, on the other hand, was more strongly correlated with the other motivational constructs. Finally, engagement in extracurricular activities in science, apart from its relations to enjoyment of science and personal value, was also highly correlated with future-oriented science motivation ($r = .540$).

TABLE 15.3

Variance–Covariance and Correlation Matrix of the Eight Motivational Constructs

	1	2	3	4	5	6	7	8
1: Enjoyment	0.389	0.233	0.306	0.123	0.187	0.109	0.147	0.184
2: Instrumental motivation	0.555	0.455	0.354	0.090	0.187	0.096	0.145	0.138
3: Future-oriented motivation	0.626	0.670	0.615	0.114	0.213	0.093	0.166	0.203
4: Self-efficacy	0.462	0.314	0.339	0.182	0.107	0.070	0.069	0.094
5: Self-concept	0.575	0.533	0.522	0.480	*0.271*	*0.070*	*0.093*	*0.114*
6: General value	0.497	0.403	0.338	0.465	*0.380*	*0.124*	*0.092*	*0.064*
7: Personal value	0.700	0.634	0.626	0.480	*0.531*	*0.771*	*0.114*	*0.092*
8: Extracurricular activities	0.617	0.426	0.540	0.459	*0.456*	*0.377*	*0.571*	*0.229*

Note: Covariances are shown in the upper triangle, correlations are shown in the lower triangle, variances are given on the diagonal. The values are based on Model 4 using the unstandardized and standardized solutions. The variance–covariance matrix and the correlation matrix are invariant across the 57 countries participating in PISA 2006.

Relation to Gender, Socioeconomic Status, and Achievement

Based on the initial invariance models, we then tested the relation of the eight motivational constructs to gender, socioeconomic status, and achievement in science. As the most defensible model of the invariance taxonomy was M4, the model in which the factor loadings and the factor variance–covariance matrix had been restricted across countries, we based all further comparisons on this model.

We first tested a model in which the factor covariance matrix of the latent variables was invariant across the 57 countries, while the three covariates—gender, SES, and achievement—could covary freely with the motivational constructs. This model (M4_pred) fit the data well (RMSEA = 0.041, TLI = 0.927, CFI = 0.934), also in comparison to M4, indicating that item intercepts within countries were not affected by the covariates that were added to the model. In the next step, we restricted the covariances between the covariates and the latent motivational variables to be invariant across the 57 countries. The model fit indices for this model (M4_pred_a) did not indicate a substantially worse fit to the data compared to model M4_pred (RMSEA = 0.042, TLI = 0.925, CFI = 0.931), as all changes in fit indices were below the prespecified thresholds. Finally, we further restricted the variances and covariances of the predictor variables to be invariant across the 57 countries (M4_pred_b). Again, the model fit indices deteriorated only slightly (RMSEA = 0.043, TLI = 0.922, CFI = 0.928) compared to M4_pred, in which the variance–covariance matrices could vary freely across the countries. The differences in the model fit indices were below the prespecified thresholds (ΔRMSEA = 0.002, ΔTLI = 0.005, ΔCFI = 0.006), indicating that it was justified to restrict the variance–covariance matrices to be completely invariant across countries. This finding mirrors the results of Marsh et al. (2006), who found a similar pattern for an even larger number of scales, but a less diverse and smaller number of countries.

How were the motivational and engagement constructs related to gender, achievement in science, and socioeconomic status? The invariance of the variance–covariance matrix across the 57 countries allowed us to assess these relations by inspecting just a single set of correlations reported in Table 15.4. Inspecting these correlations showed that gender was only marginally related to most of the motivational variables. Only the correlation with science self-concept (r = .117) was above 0.1. Boys had slightly higher academic self-concepts in science than girls, although gender only explained slightly more than one percent of the variance in science self-concept.

Unsurprisingly, science achievement was correlated higher with the motivational and engagement measures. The strongest relations emerged with science self-efficacy (r = .401) and general value of science (r = .315). Students who had higher science achievement values were also more likely to report that they would be able to solve a range of scientific problems and also tended to ascribe a higher societal value to science. Science

TABLE 15.4

Correlations of the Motivation Dimensions with Gender, Achievement, and Socioeconomic Status

	Gender	Achievement	Socio-Economic Status
Enjoyment	.017	.255	.086
Instrumental motivation	.014	.131	.063
Future-oriented motivation	.048	.146	.057
Self-efficacy	.035	.401	.252
Self-concept	.117	.222	.122
General value	.027	.315	.171
Personal value	.028	.200	.108
Extracurricular activities	.072	.109	.117
Gender		.005	.017
Achievement			.354

achievement was also correlated with enjoyment of science ($r = .255$), science self-concept ($r = .222$), and personal value of science ($r = .200$). The difference in the bivariate relations of science achievement and science self-efficacy and science self-concept is in line with self-concept theory (Marsh 2006): The items used to measure science self-efficacy were criterion-referenced and assessed students' confidence in being able to solve specific science problems similar to the actual test items. The items used to measure science self-concept, however, did not contain a specific criterion-reference, so students had to rely on other frames of reference, typically their peers, to judge their achievement. As there are considerable school-to-school differences in achievement and students tend to judge their achievement relative to their peers, the relations between science self-concept and individual achievement were diluted relative to the relationship between achievement and science self-efficacy (see Marsh et al. 2008, for a review of the underlying theory and mechanisms; also see Nagengast and Marsh 2012, for an application to these data). The correlations of science achievement with future-oriented science motivation ($r = .146$), instrumental motivation in science ($r = .131$), and engagement in extracurricular science activities ($r = .109$) were considerably smaller, albeit still above 0.1.

Finally, self-efficacy in science was most strongly correlated with socioeconomic status ($r = .252$), followed by general value of science ($r = .171$). There were only three other correlations larger than .1: science self-concept and socioeconomic status were correlated with $r = .122$, engagement in science activities and socioeconomic status correlated with $r = .117$, and personal value of science correlated with socioeconomic status $r = .108$.

After investigating the univariate relations between the covariates and the eight motivational constructs, we also fitted MIMIC models to study how gender, science achievement, and socioeconomic status jointly predicted motivation and engagement. As socioeconomic status and achievement were

TABLE 15.5

Multiple Regressions of the Motivation Dimensions on Gender, Achievement, and Socioeconomic Status (MIMIC Model)

	Gender		Achievement		Socioeconomic Status	
	β	se	β	se	β	se
Enjoyment	0.016	(0.002)	0.257	(0.002)	–0.005	(0.002)
Instrumental motivation	0.013	(0.002)	0.124	(0.002)	0.018	(0.002)
Future-oriented motivation	0.047	(0.002)	0.143	(0.002)	0.006	(0.002)
Self-efficacy	0.031	(0.002)	0.356	(0.002)	0.126	(0.002)
Self-concept	0.115	(0.002)	0.205	(0.002)	0.047	(0.002)
General value	0.024	(0.002)	0.291	(0.002)	0.067	(0.002)
Personal value	0.026	(0.002)	0.184	(0.002)	0.043	(0.002)
Extracurricular activities	0.070	(0.002)	0.077	(0.002)	0.089	(0.002)

Note: Regression coefficients from the completely standardized solution. All coefficients are statistically significant at $\alpha = 0.05$; se = standard error.

moderately correlated ($r = 0.354$) and both showed relatively strong univariate relations to the eight motivational constructs, it was important to determine their unique contribution for explaining engagement and motivation. The standardized path coefficients and their standard errors are presented in Table 15.5. Owing to the large sample size, all path coefficients were statistically significant at $p < 0.05$, hence we focused on interpreting only those standardized coefficients larger than 0.1. Enjoyment of science, instrumental motivation, career aspirations in science, general value of science, and personal value of science were only moderately predicted by science achievement. Science self-efficacy was substantially related to both science achievement ($\beta = 0.356$, se = 0.002) and socioeconomic status ($\beta = 0.126$, se = 0.002). Science self-concept was substantially related to gender ($\beta = 0.115$, se = 0.002) and to science achievement ($\beta = 0.205$, se = 0.002). For engagement in extracurricular activities in science, none of the predictor variables had a standardized path coefficient larger than 0.1. However, the coefficient for socioeconomic status was the largest ($\beta = 0.089$, se = 0.002), although gender and science achievement followed closely behind.

Discussion

Summary of Results

Overall, we found that the factor loadings and factor variance–covariance matrices of the eight factors were invariant across the 57 countries. Setting item intercepts to be invariant across countries reduced the fit substantially, although the overall model fit was still marginally acceptable when compared to the prespecified cut-offs. This was not the case for the residual variances that were clearly not invariant across countries as was also obvious from inspection of the reliability estimates for each country reported in the PISA technical manual (OECD 2009). Hence, the overall invariance of the measurement model was not as strong as in the similar investigation by Marsh et al. (2006) who—based on a smaller and less diverse sample of countries and other motivational constructs—found invariance of factor loadings, the factor variance–covariance matrix and residual variances, but did not look at intercepts or the mean structure more generally.

Overall, the eight scales were moderately to strongly correlated when correlations were evaluated using the latent variables in the CFA model. The correlations between science self-efficacy and general value of science and the other constructs tended to be somewhat smaller. Involvement in extracurricular science activities also had somewhat weaker relations to the other constructs. The other motivational and engagement constructs were not differentially related to one another. This is not surprising given that all motivational constructs were assessed in the subject area of science. It has been often shown that self-perceptions of abilities such as academic self-concept and scales that assess the perceived importance, value, and enjoyment within a domain tend to be highly correlated (e.g., Bong 2001; Marsh et al. 2006). In fact, while there is strong support for the domain specificity of many motivational constructs (correlations are low between domains)—particularly self-concept—there is less support for their specificity and construct validity within domains (correlations between different motivational constructs tend to be high, Bong 2001; Marsh et al. 2003). As the background questionnaire in PISA 2006 was limited to science, we could not test predictions of the multidimensional hierarchical model of self-concept (Marsh and Shavelson 1985; Shavelson et al. 1976) and the internal/external frame-of-reference model (Marsh 1986) that predicts low correlations between self-concept (and by extension, some other motivational) scales assessed in different subject areas due to the domain specificity of these scales. Typically, self-concept scales in different academic domains are correlated close to zero, while the corresponding correlations between the achievement scores tend to be moderate to high (e.g., Marsh et al. 1998, 2006, also see the review in Marsh 2006).

The correlations of the motivation and engagement scales with gender, achievement, and socioeconomic status (and by implication the multiple regressions on these covariates) were also reasonably invariant across the 57 countries. These findings showed that the only substantial gender differences were found for science self-concept with boys having higher self-perceptions of their abilities than girls. However, in all motivational constructs, these differences tended to be very small. Achievement was positively related to all motivation and engagement scales with the strongest relations to self-efficacy in science, enjoyment of science, general and personal value of science, and academic self-concept in science. Socioeconomic status was most strongly related to self-efficacy in science and had only weak associations with the other motivational and engagement constructs once achievement differences were controlled in a multiple-regression model.

In line with self-concept theory (Marsh 2006), but in contrast to Marsh et al. (2006), we found that self-efficacy in science was more strongly related to science achievement than academic self-concept was related to science achievement. The reasons for this difference are as follows: In Marsh et al.'s (2006) evaluation of the SAL questionnaire for PISA 2000, academic self-concept was assessed in the domains of reading and math. Each of the corresponding domain-specific self-concepts correlated more strongly with the achievement in the domain and less strongly with achievement in the other domain, thus confirming predictions of the hierarchical model of self-concept (see Marsh 2006). However, self-efficacy was assessed in a domain-general way, referring to general beliefs about the ability to master academic tasks and without a criterion of success implicit in the item so students had to use external frames of reference when judging their self-efficacy. The resulting self-efficacy measure correlated about as strongly with achievement in math and reading as the corresponding domain-specific academic self-concepts. In PISA 2006, self-efficacy and academic self-concept were both assessed in a domain-specific way for science. While the items for academic self-concept were similar though not identical to the items analyzed in PISA 2000, the self-efficacy items were very different. Instead of assessing general perceptions of the ability to succeed in a variety of academic tasks, they referred to very specific applied science-related problems (e.g., recognizing the science question underlying a newspaper report on a health issue or interpreting the scientific information provided on the labeling of food items; see OECD 2009) and asked students how easy they would find it to perform these tasks on their own. Obviously these questions were much more closely aligned with the actual achievement items that also assessed science competencies in real-world contexts. Hence, these items should be less influenced by frame-of-reference effects that impact academic self-concept scales and deflate their correlations with achievement measures (see Marsh 2006) and the resulting correlations with the achievement measures are expected to be higher.

The difference in the assessment strategy for the self-efficacy items between PISA 2000 and PISA 2006 is a good example of a specific problem in the assessment of motivational constructs that Marsh (1994) (Marsh et al. 2003; see also Pintrich 2003) referred to as "jingle-jangle fallacies" (Marsh et al. 2003, p. 189): Scales with the same name assess rather different constructs; scales with different names assess essentially the same constructs. Similarly, while there is good evidence for domain-specificity of motivational constructs, support for the construct-specificity of motivational scales within a domain is equivocal. Apart from testing measurement properties of scales, a careful inspection of item contents is always warranted to avoid falling prey to these all too common problems.

Applications of the Motivational Scales in the Background Questionnaire

Measurement invariance of the eight motivational scales is an important precondition for their use in addressing substantive research questions on motivational processes in cross-cultural research. The PISA research program offers unprecedented opportunities for testing the cross-cultural generalizability of predictions from educational psychological theories and researchers in motivation and self-concept research have already taken advantage of these opportunities also using the PISA 2006 data.

For example, Nagengast and Marsh (2011, 2012) tested the big-fish-little-pond effect (BFLPE) using data from PISA 2006. The BFLPE posits that students use the achievement of their peers as a frame of reference for judging their achievement (see Marsh et al., 2008 for an overview of research and theory of the BFLPE). BFLPE theory predicts that individual achievement is positively related to academic self-concept, while school- or class-average achievement is negatively related to academic self-concept. Nagengast and Marsh (2011) tested the BFLPE hypothesis with data from PISA 2006 in the United Kingdom and found support for the BFLPE in science and for invariance of these effects across the four U.K. countries. In an extension of these findings, Nagengast and Marsh (2012) studied the relations of achievement in science, academic self-concept in science, and career aspirations (as measured by the scale future-oriented motivation) in the international sample of PISA 2006. They found substantial BFLPEs for both academic self-concept and career aspirations in science that generalized across countries participating in PISA 2006. Individual achievement positively predicted self-perceptions of ability and career aspirations in science, while school-average ability had a negative effect on these variables once individual differences in achievement were controlled. Using a multilevel mediation model (Pituch and Stapleton 2011; Preacher et al. 2010), they found that both the positive individual level effect and the negative contextual effect of science achievement on career aspirations were mediated by individual academic self-concept.

Similarly, Nagengast et al. (2011) used data from PISA 2006 to test predictions of modern expectancy-value theory of achievement motivation (Eccles 1983; Wigfield and Eccles 2000). Classical models of expectancy-value theory assumed that expectancy and value combined multiplicatively to determine choices, motivation, and ultimately achievement. This synergistic combination, however, had been dropped from modern approaches to expectancy-value theory. Nagengast et al. (2011) argued that this had not been a consequence of theoretical reasoning, but rather the consequence of methodological limitations: While tests of expectancy-value models had classically used experiments that manipulated expectancy of success and value of the outcomes, modern approaches had relied on modeling unreliable scale scores using linear structural equation models. This considerably reduced chances of detecting interaction effects, leading to the silent abandonment of the multiplicative relation. Using structural equation models with latent interactions, Nagengast et al. were able to return the "x" to expectancy-value theory: They showed that there was an interaction between academic self-concept in science (expectancy of success) and science enjoyment (intrinsic value) in predicting career aspirations (a proxy for career choice) and extracurricular activities (engagement), providing evidence of synergistic relations between expectancy and value. In the strongest cross-cultural test of prediction of expectancy-value theory, this effect generalized across the 57 countries partaking in PISA 2006, providing strong evidence for its cross-cultural generalizability.

Limitations and Directions for Further Research

In this chapter, we analyzed the measurement invariance of science-related motivation and engagement constructs in PISA 2006 using MG-SEM. Our investigation served two purposes: Methodologically, we used it to demonstrate the implementation of tests of cross-cultural measurement invariance of constructs in the background questionnaires of large-scale assessments (see also Marsh et al. 2006; OECD 2010). Substantively, we were interested in evaluating engagement and motivational measures in science, a subject with high economic importance, but less systematic research efforts with respect to motivational and engagement scales. Given the limited space in this chapter, we can only point out several directions for further substantive and methodological research.

Our analyses did not find good support for the invariance of item intercepts across the 57 countries participating in PISA 2006. This violation is of no consequence if the analytical focus is on relations among variables and theoretical models of these relations within each country (e.g., in the substantive applications discussed above and many applications using the background questionnaire). However, invariance of item intercepts is a critical issue if the focus is on rank-ordering countries (i.e., constructing league tables) in relation to mean levels of these constructs. Even if there was good support for the invariance of item intercepts, Marsh et al. (2006) argued that the interpretation of mean-level differences in countries would not be appropriate: Cultural differences in modesty and self-enhancement might affect the mean responses of items while leaving their interrelations and relations to the underlying constructs intact. Although clearly beyond the scope of this chapter, it might useful to consider pursuing a strategy of partial invariance of item intercepts (Byrne et al. 1989). The problematic nature of between-country comparisons on self-report motivational measures notwithstanding, this strategy might be useful to determine those items which were relatively culturally stable with respect to their mean levels. Given the large number of countries and items to test, a systematic analysis would be very challenging and time consuming to undertake.

A second methodological limitation concerns the treatment of item indicators as continuous rather than as ordinal variables. As the underlying items were answered on a four-point Likert scale, it was justifiable to treat them as continuous variables (e.g., Beauducel and Herzberg 2006; Dolan 1994; Lei 2009; Lubke and Muthén 2004; Rhemtulla et al. 2012). Again, this decision was partly based on computational limitations: Treating items as ordinal would have increased the computational burden considerably by increasing the number of parameters to be estimated and would have made the estimation of the MG model with 57 countries impossible. Owing to computational constraints, we were not able to analyze the raw data from all countries, but had to rely on an MG analysis based on the weighted covariance matrices and mean vectors. Future hardware developments, however, will undoubtedly make the estimation of complex MG models more easily feasible.

Substantively, the set of motivational constructs included in the background questionnaire of PISA 2006 was limited compared to the SAL approach tested by Marsh et al. (2006). The number of constructs was considerably smaller and all scales referred only to the single subject area of science that precluded tests of the internal/external frame-of-reference model of academic self-concept and other models that make hypotheses about more than one academic domain. While the aforementioned limitations might be easily addressed in future large-scale assessment studies, the final substantive limitation is likely to hamper research on motivational constructs based on international large-scale assessment: As it is often the case in these studies, the data were strictly cross-sectional. Hence, it is not possible to disentangle the causal ordering of the motivational constructs and achievement. It is well known that academic self-concept and achievement are reciprocally related (e.g., Marsh and Craven 2006; Marsh and Yeung 1997; Valentine and DuBois 2005), so that academic self-concept benefits achievement over and above prior achievement differences and vice versa. However, this limitation should not detract from their usefulness for comparing motivational processes in large representative cross-cultural samples that offer unprecedented opportunities for researchers trying to uncover the structure and mechanisms driving student motivation and engagement.

References

Artelt, C. 2005. Cross-cultural approaches to measuring motivation. *Educational Assessment*, 10, 231–255.

Artelt, C., Baumert, J., Julius-McElvany, N., and Peschar, J. 2003. *Learners for Life: Student Approaches to Learning. Results from PISA 2000*. Paris: Organisation for Economic Co-operation and Development.

Beauducel, A. and Herzberg, P. Y. 2006. On the performance of maximum likelihood versus means and variance adjusted weighted least squares estimation in CFA. *Structural equation Modeling*, 13, 186–203. doi: 10.1207/s15328007sem1302_2

Bentler, P. M. 1990. Comparative fit indexes in structural models. *Psychological Bulletin*, 107, 238–246. doi:10.1037/0033–2909.107.2.238

Bentler, P. M. and Bonett, D. G. 1980. Significance tests and goodness-of-fit in the analysis of covariance structures. *Psychological Bulletin*, 88, 588–606.

Bong, M. 2001. Between- and within-domain relations of academic motivation among middle and high school students: Self-efficacy, task value, and achievement goals. *Journal of Educational Psychology*, 93, 23–34. doi:10.1037/0022–0663.93.1.23

Browne, M. W. and Cudeck, R. 1993. Alternative ways of assessing model fit. In K. A. Bollen and J. S. Long (Eds.), *Testing Structural Equation Models* (pp. 136–162). Newbury Park, CA: Sage.

Brunner, M., Keller, U., Hornung, C., Reichert, M., and Martin, R. 2009. The cross-cultural generalizability of a new structural model of academic self-concepts. *Learning and Individual Differences*, 19, 387–403.

Byrne, B. M., Shavelson, R. J., and Muthén, B. O. 1989. Testing for the equivalence of factor covariance and mean structures—the issue of partial measurement invariance. *Psychological Bulletin*, 105, 456–466.

Chen, F. F. 2007. Sensitivity of goodness of fit indexes to lack of measurement invariance. *Structural Equation Modeling*, 14, 464–504.

Chen, F. F. 2008. What happens if we compare chopsticks with forks? The impact of making inappropriate comparisons in cross-cultural research. *Journal of Personality and Social Psychology*, 95, 1005–1018.

Chen, F. F., Curran, P. J., Bollen, K. A., Kirby, J., and Paxton, P. 2008. An empirical evaluation of the use of fixed cutoff points in RMSEA test statistic in structural equation models. *Sociological Methods & Research*, 36, 462–494. doi: 10.1177/0049124108314720.

Cheung, G. W. and Rensvold, R. B. 2002. Evaluating goodness-of-fit indexes for testing measurement invariance. *Structural Equation Modeling*, 9, 233–255. doi:10.1207/S15328007SEM0902_5

Chiu, M. M. and Klassen, R. M. 2010. Relations of mathematics self-concept and its calibration with mathematics achievement: Cultural differences among fifteen-year-olds in 34 countries. *Learning and Instruction*, 20, 2–17. doi:10.1016/j.learninstruc.2008.11.002

Dolan, C. V. 1994. Factor analysis of variables with 2, 3, 5 and 7 response categories: A comparison of categorical variable estimators using simulated data. *British Journal of Mathematical and Statistical Psychology*, 47, 309–326. doi:10.1111/j.2044–8317.1994.tb01039.x

Eccles (Parsons), J. S. 1983. Expectancies, values, and academic behaviours. In J. T. Spence (Ed.), *Achievement and Achievement Motivation* (pp. 75–146). San Francisco, CA: W. H. Freeman.

Enders, C. K. 2010. *Applied Missing Data Analysis*. New York: Guilford.

Graham, J. W. 2009. Missing data analysis: Making it work in the real world. *Annual Review of Psychology*, 60, 549–576.

Heene, M., Hilbert, S., Draxler, C., Ziegler, M., and Bühner, M. 2011. Masking misfit in confirmatory factor analysis by increasing unique variances: A cautionary note on the usefulness of cutoff values of fit indices. *Psychological Methods*, 16, 319–336. doi:10.1037/a0024917

Hu, L. and Bentler, P. M. 1999. Cutoff criteria for fit indexes in covariance structure analysis: Conventional criteria versus new alternatives. *Structural Equation Modeling*, 6, 1–55. doi:10.1080/10705519909540118.

IBM. 2011. IBM SPSS Statistics. Version 20 [Computer Software].

Jöreskog, K. G. and Goldberger, A. S. 1975. Estimation of a model with multiple indicators and multiple causes of a single latent variable. *Journal of the American Statistical Association*, 70, 631–639.

Kaplan, D. 2009. *Structural Equation Models. Foundations and Extensions (2nd ed.).* Thousand Oaks, CA: Sage.

Lee, J., Little, T. D., and Preacher, K. J. 2011. Methodological issues in using structural equation models for testing differential item functioning. In E. Davidov, P. Schmitt, and J. Billiet (Eds.), *Cross-Cultural Analysis. Methods and Applications* (Vol. 1, pp. 55–85). New York: Routledge. doi:10.1109/TSC.2008.18

Lei, P.-W. 2009. Evaluating estimation methods for ordinal data in structural equation modeling. *Quality & Quantity*, 43, 495–507. doi: 10.1007/s11135-007-9133-z.

Little, T. D. 1997. Mean and covariance structures (MACS) analyses of cross-cultural data: Practical and theoretical issues. *Multivariate Behavioral Research*, 32, 53–76.

Lubke, G. H. and Muthén, B. O. 2004. Applying MG confirmatory factor models for continuous outcomes to Likert scale data complicates meaningful group comparisons. *Structural equation Modeling*, 11, 514–534. doi:10.1207/s15328007sem1104_2

Lumley, T. 2004. Analysis of complex survey samples. *Journal of Statistical Software*, 9, 1–19.

Lumley, T. 2011. *Survey: Analysis of Complex Survey Samples*. R package version 3.26.

Marsh, H. W. 1986. Verbal and math self-concepts: An internal/external frame of reference model. *American Educational Research Journal*, 23, 129–149.

Marsh, H. W. 1994. Sport motivation orientations: Beware of the jingle-jangle fallacies. *Journal of Sport and Exercise Psychology*, 16, 365–380.

Marsh, H. W. 2006. *Self-Concept Theory, Measurement and Research into Practice: The Role of Self-Concept in Educational Psychology*. Leicester, UK: British Psychological Society.

Marsh, H. W., Balla, J. R., and Hau, K. T. 1996. An evaluation of incremental fit indexes: A clarification of mathematical and empirical properties. In G. A. Marcoulides and R. E. Schumacker (Eds.), *Advanced Structural Equation Modeling Techniques* (pp. 315–353. Mahwah, NJ : Lawrence Erlbaum.

Marsh, H. W., Balla, J. R., and McDonald, R. P. 1988. Goodness-of-fit indexes in confirmatory factor analysis: The effect of sample size. *Psychological Bulletin*, 103, 391–410.

Marsh, H. W. and Craven, R. G. 2006. Reciprocal effects of self-concept and performance from a multidimensional perspective: Beyond seductive pleasure and unidimensional perspectives. *Perspectives on Psychological Science*, 1, 133–163.

Marsh, H. W., Craven, R. G., and Debus, R. L. 1998. Structure, stability, and development of young children's self-concepts. *Child Development*, 69, 1030–1053.

Marsh, H. W., Craven, R. G., Hinkley, J. W., and Debus, R. L. 2003. Evaluation of the Big-Two-Factor Theory of academic motivation orientations: An evaluation of jingle-jangle fallacies. *Multivariate Behavioral Research*, 38, 189–224.

Marsh, H. W. and Hau, K.-T. 2004. Explaining paradoxical relations between academic self-concepts and achievements: Cross-cultural generalizability of the internal/external frame of reference predictions across 26 countries. *Journal of Educational Psychology*, 96, 56–67.

Marsh, H. W., Hau, K. T., and Wen, Z. 2004. In search of golden rules: Comment on hypothesis-testing approaches to setting cutoff values for fit indices and dangers in overgeneralizing Hu and Bentler's 1999 findings. *Structural Equation Modeling*, 11, 320–341.

Marsh, H. W. and Grayson, D. 1994. Longitudinal stability of latent means and individual differences: A unified approach. *Structural Equation Modeling*, 1, 317–359.

Marsh, H. W., Hau, K.-T., Artelt, C., Baumert, J., and Peschar, J. 2006. OECD's brief self-report measure of educational psychology's most useful affective constructs: Cross-cultural, psychometric comparisons across 25 countries. *International Journal of Testing*, 6, 311–360. doi:10.1207/s15327574ijt0604_1

Marsh, H. W., Muthén, B. O., Asparouhov, T., Lüdtke, O., Robitzsch, A., Morin, A. J. S., and Trautwein, U. 2009. Exploratory structural equation modeling, integrating CFA and EFA: Application to students' evaluations of university teaching. *Structural Equation Modeling*, 16, 439–476.

Marsh, H. W., Seaton, M., Trautwein, U., Lüdtke, O., Hau, K.-T., O'Mara, A. J., and Craven, R. G. 2008. The big-fish-little-pond effect stands up to critical scrutiny: Implications for theory, methodology, and future research. *Educational Psychology Review*, 3, 319–350.

Marsh, H. W. and Shavelson, R. 1985. Self-concept: Its multifaceted, hierarchical structure. *Educational Psychologist*, 20, 107–125.

Marsh, H. W., Trautwein, U., Lüdtke, O., Köller, O., and Baumert, J. 2006. Integration of multidimensional self-concept and core personality constructs: Construct validation and relations to well-being and achievement. *Journal of Personality*, 74, 403–456.

Marsh, H. W. and Yeung, A. S. 1997. Causal effects of academic self-concept on academic achievement: Structural equation models of longitudinal data. *Journal of Educational Psychology*, 89, 41–54.

Muthén, L. K. and Muthén, B. O. (1998–2010). *Mplus User's Guide* (6th ed.). Los Angeles, CA: Muthén & Muthén.

Meredith, W. 1993. Measurement invariance, factor analysis and factorial invariance. *Psychometrika*, 58, 525–543.

Nagengast, B. and Marsh, H. W. 2011. The negative effect of school-average ability on science self-concept in the UK, the UK countries and the world: The Big-Fish-Little-Pond-Effect for PISA 2006. *Educational Psychology*, 31, 629–656.

Nagengast, B. and Marsh, H. W. 2012. Big fish in little ponds aspire more: Mediation and cross-cultural generalizability of school-average ability effects on self-concept and career aspirations in science. *Journal of Educational Psychology*, 104, 1033–1053. doi:10.1037/a0027697

Nagengast, B., Marsh, H. W., Scalas, L. F., Xu, M. K., Hau, K.-T., and Trautwein, U. 2011. Who took the "x" out of expectancy-value theory? A psychological mystery, a substantive-methodological synergy, and a cross-national Generalization. *Psychological Science*, 22, 1058–1066. doi:10.1177/0956797611415540

OECD. 2007. *PISA 2006 Science Competencies for Tomorrow's World*. Paris: OECD.

OECD. 2009. *PISA 2006. Technical Report*. Paris, France: OECD.

OECD. 2010. *TALIS 2008. Technical Report*. Paris, France: OECD.

Peschar, J. L., Veenstra, R., and Molenaar I. W. 1999. *Self Regulated Learning as a Cross-Curricular Competency. The Construction of Instruments in 22 Countries for the PISA Main Study 2000*. Paris: Organisation for Economic Co-operation and Development.

Pintrich, P. R. 2003. A motivational science perspective on the role of student motivation in learning and teaching contexts. *Journal of Educational Psychology*, 95, 667–686.

Pituch, K. A. and Stapleton, L. M. 2011. Hierarchical linear and structural equation modeling approaches to mediation analysis in randomized field experiments. In M. Williams and P. Vogt (Eds.), *The Sage Handbook of Innovation in Social Research Methods* (pp. 590 – 619). Thousand Oaks, CA: Sage.

Preacher, K. J., Zyphur, M. J., and Zhang, Z. 2010. A general multilevel SEM framework for assessing multilevel mediation. *Psychological Methods*, 15, 209–233. doi:10.1037/a0020141

R Development Core Team. 2011. *R: A language and environment for statistical computing*. R Foundation for Statistical Computing, Vienna, Austria. Retrieved from http://www.R-project.org/.

Raykov, T. 2009. Evaluation of scale reliability for unidimensional measures using latent variable modeling. *Measurement and Evaluation in Counseling and Development*, 42, 223–232.

Raykov, T. and Shrout, P. E. 2002. Reliability of scales with general structure: Point and interval estimation using a structural equation modeling approach. *Structural Equation Modeling*, 9, 195–212. doi:10.1207/S15328007SEM0902_3

Revelle, W. and Zinbarg, R. E. 2009. Coefficients alpha, beta, omega, and the glb: Comments on Sijtsma. *Psychometrika*, 74, 145–154. doi:10.1007/s11336–008–9102-z

Rhemtulla, M., Brosseau-Liard, P. E., and Savalei, V. 2012. When can categorical variables be treated as continuous? A comparison of robust continuous and categorical SEM estimation methods under suboptimal conditions. *Psychological Methods*, 17, 354–373. doi: 10.1037/a0029315

Rubin, D. B. 1987. *Multiple Imputation for Nonresponse in Surveys*. Hoboken, NJ: Wiley.

Schafer, J. L. 1997. *Analysis of Incomplete Multivariate Data*. Boca Raton, FL: Chapman & Hall/CRC.

Seaton, M., Marsh, H. W., and Craven, R. G. 2009. Earning its place as a pan-human theory: Universality of the big-fish-little-pond effect across 41 culturally and economically diverse countries. *Journal of Educational Psychology*, 101, 403–419. doi:10.1037/a0013838

Shavelson, R. J., Hubner, J. J., and Stanton, G. C. 1976. Validation of construct interpretations. *Review of Educational Research*, 46, 407–441.

Sijtsma, K. 2009. On the use, the misuse, and the very limited usefulness of Cronbach's alpha. *Psychometrika*, 74, 107–120.

Sörbom, D. 1974. A general method for studying differences in factor means and factor structures between groups. *British Journal of Mathematical and Statistical Psychology*, 27, 229–239.

Steiger, J. H. and Lind, J. C. 1980, May. *Statistically Based Tests for the Number of Common Factors*. Paper presented at the annual meeting of the Psychometric Society, Iowa City, IA.

Valentine, J. C. and DuBois, D.L. 2005. Effects of self-beliefs on academic achievement and vice versa. In H.W. Marsh, R.G. Craven, and D.M. McInerney (Eds.), *The New Frontiers of Self Research* (pp. 53–77). Greenwich, CT: Information Age.

van de Vijver, F. and Leung, K. 2000. Methodological issues in psychological research on culture. *Journal of Cross-Cultural Psychology*, 31, 33–51.

Vandenberg, R. J. and Lance, C. E. 2000. A review and synthesis of the measurement invariance literature: Suggestions, practices, and recommendations for organizational research. *Organizational Research Methods*, 3, 4–70.

Widaman, K. F. and Reise, S. P. 1997. Exploring the measurement invariance of psychological instruments: Applications in the substance use domain. In K.J. Bryant, M. Windle, and S.G. West (Eds.), *The Science of Prevention: Methodological Advances from Alcohol and Substance Abuse Research* (pp. 281–324). Washington, DC: American Psychological Association.

Widaman, K. F. and Thompson, J. S. 2003. On specifying the null model for incremental fit indices in structural equation modeling. *Psychological Methods*, 8, 16–37. doi:10.1037/1082–989X.8.1.16

Wigfield, A. and Eccles, J. S. 2000. Expectancy–value theory of achievement motivation. *Contemporary Educational Psychology*, 25, 65–81.

16

Contextual Indicators in Adult Literacy Studies: The Case of PIAAC

Jim Allen
Maastricht University

Rolf van der Velden
Maastricht University

CONTENTS

What Is the PIAAC Project About?

Need to Monitor Human Capital

The last few decades have seen an increased awareness of human capital as one of the driving forces of economic development. Policymakers are beginning to realize the importance of monitoring, and if necessary investing in, education and training as a way of improving the existing stock of skills. During the 1980s, the Organisation for Economic Co-operation and Development (OECD) started the large-scale project Indicators of Education Systems (INES) to measure input, process, and output of education (OECD 1994). This project resulted in the annual publication *Education at a Glance.* What soon became clear is that education and training as such are only weak indicators of human capital. Countries that have comparable levels of education can nevertheless differ significantly in the level of skills that affect important economic and social outcomes.

Development of ILSAs

This awareness triggered the development of several international large-scale assessments (ILSAs), such as the Programme for International Student Assessment (PISA) aimed at students in secondary education, as well as surveys aimed at measuring the skills of adults: the International Adult Literacy Survey (IALS) carried out in 1994/1998 (OECD/Statistics Canada 2000) and the Adult Literacy and Life Skills Survey (ALL) carried out in 2003/2008 (OECD/Statistics Canada 2005). The Programme for the International Assessment of Adult Competencies (PIAAC) is the successor of these two adult literacy surveys. The survey aims at assessing key skills of the 16–65-year-olds. PIAAC is conducted by a group of 24 countries together with an international consortium consisting of several organizations (CapStan—Belgium; CRP—Luxembourg; DIPF—Germany; ETS—USA; GESIS—Germany; Maastricht University—the Netherlands; WESTAT—USA) working for OECD.

Quest for Key Skills

A major project supporting the above-mentioned ILSAs was the Definition and Selection of Competencies (DeSeCo) project, which was initiated by the OECD to provide an overarching framework for international skills assessments. Competencies are defined in this project as "the ability to successfully meet complex demands in a particular context through the mobilization of psychosocial prerequisites (including both cognitive and non-cognitive aspects)" (Rychen and Salganik 2003, p. 43).

The theoretical framework provided by the DeSeCo project indicates the main underlying competencies that give skills their significance, but does not in itself directly give rise to clear recommendations as to the skills to be measured. Binkley et al. (2003) developed a framework that provides more detailed guidance for the development of skills measurements for adults. This work concentrated on two strands of research: the skills necessary in the workplace and cognitive functioning. The choice of direct assessments in the ALL survey was based partly on these theoretical notions, but also on practical considerations, such as an established tradition of measurements that are sufficiently short to be used during a household survey (Murray 2003). As a consequence, ALL focused on literacy and numeracy skills. PIAAC builds on the direct assessments used in ALL, extending these to the area of problem solving in technology-rich environments (OECD 2012).

In the adult literacy surveys the skills of the population are not only measured directly in the assessment, but also indirectly in the background questionnaire (BQ) by asking about the use of these skills both inside and outside working life. This engagement or use of skills is of course strongly correlated with the actual performance, but it also allows examining whether there are skill mismatches (e.g., people who have good reading skills, but do not use them in working life).

Other skill areas, particularly those involving intrapersonal, interpersonal, and other generic "soft" skills, are not included in the direct assessment, but the BQ does contain several indicators related to skill use or skill requirements in these areas. This comprises a module that has been specifically developed for the PIAAC project: the Job Requirements Approach (JRA) Module (Green 2008). The main arguments for developing a separate JRA module for PIAAC were the following:

- To provide a cost-effective way of assessing the relevance of skills not covered by the direct assessments
- To provide some information on the demand side for skills, in order to supplement the information provided by the direct assessments on the supply side

The JRA module comprises questions on problem solving, a range of interaction/social skills (influence, managerial skills, communication skills), and physical skills (strength and manual skill).

Design of the PIAAC Project

The PIAAC project is currently carried out in 23 OECD and 2 partner countries. In each country, some 5000 adults were surveyed in 2011/2012 and results are expected in 2013. A second round with an additional eight countries has recently started and will finish in 2015.

Like PISA, PIAAC is designed as multicycle research. New waves are expected to start every 5–7 years. Moreover, the tests and questionnaires are designed such that the main results can also be compared with the earlier IALS and ALL surveys. This means that PIAAC will not only allow for comparisons across countries, but also for comparisons over time. This is important, as many of the policy-relevant issues deal with evaluations of changes in the educational or labor market policy. The OECD intends to have a complementary policy survey to document relevant policy initiatives that might affect the skills of the population (Schleicher 2008). This allows for the identification of differences in policies across countries that might serve as an explanation for observed skill differences.

Although PIAAC is designed as a cross-sectional survey, there are a couple of longitudinal elements as well. The multicycle nature and the earlier adult surveys allow for so-called synthetic cohort constructions (see, e.g., Green and Riddell 2012; Willms and Murray 2007). Basically, this approach "follows" a certain age cohort over time: for example, the 30–40-year-olds in the IALS study of 1994 are compared with the 39–49-year-olds in the ALL study of 2003. In some countries, the PIAAC project intends to adopt a specific form of a longitudinal study by linking the individual data to registry data. This will be done in those countries where such national registries are available; the Scandinavian countries and the Netherlands are examples. These national registries contain information on, for example, labor force participation, occupation, earnings, educational attainment, and social security. This will enable the researchers to link the PIAAC data to later outcomes in life and thus construct a longitudinal design.

Development of the BQ

The initial developmental work for PIAAC was undertaken by the OECD as early as 2006. Several working groups from Network B of the OECD were involved in developing parts of the questionnaire. A separate pilot was set up to develop the JRA module. This module was based on the U.K. Skills Survey (Felstead et al. 2002) and tested in four different countries in 2008 (Green 2008). Based on the results of the developmental work, the OECD set up a country consulting process to make a priority rating of what should be covered in the BQ for PIAAC.

This early developmental work and priority rating formed the input of the development of the BQ by the PIAAC Consortium* that was responsible for the actual PIAAC project. The consortium was assisted by a group of experts who advised on the content of the questionnaire. Several drafts of the BQ were developed and discussed with the countries. Finally, a questionnaire with a total length of some 55 min was tested in a field test carried out in spring 2010. Based on the results of this field test and a country priority rating concerning the policy relevance of the different items, a final version was developed for the main survey with a total length of 40–45 min.

Table 16.1 presents an overview of the variables collected in the BQ. These can basically be distinguished into variables that are needed to perform an adequate descriptive analysis of the distribution of skills in the population, variables that can be seen as an important determinant of skills development, variables that are related to the outcomes of skills, and variables that

---

* This consortium was led by ETS and consisted of the following organisations: cApStAn, DIPF, ETS, GESIS, IEA, ROA, TUDOR, and WESTAT. ROA was responsible for the development of the BQ.

TABLE 16.1

Overview of Variables

Concept	Descriptive Analysis	Determinants of Skills	Economic and Social Outcomes	Control Variables
Direct assessment literacy, numeracy, problem solving				
Skill use literacy, numeracy, ICT				
Skill use in other domains (JRA)				
Gender	(Reporting category)			
Age	(Reporting category)			
Education parents	(Reporting category)			
Migration status	(Reporting category)			
Cultural capital parental home				
Mother tongue				
Language spoken at home				
Learning style				
Region	(Reporting category)			
Highest level of education	(Reporting category)			
Qualifications not completed				
Field of study				
Current/recent training				
Employers' contribution to training costs				
Training undertaken in working time				
Reasons for training (work-related versus non-work-related)				
Work experience (total duration, changes of employer)				
On-the-job training				
Informal learning at work				
Current main status (labour force status and self-declared status)	(Reporting category)			

continued

TABLE 16.1 (continued)

Overview of Variables

Concept	Descriptive Analysis	Determinants of Skills	Economic and Social Outcomes	Control Variables
Occupation	(Reporting category)			
Economic sector	(Reporting category)			
Employed versus self-employed				
Supervisory status				
Tenure current job				
Firm size	(Reporting category)			
Job security				
Working hours				
Earnings				
Education–job match				
General health status				
Voluntary work				
Political efficacy				
Social trust				

are needed as control variables. We will describe these variables and their purpose more specifically in the next section.

What Are the Policy Questions That PIAAC Seeks to Answer?

The PIAAC project focuses on three broad policy questions:

1. How are skills distributed?
2. Why are skills important?
3. What factors are related to skill acquisition and decline?

We will elaborate on how PIAAC addresses these questions in more detail in the following subsections. Note that the three questions follow the same logic as displayed in Table 16.1.

How Are Skills Distributed?

Human capital is considered as the driving force of economic growth. Investments in skills are essential in order to keep up with technological change (so-called skill-biased technological change: Levy 2010), as well as with other changes as a result of market trends or organizational developments (e.g., the introduction of high-performance workplace practices: OECD 1999). Policymakers have an interest in monitoring the stock of human capital in their country and in identifying the different levels among relevant subgroups. PIAAC assesses the stock of human capital in a society by providing a descriptive analysis of the distribution of skills proficiencies and skills use in the adult population. The survey will enable countries to answer questions such as

- How are skills proficiencies distributed across countries, or across regions, occupations, or sectors of industry within countries?
- Are skills equally distributed among relevant subgroups based on gender, age group, or migration status?
- Are skills being underdeveloped for particular levels of education?
- What is the level of investment in education and training, and are particular population subgroups excluded from adult learning systems?

In order to answer these questions, the survey assesses the level of skill proficiency in the three different domains (literacy, numeracy, and problem-solving in technology-rich environments), the degree of use of these skills both at work and in everyday life, as well as the degree of use of skills at work in other domains (the JRA module). The BQ also contains several indicators of investments in education and training, such as the level and field of the highest degree attained in education, the volume and intensity of formal and nonformal training received in the past 12 months, the extent to which this training has taken place for work-related reasons, who is financing such investments, and reasons for not participating in further training.

In order to effectively address questions concerning the distribution of skills across relevant subgroups, the BQ contains questions that allow us to derive relevant reporting categories, such as those based on gender, age, socioeconomic background, migration status, labor market status, region, occupation, industry sector, firm size, and level of education.

Why Are Skills Important?

Skills and Economic Outcomes

Policymakers would have little interest in investments in skills if skills showed no relation with relevant economic and social outcomes. Other services compete with education and training for a share in budgets, so the case for returns to educational investments needs to be based on a secure and sophisticated evidence base. Moreover, governments and the public have a legitimate desire to hold education accountable for the effects of the efforts put into this sector. For that reason, one of the key goals of the BQ is to provide indicators that can be used to show whether differences in skill levels matter, economically and socially. Figure 16.1 shows the underlying mechanisms how skills are thought to affect these outcomes.

The most obvious concern is to establish that skill levels are related to economic outcomes of individuals. Cognitive skills are thought to be a key determinant of individuals' productivity, so it is not surprising to find that cognitive skills are related to economic success. There is a large body of evidence showing that higher cognitive skills are indeed associated with better labor market outcomes (e.g., Heckman et al. 2006). Relevant questions to be addressed are

- How are skills related to individual employment opportunities, job security, earnings, or other indicators of labor market success?
- Do low skill proficiencies form a barrier to individuals entering the labor force?
- Is there a minimum level of skills that is needed to be employable?
- Can skills compensate for low educational qualifications?

One of the interesting questions in this respect is the precise role of education and skills in producing these outcomes. There are rivaling hypotheses on this point. The human capital theory (Becker 1964) claims that people with more years of schooling earn more because they are more productive. Scholars such as Spence (1973) and Arrow (1973) have pointed out that rewards are

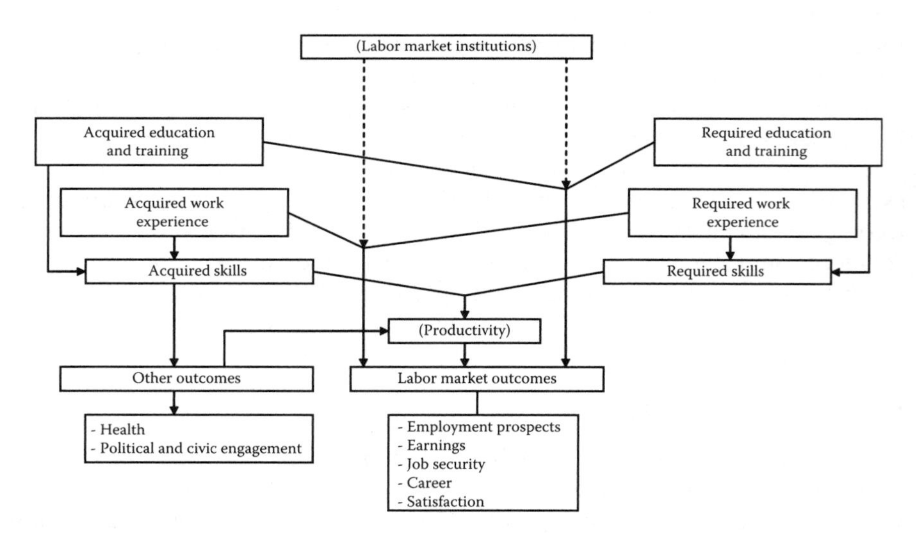

FIGURE 16.1
Schematic representation of economic and social outcomes (variables in brackets () not measured).

often based on signals such as formal qualifications as well as on the basis of productivity. The labor queue theory (Thurow 1975) points out that many relevant skills are learned on the job rather than in education. Credentialists such as Collins (1979) claim that more education does not lead to superior skills at all, but is used by "gatekeepers" to legitimize the rationing of access to high-status, highly paid jobs. There is probably an element of truth in all these theories; the crucial point is to specify the contexts under which one or the other mechanism prevails. In a study like PIAAC, we might expect large differences between countries and between sectors of industry in the relative effect of skills versus educational credentials on labor market outcomes.

Key economic outcome indicators included in the BQ are current labor market status (employed, unemployed, inactive), self-declared main status (allowing a better understanding of the situation of those not in the labor force), working hours, individual earnings, job security, occupational status, and the quality of the match between education and work.

Skills and Other Outcomes

Apart from economic outcomes, the relation of skills with other outcomes, such as health status and civic participation, is of interest as well (Schuller and Desjardins 2007). Adverse outcomes in such areas place large burdens on governments, businesses, and individuals, both in terms of direct expenditures (such as public health care budgets) and indirect costs (such as productivity loss due to worker illnesses). Relevant questions are

- To what extent is literacy related to the health status of individuals?
- To what extent do individuals with low skills appear to be less engaged in broader society (voluntary work, social trust)?
- To what extent does the engagement of migration groups or linguistic minorities appear to be inhibited by their lack of skill in the main language(s) in their country of residence?

Education not only affects individual outcomes in these domains, but also social returns due to spillover effects. As with the effects of education on labor market outcomes, the effects of education on other outcomes are still not fully understood. Broadly, two mechanisms can be distinguished (Pallas 2000): an effect on skills—education directly affects knowledge and skills that are relevant for healthy behavior, civic engagement, and so on—and an effect on allocation—higher education increases the chance of ending up in healthier jobs or in social networks in which civic engagement is higher. The PIAAC data should provide clues as to the extent to which of these mechanisms is dominant, and how this varies both between and within countries.

The BQ includes indicators of subjective health status, involvement in voluntary work, political efficacy, and social trust.

What Factors Are Related to Skill Acquisition and Decline?

Assuming that skills matter economically and socially, policymakers have an interest in knowing what factors are related to the acquisition but also to the decline of skills over the life course. Figure 16.2 shows the assumed interplay of the main determinants of skill acquisition and decline.

The primary interest is in factors that can be directly influenced by policy, such as the provision of education and training. However, it is also relevant to compare the efficiency of such skill production routes with that of other routes not directly under the control of policymakers, such as informal learning activities in which people may engage at work or outside of work. Assessing the overall relation between education and training and skill levels is only a first step in unraveling the determinants of skills acquisition. We can assume that not all education and training activities have the same impact on skills development. Nor can we assume that the impact is the same for all relevant subgroups. Policymakers have an interest in seeing which characteristics of education and training are most strongly related to higher skill levels in the population and which subgroups appear to gain most from which type of intervention. Finally, we need to be aware that skills cannot only be acquired, but skills can also be lost. Preventing skill decline is probably just as important as promoting skill acquisition, but the underlying factors affecting these processes may be quite different, and it is important to have good insight into both processes. This means that the survey needs to

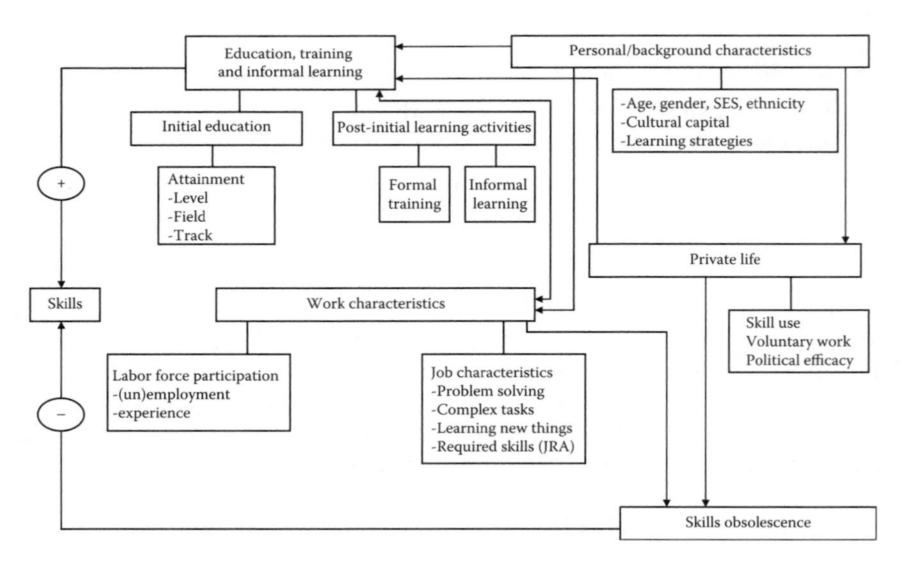

FIGURE 16.2
Schematic representation of skill acquisition and decline.

enable countries to answer questions such as

- What is the relation between education, training, and other learning activities (such as informal on-the-job learning) and the skill development of people?
- Are there subgroups that appear to profit more or less from investments in education and training?
- What is the relation between underinvestment in work-related training and adult skill levels?
- How are characteristics of the work environment related to skill levels?
- Is informal learning on the job a substitute for work-related training?
- How do processes of skill acquisition and decline vary with age?
- What are the factors that are related to skill decline?
- Are these the same as the factors that are related to skill acquisition?

Skill Acquisition

There is little doubt that education is the single most important factor that contributes to the acquisition of skills. However, people do not only learn during initial education, but also later in life. Several studies have found high returns to workers' participation in training (for a review, see Bassanini et al. 2005), but others have questioned whether it is formal training that really makes the difference. Employees spend much more time on informal learning activities than on formal learning, and the activities are often interrelated (Borghans et al. 2006). We have already indicated that the BQ contains several indicators related to the formal and informal training in which adults have been engaged, like the level and field of the highest degree that was attained in education, and the volume and intensity of the formal and informal training in the past 12 months. The BQ also explores other routes through which skills can be acquired, such as informal learning on the job.

Regardless of the specifics of the training and learning practices available through the organization in which individuals work, the amount of work experience acquired can be expected to have a strong effect on skills development. However, one is most likely to be exposed to learning situations early in one's career. As the career develops, the chance that one will be exposed to new stimuli is decreased, a pattern that is reinforced by typical patterns of brain development over the life cycle. In addition to total work experience, the number of changes of employer and/or career breaks is therefore also important. The PIAAC BQ contains items on each of these factors.

There are also factors outside the world of work that can affect development and retention of competencies. The questionnaire therefore covers some activities in which respondents may be involved in in everyday life from which they can learn as well.

Not everybody is equally likely to profit from investments in these forms of learning. Some groups may face constraints in their investments, or some may have an inappropriate learning style that prevents optimal development of their skills in education or training. The BQ contains indicators of constraints that are expected to affect the differential investment in education and training, which may be related to gender, marital status, family formation, socioeconomic status, and ethnic status. Moreover, the BQ contains a scale that measures the kind of learning strategy that adults apply when learning something new.

Skill Loss

The single most important finding of the IALS and ALL surveys was that skill loss was sufficient to offset all expected gains from increasing educational quality and quantity. Higher-order brain functions follow a predictable developmental pattern, and reach a plateau of optimal functioning in young adulthood. Most cognitive abilities tend to decline with advancing age. However, large individual differences exist in the onset and rate of decline of specific cognitive functions. Fluid abilities (functions that involve controlled and effortful processing of novel information) typically start declining in the mid-20s, while crystallized skills (the representation of learned skills and access to knowledge) may improve until and even beyond the age of 60. One may hypothesize that numeracy and literacy skills relate more strongly to crystallized abilities, and are therefore less susceptible to decline than dynamic problem solving in a technology-rich environment, which may relate more to fluid abilities.

Building partly on insights from cognitive and neuropsychology, De Grip and Van Loo (2002) developed a typology of different types of skills obsolescence, of which so-called skill *wear*—which results from the natural aging process and may be accelerated by physically or mentally challenging working conditions—and skill *atrophy*—which is due to the lack or insufficient use of skills and may be brought on by unemployment, career interruptions, or working below one's level of education—are particularly relevant from the point of view of PIAAC. The BQ contains indicators that are thought to be related to *wear* (age, health) and *atrophy* (unemployment, working below one's level, tenure, and sector of industry). In addition, several of the abovementioned indicators of workplace learning have a direct bearing on not only the job learning but also the risk of skills obsolescence in the form of atrophy.

Advancements and Caveats

Without any doubt, the PIAAC will become one of the most important ILSAs of the coming decade. Part of its strength lies in the innovative character of the assessment, that is, the development of the new domain of problem solving in technology-rich environments, as well as the development of the JRA module to assess a broad array of skills used at work. The strength also lies in the coherence of the conceptual framework underlying both the direct assessment and the BQ. The survey is also innovative in using a specifically developed computer platform that allows for both the BQ and the direct assessment to take place on the computer (only respondents who lack the necessary computer skills will take a paper-and-pencil test).

Compared to many international comparative surveys, PIAAC is also rather unique in terms of the ex-ante harmonization of the instruments. All items were centrally determined and national adaptations needed to be approved by the consortium. In addition, the translations were checked by the consortium to ensure the overall comparability of the items. The use of a computer platform also greatly improved the quality of the data and kept nonresponse to a minimum.

Despite these advancements made in PIAAC, there are some caveats that we would like to point out.

No Claims for Causality

For evidence-based policy, it is important that the estimates of the relationships between various factors and skills are as free as possible from bias. Policymakers would be misinformed if the effects of, for example, investments in education or training were seriously overestimated (in which case some of the investment might be a waste of money) or underestimated (in which case policymakers might decide to withhold investments in education or training that would have yielded large returns in terms of skills). Estimates based on nonexperimental data collections such as the ones obtained through PIAAC may be biased. The problem occurs when there is some other variable that affects both the decision of an individual to invest in education or training (the variable of interest) as well as his or her skills levels (the dependent variable). When this variable is not controlled for in the analysis (unobserved heterogeneity), the parameter estimates of the other variables in the model may be biased: the so-called omitted variable bias. Another problem occurs when the direction of causality is not clear (endogeneity). For example, people may follow training because they were promoted to a better job rather than having been promoted as a result of their investment in training. In that case, the causal link between training and wages is reversed.

This is not a new problem, and in the past social scientists have used panel designs and control variables to obtain better estimates of the statistical associations of interest. However, this will never rule out the possibility that some omitted variable may bias the true relation. In the past decade, new statistical techniques have been developed to deal with this problem of unobserved heterogeneity, such as the use of instrumental variables or the use of selection models (e.g., Heckman 1979). The basic idea is to use other variables or other models to estimate the "true" effect of the variables of interest. It is now a common understanding among social scientists that the use of these additional techniques are essential to improving our knowledge of the effects of the variables of interest, and to giving policymakers the unbiased estimates they need to make valid decisions.

For the PIAAC project, we have tried to develop instrumental variables that could be used in analyses. This is not an easy task, but we did come up with some instruments that could be used, for example, birth order, proximity of potential schooling institutes or discount rate to instrument education, and organizational restructuring and distance to current and alternative jobs to instrument non-formal training. The main goal of including such instrumental variables is to get an unbiased estimate of the effect of, say, education on skills. These variables should therefore be truly exogenous in the sense that they are unrelated to the dependent variable (skills levels) or to other unobserved explanatory variables. By their very nature, these variables usually have little policy relevance in themselves. That was one of the main reasons why most of the instrumental variables that were proposed were not included in the final questionnaire for the field test. In the face of severe space limitations, it was hard to justify retaining variables that have little policy relevance in themselves, even if they could have helped in obtaining the unbiased estimates that policymakers need to make evidence-based policy.

Underestimation of Specific Professional Skills

Despite the fact that employers often list generic cognitive skills and personal traits as the most important skills required in the workplace, professional expertise is a *conditio sine qua non* for success in many occupations. For example, nobody would doubt that in order to become a good medical doctor, architect, or car mechanic, one needs to acquire the domain-specific knowledge and skills that make up the professional domains of these occupations. The German psychologist Weinert formulated this as follows: "Over the last decades, the cognitive sciences have convincingly demonstrated that context-specific skills and knowledge play a crucial role in solving difficult tasks. Generally, key competencies cannot adequately compensate for a lack of content-specific competencies" (Weinert 2001, p. 53).

There is, however, a plethora of specific professional skills, and usually they are not taken up in ILSAs such as PIAAC, simply because there is no common assessment instrument that allows all different types of professional skills to be measured in a meaningful way for large populations.* The absence of direct measures of specific professional skills will automatically lead to an overestimation of the relevance of the generic skills, such as the ones measured in the

* There are some exceptions though. In Germany, there is an attempt to develop an assessment for vocational education, called PISA-VET (Baethge et al. 2006). The OECD is currently carrying out a pilot to measure specific skills of higher educated in the so-called AHELO project.

direct assessment. It underscores the importance of obtaining detailed information on the occupation of working respondents, as is done in the BQ. Since the differences among occupations in the skills measured in the direct assessments is likely to be at least matched and probably eclipsed by differences in level and type of specific skills, the residual occupation-level variance in economic outcomes should provide a rough indication of the economic importance of specific skills relative to the generic skills measured.

References

Arrow, K.J. 1973. Higher education as a filter. *Journal of Political Economy*, 2(3), 193–216.

Baethge, M., F. Achtenhagen, L. Arends, E. Babic, V. Baethge-Kinsky, and S. Weber 2006. *PISA-VET: A Feasibility-Study*, Stuttgart: Frans Steiner Verlag.

Bassanini, A., A.L. Booth, G. Brunello, M. De Paola, and E. Leuven. 2005. Workplace Training in Europe. *IZA-Discussion paper 1640*, Bonn: IZA.

Becker, G.S. 1964. *Human Capital: A Theoretical and Empirical Analysis with Special Reference to Education*, New York: Columbia University Press.

Binkley, M.R., R. Sternberg, S. Jones, D. Nohara, T.S. Murray and Y. Clermont. 2003. Moving towards measurement: The overarching conceptual framework for the ALL study, In: T.S. Murray, Y. Clermont, and M. Binkley (eds.) *Measuring Adult Literacy and Life Skills: New Frameworks for Assessment*, Ottawa, Canada: Statcan.

Borghans, L., B. Golsteyn, and A. de Grip. 2006. *Meer Werken is Meer Leren: Determinanten van Kennisontwikkeling*, 's-Hertogenbosch: CINOP.

Collins, R. 1979. *The Credential Society: An Historical Sociology of Education and Stratification*, New York: Academic Press.

De Grip, A. and J. Van Loo. 2002. The economics of skills obsolescence: A review, In: A. de Grip, J. van Loo, and K. Mayhew (Eds.), *The Economics of Skills Obsolescence*, Research in Labour Economics, vol. 21, Amsterdam/Boston: JAI Press, 1–26.

Felstead, A., D. Gallie, and F. Green. 2002. *Work Skills in Britain 1986-2001*, Oxford: SKOPE, Oxford and Warwick Universities.

Green, D.A. and W.C. Riddell. 2012. Ageing and Literacy Skills: Evidence from Canada, Norway and the United States, *IZA-Discussion Paper 6424*, Bonn: IZA Bonn: IZA.

Green, F. 2008. Draft Report on the Validation of the Results of the Job Requirements Approach (JRA) Pilot Survey. Mimeo.

Heckman, J.J. 1979. Sample selection bias as a specification error. *Econometrica* 47, 1, 153–161.

Heckman, J.J., J. Stixrud, and S. Urzua. 2006. The effects of cognitive and noncognitive abilities on labour market outcomes and social behavior. *Journal of Labour Economics*, 24(3), 411–482.

Levy, F. 2010. How Technology Changes Demands for Human Skills, *OECD Education Working Papers 45*, Paris: OECD.

Murray, T.S. 2003. Reflections on international competence assessments, In: D.S. Rychen and L.H. Salganik (eds.), *Key Competencies for a Successful Life and a Well-Functioning Society*, Göttingen: Hogrefe & Huber, 135–159.

OECD. 1994. *Making Education Count: Developing and Using International Indicators*, Paris: OECD.

OECD. 1999. *Economic Outlook*, Paris: OECD.

OECD/Statistics Canada. 2000. *Literacy in the Information Age: Final Report of the International Adult Literacy Survey*, Paris: OECD and Ottawa: Statistics Canada.

OECD/Statistics Canada. 2005. *Learning a Living: First results of the Adult Literacy and Life Skills Survey*, Paris: OECD and Ottawa: Statistics Canada.

OECD. 2012. *Literacy, Numeracy and Problem Solving in Technology-Rich Environments: Framework for the OECD Survey of Adult Skills*, Paris: OECD.

Pallas, A.M. 2000. The effects of schooling on individual lives. In: M.T. Hallinan (ed.), *Handbook of the Sociology of Education*, New York: Kluwer Academic/Plenum Publishers, 499–525.

Rychen, D.S. and L.H. Salganik. 2003. A holistic model of competence, In: D.S. Rychen and L.H. Salganik (eds.) *Key Competencies for a Successful Life and a Well-Functioning Society*, Göttingen: Hogrefe & Huber, 41–62.

Schleicher, A. 2008. PIAAC: A new strategy for assessing adult competencies. *International Review of Education*, 54, 627–650.

Schuller, T. and R. Desjardins. 2007. *Understanding the Social Outcomes of Learning*, Paris: OECD.

Spence, M. 1973. Job market signalling, *Quarterly Journal of Economics*, 87, 1, 355–374.

Thurow, L.C. 1975. *Generating Inequality*, New York: Basic Books.

Weinert, F.E. 2001. Concept of competence: A conceptual clarification, In: D.S. Rychen and L.H. Salganik (eds.), *Defining and Selecting Key Competencies*, Göttingen: Hogrefe & Huber, 45–65.

Willms, J.D. and T.S. Murray. 2007. *Gaining and Losing Literacy Skills Over the Lifecourse*, Ottawa: Statistics Canada.

Section III

Advanced Analytic Methods for Analyzing International Large-Scale Assessment Data

17

Incorporating Sampling Weights into Single- and Multilevel Analyses

Laura M. Stapleton
University of Maryland

CONTENTS

Introduction

Sampling weights are often a source of confusion for data analysts who are unfamiliar with sampling theory. This chapter will hopefully clarify the issues and demonstrate how and why sampling weights are incorporated at the analysis stage. The reader is encouraged to first read Chapter 6 of this volume (Rust 2013) to understand basic sampling procedures in international large-scale surveys as these procedures will be referenced throughout this chapter.

In general, the incorporation of sampling weights into analytic models allows for unbiased estimates of population parameters. Because the sample that is obtained may not reflect the population in known ways, sampling weights allow the statistical analysis to more appropriately reflect the population characteristics. As a simple example, suppose a population consisted of 20 people, 10 of whom were female and 10 were male. Further suppose that two females were randomly selected for the sample, which reflects a probability of selection ($\pi_{females}$) of 2/10 = .2. Also, four males are randomly selected for the sample reflecting a probability of selection of π_{males} = 4/10 = .4. If the height of all individuals in the sample is measured and the sample mean calculated, a biased estimate of the population average height would likely result because the sample contains proportionately more males than it should (or proportionately fewer females than it should). In general, the sampling weight is the inverse of the selection probability (a discussion of adjustments to this base weight is provided in the section "Types of Weights"). Therefore, for females, the sampling weight, $w_{females}$, is 1/.2, or 5. Note that the sum of the weights for the females in the sample is equal to the number of females in the population. Similarly, for males, the sampling weight is w_{males} = 1/.4, or 2.5. In essence, the sampling weight provides the number of units (in this case, people) that any given observation represents. In this example, each female in the sample represents five females and each male represents 2.5 males.

Given this extremely simple introduction, the remainder of this chapter presents information on different definitions of weights available with international large-scale assessments and examples of how the use of the different weights (or lack of use of them) can affect estimates from multivariate models. Information on the capabilities of current versions of statistical software to incorporate weights into analyses is provided.

An important caveat to this chapter is that the inclusion of sampling weights into an analysis does not, by itself, accommodate the fact that data were obtained from a complex sampling process. An appropriate variance estimation technique must be used in the analysis. Variance estimation is a phrase that encompasses estimation of standard errors of parameter estimates. Refer to Chapter 6 (Rust 2013) and Wolter (2007) to obtain more information on variance estimation; in addition, refer to the subsection on replicate weights in the section "Types of Weights."

Types of Weights

In any large-scale dataset, it is possible that the researcher may find a number of different weight variables. To be prepared to use these weights and select the appropriate one(s) for a particular analysis, the researcher should familiarize him- or herself with the types of weights that may be provided. In this section, the definitions for a variety of weights and weight adjustment variables are presented.

Base Weights

As addressed in Chapter 6 on sampling (Rust 2013), selection of observations from a sampling frame often occurs in different stages and there may be weights provided for each of these stages of selection. While many surveys in the education realm are two-stage surveys of schools and then of students, others may be three-stage, including the selection of geographic areas, followed by selection of schools, and then of students. At each of these stages of selection, the unit being selected (e.g., the school or the student), is associated with some known probability of selection, π. Assuming a two-stage sample, the school has a given sampling weight that differs across the schools; it is rare that schools are selected using a simple random sample. As described in Chapter 6, the frequent use of probability proportionate to size sampling results in large schools having relatively larger selection probabilities, π_j, as compared to smaller schools. Therefore, taking the inverse $1/\pi_j = w_j$, the sampling weights for large schools will tend to be smaller than the weights for small schools. Because the selection is being undertaken on schools in the sampling frame of schools, the school sampling weight represents the number of schools that a given selected school represents.

Once schools are selected, either students of a given age are selected or classrooms are selected and all students in the classroom are included (or students may further be selected within classrooms in a three-stage sample). These second stage selection probabilities, $\pi_{i|j}$, reflect the conditional selection probability, that is, given that its school was selected, it is the selection probability of the classroom or student within the school. The within-school, conditional sampling weight is thus $1/\pi_{i|j} = w_{i|j}$.

For a two-stage sample, these two base weights (the inverse of the selection probability of the school and the inverse of the conditional selection probability of the classroom or student) are used to determine an overall base weight. Multiplying the two together provides the overall, or unconditional, sampling weight of the observation: $w_{ij} = w_j \times w_{i|j}$.

Weight Adjustments

It is possible that some schools or some classrooms or students did not agree to participate in the study or did not respond. This nonparticipation or nonresponse needs to be accommodated such that the initial sampling weights are adjusted. For example, suppose that in the simple example above initially four females had been selected for the sample (out of the 10 in the population). The initial selection probability for females was .4 and the initial sampling weight for each female was 2.5. However, suppose that two of the selected females chose not to participate. A nonresponse adjustment should be applied to the initial sampling weight. In this case, the nonresponse adjustment would be 4/2; within the female stratum, each of the sampling weights needs to be inflated to represent the two females who did not respond. These adjustment factors may exist for both the school-level weights and the classroom or student-level weights. Note that nonresponse adjustment factors might have been derived in a very complex way, taking advantage of auxiliary information about the responders as well as the sampling frame information and not based on just the explicit stratum information as shown in the example here with gender. This simple adjustment approach is referred to as a weighting class adjustment method (Little and Rubin 2002). Other examples of nonresponse adjustments include the use of response propensity models and propensity score weighting (Heeringa et al. 2010).

Total, House, and Senate Weights

In international large-scale assessments, a complication arises in that comparisons of parameter estimates are made across countries that may be very different in size. For this reason, a set of transformed weights may be provided on the dataset to aid in country comparisons. Therefore, a researcher may find weights referred to as Total, House, and Senate weights.

First, the Total student weight is as described above and will be identified in this chapter as w_{ij}, where the i indexes the individual student and j indexes the school. It is the unconditional overall weight after each of the selection stage weights have been multiplied along with their nonresponse adjustments. The sum of these Total weights is intended to be an estimate of the population size in each country. This sampling weight should be used, for example, if the user is interested in obtaining estimates of the total number of students with a computer in the home in each country. However, note that if the software that the researcher is using does not appropriately estimate standard errors given the complex sampling design, the standard errors may be calculated as if the sample size were the size of the population and thus result in the estimation of extremely small standard errors! An example of this sort of problem is provided in the section "Example Analyses." Another problem with using the Total student weight in analyses occurs in those analyses where data are combined across countries. For example, suppose a researcher wanted to find the average performance of students in several Western countries, including the United States and England. The population size of the United States is much larger than that of England and use of the Total weight in a combined analysis will result in the estimated combined mean being closer to the U.S. sample mean than the England sample mean. This estimated combined mean is an unbiased estimate of the mean of all students from these countries but if the researcher were interested in the average of the country means, then the estimate would be inappropriate.

A proportional transformation can be applied to the Total weights within each country to result in weights that sum up to a given size within each country. Use of a sampling weight based on a proportional transformation will not affect estimates of means, proportions, regression coefficients, and so on, but it will affect the estimate of the sum of the variable and the sum of the sampling weight itself. A proportional transformation is used to create new weights, referred to as Senate and House weights, typically provided on international large-scale data files. Each type of weight, Senate and House, is appropriate for different analyses. As will be seen, these weights were so named to reflect the approach in the U.S. Congress where, in the U.S. Senate, states are given equal "weight" regardless of their population size but, in the House of Representatives, states are given weight as a function of their population size. This same approach is taken to equally weight or proportionately weight data from countries when their data are combined.

The Senate weight might be defined within each country as

$$\text{Senate } w_{ij} = w_{ij} \times \frac{500}{\sum_{j=1}^{J} \sum_{i=1}^{n_j} w_{ij}} \tag{17.1}$$

where J represents the number of schools in the sample in the country of interest and n_j represents the number of students within school j. The resulting Senate weights will sum to 500 within each country; the number 500 is arbitrary and may be different across different large-scale datasets. Because this weight sums to the same size across countries, this weight could be used when the analyst wishes to combine data across countries but treat each country as an equal contributor to an estimate.

In contrast, the House weight is defined within each country as

$$\text{House } w_{ij} = w_{ij} \times \frac{n}{\sum_{j=1}^{J} \sum_{i=1}^{n_j} w_{ij}} \tag{17.2}$$

where n represents the total number of students in the sample across all schools in the country of interest. Within each country, the House weight will sum to the sample size in that country. Use of this weight in analyses will not reflect equal contribution of countries to the parameter estimate of interest (unless countries coincidentally happen to have the same sample size) but will reflect the contribution based on sample size. The only time that the House weight should be used is when the software bases standard error calculation on the sum of the sampling weights and not on the actual sample size itself, as shown in the example in the section "Example Analyses." Given rapid advances in the ability of software to accommodate survey data with sampling weights appropriately, the need for the use of the House weight is lessening.

Note that if one were conducting an analysis within a single country only, any of these three types of weights (Total, Senate, and House) would provide the same point estimates of most parameters of interest such as means, proportions, regression coefficients, and differences in means.

Other Weight Transformations

Outside of international large-scale assessments, the House weight transformation discussed above is referred to as "normalizing the weight" or as a "relative sample size" weight and is applicable in any probability survey sample context and not just in international assessment programs. A final transformation that may be conducted but typically not provided on large-scale datasets is referred to as an "effective sample size" weight (Pfeffermann et al. 1998; Potthoff et al. 1992). This effective sample size weight sums to less than the sample size and reflects the potential loss of precision in estimates due to clustering and/or disproportionate selection mechanisms. This weight is not often used in practice but has sometimes been shown to be advantageous in multilevel analyses as is discussed in the section "More Advanced Estimates."

Replicate Weights

A final set of weights that may be found on international large-scale assessment data files, which are of considerable confusion to those who are not familiar with variance estimation, are replicate weights. These replicate weights typically are versions of the unconditional sampling weight but for each observation, the weight has either been set to 0 or inflated to account for their neighbor observation's weights (e.g., those observations from the same primary sampling unit) that were set to zero. In some datasets, these weights are not already provided, but status indicators of variance stratum and primary sampling unit are provided so that software can create the replicate weights as needed. The replicate weights should not be used within a single analysis; they should only be used in a sampling variance estimation procedure, where the analysis is run each time for each set of replicate weights and then combined to provide an empirical estimate of the sampling variance of the model parameter estimates. More information about this replication estimation procedure is available in Chapter 6 (Rust 2013) and Lohr (2010).

Summary

The weights discussed above (first-stage, conditional, Total, House, Senate, and replicate) are the most frequently encountered weights on large-scale assessment datasets but it is not a comprehensive list. Other weights that may be found with complex survey programs include panel weights and linkage weights. Panel weights are used in longitudinal surveys where some participants may have dropped out or not completed a specific wave of the study. Panel weights allow the user to appropriately weight the sample to address the fact that the data used for the analysis may only include some of the original participants in the study. Linkage weights, such as "student–teacher" linkage weights can be used when more than one informant may be associated with a student. For example, if some students in the assessment program have reports from more than one teacher, that student's weight would need to be partitioned to each of the multiple teachers in order for that student not to count more than he/she should when using the teacher report data to estimate student characteristics.

With this brief overview of the types of weights that may be included on data files, a user should now be in a good position to understand which weights might be appropriate for his or her analysis. The next section will present examples of the effects of inclusion of weights into analyses.

Example Analyses

Simple Means

To demonstrate the differences in estimates that are obtained when using some of the different weights defined above, a simple analysis was conducted here for two countries selected from the 1999 TIMSS dataset (made publicly available in the spring of 2012 at http://timss.bc.edu/timss1999i/database.html). Data from the United States and from Turkey were selected for the example. These two countries had similar sampling procedures but represent very different population sizes. In each of these countries, a three-stage sample was drawn. In Turkey, 40 provinces were initially selected from 80 in the country. Within each selected province, at least four schools were sampled with implicit stratification by county. Finally, within schools, one classroom was selected. The data for Turkey come from 7841 students within 204 schools. In the United States, 52 geographic areas were first selected out of 1027 in the sampling frame. Within these 52 PSUs, schools were selected within explicit strata defined by school type and primary sampling unit (PSU) size. Schools were also sampled using implicit stratification by religious denomination in the private school strata and percentage minority within the public school strata. From one to eight classrooms were selected within each school with an average of 2.04 classrooms per school. The data for the United States come from 9072 students within 221 schools. Sampling information for other countries is available in Appendix C of the TIMSS 1999 Technical Report (Martin et al. 2000).

For the example analysis, the variable "Do you have a computer at your home?" (*BSBGPS02*) was selected to show the differences in estimates that are obtained with the various weighting options with the software SPSS. The variable *BSBGPS02* was recoded such that 1 represents "yes" and 0 represents "no." Also note that this response was missing for 1.2% of students from Turkey and 3.1% respondents from the United States and therefore the analyses included 8789 students from the United States and 7748 students from Turkey. While listwise deletion has been used for this example, there are more appropriate ways to address the missingness (e.g., with multiple imputation) that is beyond the scope of this chapter. Chapter 20 of this volume (Shin 2013) includes a discussion of one approach to modeling with missing data within the context of the HLM software. Additionally, Chapter 19 (Rutkowski 2013) and Chapter 21 (Anderson et al. 2013) provide a different approach. Combined, the reader will be exposed to some of the best practices in modeling with missing data.

For this example, three sets of analyses were run. Each set included the estimation of the population mean of *BSBGPS02* without a weight variable and with the *TOTWGT, HOUWGT,* and *SENWGT* variables, which represent the unconditional overall weight, the House weight, and the Senate weight, respectively (defined above in the section "Total, House, and Senate Weights"). This set of analyses was run three times: on the data from the United States only, on the data from Turkey only, and on the combined sample of the data from the United States and Turkey. The latter would be of interest if a researcher wanted to report out average characteristics of the students from a set of countries. Table 17.1 contains the estimated mean and standard error using SPSS with the respective weight variable identified using the pull down menu options "DATA..." and then "WEIGHT CASES..." and by selecting the appropriate weight (*TOTWGT, HOUWGT,* or *SENWGT*). Given that *BSBGPS02* is coded as 0 and 1, the mean represents the proportion of students having a computer in the home. It should be noted that the standard error estimates shown in Table 17.1 are not correct (as will be discussed in a later section).

TABLE 17.1

Example SPSS Estimates from TIMSS 1999 Data for Turkey and the United States Using Three Different Types of Sampling Weights

		No Weight	*TOTWGT*	*HOUWGT*	*SENWGT*
United States	Mean	0.79	0.80	0.80	0.80
	se(Mean)	0.004	0.000	0.004	0.018
	N	8789	3,252,802	8845	487
Turkey	Mean	0.10	0.10	0.10	0.10
	se(Mean)	0.003	0.000	0.003	0.013
	N	7748	611,472	7757	495
United States and Turkey combined	Mean	0.46	0.69	0.47	0.44
	se(Mean)	0.004	0.000	0.004	0.016
	N	16,537	3,864,273	16,602	982

Focusing on the first row of estimates, note that for the United States, the use of weights as compared to no weights in the analysis resulted in a slightly higher estimate of the proportion of students who had a computer in the home. Without weights, one would estimate that 79% of the students had a computer but with weights that estimate increases to 80%. This means that students with higher sampling weights must have been more likely to have a computer in the home. In fact, in the United States sample, there is a .05 correlation between *BSBGPS02* and the overall student sampling weight. In Turkey, there was nearly no relation between the sampling weight and *BSBGPS02* ($r = -.02$) and, therefore, there was basically no difference in the estimate of the mean when using weights versus not using weights. When a difference in the estimate exists under weighted and unweighted analyses, it is referred to as informativeness of the weights. Tests of informativeness of weights have been proposed (Asparouhov and Muthén 2007; Pfeffermann 1993) and procedures to evaluate the effects of inclusion of weighting variables are discussed by Korn and Graubard (1999), Snijders and Bosker (2012), and Carle (2009). These procedures will be summarized in the section "Unweighted Approaches to Analyses" of this chapter.

Of utmost importance in Table 17.1, note how the *se(Mean)* and *N* rows differ for the United States and for Turkey across the different weighting options. Because the SPSS software uses the WEIGHT CASES as a frequency weight as opposed to a sampling or probability weight, the weight is treated as if it should replicate the data that many times and functionally create a bigger dataset. As an example, note that the *N* for the analysis that includes the *TOTWGT* variable as the weight variable is quite large. Therefore, the standard error, calculated by SPSS as the estimate of the standard deviation divided by the square root of *N*, is extremely small (so small that it is less than .0005 and appears as .000 on the output). For this reason, of popular software using sampling weights as frequency weights, the concept of the House weight was created. This transformation scales the weights so that they sum to the sample size instead of the population size. Note that the *N* associated with the analysis with the *HOUWGT* variable identified as the weighting variable is 8845 for the United States and 7757 for Turkey. The sum of the House weights typically would be expected to be the same as the sample size but due to the missing data on *BSBGPS02*, the numbers are not quite equal. It should be clear that because the House weight is just a proportional transformation of the overall weight (*TOTWGT*), the point estimate of the mean does not differ between the two methods. Finally, examine the Senate weight column. Note that the *N* for the United States and Turkey are both quite low. In this transformation approach, the weight variable was scaled such that the sum of the weights within a country would be 500 (again, due to the use of listwise deletion given the missing data, the weights do not sum up to 500). For the United States, the standard error as calculated by the SPSS software is based on the assumption that just under 500 respondents contributed to the estimate; therefore, the standard error is much larger than it likely should be (given that the actual estimate is based on nearly 9000 students). The Senate weight is really only appropriate to use when the analyst wishes to obtain an estimate of the average for students from two or more countries and wants each country to contribute equally to the estimate.

The last block of analysis results in the table shows the combined estimate of data from the United States and Turkey. This set of analyses reveals great differences in the point estimate of the proportion of students with a computer in the home when using the different weighting options. Using the Senate weight, the point estimate should represent the United States and Turkey equally and thus the overall point estimate should be halfway between the estimate for the United States and the estimate for Turkey. The estimate of 44% of students having a computer in the home is an equally weighted combination of the 80% for the United States and the 10% for Turkey (again, given that the sum of the Senate weights did not equal 500 for each country due to missing data, this overall estimate differs a little from equal weighting of countries as Turkey has a higher sum of the Senate weights than does the United States). The point estimate when using the *TOTWGT* variable represents a mixture of students from the United States and Turkey in terms of the population sizes. Given the N for the United States is over 3 million and Turkey just over 600,000, the point estimate of the proportion of students with computers in the home is more heavily weighted toward the United States value at 69% (which is an unbiased estimate of the proportion of all students in the United States and Turkey having a computer—but it does not allow each country to weigh equally in the estimate). Note, however, that the standard error for this estimate of 69% as calculated in SPSS is quite low and assumes that the sample size equals almost 4 million! Unfortunately, the House weight cannot be used in this scenario to obtain an appropriate standard error. Because the House weights are scaled *within* country to sum to the sample size, if they are used in the analysis then the comparative size of the United States population and the Turkey population will not be reflected in the overall estimate—the comparative sizes of the *samples* will be reflected in the overall estimate; the point estimate of the mean from the unweighted analysis and the House weight analysis has little interpretable meaning apart from being the mean of the combined sample. If an analyst wished to obtain an appropriate standard error for the average *BSBGPS02* variable of the combined sample of the United States and Turkey with each country being represented in proportion to its actual size, a new weight would need to be created. This weight would be

$$\text{Multicountry } w_{ijk} = w_{ijk} \frac{\sum_{k=1}^{K} n_k}{\sum_{k=1}^{K} \sum_{j=1}^{J} \sum_{i=1}^{n_j} w_{ijk}} \tag{17.3}$$

where K is the number of countries in the combined group, k indexes the country, and n_k represents the sample size within country k. In essence, this formula would provide a House-type weight for the combined sample.

Note that this problem (of the standard error being associated with the sum of the weights and therefore the selection of weight is crucial) has become less of an issue in applied analyses. Popular programs for general statistical analysis, such as SAS and SPSS, now have capabilities to appropriately estimate point and variance estimates with data obtained from complex sampling procedures. For example, if one were to run the same analyses as shown in Table 17.1 but use the SPSS *Complex Samples* option with the *TOTWGT* variable as the sampling weight variable, an estimate of the mean for the U.S. sample of .80 would be obtained and the estimate of the standard error would be .005. This standard error estimate takes into account the coefficient of variation of the weights, and also the clustered nature of the sample design. The variability in the weights (the departure from an equal probability of selection approach) can lead to some loss in precision depending on the relation between probability of selection and distribution of residuals (Heeringa et al. 2010). As of version 20.0 for IBM SPSS, the SPSS analysis options includes linearized estimates of standard errors (see Chapter 6 for an explanation) and can accommodate estimates of means, frequencies, the general linear model, logistic regression, and Cox regression. Similarly, the SAS version 9 software has *SURVEY* procedures that accommodate the same estimators: means, frequencies, general linear model, logistic regression, and Cox proportional hazards models.

Note that this simple example of the differences in estimates based on different weight selections has ignored the issue of variance estimation to account for the three-stage sampling design that was utilized in the United States and in Turkey and therefore the standard errors shown in this example may not be appropriate. To address that issue, see Chapter 6 (Rust 2013).

More Advanced Estimates

Sample Dataset

For the next two examples, a hypothetical sample dataset was created to demonstrate the potential issues in more advanced analysis with weighted responses. Data were randomly generated such that two variables, hours spent on homework each week and ability score from a multi-item assessment, would have a specific relation to each other. A population of 100 schools was generated, randomly varying in size, with an average size of 500 students and a standard deviation of 100 students. Within each of these schools, individual observations of homework hours and ability score were generated in a number to match the school size. These individual observations were generated within two different strata: one stratum contained students of one type (such as English as a Second Language [ESL] status) and the other stratum contained students of another (native English speakers), resulting in a total of 50,547 observations across the 100 schools. The data were generated such that about one-third of the students in the population were ESL students and two-thirds were native English speakers. Additionally, the generation mechanism for the data was designed such that schools would differ in their average amounts of homework hours and ability score, conditional on size of school, and students in the two strata differed, on average, in their levels of ability score. In this hypothetical dataset, the population mean ability score for ESL students was 138.4 while the mean for native English speaking students was 156.8.

Using this population of data, a small dataset was selected to demonstrate differences in weighting approaches. Specifically, a sample was drawn of 10 schools, using probability proportionate to size sampling; the selected schools are shown in Table 17.2 with their sampling information.

The first columns were data from the sampling frame, including the identification number of the school and the measure of size or the approximate number of students of interest at the school. One can note that the selection probability is a function of the size of the school; those schools with greater enrollments, such as school ID 36, have the highest probability of selection into the sample. The school sampling weight is the inverse of the selection probability and thus school ID 36 has the smallest sampling weight (1/.1383 = 7.23). Note that the sum of the school sampling weights in the final column would be an estimate of the population number of schools.

Suppose that each of these schools had agreed to participate in the study. If they had not, the sampling weights of the participating schools would need to be adjusted upward to represent those schools that did not, using either a weighting class model or some other propensity model. Students were then sampled within each participating school using a sampling by age (as opposed to entire school or classroom) strategy. Suppose that there was also an interest in obtaining subpopulation estimates for students who

TABLE 17.2

Example Sample of Selected Schools

School ID	Measure of Size	Selection Probability (π_j)	School Sampling Weight (w_j)
14	489	.0967	10.34
20	578	.1144	8.75
25	492	.0973	10.27
35	423	.0837	11.95
36	699	.1383	7.23
38	537	.1062	9.41
56	446	.0882	11.33
79	518	.1025	9.76
90	575	.1138	8.79
95	549	.1086	9.21

were nonnative English speakers. If a random sample of students within each school were selected, then there would be no guarantee that a sufficient number of ESL students would be in the sample to provide subgroup estimates. Therefore, the student population in each school was stratified into two strata: those students who were native English speakers and ESL students. Within each of the participating schools, two students were randomly selected from each stratum. Example data for the 40 students in the final sample are shown in Table 17.3.

There are several things to note about the data in Table 17.3. In the first two columns are the IDs of the school and of the student. The third column contains the stratum information for the student where 1 = ESL and 2 = native English speaker. Homework hours and Ability score are the two measured variables of interest for each student. In the sixth column is the student's probability of selection within the selected school, referred to as the conditional selection probability. This probability is a function of the number of students sampled from within the stratum (2 in this example) and the number of students in the school in the specific stratum ($\pi_{i|j} = (2/n_{jk})$), where the k subscript denotes stratum (1 or 2 in this case). Taking the first four records as an example, School ID 14 enrolled 489 students of the population of interest, 486 of these were in stratum 1 and 3 of these were in stratum 2; School ID 14 is primarily serving nonnative English speaking students. Sampling two students out of each stratum gave a very small selection probability for students in stratum 1 (just .00412) and a very large selection probability for students in stratum 2 (.66667). Taking the inverse of these selection probabilities, the conditional sampling weight for each student is obtained as shown in the seventh column. Given this sampling process, both of the students who were sampled in stratum 1 in School ID 14 have conditional (within-school) sampling weights of ($w_{i|j} = (1/\pi_{i|j}) = (1/0.00412) = 243$. Each student in this stratum and school represents 243 students in school ID 14 in stratum 1. In order to generalize to all students in the population and not just those students from the particular sampled schools, the total sampling weight is created as shown in the eighth column. The total sampling weight is the product of the school-level weight and the conditional within-school weight. Students in the first stratum of School ID 14 have a total sampling weight of $w_{ij} = w_j \times w_{i|j} = 10.34 \times 243.0 = 2511.84$. Each of these students represent 2511.84 students in the population.

Note two important concepts regarding the information provided in Table 17.3. The sum of the weights in the seventh column, the conditional sampling weights, within each school is the population size within the school. For example, for School ID 14, 243 + 243 + 1.5 + 1.5 = 489. Also, it should be noted that the sum of the weights in the eighth column, the total sampling weights, is an estimate of the total population size and will, in fact, be the population size if the measure of size used for each school is the true enrollment and each school had sufficient numbers of students to meet the sampling requirement (e.g., in this case had at least two students in each stratum).

TABLE 17.3

Example of the Sample of Student-Level Data

School ID	Student ID	Student Stratum	Homework Hours	Ability Score	Conditional Selection Probability ($\pi_{i\|j}$)	Conditional Sampling Weight ($w_{i\|j}$)	Total Sampling Weight (w_{ij})
14	211	1	13	142	0.00412	243	2511.84
14	470	1	11	138	0.00412	243	2511.84
14	243	2	15	147	0.66667	1.5	15.51
14	397	2	16	149	0.66667	1.5	15.51
20	444	1	4	135	0.00557	179.5	1569.76
20	542	1	10	144	0.00557	179.5	1569.76
20	42	2	10	148	0.00913	109.5	957.59
20	88	2	11	147	0.00913	109.5	957.59
25	91	1	7	144	0.03125	32	328.76
25	370	1	7	145	0.03125	32	328.76
25	251	2	8	147	0.00467	214	2198.59
25	387	2	11	154	0.00467	214	2198.59
35	98	1	10	144	0.00913	109.5	1308.49
35	205	1	10	144	0.00913	109.5	1308.49
35	23	2	11	147	0.0098	102	1218.86
35	403	2	10	147	0.0098	102	1218.86
36	258	1	4	146	0.01307	76.5	553.2
36	544	1	3	143	0.01307	76.5	553.2
36	268	2	5	149	0.00366	273	1974.15
36	527	2	6	151	0.00366	273	1974.15
38	252	1	7	146	0.03125	32	301.21
38	368	1	7	146	0.03125	32	301.21
38	230	2	10	152	0.00423	236.5	2226.14
38	439	2	13	162	0.00423	236.5	2226.14
56	278	1	10	156	0.0625	16	181.33
56	297	1	11	156	0.0625	16	181.33
56	53	2	11	161	0.00483	207	2346.02
56	141	2	16	168	0.00483	207	2346.02
79	43	1	8	142	0.00595	168	1639.36
79	506	1	10	144	0.00595	168	1639.36
79	330	2	13	153	0.01099	91	887.99
79	367	2	10	149	0.01099	91	887.99
90	43	1	7	134	0.00383	261	2294.39
90	469	1	7	139	0.00383	261	2294.39
90	387	2	12	149	0.03774	26.5	232.96
90	444	2	11	148	0.03774	26.5	232.96
95	50	1	6	145	0.14286	7	64.45
95	115	1	5	142	0.14286	7	64.45
95	9	2	7	153	0.00374	267.5	2462.9
95	210	2	13	166	0.00374	267.5	2462.9

With these data, a regression analysis of Ability score on Homework hours per week was conducted where the following model was of interest:

$$Y_{ij} = \beta_0 + \beta_1 X_{ij} + r_{ij} \tag{17.4}$$

Table 17.4 contains the population values for the coefficients as well as the results of the analysis using three different approaches with SAS version 9.1 software. The first analysis was run assuming that the 40 observations represented a simple random sample from the population of interest. Given a simple random sample, each observation has an equivalent selection probability and thus the use of sampling weights would be moot. The following SAS syntax was used: PROC REG; MODEL Y = X, where X is defined as homework hours and Y is defined as ability score. The second analysis incorporated the total sampling weight into the analysis and thus should differentially weight observations from strata 1 and 2 so that the overall regression coefficient represents the disproportionality in the population. This second analysis, however, did not use appropriate sampling variance estimators: PROC REG; MODEL Y = X; WEIGHT TOTWGT. The final analysis included both the sampling weight as well as an estimator that accounted for the clustering of the student data within school. A Taylor series (TS) linearization was used to obtain sampling variances (standard errors): PROC SURVEYREG; CLUSTER J; MODEL Y = X; WEIGHT TOTWGT (SAS Institute Inc. 2004).

Given an unweighted analysis, it would have been concluded that students who study for 0 h a week, on average, have an assessment score of 136.80. Given the standard error estimate, this value is significantly different from zero. Of more interest, one would conclude that, in this sample, for each hour of homework per week assessment scores would be predicted to be 1.20 units higher with a standard error of 0.32. As shown in the first column, the data were generated such that, in the population, the intercept is 133.6 and the slope is 1.8. Note that a 95% confidence interval for the intercept based on this unweighted analysis would include the value of 133.6 and the 95% confidence interval for the slope would just barely contain the true population value of 1.8.

Without weighting the analysis, the fact that some of the schools had greater probabilities of being in the sample (the larger schools) and some of the

TABLE 17.4

SAS Results from a Simple Regression Using Unweighted and Weighted Data

	Population Values	Unweighted Estimates	Weighted Estimates	Weighted, TS Linearized Variance Estimates
β_0	133.6	136.80	131.66	131.66
$se(\beta_0)$		3.15	4.13	6.95
β_1	1.8	1.20	1.79	1.79
$se(\beta_1)$		0.32	0.41	0.74

students had greater probabilities of being in the sample (those who had a status of studying ESL) was being ignored. Therefore, the estimates are disproportionately representing ESL students, especially those from larger schools. If the total sampling weight is incorporated into the analysis, allowing some students to represent as few as 17 students and some to represent over 2500 students, then one can expect the estimates to change. The second column of weighted estimates shows that on average, for students who study 0 h of homework per week, the expected score is 131.66 and with each hour of homework, one would expect the score to increase by 1.79 units. These weighted estimates are closer to the true population values and the 95% confidence interval for each parameter estimate contains the population values in this example. The standard errors are still not appropriate in this analysis because the data were drawn using a two-stage sampling technique. If schools differ in their average amount of X and Y, then a dependency exists. This dependency, or clustering, typically leads to less precise estimates of model parameters (Kish 1965). For this example, the level of clustering homogeneity (measured by the intraclass correlation or ICC) is .144 for the hours of homework variable and .321 for the ability score variable. The final analysis uses an appropriate procedure to account for the imprecision introduced by two-stage nature of the sampling design. Note that the weighted point estimates are the same in the last two analyses and it is only the estimates of the standard errors that are adjusted. As discussed in Chapter 6 (Rust 2013), these appropriate standard errors can be obtained with a variety of software including SAS (as done here), the Complex Samples function of SPSS, Stata, M*plus*, and WesVar, among others.

Turning now to a more complex situation, suppose it was of interest to run a multilevel analysis of the relation between homework hours and ability scores within schools and between schools (Raudenbush and Bryk 2002). Specifically, suppose the following two-level model was hypothesized for these data:

$$
\begin{aligned}
Y_{ij} &= \beta_{0j} + \beta_{1j}(X_{ij} - \bar{X}_j) + r_{ij} \\
\beta_{0j} &= \gamma_{00} + \gamma_{01}\bar{X}_j + u_{0j} \\
\beta_{1j} &= \gamma_{10}
\end{aligned}
\qquad (17.5)
$$

where X_{ij} represents the homework hours score for individual i in school j, $\bar{X}_j$ represents the school average amount of homework (based on an aggregation of individual student values), and Y_{ij} represents the ability score of student i in school j. Note that this model posits that there is a random effect of school for ability scores (u_{0j}) but the effect of hours of homework is not modeled to differ across school. Also note that at level-1, the hours of homework variable is being modeled as cluster-mean centered. This model was run using M*plus* version 6.1 on the example data in Table 17.3.

As can be seen in Table 17.5, the weighted and unweighted point estimates differ slightly with the weighted estimates being closer to the population

TABLE 17.5

M*plus* Results from a Multilevel Regression Using Unweighted and Weighted Data

	Population Values	Unweighted	Weighted at Each Level
γ_{00}	132.67	145.825	146.845
$se(\gamma_{00})$		5.441	6.858
γ_{01}	1.884	0.237	0.181
$se(\gamma_{01})$		0.685	0.826
γ_{10}	1.955	2.378	2.116
$se(\gamma_{10})$		0.167	0.173

values in one of the three cases. Importantly, it should be noted that the standard errors for the weighted estimates are (appropriately) larger than those provided in an unweighted analysis. If one were to compute confidence intervals around each estimate using a formula appropriate given the degrees of freedom of each parameter estimate ($df = 8$ for level-2 estimates and $df = 29$ for the level-1 estimate), the unweighted confidence interval for the two slope parameter estimates would not contain the population value. Alternatively, all three 95% confidence intervals for the weighted estimates would contain the population values.

Note that there are different estimation methods for the inclusion of sampling weights into a multilevel analysis. One method, probability-weighted iterative generalized least squares (PWIGLS), was developed by Pfeffermann et al. (1998) and currently is implemented in the LISREL, HLM, and MLwiN software packages. Alternately, a multilevel pseudo maximum likelihood method (MPML; Asparouhov and Muthén 2006; Rabe-Hesketh and Skrondal 2006) is used within the M*plus* software and the Stata *gllamm* procedure. For single-level latent variable models, a pseudo maximum likelihood estimation is also available in M*plus* and Stata. Both PWIGLS and MPML estimation methods include a sandwich estimator (Huber 1967) for sampling variance estimation whereby the sampling variances are adjusted given the sampling design (see Asparouhov and Muthén [2006] or Rabe-Hesketh and Skrondal [2006] for details on how it is accomplished within the two-level model). For the multilevel example shown here, results from HLM and M*plus* differed very minimally.

In multilevel analyses, whether the PWIGLS or the MPML estimation is used, two weights are required for the estimation, the conditional sampling weights within clusters at level 1 and the inverse of the selection probability of the cluster as the sampling weight at level 2. Unlike in single-level models, in multilevel models, the scaling of the weight at level 1 can affect point estimates in the model (Pfeffermann et al. 1998). The distribution of the level-2 random effect estimate is affected by the ratio of the actual cluster sample size and the sum of the weights within the cluster (Asparouhov 2006) especially for dichotomous outcomes (Rabe-Hesketh and Skrondal 2006). Within some software programs, the $w_{i|j}$ weights are typically scaled so that they sum to the actual cluster size, n_j (called a relative sample size weight or normalized weight), although a weight scaling to the effective sample size is also possible. In this alternate weight scaling, the weights are adjusted to reflect the effective sample size (typically lower than the actual sample size to reflect the loss in efficiency due to nonoptimal sampling proportions) given the sampling design. Specifically, the effective sample size weights are determined as follows, assuming a two-stage sampling design:

$$J^* = \frac{\left(\sum w_j\right)^2}{\sum w_j^2} \tag{17.6}$$

$$n_j^* = \frac{\left(\sum w_{i|j}\right)^2}{\sum w_{i|j}^2} \tag{17.7}$$

where J^* in Equation 17.6 is the effective sample size at level 2 and n_j^* in Equation 17.7 is the effective sample size at level 1. At each level, as weights depart from equality, the ratio lessens. If all weights were equal, then the ratio would be the same as the sample size at that level. For the example data in Table 17.3, the school effective sample sizes ranged from 2.02 (School ID 14) to 3.99 (School ID 35), although all actual sample sizes were 4. At the school level, given the weights shown in Table 17.2, the effective sample size is 9.81, very close to the actual sample size of schools.

The scaling of the weight used at level 2 of the model is not a concern as multiplying all weight values by a constant will not affect the point estimates or sampling variance estimates (assuming the software is using approaches appropriate for complex sample data). But, as stated above, the scaling of the level-1 weight can be important. One option is to rescale the conditional weight to sum to the sample size within each level-2 unit (the relative sample size weight):

$$w_{i|j}^r = w_{i|j} \frac{n_j}{\sum w_{i|j}} \tag{17.8}$$

This weight is referred to as Method A by Asparouhov (2006) and as Method 2 by Pfeffermann et al. (1998). The other option is to use the effective sample size information in the numerator instead of the actual sample size referred to alternatively as Method B and as Method 1 in the literature:

$$w_{i|j}^e = w_{i|j} \frac{n_j^*}{\sum w_{i|j}} \tag{17.9}$$

In research comparing use of the effective sample size weight scaling to use of relative weight scaling (Asparouhov 2006; Grilli and Pratesi 2004; Pfeffermann et al. 1998; Rabe-Hesketh and Skrondal 2006; Stapleton 2002), no clear preference was found between these two approaches. When cluster sizes were large, there were few differences between the methods, but when cluster sizes were small, differences emerged depending on various factors, including the informativeness of the weights and the measurement scale of the outcome. The M*plus*, MLwiN, and HLM softwares take an approach whereby the level-1 conditional weights are rescaled to sum to the cluster size (Method A or Method 2); this default can be overridden in the both the M*plus* and MLwiN software packages to provide any desired scaling. The Stata *gllamm* procedure uses the weights as scaled by the user and thus the researcher must prepare the data to reflect any desired rescaling of the conditional weights.

A practical difficulty for the applied researcher is that, on some public-release data files, only the overall, unconditional sampling weight of the ultimate sampling unit (such as students) is provided. Sometimes, no information is given regarding the sampling weight of the cluster. Some exceptions are the International Association for the Evaluation of Education Achievement (IEA) Civic Education Study and the Educational Longitudinal Study of 2002. Asparouhov (2006) suggests that if weights are not available at each level, then the researcher should not run a multilevel analysis and conduct a single-level analysis only. This might be conducted for the example above by running the following model:

$$Y_{ij} = \gamma_{00} + \gamma_{01}(\bar{X}_j) + \gamma_{10}(X_{ij} - \bar{X}_j) + e_{ij} \tag{17.10}$$

Note that the only difference between Equations 17.5 and 17.10 are that the random effects at the two levels are now combined into one residual term (e_{ij}) and therefore a parsing of variation attributable to each level cannot be undertaken. Because the degrees of freedom associated with each of the model parameters is not the same (e.g., the level-2 coefficient on the mean homework time is based on data for just 10 schools while the coefficient for the cluster-centered homework time is based on 40 individuals within 10 schools), it is imperative that an appropriate variance estimation technique be used to obtain estimates for this model as already discussed (such as with SPSS Complex Samples, SAS SURVEY procedures, etc.).

However, this solution does not allow a researcher to examine hypotheses that may be of interest, specifically, parsing variance between and within schools and allowing slopes to be random across schools. In the two-stage sampling situation, Goldstein (2003) and Kovacevic and Rai (2003) each proposed approximations to the conditional sampling weights, $w_{i|j}$, and cluster level sampling weights, w_j, if only the unconditional weights, w_{ij}, are available. The approximations provided by Kovacevic and Rai assume that the number of first-stage cluster units in the population is known and they have developed approximations for use with both simple random sampling and with probability proportionate to size (PPS) sampling. It is not atypical for international large-scale assessment sampling documentation to include an approximate number of schools that were in the sampling frame, so the use of Kovacevic and Rai's approximations by the applied researcher may be possible. Goldstein's proposed approximations assume no known information about the number of population units and provide approximations only for the simple random sampling case. Specifically, in the case with two levels of sampling and simple random sampling (SRS) used at each stage in the sample, Kovacevic and Rai propose that the school level weight, w_j, is simply estimated as

$$\hat{w}_j = \frac{G}{J} \tag{17.11}$$

where G is the number of known groups (schools) in the population and J is the number of selected schools for the sample. Goldstein (2003) suggests the following equation can be used in the SRS case when G is not known:

$$\hat{w}_j = \frac{J\left(\sum_{i=1}^{n_j} w_{ij}/n_j\right)}{\sum_{j=1}^{J}\left(\sum_{i=1}^{n_j} w_{ij}/n_j\right)} \tag{17.12}$$

where w_{ij} is the unconditional, ultimate, sampling weight for the element and n_j is the number of observations in the sample in cluster j. The numerator in Equation 17.12 contains the average sampling weight for observations in the school of interest multiplied by the number of schools in the sample and the denominator contains the sum (across all of the schools) of the average observation weight in a school. If this ratio exceeds 1.0, then the observations in the school of interest represent more of the population than other schools on average. This disproportionate representation is reflected in a higher approximated school-level weight as compared to other schools. The average school-level weights resulting from applying Equation 17.12 will be 1.0. With the more typically used PPS sampling, Kovacevic and Rai propose the following approximation for the cluster-level weight:

$$\hat{w}_j = \frac{G}{J} \times \frac{1}{\sum_{i=1}^{n_j} w_{ij}} \times \frac{1}{(1/J)\sum_{j=1}^{J}(1/\Sigma w_{ij})} \tag{17.13}$$

but note, again, that the number of schools in the population (G) is assumed known. In Equation 17.13, the first component is the base cluster sampling weight (and equivalent to Equation 17.11); this weight is adjusted by the product of the second and third components. The second component is a function of the relative weighted size of the school of interest, while the third component is a function of the average relative size of all of the schools. If this product of the second and third components of Equation 17.13 is 1.0, then the school of interest is assumed to be the same size, on average, as the others. If the product is less than 1.0, then the school is assumed to be larger than the others on average and therefore the school-level weight is adjusted downward. The Goldstein approximation is focused on the relative size of the *average* weight within a school to the *average* across all schools, while the Kovacevic and Rai approximation compares the *sum* of the weights within a school to the average *sum* across all schools, thus ignoring differences in school sizes.

For any of these approximations, once w_j is estimated, the conditional, within-cluster weight is estimated as

$$\hat{w}_{i|j} = \frac{w_{ij}}{w_j} \tag{17.14}$$

When stratification is used at the first stage of sampling, this weight scaling process could be undertaken within each explicit stratum. In the case of a three-level model, the equations could be undertaken within each secondary sampling unit then, in a second step, one can transform the approximated w_{jk} weights into conditional weights at the second-stage cluster level, $w_{j|k}$, and estimate the PSU-level weights, w_k. In a true PPS sampling design with fixed sample sizes selected at each PSU, the unconditional sampling weights should be equivalent for each observation. If this is the case for a given sample, then no approximation would be able to disentangle the appropriate conditional and cluster-level selection probabilities and therefore it is questionable that these approximations will work appropriately for many international large-scale assessment datasets. Stapleton (2012) evaluated these two approximations and concluded that neither were appropriate for use under typical conditions of public release datasets. The Goldstein approximations were appropriate for use when the selection at the first stage used a process of simple random sampling and an equal rate of sampling was used in all strata. However, the Goldstein approximations did not result in proper estimates when second-stage sampling used fixed sample sizes as opposed to fixed sampling rates. The Goldstein approximation was also accurate when probability proportionate to size sampling was used at the first stage of selection and a fixed sampling rate was used within clusters. Unfortunately, this sampling technique is not typically employed in large-scale data collection. Furthermore, when disproportionate sampling rates are used across strata at the first stage of sampling, the Goldstein approximation cannot be used. Like the Goldstein approximations, the Kovacevic and Rai approximations did not properly estimate sampling weights under probability proportionate to size sampling with fixed sample sizes within clusters. However, when disproportionate rates are used across strata at the first stage of sampling, the Kovacevic and Rai approximations are able to capture the differences in relative rates of selection (assuming that G is known).

If access to the conditional weights is impossible, it is suggested that one use the Kovacevic and Rai approximations when conducting two-level analyses with the typical PPS sampling designs, assuming that the population number of clusters is known or can be estimated. Also, users are encouraged to examine whether measures of size might be estimated from available data on the file. Equation 17.12 should yield appropriate cluster-level weights if reliable measures of size can be estimated and used to replace the sum of the weights within a cluster in the equation and approximation of conditional weights would proceed through the use of Equation 17.14.

Software for Multilevel Analysis

Current software that can handle multilevel analyses with weights include HLM 7, MLwiN 2.22, M*plus* 6.1, and some Stata procedures. Unfortunately, the MIXED procedures in SAS and SPSS do not accommodate sampling weights appropriately and there are no current R macros that address weights in multilevel models. For a full description of the capabilities of these programs, see West and Galecki (2011). For multilevel structural equation modeling, both M*plus* and Stata can be used; LISREL does not currently have the capability of including probability weights at each level of the analysis. Note that statistical software versions are quickly updated and thus the reader is encouraged to investigate the version of their preferred software to evaluate whether sampling weights are now accommodated.

Unweighted Approaches to Analyses

Given the examples presented in the section "Example Analyses," the reader may be under the impression that sampling weights should always be included in an analysis when available. This is not the case. If the selection mechanism is noninformative (e.g., if men and women in our first example are of equal height on average in the population), then the inclusion of weights in an analysis might actually pose a disadvantage. The standard errors from a weighted analysis will likely be larger than the standard errors from an unweighted analysis, leading to a loss of precision and a lowering of power of the statistical inference.

In addition, if the selection mechanism is informative, it is possible to take a model-based approach in which the analytic model contains the information that defined the sampling mechanism (Heeringa et al. 2010; Kalton 1983). For example, with the first very simple scenario presented in this chapter in the Introduction, instead of obtaining one mean overall height for the population across gender, the mean might have been calculated for women and men separately. Given that the selection probability differed only by gender, if the analysis is undertaken within each gender, there is no need for the use of sampling weights. This simple model-based example can be extended for each of the examples provided in the chapter. Note that as the sampling mechanism becomes more complex, so too would the model required for a fully model-based estimation. In the last example wherein the relation of homework hours to ability scores was examined at both the school and student level, a fully model-based solution would require several inclusions into the model: main effects of ESL-status at level 1 and the size of the school at level 2, the interaction of ESL-status with homework hours at level 1, the interaction of proportion of ESL-status with mean homework hours at level 2, the interaction of school size with mean homework hours at level 2 and three-way interactions. With only a simple sampling scenario, the model-based solution quickly became very complex.

In general, when undertaking applied analyses, researchers should consider running the analysis with and without weights as well as with different weight scalings to understand the possible implications of the inclusion of specific weights in the analysis. Given findings from multilevel simulation research (Asparouhov and Muthén 2006; Grilli and Pratesi 2004; Rabe-Hesketh and Skrondal 2006), it is clear that, to obtain unbiased estimates of coefficients under the simple conditions studied, cluster-level and conditional weights should be used in two-level analyses when the weights are informative. Unfortunately, applied researchers do not know *a priori* whether the weights are informative. By running the analysis repeatedly with different treatment of weights, some information about the role of the sampling process in the resultant estimates will be provided. Carle (2009) presents an example of the analyses one can undertake to evaluate these differences.

Snijders and Bosker (2012) suggest several methods in the investigation of whether weights should be included in multilevel models. Their suggested approaches are briefly summarized here and the interested reader is encouraged to seek out the detail in Snijders and Bosker. It should be noted, however, that empirical methods to determine whether weights should be incorporated in the analysis is dependent on sampling error—the particular sample may show no difference in inference given weighted and unweighted analysis but another sample (obtained by the same sampling method) might have shown a difference.

1. Examine the variability in the weights for the level-1 and level-2 weights separately. This examination can be conducted by calculating the effective sample sizes as shown in Equations 17.6 and 17.7. Also, the coefficient of variation (*cv*) of the weights may be calculated as the ratio of the standard deviation of the weights over the mean weight at each level. If the *cv* is close to 0 or if the effective sample

sizes do not differ greatly from the actual sample sizes, then it is unlikely that the weights are informative.

2. Add design variables to the model of interest. Adding available design variables represents a model-based approach to obtaining unbiased estimates. Both main effects and interactions should be considered. If the design variables are related to the outcome, over and above other model specification, or if the parameter estimates change substantially once design variables are added, weights are likely to be informative.
3. Apply the model to subsets of the dataset defined by design factors. Assess the degree of differences in parameter estimates across these analyses. Snijders and Bosker suggest only subsetting the sample into two or three groups at each level. Discrepancies in the model estimates would inform the researcher of the nature of the informativeness, whether at level 1 of level 2 and of possible design features that might be included as predictors as in the second approach above.
4. Include the level-1 and level-2 weights as predictors in the model as both main effects and interactions with other predictors. A multivariate test of the set of added coefficients would inform the analyst whether the weighting was informative.
5. For each cluster, compare the model-based estimates to the weighted estimates. This approach should only be done with fairly large sample sizes. A multivariate test can be done to determine whether discrepancies across all clusters are random or systematic.

Summary

International large-scale assessments include a vast amount of information that would be useful for content researchers to access in developing models to explain development of abilities and skills. While correlational in nature, the collection of assessment data (and auxiliary background information about students and the contexts in which they learn and are educated) presents an opportunity to evaluate preliminary hypotheses about proposed relations and develop new hypotheses. However, appropriate methods are required to analyze the data to obtain unbiased estimates of parameters and their sampling variances. The complex sampling designs used in the collection of international large-scale assessment data often use procedures that result in sample observations that should not be considered of equal "weight" given their disproportionate selection probabilities. In this chapter, some discussion was provided about these sampling weights and how they are accommodated in the analysis.

For basic parameter estimates, such as estimates of means and proportions, the inclusion of sampling weights is straightforward and well established in the statistical literature. Although model-based approaches can be used to obtain unbiased estimates without the inclusion of sampling weights, all the design information must be available to the analyst and appropriately modeled; such conditions are unlikely and, therefore, use of weighted estimators is the simplest approach. More advanced models, however, present challenges in estimation and methodological work in this area is currently ongoing.

References

Anderson, C., Kim, J.-S., and Keller, B. 2013. Multilevel modeling of discrete response variables. In L. Rutkowski, M. von Davier, D. Rutkowski (Eds.), *A Handbook of International Large-Scale Assessment: Background, Technical Issues, and Methods of Data Analysis*. London: Chapman Hall/CRC Press.

Asparouhov, T. 2006. General multi-level modeling with sampling weights. *Communications in Statistics: Theory and Methods*, 35, 439–460.

Asparouhov, T. and Muthén, B. 2006. Multilevel modeling of complex survey data. *Proceedings of the Joint Statistical Meeting in Seattle, August 2006. ASA Section on Survey Research Methods*, 2718–2726.

Asparouhov, T. and Muthén, B. 2007. Testing for informative weights and weights trimming in multivariate modeling with survey data. *Proceedings of the 2007 JSM Meeting in Salt Lake City, Utah, Section on Survey Research Methods*, 3394–3399.

Carle, A. C. 2009. Fitting multilevel models in complex survey data with design weights: Recommendations. *BMC Medical Research Methodology*, 9, 49.

Goldstein, H. 2003. *Multilevel Statistical Models*, 3rd edition. London: Arnold.

Grilli, L. and Pratesi, M. 2004. Weighted estimation in multilevel ordinal and binary models in the presence of informative sampling designs. *Survey Methodology*, 30, 93–103.

Heeringa, S. G., West, B. T., and Berglund, P. A. 2010. *Applied Survey Data Analysis*. Boca Raton, FL: CRC Press.

Huber, P. J. 1967. The behavior of maximum likelihood estimation under nonstandard conditions. *Proceedings of the Fifth Berkeley Symposium on Mathematical Statistics and Probability*. pp. 221–233. Berkeley, CA: University of California Press.

Kalton, G. 1983. Models in the practice of survey sampling. *International Statistical Review*, 51, 175–188.

Kish, L. 1965. *Survey Sampling*. New York, NY: John Wiley & Sons, Inc.

Korn, E. L. and Graubard, B. I. 1999. *Analysis of Health Surveys*. New York, NY: Wiley.

Kovacevic, M. S. and Rai, S. N. 2003. A pseudo maximum likelihood approach to multilevel modeling of survey data. *Communications in Statistics*, 32, 103–121.

Little, R. J. A. and Rubin, D. B. 2002. *Statistical Analysis with Missing Data, Second Edition*. Hoboken, NJ: Wiley-Interscience.

Lohr, S. 2010. *Sampling: Design and Analysis*, 2nd edition. Pacific Grove, CA: Duxbury Press.

Martin, M. O., Gregory, K. D., and Stemler, S. E. 2000. *TIMSS 1999 Technical Report*. Chestnut Hill, MA: International Study Center Boston College.
Pfeffermann, D. 1993. The role of sampling weights when modeling survey data. *International Statistical Review*, 61, 317–337.
Pfeffermann, D., Skinner, C. J., Holmes, D. J., Goldstein, H., and Rasbash, J. 1998. Weighting for unequal selection probabilities in multilevel models. *Journal of the Royal Statistical Society, Series B*, 60, 23–40.
Potthoff, R. F., Woodbury, M. A., and Manton, K. G. 1992. "Equivalent sample size" and "equivalent degrees of freedom" refinements for inference using survey weights under superpopulation models. *Journal of the American Statistical Association*, 87, 383–396.
Rabe-Hesketh, S. and Skrondal, A. 2006. Multilevel modeling of complex survey data. *Journal of the Royal Statistical Society, Series B*, 60, 23–56.
Raudenbush, S. W. and Bryk, A. S. 2002. *Hierarchical Linear Models: Applications and Data Analysis Methods, Second Edition*. Newbury Park, CA: Sage.
Rust, K. 2013. Sampling, weighting, and variance estimation in international large-scale assessments. In L. Rutkowski, M. von Davier, D. Rutkowski (Eds.), *A Handbook of International Large-Scale Assessment: Background, Technical Issues, and Methods of Data Analysis*. London: Chapman Hall/CRC Press.
Rutkowski, L. 2013. Structural models for ILSA data. In L. Rutkowski, M. von Davier, D. Rutkowski (Eds.), *A Handbook of International Large-Scale Assessment: Background, Technical Issues, and Methods of Data Analysis*. London: Chapman Hall/CRC Press.
SAS Institute Inc. 2004. *SAS/STAT 9.1 User's Guide*, Volumes 1–7. Cary, NC: SAS Institute, Inc.
Shin, S.-H. 2013. Efficient handling of predictors and outcomes having missing data. In L. Rutkowski, M. von Davier, D. Rutkowski (Eds.), *A Handbook of International Large-Scale Assessment: Background, Technical Issues, and Methods of Data Analysis*. London: Chapman Hall/CRC Press.
Snijders, T. A. B. and Bosker, R. J. 2012. *Multilevel Analysis: An Introduction to Basic and Advanced Multilevel Modeling*, 2nd edition. London: Sage Publications Ltd.
Stapleton, L. M. 2002. The incorporation of sample weights into multilevel structural equation models. *Structural Equation Modeling*, 9, 475–502.
Stapleton, L. M. 2012. Evaluation of conditional weight approximations for two-level models. *Communications in Statistics: Simulation and Computation*, 41, 182–204.
West, B. and Galecki, A. 2011. An overview of current software procedures for fitting linear mixed models. *American Statistician*, 65, 274–282.
Wolter, K. M. 2007. *Introduction to Variance Estimation*. 2nd edition. New York: Springer.

18

Multilevel Analysis of Assessment Data

Jee-Seon Kim
University of Wisconsin-Madison

Carolyn J. Anderson
University of Illinois at Urbana-Champaign

Bryan Keller
Teachers College, Columbia University

CONTENTS

Introduction

Large-scale assessment data often exhibit a multilevel structure as a result of either sampling procedures, such as stratified sampling, or contextual factors, such as school settings where students are nested within schools, or cross-cultural settings where individuals are nested within countries. Observations within a cluster are likely to be correlated with one another and their dependency should be accounted for in the data analysis to permit valid statistical inferences. Moreover, relationships among variables may vary within clusters allowing for more detailed and informative study of contextual effects and their correlates.

The notion of clustered data and related issues of dependence are not new and have been studied for a long time. For example, the concept of a *design effect* was already used in 1965 by Kish as a required adjustment to account for correlations among observations within clusters in randomized trials. However, major advances in multilevel analysis have been made during the past couple of decades, and more recently, the methodology has become increasingly popular across disciplines owing to increased computing power and the subsequent emergence of user-friendly software. There now exist several software packages developed specifically for multilevel models, including HLM, MLwiN, and SuperMix (intermixture of MIXOR, MIXREG, MIXNO, and MIXPREG), as well as modules available for multilevel analysis in several general-purpose software packages, such as lme4 and nlme in R, MIXED, NLMIXED, and GLIMMIX in SAS, and xtreg and GLLAMM in STATA. Links to multilevel modeling software are available at the *Centre for Multilevel Modelling* website (www.bristol.ac.uk/cmm/learning/mmsoftware/). Other general statistical programs that enable some form of multilevel modeling include WinBUGS, Latent Gold, and Mplus. Although software for multilevel analysis may not be difficult to use for most researchers and practitioners with basic computer skills and fundamental knowledge of regression, the implementation of the methodology with real data is not always straightforward due to the intrinsic complexity of multilevel models and the complicated structures often present in multilevel data. Many national or international datasets are also observational or quasi-experimental, have sampling weights for primary and secondary sampling units, and/or may include missing data across levels.

This chapter demonstrates a multilevel analysis of large-scale assessment data using reading achievement scores and related variables from the 2006 Progress in International Reading Literacy Study (PIRLS) within the United States (Martin et al., 2007). Often a goal of research using such datasets is to study factors that potentially impact reading literacy and related issues. Initially, we examine descriptive statistics at each level. We then consider a multilevel model specification for the data, along with associated assumptions, diagnostics, and interpretation. We also address the presence of sampling weights across levels and the need to impute missing data at different levels.

The next section explains the PIRLS 2006 data used for this analysis and describes the variables of interest. Such examinations reflect *exploratory data analysis* (EDA) and can be conducted in various ways, for example, by inspecting distributions and/or patterns across variables and visualizing relationships among them. This step is critical to understand the data and also to check whether the planned modeling technique appears appropriate. Subsequent sections discuss steps in applying and evaluating multilevel models with large-scale data. The chapter concludes by presenting some advanced topics and remaining methodological issues in multilevel modeling, suggesting further readings, and cross-referencing other chapters in this volume for specific topics such as nonlinear multilevel models for discrete data, imputing multilevel missing data, and making causal inferences based on large-scale assessment data.

Data

PIRLS

PIRLS is an international comparative study of the reading literacy of young students. It focuses on the reading achievement and reading behaviors and attitudes of fourth-grade students in the United States and students in the equivalent of fourth grade across other participating countries. PIRLS was first administered in 2001 and included 35 countries, but expanded to 40 countries in 2006. The assessment includes a written test of reading comprehension and a series of questionnaires focusing on the factors associated with the development of reading literacy. PIRLS is coordinated by the International Association for the Evaluation of Educational Achievement.

In PIRLS, two types of reading are assumed that account for most of the reading young students do: reading for literary experience and reading to acquire and use information. In the assessment, narrative fiction is used to assess students' abilities to read for literary experience, while a variety of informational texts are used to assess students' abilities to acquire and use information while reading. The PIRLS assessment contains an equal proportion of texts related to each purpose.

The multilevel analysis in this chapter is illustrated using PIRLS 2006 data only from the United States. The data are from a two-stage cluster sample with sampling probability weights. In 2006, a nationally representative sample of fourth-grade students was selected resulting in 5190 students from 183 schools. Schools were randomly selected first, and then one or two classrooms were randomly selected within each school. Although some schools have students from two classrooms, most schools have only one classroom in the study, and we therefore treated the data as consisting of two levels: students within schools. The number of students per school ranges from 8 to 68 with an average equal to 28. Data from both the fourth-grade student questionnaire and the school questionnaire are used.

Plausible Value Outcomes

In PIRLS 2006, students were asked to engage in a full repertoire of reading skills and strategies, including retrieving and focusing on specific ideas, making simple to more complex inferences, and examining and evaluating text features. To provide good coverage of each skill domain, the test items required a total of 5 h and 20 min of testing time. However, testing time was kept to 1 h and 20 min for each student by clustering items in blocks and randomly rotating the blocks of items through the nine student test booklets. As a consequence, no student received all items but each item was answered by a randomly equivalent sample of students.

The matrix-sampling design used in PIRLS 2006 solicits relatively few responses from each sampled student while maintaining a wide range of content representation when responses are aggregated across all students. In such designs, the advantage of estimating population characteristics efficiently comes at the cost of an inability to make precise statements about individuals. More significantly, the amount of uncertainty may be different for different individuals. The uncertainty associated with individual proficiency estimates becomes too large to be ignored and a simple aggregation of individual student scores can lead to seriously biased estimates of population characteristics (Wingersky et al. 1987).

To address this issue, a *plausible values methodology* by Mislevy (1991) was implemented in the current analysis to represent the measured reading proficiency of each student. In short, the method uses all available data to estimate the characteristics of student populations and subpopulations and uses multiply imputed scores, called *plausible values*, to account for the uncertainty. The PIRLS 2006 data consists of five plausible values representing an overall reading score for each student. We refer to them as `readingPV1` to `readingPV5` in this chapter. All five plausible values were used as outcomes in our analysis. We explain in the section "Multiple Imputation" how results from the five analyses are combined. The distributions of the five plausible values are shown in Figure 18.1. The five empirical distributions suggest that the normality assumption appears reasonable. The figure also shows that, although plausible values fluctuate within persons (e.g., five values for a subject were 525.22, 603.41, 591.00, 621.33, and 583.37), the means and standard deviations of the five distributions across persons are very close to each other, as well as their general distributional form, as one would expect given their generation from the same distributions.

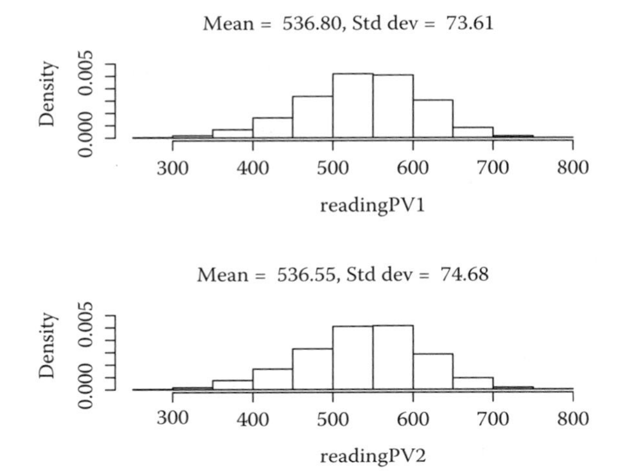

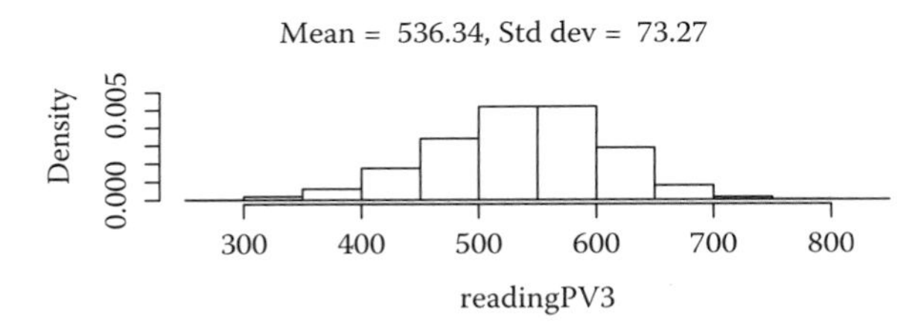

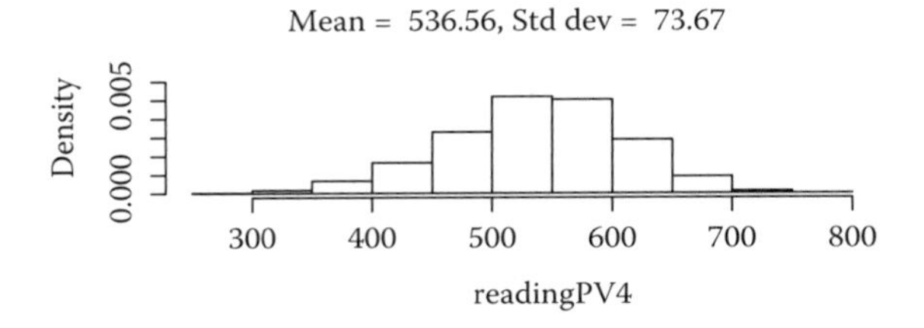

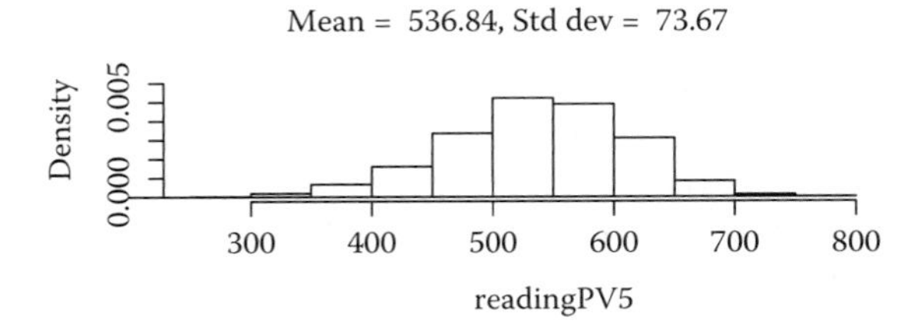

FIGURE 18.1
Histograms of plausible values 1 through 5 with sample means and standard deviations for the five reading plausible value scores.

Predictors

Background questionnaires were administered to collect information about student home and school experiences in learning to read. A student questionnaire addressed student attitudes toward reading and their reading habits. In addition, questionnaires were given to students' teachers and school principals to gather information about school experiences in developing reading literacy. In countries other than the United States, a parent questionnaire was also administered.

Along with the outcome variables, a number of explanatory variables were selected from both the student and school levels, as they are hypothesized to affect or be related to reading performance. Basic summary statistics are shown in Table 18.1. In the table, the means and standard deviations for subgroups (e.g., female vs. male) were calculated for `readingPV1` to `readingPV5` separately and then averaged. The differences of the statistics across the five plausible values are not reported here as the outcome distributions are very similar across plausible values, as shown in Figure 18.1, and the group means and standard deviations were also stable across the five outcome variables. For example, the means of `readingPV1` to `readingPV5` for females ($n = 2582$) were 541.42, 541.99, 541.54, 541.29, and 542.45. If the descriptive statistics were substantially different across the plausible values, we would have reported the variability across the plausible values.

Exploratory Data Analysis

EDA is an important part of data analysis. In this section, we describe and illustrate various graphical methods that are especially useful for multilevel analysis. The particular figures drawn depend on the nature of the predictor variables.

We start by examining the relationship between reading scores and possible microlevel predictor variables. If predictors are numerical (continuous), a plot of student data where each cell of the panel contains a plot of the response variable by a predictor variable can be very informative. Linear (alternatively quadratic, cubic, or spline) regression curves fit to each school's data can be overlaid in each cell of the panel.* Features to look for in such plots are the nature of the relationship between the predictor and response (e.g., Does a linear relationship seem reasonable? Is the relationship positive, negative, or nonexistent? Is the relationship basically the same for all schools?); the variability of the data for each school; and anything anomalous (e.g., Is a predictor that was thought to be numerical actually discrete? Is there a school that is very different from all the others? Are there outliers in the response or predictor variable?). Information about similarities and differences between schools can often be better seen by overlaying the school regression lines in a single figure.

* If there are a large number of clusters, a random sample of groups can be used.

TABLE 18.1

Information about Explanatory Variables

Variable	Levels	Percentage (%)	Mean	Std Dev
Level-1 Variables (N = 5190 Students)				
female	Female	49.78	541.74	70.59
	Male		531.51	76.46
enjoyreading	Disagree a lot	11.50	504.42	67.86
	Disagree a little	9.72	516.51	68.79
	Agree a little	24.95	536.77	68.53
	Agree a lot	53.83	549.30	74.26
bornUS	Student born in US	92.12	540.71	72.29
	Student not born in US		507.24	75.04
mombornUS	Mother born in US	67.57	546.42	71.65
	Mother not born in US	21.55	523.46	72.92
	Don't know	10.88	515.62	72.54
dadbornUS	Father born in US	62.67	547.84	71.41
	Father not born in US	21.82	524.17	72.59
	Don't know	15.51	518.24	73.14
home25books	25 or more books at home	64.50	552.01	72.09
	Less than 25 books at home		517.22	67.46
watchTV5H	Watch TV 5+ h on normal school day	28.37	513.14	70.16
	Watch TV <5 h on normal school day		549.51	71.67
comgames5H	Play computer/video games 5+ h/day	30.29	513.12	69.39
	Play computer/video games <5 h/day		550.79	71.49
Level-2 Variables (M = 183 Schools)				
econDisadv	>50% economically disadvantaged	41.67	511.99	71.88
	26–50% economically disadvantaged	23.33	542.29	72.07
	11–25% economically disadvantaged	10.00	557.84	68.39
	<10% economically disadvantaged	25.00	567.05	65.61
schoolLIB	School has a library	98.29	537.89	73.60
	School does not have library		504.45	56.65

Note: Values are based on nonmissing data. Mean and standard deviation are based on the averages across the reading plausible values.

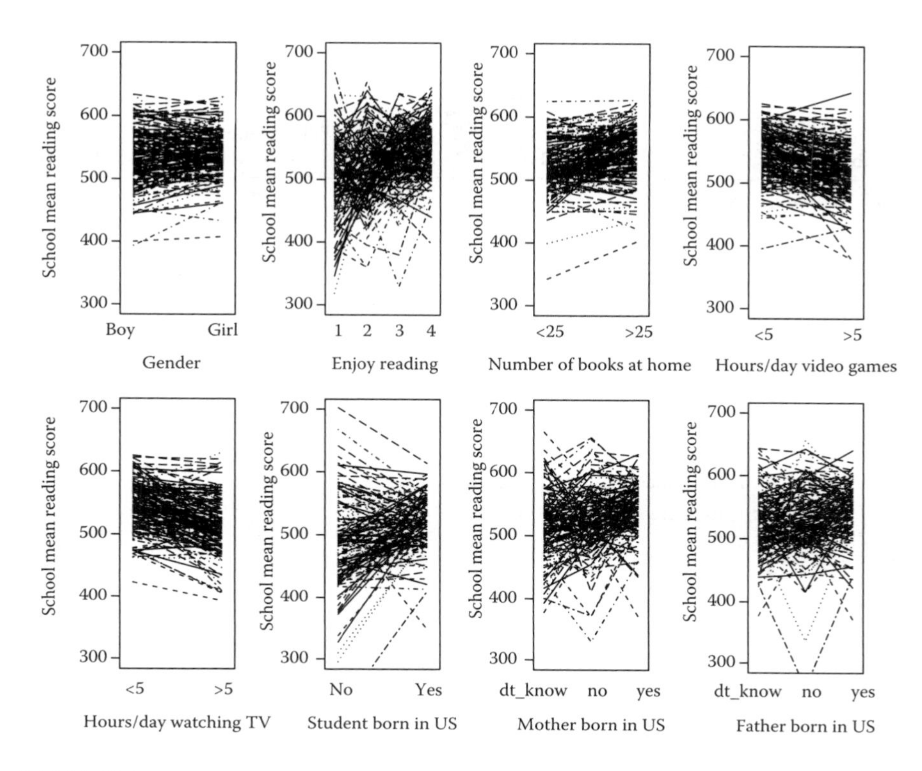

FIGURE 18.2
Panel plot of school mean reading scores plotted by each predictor variable where lines are connecting school means.

In our PIRLS example, all the predictors are discrete; therefore, we did not plot individual student data separately for each school. In Figure 18.2, we plot the school mean reading scores for each level of a predictor variable and connect the means within school by lines. The same pattern is found for each of the five plausible reading scores; therefore, we only present the figure for one of the plausible values, `readingPV1`. With the exception of the multicategory variables (`enjoyreading`, `mombornUS`, and `dadbornUS`), these lines correspond to straight lines. What is apparent in the figure is that there is considerable vertical spread among schools and this suggests that school regressions will have different intercepts. Also noticeable is the fact that some predictors have schools with vastly different slopes. For example, the slopes for `bornUS` are both negative and positive, and they show great variability among schools. Also important to note is that most students were born in the United States. The regression coefficient for `bornUS` is likely to be unstable because of the small proportion of students who were not born in the United States. The slopes for hours per day playing video games (i.e., `compgames5H`) also show considerable variability; whereas, the lines for gender are basically flat, which suggests that gender may not be a good predictor of reading scores. Since the means for both levels of `compgames5` (more or less than 5 h per day) are based on a relatively large number of students, the model fit to the data may require different slopes for `compgames5H` for each school.

Another feature of the data that can be seen in Figure 18.2 is that all the predictors should be treated as discrete even if they have three or more categories. For example, the relationships between reading scores and `mombornUS`, `dadbornUS`, and `enjoyreading` are not linear. Also from Figure 18.2, we can get a sense of the general direction of the relationship. Reading scores on average are shown to be higher for girls, larger values of enjoyment of reading, more books in the home, fewer hours playing video games, fewer hours watching TV, students born in the United States, and students whose mothers and fathers were born in the United States. To see the direction of the relationship more clearly, in Figure 18.3, the means taken over all schools and plausible values are plotted by each predictor variable, also shown in Table 18.1. Although the data are clustered, these means are unbiased estimates of

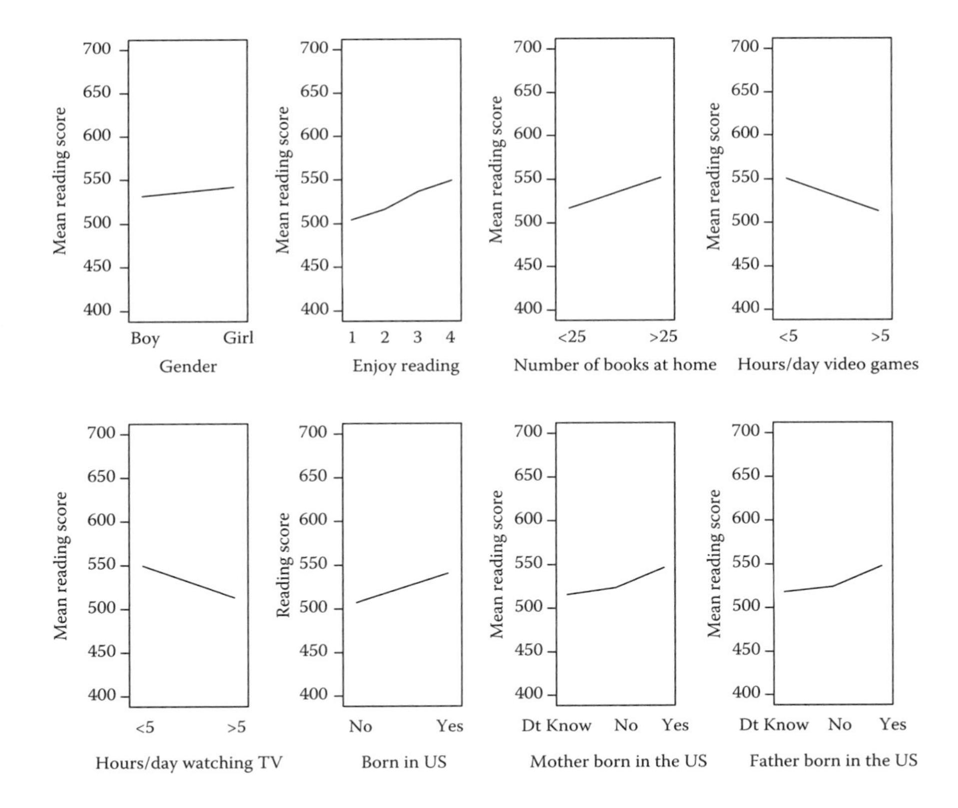

FIGURE 18.3
Panel plot of marginal mean reading scores by each predictor variable.

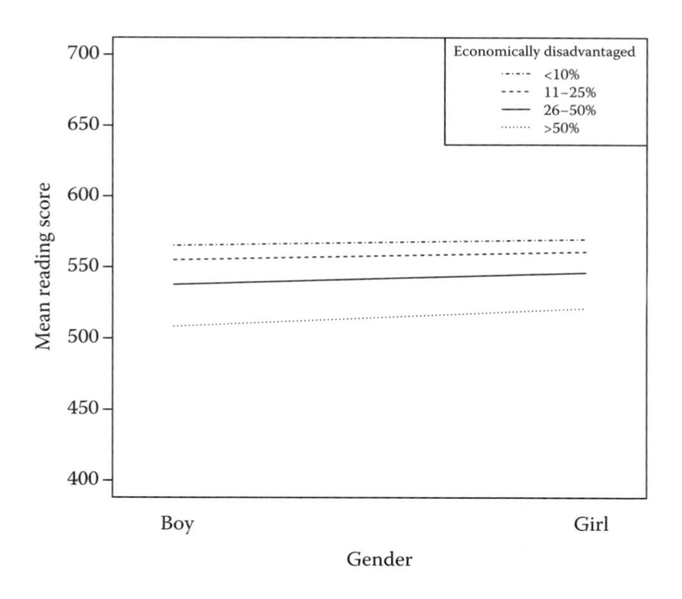

FIGURE 18.4
Mean reading scores plotted by gender with different lines (means) for different school percentages of students from economically disadvantaged families.

the effects of each predictor on the response (Diggle et al. 1994; Verbeke and Molenberghs 2000). In the models, the (fixed) regression parameters that are the same for all schools (and the average school) are expected to be in the same direction as the means illustrated in the figure.

Figures 18.1 and 18.2 focused only on effects at the level of the student. In the next figure, we illustrate how effects at the school level can be examined. In Figure 18.4, the mean reading scores are plotted for boys and girls at each level of `econDisadv`.* The lines for `econDisadv` are parallel and suggest there is no interaction between gender and `econDisadv`; however, the vertical spread between the level of economically disadvantaged students at a school suggests that this predictor may help account for differences between school intercepts. The schools with the lowest percentage of economically disadvantaged families (top curve, <10%) have the highest mean reading scores, followed by levels of 11–25% and 26–50%, while the lowest reading scores are for schools where >50% of families are disadvantaged.

As a final form of EDA, we fit normal least squares regression models with only the student (micro)-level predictor variables to each school's data. Since the intercept and regression coefficients are school specific, these regressions represent the best possible model fit for the data where goodness of fit can

* If a school-level predictor is continuous, a figure similar to Figure 18.4 can be created by discretizing the predictor and computing means for the artificial levels. However, when the time to model the data comes, the predictor should be treated as a continuous variable.

be measured by multiple R^2. Since schools with smaller numbers of students will tend to be fit better by the model than schools with larger numbers of students, the multiple R^2s for each school are plotted against school sample size. For our PIRLS data, these are plotted in Figure 18.5. Most of the schools' multiple R^2s are between .30 and .90, indicating that this set of student-level predictors could lead to a good level 1 or student-specific model for the data.

As a global summary statistic, we also computed meta-R^2 as

$$\text{meta-}R^2 = \frac{\sum_{j=1}^{M}(\text{SSTOT}_j - \text{SSE}_j)}{\sum_{j=1}^{M}\text{SSTOT}_j}$$

where SSTOT_j and SSE_j are the total sum of squares and the error sum of squares, respectively, for school j (Verbeke and Molenberghs 2000). Meta-R^2 measures the proportion of total within-school variability that can be explained by linear regression. For our PIRLS data, meta-R^2 equals .49. Both the R^2 and meta-R^2 can be used to compare plausible, student-specific models for the data.

Other exploratory analyses are possible that focus on the random aspects of the data and measure how well alternative models may fit the data (e.g.,

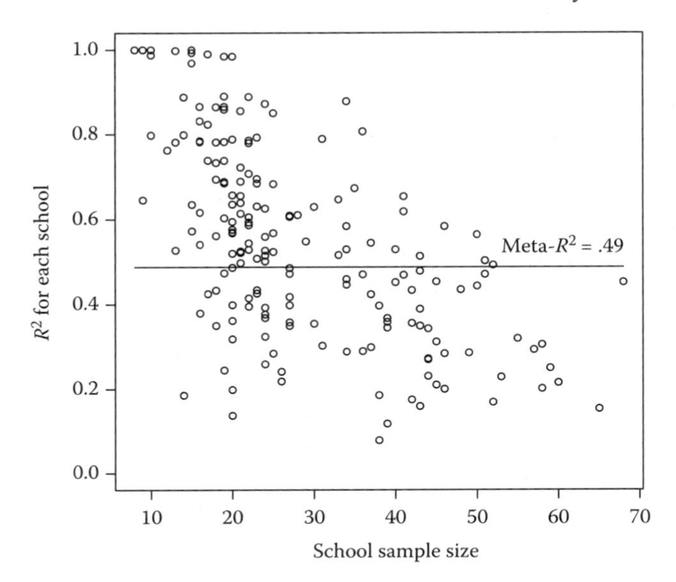

FIGURE 18.5
Multiple R^2s from linear regressions fit to each school's data plotted against school size with meta-R^2 indicated by the horizontal line.

see Diggle et al. 1994; Verbeke and Molenberghs 2000); however, a better understanding of multilevel modeling is required to explain such analyses. Therefore, we forgo these in favor of presenting more about the basic models and their properties. While conducting our EDA, we noticed that there are missing values on both student and school predictors. In the next section, we discuss how we dealt with missing data.

Missing Data

Although the percentages of missingness are not alarmingly high for many variables in PIRLS, most variables have missing values, as is often the case in large-scale assessment data. The patterns and percentages of missingness at the student and school levels are summarized in Tables 18.2 and 18.3. Each row represents a pattern of missing data where an "X" indicates that the column variable does not have missing values and "." indicates that the variable has missing values. The last two columns give the frequency and percentages with which each of the row patterns occur. The variables are ordered from those with the least missingness to those with the most missingness. For example, `female` has the least missing with three missing values (0.06%) while `comgames5H` has the most missing with 265 missing values (5.11%) at the student level. Among the 5190 students and 183 schools, 4570 (88.05%) students and 173 (94.54%) schools have no missing values. All of the variables in Table 18.1 as well as the five reading plausible values were used for imputing missing predictor values.

Multiple Imputation

Missing data are a major problem in most large-scale datasets and the clustered structure of multilevel data adds another layer of complexity to the treatment of missing observations. Simply removing students or whole schools due to missing data is not only a waste of information but also can result in biased parameter estimates (Allison 2002; Enders 2010; Schafer 1997; Van Buuren 2011, 2012). The program Mplus (Muthén and Muthén 2010) can fit models using maximum likelihood estimation (MLE) with missing values on the response variable, but MLE cannot handle missing data for the predictor or explanatory variables. A multiple imputation approach can impute missing response and predictor variables. The missing data mechanism we assume is *missing at random (MAR)*, which implies that conditional on the observed data, the missing values are independent of the unobserved data or, in other words, the missingness itself contains no information about the unobserved values that is not also contained in the observed data (Rubin 1976; Snijders and Bosker 2012).

Plausible values are provided for reading proficiency and thus no imputation is needed for outcomes, but we have missing data at both student and school levels with discrete predictors. Relatively little research has been conducted regarding missing clustered data. Most of the work pertains to either normally distributed variables (e.g., Chapter 20 of this volume by Shin; Shin

TABLE 18.2

Missing Data on Student Questionnaire

Pattern	female	enjoyreading	dadbornUS	mombornUS	bornUS	home25books	watchTV5H	comgames5H	Frequency	Percentage
1	X	X	X	X	X	X	X	X	4570	88.05
2	.	X	X	X	X	X	X	X	2	0.04
3	X	.	X	X	X	X	X	X	62	1.19
4	X	X	X	X	.	X	X	X	6	0.12
5	X	X	X	.	X	X	X	X	1	0.02
6	X	X	.	X	X	X	X	X	1	0.02
7	X	X	X	X	X	.	X	X	116	2.24
8	X	X	X	X	X	X	.	X	13	0.25
9	X	X	X	X	X	X	X	.	34	0.66
10	X	X	.	.	X	X	X	X	2	0.04
11	X	.	X	X	X	.	X	X	7	0.13
12	X	X	X	X	.	.	X	X	1	0.02
13	X	.	X	X	X	X	.	X	17	0.33
14	X	X	X	X	X	.	.	X	2	0.04
15	X	.	X	X	X	X	X	.	9	0.17
16	X	X	X	X	X	.	X	.	2	0.04
17	X	X	X	X	X	X	.	.	171	3.29
18	X	X	.	.	.	X	X	X	38	0.73
19	X	.	X	.	X	X	X	.	1	0.02
20	X	.	X	X	X	.	X	.	2	0.04

continued

TABLE 18.2 (continued)

Missing Data on Student Questionnaire

Pattern	female	enjoyreading	dadbornUS	mombornUS	bornUS	home25books	watchTV5H	comgames5H	Frequency	Percentage
21	X	.	X	X	X	X	.	.	8	0.15
22	X	X	X	.	X	X	.	.	1	0.02
23	X	X	X	X	X	.	.	.	15	0.29
24	X	.	.	.	.	X	X	X	1	0.02
25	X	X	.	.	.	.	X	X	57	1.10
26	X	.	X	X	X	.	.	.	1	0.02
27	.	X	.	.	.	.	X	X	1	0.02
28	X	.	.	.	.	.	X	X	25	0.48
29	X	.	.	.	.	X	.	X	2	0.04
30	X	X	.	.	.	.	.	X	1	0.02
31	X	X	.	.	.	X	.	.	4	0.08
32	X	X	.	.	.	.	.	.	4	0.08
33	X	.	.	.	.	.	.	.	13	0.25
Frequency	3	148	149	151	153	247	252	265	5190	
Percentage	0.06	2.85	2.87	2.91	2.95	4.76	4.86	5.11		

Note: "X" indicates a nonmissing value and "." indicates a missing value.

TABLE 18.3

Missing School-Level Data

Pattern	econDisadvN	schoolLIB	Frequency	Percentage
1	X	X	173	94.54
2	.	X	2	1.09
3	X	.	7	3.83
4	.	.	1	0.55
Frequency	3	8	183	
Percentage	1.64	4.37		

Note: "X" indicates a nonmissing value and "." indicates a missing value.

and Raudenbush 2007, 2010) or uses simple cases without missing level-2 variables (Van Buuren 2011, 2012). Neither of these two solutions works for our data. Some proposals have been put forth that include dummy variables for each cluster (Reiter et al. 2006); however, this presupposes that clusters only differ in terms of their intercepts. In our case, we do not want to make this assumption and due to the large number of clusters, this method is not pursued here. Also rare are proposals for incorporating sampling weights into the imputation model for missing data. An exception is Amer (2009), but this example only deals with two clusters.

There are two general approaches to impute multivariate missing data: joint modeling (JM) and fully conditional specification (FCS) (Van Buuren 2012). JM imputes missing values in multiple variables simultaneously based on the assumption that the data can be described by a multivariate distribution. The multivariate normal distribution is the mostly widely applied, although any multivariate distribution can be assumed in theory. On the other hand, FCS imputes data on a variable-by-variable basis. FCS requires an imputation model for each incomplete variable as the outcome, and creates imputations per variable in an iterative fashion. As FCS directly specifies the conditional distributions from which draws should be made, the method does not require specification of a multivariate model for multiple variables. An overview of similarities and differences between JM and FCS is provided in Van Buuren (2007).

For missing values among binary and ordinal predictors that do not likely follow a multivariate normal distribution, we used FCS as implemented in the package mice in R (Van Buuren 2011). We used two datasets for imputation; the school data and student data. The student data ($N = 5190$) consist of all level-1 and level-2 variables, where the level-2 values are constant for students within the same schools. The school data ($M = 183$) consists of the same number of variables, where the means of the student-level variables are used as the corresponding school-level variables. Although we acknowledge that our approach of imputing missing values at each level separately is not optimal, imputing missing values while preserving the multilevel structure is not simple; more complex approaches that may be superior are described in Swoboda (2011) and Kim and Swoboda (2012). However, we at least use information across levels to improve the quality of the multiple imputation. For example, the five plausible values at the student level are used to impute the missing values of the school-level predictors, and the school-level predictors are used to impute missing values of the student-level predictors.

There appears to be no consensus on the appropriate number of imputed datasets (Allison 2002; Enders 2010; Schafer 1997; Van Buuren 2012), and the answer may depend on the missing patterns and frequencies. We imputed five datasets for the predictors and crossed them with the five plausible value outcomes, resulting in 25 datasets for further analysis. The different models were all fit to each of the 25 imputed datasets, and we thus obtained 25 different outcomes for each model.

To synthesize results from the multiple datasets for each model, we used Little and Rubin's procedures for combining multiple results, which appears to be the generally accepted standard in the imputation literature (Little and Rubin 2002; Snijders and Bosker 2012). The Little and Rubin procedure accounts for two sources of imprecision in the parameter estimates: *the within-dataset uncertainty* and *the imputation uncertainty*. The within-dataset uncertainty is accounted for by the standard error of estimate, and the imputation uncertainty is accounted for by the variance of the estimates across datasets. This procedure can be presented as follows: Let θ be a parameter of interest. The average estimate is obtained across M_{imp} imputed datasets:

$$\bar{\theta} = \frac{1}{M_{\text{imp}}} \sum_{m=1}^{M_{\text{imp}}} \hat{\theta}_m. \tag{18.1}$$

The within-dataset uncertainty is the average of the squared standard errors,

$$\bar{W} = \frac{1}{M_{\text{imp}}} \sum_{m=1}^{M_{\text{imp}}} \text{SE}\left(\hat{\theta}_m\right)^2,$$

and the between-imputation uncertainty is the variance of the estimate across the multiple datasets:

$$B = \frac{1}{M_{\text{imp}} - 1} \sum_{m=1}^{M_{\text{imp}}} \left(\hat{\theta}_m - \bar{\theta}\right)^2.$$

These two sources of uncertainty are used to compute the standard errors of the average estimate by taking the square root of a weighted sum of the two sources as follows:

$$\mathrm{SE}(\bar{\theta}) = \sqrt{\bar{W} + \left(1 + \frac{1}{M_{\mathrm{imp}}}\right)B}. \tag{18.2}$$

When we later fit multilevel models, Equations 18.1 and 18.2 are used to report results in the tables.

Multilevel Modeling

Multilevel modeling is a statistical methodology with many alternative names, including hierarchical linear modeling, mixed modeling, random-effects modeling, nested modeling, random coefficient modeling, and variance component analysis. Not surprisingly, the notation for multilevel models varies across the literature and software used in various disciplines. While we attempt to use notation that is largely consistent with those of the other chapters in the volume, readers should expect some differences across other resources for multilevel analysis.

Despite subtle or sometimes not-so-subtle differences in notation and terminology across disciplines, there are a number of core concepts that are consistent, such as intraclass correlation (ICC), fixed and random effects, random intercepts and random slopes (a.k.a. random coefficients), centering, reliability of aggregated variables, within- and between-group covariability, cross-level interactions, and slopes as outcomes models. Owing to space limits, we provide a list of multilevel modeling textbooks at the end of this chapter and refer readers to these books for definitions and examples of these concepts.

Presentation of the Model

A single-level regression model can be generalized to a multilevel model by allowing a regression coefficient to be random over clusters. Let the outcome variable (e.g., reading proficiency) for an individual i in cluster (e.g., school) j be Y_{ij}, where $i = 1,\ldots,n_j$ and $j = 1,\ldots,M$. The distribution of Y_{ij} within a cluster is assumed to be normal. The level-1 cluster-specific model can be written as

$$Y_{ij} = \beta_{0j} + \sum_{p=1}^{P} \beta_{pj} x_{pij} + \varepsilon_{ij}, \tag{18.3}$$

where β_{0j} is the intercept for cluster j, β_{pj} is the regression coefficient of cluster j for predictor variable x_{pij}, $p = 1,\ldots,P$, and ε_{ij} is the student-level random effect or error term, usually assumed to be independently and normally distributed with a zero mean and an unknown variance σ^2 (i.e., $\varepsilon_{ij} \sim N(0, \sigma^2)$ *i.i.d.*).

While the level-1 model accounts for variability at the student level, the level-2 model describes variability between clusters (schools in our analysis). The level-2 models are linear models for each of the level-1 regression coefficients:

$$\begin{aligned}
\beta_{0j} &= \gamma_{00} + \sum_{q=1}^{Q} \gamma_{0q} z_{qj} + U_{0j} \\
\beta_{1j} &= \gamma_{10} + \sum_{q=1}^{Q} \gamma_{1q} z_{qj} + U_{1j} \\
&\vdots \\
\beta_{Pj} &= \gamma_{P0} + \sum_{q=1}^{Q} \gamma_{Pq} z_{qj} + U_{Pj},
\end{aligned} \tag{18.4}$$

where the z_{qj}'s are predictors or explanatory variables that model systematic differences between clusters, $q = 1,\ldots,Q$, the γ_{p0}'s are level-2 intercepts, the other γ's are fixed regression coefficients for level-2 predictors, and the U_{pj}'s are unobserved random effects (residuals).

The unexplained, random, or stochastic between-cluster differences are modeled as random effects. The distributional assumption for the U_{pj}'s in Equation 18.4 is

$$\boldsymbol{U}_j = \begin{pmatrix} U_{0j} \\ U_{1j} \\ \vdots \\ U_{Pj} \end{pmatrix} \sim MVN\left(\begin{pmatrix} 0 \\ 0 \\ \vdots \\ 0 \end{pmatrix}, \begin{pmatrix} \tau_{00} & \tau_{10} & \cdots & \tau_{p0} \\ \tau_{10} & \tau_{11} & \cdots & \tau_{p1} \\ \vdots & \vdots & \ddots & \vdots \\ \tau_{p0} & \tau_{p1} & \ldots & \tau_{pp} \end{pmatrix} \right) i.i.d., \tag{18.5}$$

where MVN stands for multivariate normal and *i.i.d.* is for independent and identically distributed. For short, $\boldsymbol{U}_j \sim MVN(\boldsymbol{0}, \boldsymbol{T})$ *i.i.d.*

Substituting the level-2 equations into the regression coefficients of the level-1 model yields a linear mixed-effects model

$$Y_{ij} = \gamma_{00} + \sum_{p=1}^{P} \gamma_{p0} x_{pij} + \sum_{q=1}^{Q} \gamma_{0q} z_{qj} + \sum_{p=1}^{P} \sum_{q=1}^{Q} (\gamma_{pq} z_{qj} + U_{pj}) x_{pij} + U_{0j} + \varepsilon_{ij}. \tag{18.6}$$

The mixed-effects model shows that γ_{00} is the overall intercept, the γ_{p0}'s are fixed regression coefficients for level-1 effects, the γ_{0q}'s are fixed regression coefficients for level-2 effects, and the γ_{pq}'s are fixed effect regression coefficients for cross-level interaction terms. The first subscript on γ_{pq} corresponds to the predictor in the level-1 model and the second subscript corresponds to the predictor in the level-2 model. For example, γ_{32} is the regression coefficient for the interaction between x_{3ij} and z_{2j}. Random effects consist of the random intercept U_{0j}, random slopes U_{pj}, and the error term ε_{ij}.

Empty Model

Whereas multilevel analysis investigates relationships between predictors and outcomes in the context of hierarchical data, it is often useful to consider *the empty model* before entering predictors into the model. The empty model consists of the outcome variable (e.g., reading proficiency) and the group membership identification variable (e.g., school ID#) as the only variables in the model, and is also known as *the null model* or *the unconditional means model*. The empty model can be presented as

$$Y_{ij} = \gamma_{00} + U_{0j} + \varepsilon_{ij}. \tag{18.7}$$

Equation 18.7 represents the simplest multilevel model in the two-level case, and is identical to a one-way random effects ANOVA model. In addition to the grand mean of the outcome value, γ_{00}, the model reflects the variability of the outcome variable as the sum of the between- and within-group variance components:

$$\text{var}(Y_{ij}) = \text{var}(U_{0j}) + \text{var}(\varepsilon_{ij}) = \tau_{00} + \sigma^2. \tag{18.8}$$

The covariance between two observations within the same cluster j is equal to the variance of U_{0j}, and can be viewed as the amount of variance shared by the two units (e.g., two students within the same school) as a result of context:

$$\text{cov}(Y_{ij}, Y_{i'j}) = \text{var}(U_{0j}) = \tau_{00}. \tag{18.9}$$

Equation 18.8 shows that the model assumes a constant variance within each cluster. This assumption is not required for multilevel models with random slopes (see the section "Model Specification and Comparison"). Equation 18.9 shows there exists dependency within clusters, unlike the independence assumption in single-level models. It also demonstrates that the covariance between observations within each cluster is equal to the between-group variance.

Based on the between- and within-group variance components, we can calculate the correlation between two observations from the same cluster:

$$\text{corr}(Y_{ij}, Y_{i'j}) = \frac{\tau_{00}}{\tau_{00} + \sigma^2}, \tag{18.10}$$

which is known as the *intraclass correlation coefficient* and denoted as $\rho_I(Y)$. The intraclass correlation (ICC) can be interpreted as the correlation between two observations within a cluster (as defined in Equation 18.10) and also as the proportion of the variance accounted for by groups. In other words, the ICC indicates the degree of association among observations within clusters. The value can be as small as zero (when $\tau_{00} = 0$) and as large as one (when $\sigma^2 = 0$). A large ICC reflects a high degree of resemblance within clusters and strong dependence among observations. The ICC can also be calculated after entering predictors to the model, in which case it is referred to as a *residual* ICC.

Note that although $\rho_I(Y)$ is often simplified to ρ_I, the ICC depends on the variable of interest and different ICCs would be calculated for different variables. For example, an ICC of mathematics scores can be higher than an ICC of history scores, indicating a stronger correlation among mathematics scores within the same classes or schools than history scores. In the PIRLS 2006 US data, we fit empty models using `readingPV1` to `readingPV5` separately and combined the results (see the section "Multiple Imputation"). The average ICC for reading proficiency was 1244.05/(1244.05 + 4282.11) = .225 with values of .223 to .230 across the five plausible values. The estimated ICC suggests a total variance of 5526.16, about 22.5% of which is attributable to differences between schools.

Modeling Strategy

Statistical modeling is a process that is guided by substantive theory, the results of exploratory analysis, and results from fitting various models to the data. The two most common approaches advocated in the literature on linear mixed models are "step-up" and "top-down" methods (Ryoo 2011). The step-up method starts with a simple level-1 model including a random intercept to which fixed effects are successively added followed by random effects (Pinheiro and Bates 2000; Raudenbush and Bryk 2002). The top-down approach starts with the most complex polynomial model representing (level 1) effects and a random intercept, where the first step is to determine the correct order of the polynomial and the second step is to build the random effects part of the model (Diggle et al. 1994; Verbeke and Molenberghs 2000). The top-down approach lends itself better than the step-up approach to longitudinal data where change is not always linear. The little research that exists on the subject of modeling approaches with linear mixed models has found that the step-up approach tends to identify the true model in simulations more effectively than the top-down method (Ryoo 2011). In this section, a version of the step-up approach is used where we start with simple models and work toward more complex ones. Besides the work by Ryoo (2011), a reason for preferring this approach is that a model with all potentially interesting fixed and random effects suggested by the exploratory analysis may fail as a starting model because complex models are often not supported by the data; that is, such models frequently fail to converge or yield improper solutions (such as an improper covariance matrix $\boldsymbol{T}$).

Model building in multilevel analysis consists of selecting fixed effects for the mean structure and random effects for the covariance structure. The two structures are not independent. The mean structure aims to explain the systematic part of the variability of the outcome (e.g., reading proficiency in our example) and the covariance structure helps to account for the random or unexplained variance in the data. Whereas traditional single-level regression such as ANCOVA models focus mainly on the mean structure, it is critical to specify the appropriate covariance structure in multilevel models because an underparameterized covariance structure invalidates statistical inferences and an overparameterized covariance structure leads to inefficient estimation and poor standard errors (Verbeke and Molenberghs 2000). Therefore, both appropriate mean and covariance structures are essential for making valid conclusions and proper predictions.

Although no strategy guarantees the optimal model or a model that satisfies all assumptions, a combination of general modeling guidelines and EDA can help find a good model with appropriate mean and covariance structures. We propose an eight-step procedure for model building as follows. First, examine the data using EDA, including the strategies in the section "Exploratory Data Analysis." Second, specify systematic (fixed effects) that are of theoretical importance. Third, build a random-effects structure (possibly based on results of EDA). Fourth, revisit and revise fixed effects based on EDA and conduct tests for the fixed effects. Fifth, retest random effects and possibly revise. Sixth, repeat/cycle through the fourth and fifth steps. Seventh, conduct model diagnostics and residual analysis. Eighth, and finally, interpret results. We followed this eight-step strategy for our analysis in this chapter.

Model Specification and Comparison

After the examination of the empty model, we included a set of theoretically important fixed effects (Model 1), examined potential random effects (Models 2 and 3), and considered other predictors that might also be important (Model 4). The parameter estimates and model fit indexes of Models 1 through 4 are shown in Table 18.4. The fixed effect estimates are the means of the estimates taken across the 25 imputed datasets using Equation 18.1, and the corresponding standard errors are calculated using Equation 18.2 in the section "Multiple Imputation".

One can use either model-based or empirical standard errors to compute test statistics for the fixed effects. The model-based standard errors are obtained under the assumption that the covariance matrix of the observations is specified correctly, and the empirical standard errors are based on the residuals. The latter are also referred to as Huber–White's robust sandwich standard errors (Huber 1967; White 1980). It is known that model-based standard errors tend to be underestimated when a model is misspecified, whereas the empirical standard errors are relatively robust to model misspecification (Diggle et al. 1994; Raudenbush and Bryk 2002). Kim and Frees

TABLE 18.4

Model Coefficients, Standard Errors, and Fit Indexes, Averaged across 25 (5 Plausible Outcome Values × 5 Independent Variable Imputations) Datasets

Model	Model 1	Model 2	Model 3	Model 4
Fixed Effects		**Coefficient (SE)**		
Intercept	544.93 (6.08)	545.29 (6.14)	544.19 (6.15)	523.72 (11.95)
female	2.05 (2.20)	2.10 (2.19)	2.08 (2.19)	−3.00 (3.64)
enjoyreading (agree a little)	−9.32 (2.29)	−9.31 (2.29)	−9.18 (2.29)	−9.13 (2.30)
enjoyreading (disagree a little)	−22.78 (3.70)	−22.90 (3.70)	−22.56 (3.67)	−22.28 (3.66)
enjoyreading (disagree a lot)	−28.53 (3.60)	−28.57 (3.61)	−28.65 (3.59)	−28.73 (3.58)
bornUS	18.30 (4.10)	18.00 (4.15)	18.01 (4.17)	17.27 (4.25)
dadbornUS (don't know)	−8.40 (3.43)	−8.26 (3.45)	−8.21 (3.46)	−5.55 (4.31)
dadbornUS (yes)	6.84 (3.13)	6.92 (3.13)	7.04 (3.12)	5.32 (3.49)
home25books	16.12 (2.22)	15.99 (2.20)	16.01 (2.21)	15.93 (2.22)
watchTV5H	−15.15 (2.52)	−15.22 (2.53)	−15.09 (2.54)	−14.93 (2.55)
comgames5H	−16.05 (2.79)	−16.14 (2.80)	−16.02 (2.85)	−15.75 (2.86)
econDisadv (>50%)	−42.15 (5.09)	−42.45 (5.11)	−40.80 (5.05)	−45.68 (6.27)
econDisadv (26–50%)	−14.62 (5.18)	−14.53 (5.23)	−14.00 (5.22)	−17.19 (6.24)
econDisadv (11–25%)	−6.11 (5.93)	−6.27 (5.96)	−5.82 (5.71)	−8.17 (7.77)
mombornUS (don't know)				−5.48 (5.12)
mombornUS (yes)				2.83 (3.51)
schoolLIB				23.82 (9.92)
female × econDisadv (>50%)				8.13 (4.67)
female × econDisadv (26–50%)				5.61 (5.80)
female × econDisadv (11–25%)				3.47 (7.26)
Variance Components		**Estimate (SE)**		
σ^2	3834.80 (117.33)	3804.26 (116.431)	3773.05 (120.57)	3768.94 (119.92)
Intercept τ_{00}	513.84 (76.12)	520.11 (85.64)	712.09 (131.62)	700.66 (133.84)
Slope (comgames5H) τ_{11}		152.87 (91.31)	165.53 (110.43)	162.12 (110.82)
Slope (female) τ_{22}			125.33 (83.38)	114.59 (79.55)
Covariance τ_{01}		−35.50 (72.43)	−82.73 (103.61)	−75.37 (105.34)
Covariance τ_{02}			−201.15 (86.79)	−197.11 (88.31)
Covariance τ_{12}			14.43 (71.42)	2.52 (72.54)
Fit Indexes		**Mean (SD)**		
−2 × log-likelihood	57831.47 (83.13)	57826.33 (83.04)	57814.53 (81.98)	57803.22 (83.10)
AIC	57863.47 (83.13)	57862.33 (83.04)	57856.53 (81.98)	57857.22 (83.10)
BIC	57968.35 (83.13)	57980.32 (83.04)	57994.18 (81.98)	58034.19 (83.10)

Note: Between-imputation standard deviations for the fit indexes are also presented.

(2006, 2007) also showed through simulations that model-based standard errors are particularly sensitive to misspecification due to omitted variables. We used empirical standard errors in our analysis.

In addition to statistical tests for individual regression coefficients using t-test statistics, we compared the models by means of likelihood values, Akaike Information Criterion (AIC = −2 log-likelihood + 2 × the number of parameters), and Bayesian Information Criterion (BIC = −2 log-likelihood + 2 × the number of parameters × log(N)). We obtained these model fit indexes for each model in the 25 datasets and reported the means and standard deviations of the three indexes across the imputed datasets. As only the likelihood function varies across the imputed datasets while the number of parameters and sample size stay constant, the standard deviations are the same for the three fit indexes for each model.

With respect to evaluating the random effects, standard errors of variance components are known to be unreliable in multilevel models, and it is advised not to use standard errors to evaluate the significance of variance components. Some multilevel software (e.g., package lme4 in R) do not provide standard errors for variance components. Instead, a likelihood ratio test statistic based on the full and reduced models is used to compare two models with and without a random effect. Moreover, the likelihood ratio test for comparing fixed effect parameters needs to be modified to test the significance of a random effect, as testing variance components deals with a null hypothesis at the boundary value (e.g., H_0: $\tau_{11} = 0$) and a directional alternative hypothesis (e.g., H_1: $\tau_{11} > 0$).

A simple modification of the χ^2 critical value can account for the otherwise overly conservative likelihood ratio test for variance components. Specifically, results from simulation studies (Self and Liang 1987; Stram and Lee 1994, 1995) have shown that when testing H_0: $\tau_{00} = 0$ versus H_1: $\tau_{00} > 0$, the likelihood ratio test statistic asymptotically follows a distribution of $\frac{1}{2}\chi_1^2$ rather than χ_1^2. The asymptotic distribution of the likelihood ratio test statistic for testing an additional random effect (e.g., random slope) is shown to be a mixture of χ_1^2 and χ_2^2 distributions. This implies that for testing the significance of the first random slope by comparing −2log-likelihood of Model 1 and Model 2 (which has two more parameters, τ_{11} and τ_{01}, than Model 1) at the type I error rate of 0.05, the proper critical value is $\frac{1}{2}\chi_1^2 + \frac{1}{2}\chi_2^2 = \frac{1}{2}(3.84 + 5.99) = 4.92$, which is smaller than the standard likelihood ratio test critical value with two degrees of freedom. For testing the second random slope by comparing Model 2 and Model 3 (with three more parameters τ_{22}, τ_{02}, and τ_{12}), we can use the critical value of $\frac{1}{2}\chi_2^2 + \frac{1}{2}\chi_3^2 = \frac{1}{2}(5.99 + 7.81) = 6.90$. This downward adjustment in the critical values using mixture distributions makes the likelihood ratio test for random effects more accurate.

Table 18.4 shows that AIC and BIC disagree in regard to the best fitting model for this example. While Model 3 (random slopes of `comgames5H` and female) has the smallest AIC, Model 1 (random intercept only) has the smallest BIC. At the same time, the four models are nested within each other and we can also directly compare them using the likelihood ratio test. To compare nested models with different fixed effects, such as Model 3 versus Model 4, it is important to estimate models using the full information maximum likelihood (FIML) method, not the residual maximum likelihood (REML) method. Models 1 through 3 are different only in random effects, and thus both FIML and REML methods can be used in comparing them via likelihood values. All models in this chapter were estimated by the FIML method.

In the comparison of Model 1 versus Model 2 with the random slope of `comgames5H`, the difference in –2log-likelihood is 5.14, which is greater than the critical value of $\frac{1}{2}\chi_1^2 + \frac{1}{2}\chi_2^2 = 4.92$, suggesting the effects of comgames5H vary across schools. In the comparison of Model 2 versus Model 3 with the additional random slope of female, the difference in –2log-likelihood is 11.80 and is greater than the critical value of $\frac{1}{2}\chi_2^2 + \frac{1}{2}\chi_3^2 = 6.90$, suggesting the effect of gender is also different across schools. Finally, in comparing Model 3 to Model 4 with the additional predictors of `mombornUS`, `schoolLIB`, and the interaction `female` × `econDisadv`, the standard critical value for the likelihood ratio test was used because the difference between the two models involves only fixed effects. The difference in –2log-likelihood is 11.31 between Model 3 and Model 4 with six additional parameters, and this difference is smaller than the critical value of $\chi_6^2 = 14.07$. Therefore, based on the series of likelihood ratio tests, we chose Model 3 as the best-fitting model in our analysis.

Although the summary statistics in Table 18.1 suggest that whether a student's mother was born in the United States is important with respect to that student's reading performance, the effect was not significant in Model 4. We suspect this is due to the strong association between `dadbornUS` and `mombornUS`. Among 5038 students who answered each question as "yes," "no," or "don't know," 84% chose the same category for the two questions. Therefore, there is little unique information in `mombornUS` when `dadbornUS` is already in the model, and the two highly correlated variables lead to multicollinearity in estimation. For a similar reason, we speculate whether the school has a library or not seems important based on the mean reading performance, but the effect of `schoolLIB` is not significant when *econDisadv* is already in the model. Those schools without a library all had high percentages of economically disadvantaged students. We also examined a number of interactions, including the cross-level interaction `female` × `econDisadv` but did not observe a significant interaction effect.

R^2-Type Measures for Explained Variance

In standard regression models, the R^2 index represents the percentage of variance in the outcome variable that is accounted for by the explanatory variables. In multilevel models, there are complications with the use of R^2 to assess prediction. One complication is that there are models at multiple levels that represent different sources of variance in the data; within-cluster and between-cluster models. Therefore, different R^2-type measures can be calculated at each level, often referred to as R_1^2 for a level-1 model and R_2^2 for a level-2 model.

In single-level linear regression, R^2 has a number of different interpretations, including the squared correlation between observed and predicted values of the response, the proportional decrease in variance of the response given the predictor variables, and the proportional reduction of prediction error variance. The definition of the proportional reduction in explained variance is problematic in multilevel models because it is theoretically and empirically possible that a residual variance at some level of the model will increase after adding an explanatory variable, and thus interpreting R^2 as a simple percentage of variance accounted for (analogous to single-level regression analysis) is not recommended.

Snijders and Bosker (1994, 1999) proposed some alternative ways to define explained variation that provide less problematic interpretations based on the proportional decrease in prediction error variance (i.e., the mean squared error of prediction). Specifically, consider a two-level random intercept model with variance components at levels 1 and 2, denoted σ^2 and τ_{00}, respectively. We will obtain different estimates of the two variance components under an empty unconditional model (Equation 18.7) and a conditional model with explanatory variables. Denote the estimates from the unconditional model as $(\hat{\sigma}^2 \text{ and } \hat{\tau}_{00})_{\text{unconditional}}$ and the conditional model as $(\hat{\sigma}^2 \text{ and } \hat{\tau}_{00})_{\text{conditional}}$. Snijders and Bosker (1999) defined their alternative R^2 indices as follows. The first index is the proportional reduction of error for predicting a level-1 outcome, and can be computed as

$$R_1^2 = 1 - \frac{(\hat{\sigma}^2 + \hat{\tau}_{00})_{\text{conditional}}}{(\hat{\sigma}^2 + \hat{\tau}_{00})_{\text{unconditional}}}. \tag{18.11}$$

The second index is the proportional reduction of error for predicting a group mean, which can be computed as

$$R_2^2 = 1 - \frac{(\hat{\sigma}^2 / \tilde{n} + \hat{\tau}_{00})_{\text{conditional}}}{(\hat{\sigma}^2 / \tilde{n} + \hat{\tau}_{00})_{\text{unconditional}}}, \tag{18.12}$$

where $\tilde{n}$ is the expected number of level-1 units per level-2 unit. When the number of level-1 units, n_j, varies greatly across level-2 units, $\tilde{n}$ can be substituted by the harmonic mean

$$\tilde{n} = \frac{M}{\sum_{j=1}^{M} (1/n_j)}.$$

Relative to the arithmetic mean, the harmonic mean gives less weight to clusters with much larger sample sizes and thus is more appropriate as the expected number of level-1 units per level-2 unit in the population. In the

PIRLS 2006 example, the harmonic mean for the number of students per school is 23.39 (as opposed to the arithmetic mean of 28.36).

Note that Equations 18.11 and 18.12 represent R_1^2 and R_2^2 for a random intercept model as the conditional model. However, our final model (Model 3 in Table 18.4) consists of two random slopes. Although R_1^2 and R_2^2 can be calculated for random slope models, the process is considerably more complicated than for random intercept models, because level-2 variances are not constant but functions of explanatory variables. Recchia (2010) developed a SAS macro to calculate R_1^2 and R_2^2 for multilevel models. Alternatively, Snijders and Bosker (1999) suggested using approximate R_1^2 and R_2^2 values based on a random intercept model, because the values of R_1^2 and R_2^2 for random intercept and random slope models are similar when they have the same fixed effects specification.

Following the suggestion by Snijders and Bosker (1999), we calculated approximate R_1^2 and R_2^2 in our example by comparing the empty unconditional model to the random intercept conditional model, Model 1 in Table 18.4, which is identical to our final model except for the two random slopes. We obtained R_1^2 and R_2^2 for each of the 25 datasets and the values are reasonably stable across the imputed datasets. At level 1, R_1^2 varies from .206 to .225, with a mean of .214 (SD = .005), implying that there is a 21.4% reduction in the mean square prediction error within schools. At level 2, R_2^2 varies from .509 to .544, with a mean of .525 (SD = .009), implying that we reduced predictive error 52.5% for the school means by including explanatory variables.

Incorporating Design Weights

Often with complex, large-scale surveys, the probability of selection of clusters and observations within cluster are unequal. Typically, weights are given by the organization conducting the survey that reflect the design of the study. When weights are excluded from the modeling, parameter estimates may be biased but efficient; including weights leads to unbiased estimates but less efficient ones. The decision to include design weights or not should be based on whether the weights are likely to have an impact or are *informative*, and whether the probability of selection (the sampling model) is related to the probability model for the data (i.e., the HLM). To determine whether weights are informative, we need to compute them and examine their distribution.

We adopted recommendations by Rutkowski et al. (2010) for computing weights in surveys such as the PIRLS. Weights should be computed for each level of sampling. Besides having unequal selection probability, all of the selected units may not respond. The weights for students (level 1 or secondary units) will be computed as the product of student and class weights:

$$w_{1ij\ell} = \underbrace{\left(WF_{ij\ell} \times WA_{ij\ell}\right)}_{\text{student}\,i}\underbrace{\left(WF_{j\ell} \times WA_{j\ell}\right)}_{\text{class}\,\ell}, \tag{18.13}$$

where $WF_{ij\ell}$ and $WF_{j\ell}$ are the inverses of the probabilities for the selection of students and classes from school j, respectively, and $WA_{ij\ell}$ and $WA_{j\ell}$ are the weight adjustments for nonresponse for student i and class ℓ from school j, respectively. The weight adjustments are for those students and classes that were selected but did not choose to participate (Rutkowski et al. 2010). The school weights used are

$$w_{2j} = WF_j \times WA_j, \tag{18.14}$$

where WF_j is the inverse of the probability of selecting school j, and WA_j is the weight adjustment for school j.

One way to assess the informativeness of the weights is to examine their distribution. If the selection probabilities are equal and all selected units respond, then the weights would all be equal and their variance would be zero. In this case, the sampling would be the same as simple random sampling and the weights could be set to one. In the PIRLS 2006 data, the mean of the weights for schools equals 305.42 and their standard deviation is 218.64, which suggests that the level 2 weights are informative. For the student level, the means of students within schools are mostly equal to one and 74% of the schools have standard deviations that equal zero. Of the remaining schools, 21% have standard deviations that are less than 0.05 and the other 5% have standard deviations less than 0.13. These standard deviations suggest that the level-1 weights are not informative and will likely have no impact on the analysis. Additional methods to assess the informativeness of weights are given in Chapter 21 by Anderson, Kim, and Keller.

The weights are typically scaled according to one of two methods that were discussed in Pfeffermann et al. (1998). The method used here is to scale the weights so that their sum equals the sample size; that is, $\Sigma_j w_{2j} = 183$ and $\Sigma_i \Sigma_\ell w_{1ij\ell} = n_j$.

Weights are incorporated into the model during estimation. The log-likelihoods for the lowest-level units (i.e., students) are multiplied by their respective weights and summed over the values within a cluster. Cluster-specific values are found by integrating out the random effects (i.e., U_{pj}'s). The cluster-specific values are summed after being multiplied by their respective weights w_{2j} to yield a function of all the data (Grilli and Pratesi 2004; Rabe-Hesketh and Skronkal 2006). The parameters that maximize the function are maximum pseudolikelihood estimates. They are not maximum likelihood estimates because the log-likelihoods are multiplied by weights. For more details, see Chapter 21 by Anderson, Kim, and Keller (see also Grilli and Pratesi 2004; Rabe-Hesketh and Skronkal 2006).

To illustrate the effect of including weights, we fit Model 3 (our best model without weights) with weights. The results are summarized in Table 18.5. Since we are using pseudolikelihood estimation, the sandwich or robust estimates of the standard errors are presented. The lower efficiency resulting from including weights is evident by comparing the standard errors of the

TABLE 18.5

Model 3 with and without Weights

	Model 3 without Weights			Model 3 with Weights		
Fixed Effects	**Coef**	**SE**	***t***	**Coef**	**SE**	***t***
Intercept	544.19	6.15	88.51	535.07	7.09	75.53
female	2.08	2.19	0.95	2.89	2.84	1.02
enjoyreading (agree a little)	−9.18	2.29	−4.01	−9.71	3.21	−3.02
enjoyreading (disagree a little)	−22.56	3.67	−6.15	−29.15	4.76	−6.13
enjoyreading (disagree a lot)	−28.65	3.59	−7.97	−29.23	4.37	−6.69
bornUS	18.01	4.17	4.32	21.00	5.29	3.97
dadbornUS (don't know)	−8.21	3.46	−2.38	−8.53	4.17	−2.04
dadbornUS (yes)	7.04	3.12	2.26	7.75	3.53	2.19
home25books	16.01	2.21	7.25	17.49	2.59	6.75
watchTV5H	−15.09	2.54	−5.94	−13.30	2.86	−4.66
comgames5H	−16.02	2.85	−5.61	−16.02	3.08	−5.20
econDisadv (>50%)	−40.80	5.05	−8.08	−37.85	6.80	−5.57
econDisadv (26–50%)	−14.00	5.22	−2.68	−4.86	6.64	−0.73
econDisadv (11–25%)	−5.82	5.71	−1.02	2.10	6.35	0.33
Variance Components	**Estimate**	**SE**		**Estimate**	**SE**	
σ^2	3773.05	120.57		3719.26	136.67	
Intercept τ_{00}	712.09	131.62		771.03	225.17	
Slope (comgames5H) τ_{11}	165.53	110.43		151.06	121.20	
Slope (female) τ_{22}	125.33	83.38		166.10	112.42	
Covariance τ_{01}	−82.73	103.61		−71.53	116.70	
Covariance τ_{02}	−201.15	86.79		−227.32	129.01	
Covariance τ_{12}	14.43	71.42		45.86	83.41	

Note: Model coefficients, standard errors, and fit indexes, averaged across 25 (5 plausible outcome values × 5 independent variable imputations) datasets are shown.

parameter estimates. The estimates of the standard errors with the weights are larger than those without weights. One of the school-level variable effects, `econDisadv (26-50%)` has very different regression coefficient estimates (−14.00 vs. −4.86) and is not significant in the model with weights.

Model Diagnostics

Before turning to interpretation, we consider an often neglected part of multilevel modeling: investigating whether there are potential violations of models assumptions or systematic misfit of the model to data. Snijders and Bosker (2012) cover diagnostics for multilevel models in great detail. We will restrict our attention to the examination of residuals and the assumption of normally distributed random effects and illustrate some of the possible methods using Model 3 without weights.

Figure 18.6 includes various model diagnostic plots. In the top row of plots, we examine the cluster-specific or conditional Pearson residuals. These residuals are based on predictions that include both the fixed effects from Table 18.4 and empirical Bayes estimates of the random effects (i.e., U_{0j}, U_{1j}, and U_{2j}). The analogous plots for marginal predictions (i.e., those that excluded estimates of the random effects) were similar to those in Figure 18.6. The plots for each imputed dataset were very similar to each other; therefore, we averaged over the 25 imputed datasets and present a single set of plots.

In Figure 18.6a, conditional Pearson residuals are plotted against predicted reading scores. If a model is a reasonable one, we should see a random collection of points, approximately equal residual variances across levels of the predicted variables, and no apparent patterns. This is the case in our example. A pattern will often be seen when a response variable is subject to either a floor or ceiling effect (i.e., scores were bounded either from below or above). In such a case, we would likely see a linear relationship. The analogous plot for the marginal effects shows a pattern in that there is a maximum pos-

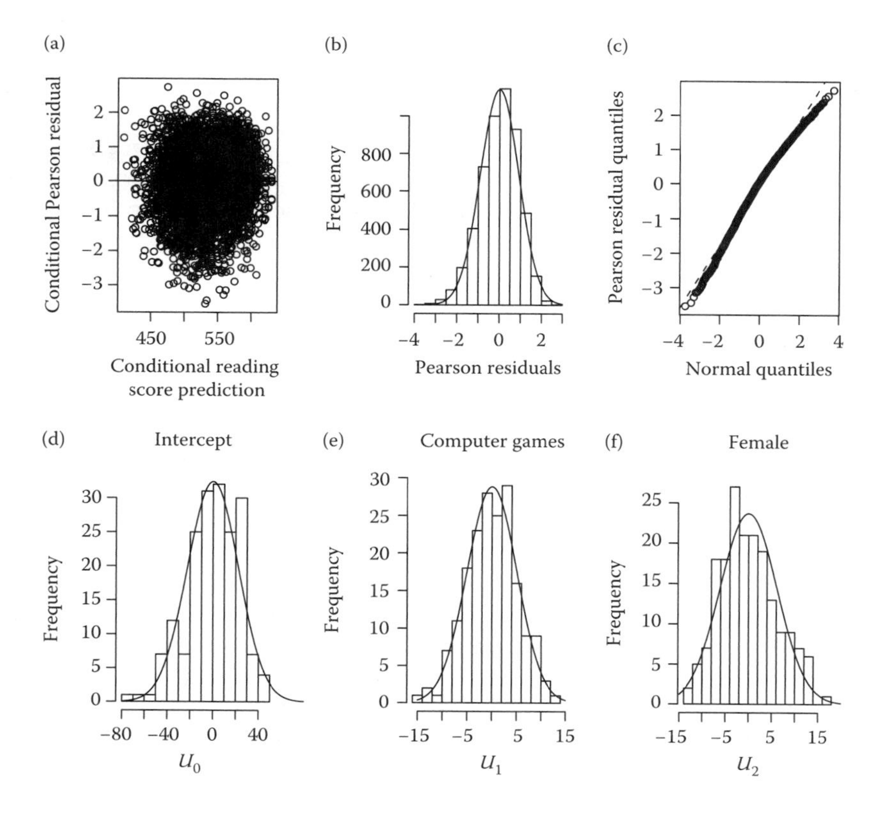

FIGURE 18.6
Various model diagnostics for Model 3 that include empirical Bayes estimates for random effects.

sible value for the predicted reading scores (i.e., a vertical line of points at predicted reading scores equal to around 575).

Figures 18.6b and 18.6c are designed to examine the normality assumption. Figure 18.6b is a histogram of the conditional Pearson residuals with the normal distribution overlaid and Figure 18.6c is a Q–Q plot. From the Q–Q plot, it appears that the tails of the distribution of residuals are a bit heavy, but overall normality seems reasonable.

Figures 18.6d through 18.6f are histograms of the estimated random effects with normal distributions overlaid. If the assumption that the random effects are multivariate normal is satisfied, then the distribution of estimated U_{pj}'s should be normal. Although the estimated U_{pj}'s look roughly normal, this only means that the assumption is tenable. It is possible that the true underlying distribution of random effects is not normal but the estimates of the U_{pj}'s may appear normal (Verbeke and Molenberghs 2000). If the distribution of $\hat{U}_{pj}$ is not normal, then the assumption is not valid.

Interpretation of Results

The final model, Model 3, has 14 regression coefficients and seven random effect parameters (variance σ^2 at level 1 and a 3×3 variance–covariance matrix at level 2) with one random intercept and two random slopes. The intercept of 544.21 is the expected reading proficiency score when all predictors are zero, implying male students who enjoy reading a lot, were not born in the United States, have a father who is also not born in the United States, have less than 25 books at home, watch TV or play computer games less than 5 h per day, and attend a school where the percentage of economically disadvantaged students is less than 10%. Female students who were born in the United States, have a father born in the United States, and/or have more books at home on average would have better reading performance. On the other hand, those who do not enjoy reading as much, watch TV or play computer games 5 h or more per day, and/or attend a school with a higher percentage of economically disadvantaged students on average would be expected to have lower reading proficiency scores.

The random intercept represents heterogeneity across schools in terms of their residual reading performance even after accounting for explainable differences by the fixed effects. Model 3 has two random slopes, both corresponding to predictors that are binary variables. The random slopes suggest that the effects of gender as well as excessive hours on computer games vary across schools. Since the two variables are both binary, we can consider four subgroups whose variances are different. For male students who spend less than 5 h on computer games (i.e., `female` = 0, `comgames5H` = 0), the variance of reading proficiency is estimated as $\hat{\tau}_{00} + \hat{\sigma}^2 = 4485.88$. For male students who spend 5 or more hours on computer games, the variance is $\hat{\tau}_{00} + 2\hat{\tau}_{02} + \hat{\tau}_{22} + \hat{\sigma}^2 = 4485.65$. For female students who spend less than 5 h on computer games, the variance is $\hat{\tau}_{00} + 2\hat{\tau}_{01} + \hat{\tau}_{11} + \hat{\sigma}^2 = 4208.72$.

Finally, for female students who spend 5 or more hours on computer games, the variance is estimated as $\hat{\tau}_{00} + 2\hat{\tau}_{01} + \hat{\tau}_{11} + 2\hat{\tau}_{02} + 2\hat{\tau}_{12} + \hat{\tau}_{22} + \hat{\sigma}^2 = 4235.31$. Therefore, the heterogeneity is the greatest among males who spend less than 5 h on computer games. Females who spend less than 5 h on computer games (the highest performance group among the four) are the most homogeneous group of students.

Discussion

Other Multilevel Models

Although the two-level models in this chapter where students are nested within schools are common in large-scale assessment data, other multilevel models are also widely used and are gaining greater interest. Three in particular are worth considering in the analysis of international assessment data: three-level models, repeated measures models, and nonlinear models for discrete outcomes. For three-level models, countries or other primary sampling units, such as states, districts, or counties, can be entered as the highest level. Depending on the number of level-3 units, distributional assumptions, and omitted factors, the highest level can be treated as *fixed* or *random*. There exists an extensive literature on this topic in econometrics (Hausman 1978; Hausman and Taylor 1981) and more recently in psychometrics (Kim and Frees, 2006, 2007). For example, we can extend the current analysis by considering another country or two and modeling results for multiple countries simultaneously. In such analyses, it should be clarified that the meanings of the variables and levels are comparable across countries. However, many variables in such analyses need to be analyzed in context and their meanings may not be exchangeable across countries (e.g., born in the country, parents born in the country, first language, economically disadvantaged, and eligible for free or reduced-price meal).

Another important type of multilevel analysis is repeated measures analysis, which includes but is not limited to longitudinal data analysis. As large-scale assessment data often consist of scores on multiple subjects or multiple aspects of a subject (e.g., reading achievement/behavior/attitude or different domains of mathematics), several related outcomes can be available for each student. In a longitudinal design, students' performances and attitudes may be recorded over time to study intraindividual change as well as interindividual differences. When multiple dependent variables are nested within students, these repeated measures can be treated as level-1 units in multilevel models. In such models, students are often entered as level-2 units, and schools are considered as level-3 units. Dependency among the level-1 units is usually very high in repeated measures models because the values belong to the same students. A flexible variance–covariance structure may be needed to account for the complex dependency among repeated measures. Multivariate multilevel models or latent growth curve models are commonly used for multiple subjects with longitudinal assessment data.

Finally, large-scale assessment data almost always include survey questionnaires which require respondents to select one of a given set of response options, thus yielding discrete outcomes. Discrete outcomes data call for the use of alternative models designed to handle their special nature. In the context of large-scale assessment studies that use complex sampling designs, complications arise because the data are typically clustered data and the probability of selection is not equal for all units. Chapter 21 by Anderson, Kim, and Keller in this volume is devoted to this issue. Multilevel models for discrete outcomes are also referred to as *generalized linear mixed models.*

Current and Future Research in Multilevel Analysis

There was a time when multilevel analysis was considered a highly advanced statistical method and only used by a limited number of trained methodologists. In the recent years, however, multilevel modeling has become available to many substantive researchers and practitioners, in our opinion largely owing to the publication of accessible textbooks, the availability of user-friendly software, and regularly offered courses and workshops. Several textbooks appear to be widely used across the disciplines, including Raudenbush and Bryk (2002), Snijders and Bosker (2012), Hox (2010), and Goldstein (2010).

Multilevel analysis has proven its importance in the social sciences, and the methodology has great potential for advancing the design and utilization of large-scale assessment data. In educational research, despite immense effort and investment, results from educational effectiveness studies remain inconsistent (Ehrenberg et al. 2001; Goldhaber and Brewer 1997; Ludwig and Bassi 1999). There are mixed findings concerning which district, school, teacher, neighborhood, family, and student variables make significant differences on educational outcomes such as improving students' academic achievement or reducing dropout rates, let alone the size of their effects. More recently, the Programme for International Student Assessment (PISA) and other international assessments have found counterintuitive relationships between achievement scores and attitudes at the country level. For example, in the PISA 2006 data, students with higher science achievement scores indicated more positive attitudes toward science as expected, when the data are analyzed within countries. However, the relationship between achievement and attitudes was reversed at the country level and it was found that high science achievement was associated with more negative attitudes based on between-country correlations (Lu and Bolt, 2012).

Puzzling or inconsistent findings in educational research are due to multiple factors including inherent difficulties in isolating the effects of variables involved in complex educational processes, omitted confounding factors, nonrandom missing data and attrition, selection bias related to school and program choice, and different dynamics among variables across clusters (e.g., countries or schools). Yet, it is often infeasible to design comprehensive experimental studies in school settings for apparent ethical and/or practical reasons. It is also impractical to expect no missing data or missing completely at random, or to collect all relevant variables to account for selection mechanisms and cultural differences.

Consequently, there is always a danger in education as well as other areas in the social sciences that decision making and policy implementation can be based on inaccurate information or in some cases misleading findings. Although we attempted to be careful in our model specification and diagnostics, the purpose of our data analysis is for demonstrating the application of multilevel analysis to large-scale assessment data rather than making practical recommendations or drawing substantive conclusions, as further investigation is necessary to provide stronger claims of causal inference between predictors and reading outcomes. Chapter 22 by Robinson in this handbook deals with making causal inferences using large-scale assessment data.

Despite many advances during the past several decades, there remain important theoretical and technical issues to be addressed in multilevel analysis. For example, although there exists a large body of literature with regard to imputation approaches for missing data, there is a lack of methodological development for effectively imputing missing values while preserving yet utilizing the multilevel structure, especially for discrete outcomes (for continuous outcomes, see Chapter 20 by Shin in this volume). Also, whereas various matching designs and strategies have been proposed in studies of educational effectiveness with observational data, little has been done on matching nonequivalent groups in the context of multilevel data. This delayed progress is largely due to the complexity of adapting statistical techniques for imputation and matching strategies into the multilevel framework, as the necessary modifications are not straightforward (Steiner et al. 2012). Nonetheless, considering the needs of these techniques and active research on these topics, we expect promising methodological developments as well as increasing applications of multilevel analysis in the coming years.

Acknowledgment

This work was supported in part by grant R305D120005 from the Institute of Education Sciences, the U.S. Department of Education awarded to the first author.

References

Allison, P. D. 2002. *Missing Data.* Sage, Newbury Park, CA.

Amer, S. R. 2009. Neural network imputation in complex survey design. *International Journal of Electrical and Electronics Engineering*, 3:52–57.

Diggle, P. J., Liang, J.-Y., and Zeger, S. L. 1994. *Analysis of Longitudinal Data.* Oxford University Press, NY.

Ehrenberg, R. G., Brewer, D. J., Gamoran, A., and Willms, J. D. 2001. Class size and student achievement. *Psychological Science in the Public Interest*, 2:1–30.

Enders, C. K. 2010. *Applied Missing Data Analysis.* Guilford Press, New York, NY.

Goldhaber, D. D. and Brewer, D. J. 1997. Why don't schools and teachers seem to matter? Assessing the impact of unobservables on educational productivity. *The Journal of Human Resources*, 32, 505–523.

Goldstein, H. 2010. *Multilevel Statistical Models.* 4th Edition, John Wiley and Sons, Ltd, Chichester, UK.

Grilli, L. and Pratesi, M. 2004. Weighted estimation in multilevel ordinal and binary models in the presence of informative sampling designs. *Survey Methodology*, 30:93–103.

Hausman, J. A. 1978. Specification tests in econometrics. *Econometrica*, 46:1251–1272.

Hausman, J. A. and Taylor, W. E. 1981. Panel data and unobservable individual effects. *Econometrica*, 49:1377–1398.

Hox, J. J. 2010. *Multilevel Analysis: Techniques and Applications.* 2nd edition, Routledge, New York, NY.

Huber, P. J. 1967. The behavior of maximum likelihood estimates under nonstandard conditions. *Proceedings of the Fifth Berkeley Symposium on Mathematical Statistics and Probability*, 1:221–231.

Kim, J.-S. and Frees, E. W. 2006. Omitted variables in multilevel models. *Psychometrika*, 71:659–690.

Kim, J.-S. and Frees, E. W. 2007. Multilevel modeling with correlated effects. *Psychometrika*, 72:505–533.

Kim, J.-S. and Swoboda, C. 2012. Strategies for imputing missing values in hierarchical data: Multilevel multiple imputation. Paper presented at the *Annual Meeting of American Educational Research Association*, Vancouver, CA.

Kish, L. 1965. *Survey Sampling.* John Wiley & Sons, Inc., New York, NY.

Little, R. J. A. and Rubin, D. B. 2002. 2nd edition, *Statistical Analysis with Missing Data.* John Wiley, New York.

Lu, Y. and Bolt, D. 2012. Application of a MIRT model to address response style differences in cross-national assessment of educational attitudes. Paper presented at the *Annual Meeting of National Council on Measurement in Education*, Vancouver, CA.

Ludwig, J. and Bassi, L. 1999. The puzzling case of school resources and student achievement. *Educational Evaluation and Policy Analysis*, 21:385–403.

Martin, M. O., Mullis, I. V. S., and Kennedy, A. M. 2007. PIRLS2006 technical report. Technical report, Chestnut Hill, MA: TIMSS & PIRLS International Study Center, Boston College.

Mislevy, R. J. 1991. Randomization-based inference about latent variables from complex samples. *Psychometrika*, 56:177–196.

Muthén, L. K. and Muthén, B. 1998–2010. *Mplus User's Guide.* 6th edition, Muthén & Muthén, Los Angeles, CA.

Pfeffermann, D., Skinner, C., Holmes, D., Goldstein, H., and Rasbash, J. 1998. Weighting for unequal selection probabilities in multilevel models (with discussion). *Journal of the Royal Statistical Society, Series B*, 60:23–56.

Pinheiro, J. and Bates, D. 2000. *Mixed-Effects Models in S and S-PLUS*. Springer Verlag, New York, NY.

Rabe-Hesketh, S. and Skronkal, A. 2006. Multilevel modeling of complex survey data. *Journal of the Royal Statistical Society, Series A*, 169:805–827.

Raudenbush, S. W. and Bryk, A. S. 2002. *Hierarchical Linear Models*. Sage, Thousand Oaks, CA.

Recchia, A. 2010. R-squared measures for two-level hierarchical linear models using SAS. *Journal of Statistical Software*, 32:1–9.

Reiter, J., Raghunathan, T., and Kinney, S. 2006. The importance of modeling smapling design in multiple imputation for missing data. *Survey Methodology*, 32:143–149.

Rubin, D. B. 1976. Inference and missing data. *Biometrika*, 63:581–592.

Rutkowski, L., Gonzalez, E., Joncas, M., and von Davier, M. 2010. International large-scale assessment data: Issues in secondary analysis and reporting. *Educational Researcher*, 39:142–151.

Ryoo, J. H. 2011. Model selection with the linear mixed model for longitudinal data. *Multivariate Behavioral Research*, 46:598–624.

Schafer, J. L. 1997. *Analysis of Incomplete Missing Data*. Chapman & Hall, London.

Self, S. G. and Liang, K. Y. 1987. Asymptotic properties of maximum likelihood estimators and likelihood tests under nonstandard conditions. *Journal of the American Statistical Association*, 82:605–610.

Shin, Y. and Raudenbush, S. W. 2007. Just-identified versus overidentified two-level hierarchical linear models with missing data. *Biometrics*, 63:1262–1268.

Shin, Y. and Raudenbush, S. W. 2010. A latent cluster-mean approach to the contextual effects model with missing data. *Journal of Educational and Behavioral Statistics*, 35:26–53.

Snijders, T. A. B. and Bosker, R. J. 1994. Modeled variance in two-level models. *Sociological Methods Research*, 22:342–363.

Snijders, T. A. B. and Bosker, R. J. 1999. *Multilevel Analysis: An Introduction to Basic and Advanced Multilevel Modeling*. 1st edition, Sage, London.

Snijders, T. A. B. and Bosker, R. J. 2012. *Multilevel Analysis*. 2nd edition, Sage, Thousand Oaks, CA.

Steiner, P., Kim, J.-S., and Thoemmes, F. 2012. Matching strategies for observational multilevel data. In *Joint Statistical Meeting Proceedings, Social Statistics Section*, San Diego, CA. American Statistical Association.

Stram, D. O. and Lee, J. W. 1994. Variance components testing in the longitudinal mixed effects model. *Biometrics*, 50:1171–1177.

Stram, D. O. and Lee, J. W. 1995. Correction to: Variance components testing in the longitudinal mixed effects model. *Biometrics*, 51:1196.

Swoboda, C. M. 2011. *A New Method for Multilevel Multiple Imputation*. Unpublished doctoral dissertation, University of Wisconsin–Madison.

Van Buuren, S. 2007. Multiple imputation of discrete and continuous data by fully conditional specification. *Statistical Methods in Medical Research*, 16:219–242.

Van Buuren, S. 2011. Multiple imputation of multilevel data. In Hox, J. and Roberts, K., eds, *Handbook of Advanced Multilevel Analysis*. Taylor & Francis, New York, NY.

Van Buuren, S. 2012. *Flexible Imputation of Missing Data*. Chapman Hall/CRC, Boca Raton, FL.

Verbeke, G. and Molenberghs, G. 2000. *Linear Mixed Models for Longitudinal Data.* Springer-Verlag, New York, NY.
White, H. 1980. A heteroskedasticity-consistent covariance matrix estimator and a direct test for heteroskedasticity. *Econometrica,* 48:817–838.
Wingersky, M., Kaplan, K. B., and Beaton, A. E. 1987. Joint estimation procedures. In Beaton, A. E., ed., *Implementing the New Design: The NAEP 1983–1984 Technical Report,* pages 285–292. Educational Testing Services, Princeton, NJ.

19

Using Structural Equation Models to Analyze ILSA Data

Leslie Rutkowski
Indiana University

Yan Zhou
Indiana University

CONTENTS

Introduction

The structural equation modeling (SEM) framework is an increasingly popular data analytic method in the social sciences (MacCallum and Austin 2000; Martens 2005). The inherent flexibility of the SEM framework and developments in theory (e.g., Bollen et al. 2010; Muthén 2002) and accompanying software (e.g., lavaan, Rosseel 2012; LISREL, Jöreskog and Sörbom 1996–2001; Mplus et al. 1998–2010) allow analysts to employ complex methods and have contributed to growth of use. Of particular interest for this chapter are advances that accommodate the peculiarities of survey data in general and educational achievement data in particular. To that end, the purpose of this chapter is to demonstrate the use of SEM in analyzing international large-scale assessment data. We address issues surrounding the treatment of missing data, deciding whether and when sampling weights should be used, estimator choices, and handling plausible values. The subjects of missing data, sampling weights, and plausible values are covered in greater detail in other chapters in this volume (see Chapters 6, 8, 17, and 20 of this volume). Thus, we avoid an in-depth overview of these matters and instead concentrate on employing methods that deal with these issues in the SEM framework. We also briefly overview estimators that are well suited for SEMs and complex data. Our examples include both single- and multilevel models.

Our approach is in the interest of providing pedagogically useful examples; however, it is important to note that any model choice should be based on the research question at hand, the underlying theory, and the structure of the data. A basic familiarity with covariance structure analysis and multilevel models are assumed; however, we very briefly review the theory and notation to orient the reader.

The first section of the current chapter concisely reviews SEM theory and notation with multilevel extensions. We then provide a short discussion of the relevant survey-related and general analytic issues, including sampling weights, missing data, incorporating plausible values, and dealing with the clustered structure inherent in ILSA data via multilevel SEM. Next, we present a number of examples of applying single- and multilevel SEM to Ireland's 2009 Programme for International Student Assessment (PISA), which emphasizes reading as the main achievement domain. The choice of Ireland here is somewhat arbitrary and reflects the fact that the first author conducted a seminar for Irish researchers on using SEM with PISA data around the time this chapter was written. Each analysis can be replicated by interested readers who visit the companion website and download the necessary data and syntax.

Brief Review of the SEM Framework

Background

As we describe later in the chapter, numerous resources exist for gaining a deeper knowledge of SEM. It will suffice for now to say that a good starting place for a thorough treatment of SEM includes Bollen's (1989) classic text. Kline's (2011) and Kaplan's (2009) recent offerings are also excellent updated resources. Here, we provide enough detail to orient the reader to the rest of the chapter. Before proceeding, we note that throughout the chapter we use the terms *latent variable* and *factor* interchangeably.

Broadly, SEM (known also as *covariance structure analysis, structural modeling, latent variable modeling,* and others) is a multivariate method that allows for the specification and testing of complex, structural relationships among variables. A clear benefit of the SEM framework is that this approach explicitly estimates and accounts for measurement error in observed variables. This is an important advantage over other multivariate methods that assume variables are measured without error, which can have serious consequences on parameter estimates if observed variables are imperfectly measured (e.g., Bollen 1989, pp. 151–178). The SEM framework also has a statistical test of overall model fit (via the χ_2 test), which is typically lacking from other multivariate methods (e.g., multiple regression or ANOVA/MANOVA). A final strength of SEM noted here is the ability to add mediational pathways to the structural parts of the model.

Generally, the SEM framework deals with three classes of models: path analytic models, measurement models, and general or structural models, the last of which combines aspects of path analytic and measurement models. In the last decade, clear links between the SEM and the multilevel modeling framework have been made (e.g., Bauer 2003; Muthén 1991, 1994; Skrondal and Rabe-Hesketh 2004), which draw on the strengths of both traditions. In particular, the noted advantages of SEM have been married with multilevel modeling's advantages of accounting for the clustered structure inherent in many kinds of data (educational data as a prime example). Multilevel modeling also brings with it the possibility to estimate effects across levels and effects that can vary across higher-level units. Both single- and multilevel SEM produce parameter estimates with measures of uncertainty and tests of overall model fit. In multilevel SEM, some or all level one parameter estimates are free to vary across level two units. One can imagine, then, that the class of models possible under this framework is broad and flexible. Further adding to the possibilities are relatively recent developments in estimators that permit distributions of dependent variables (or *endogenous* variables in SEM parlance) to deviate from the typical assumption of normality. Indeed, ordinal, binary, and multicategory outcomes are all candidates for inclusion.

Notation

Measurement Model Notation

We generally adopt the *linear structural relations* (LISREL) notation as first developed by Jöreskog (e.g., Jöreskog 1973) and well explicated in Bollen (1989). Some modifications to the notation to accommodate multilevel SEMs are made and are generally in line with Kaplan's (2009) and Bollen et al.'s (2010) approach. We begin first with the notation for measurement models, and follow with the notation for structural models (simplifications of which serve to provide notation for path analyses).

To begin, consider a typical measurement model, where a q dimensional vector of observed variables, $\boldsymbol{x}$, are said to measure some n dimensional vector of latent variables, $\boldsymbol{\xi}$; the strength of the relationship between the latent and observed variables are expressed as a $q \times n$ matrix of latent variable loadings, $\boldsymbol{\Lambda}_x$; and the vector of intercepts is given as $\boldsymbol{\alpha}_x$. We can write the model generally as

$$\boldsymbol{x} = \boldsymbol{\alpha}_x + \boldsymbol{\Lambda}_x \boldsymbol{\xi} + \boldsymbol{\delta}. \tag{19.1}$$

This specification corresponds to a measurement model where the latent variable is exogenous. In situations where the latent variable is endogenous, a y-specification is used, where the vector of observed variables are denoted as $\boldsymbol{y}$, the latent variable is given as $\boldsymbol{\eta}$, and the vector of measurement errors are denoted by $\boldsymbol{\delta}$. Using algebra of expectations and covariance, we can then express the fundamental factor analytic equation, where the covariance matrix of the observed variables is a function of the latent variables, as

$$\boldsymbol{\Sigma} = \boldsymbol{\Lambda}_x \boldsymbol{\Phi} \boldsymbol{\Lambda}_x' + \boldsymbol{\Theta}_\delta. \tag{19.2}$$

Under this specification, $\boldsymbol{\Phi}$ is an $n \times n$ matrix of latent variable variances and covariances and $\boldsymbol{\Theta}_\delta$ is a $q \times q$ (often) diagonal matrix of error variances. We differentiate this single-level measurement model from the multilevel measurement model, where the vector of observed variables for individual i ($i = 1,2,\ldots n_g$) in group g ($g = 1,2,\ldots G$) is specified as

$$\boldsymbol{x}_{ig} = \boldsymbol{\alpha}_x + \underbrace{\boldsymbol{\Lambda}_{w_{ig}} \boldsymbol{\xi}_{w_{ig}} + \boldsymbol{\delta}_{w_{ig}}}_{\text{within}} + \underbrace{\boldsymbol{\Lambda}_{b_g} \boldsymbol{\xi}_{b_g} + \boldsymbol{\delta}_{b_g}}_{\text{between}}. \tag{19.3}$$

According to this specification, $\boldsymbol{\Lambda}_{wi_g}$ is the within-groups matrix of loadings on the within-groups latent variable, $\boldsymbol{\xi}_{wi_g}$, and $\boldsymbol{\delta}_{wi_g}$ is the within-groups measurement error. The between-groups matrix of loadings, $\boldsymbol{\Lambda}_{b_g}$, expresses the relationship between the vector of observed variables and the between-groups latent variable, ξ_{b_g}. Finally, $\boldsymbol{\delta}_{b_g}$ is the between-groups measurement error. As should be apparent from the model specification, the between-groups part of the model may be different from the within-groups part of the model, pointing to latent variable structure differences at different levels of the data hierarchy. Endogenous specifications simply require substituting the appropriate y-specification notation, mentioned above.

Given a multilevel model specification, the fundamental factor analytic equation can be decomposed into a within- and between-groups part as

$$\boldsymbol{\Sigma}_T = \underbrace{\boldsymbol{\Lambda}_{x_w}\boldsymbol{\Phi}_w\boldsymbol{\Lambda}'_{x_w} + \boldsymbol{\Theta}_{\delta_w}}_{\text{within}} + \underbrace{\boldsymbol{\Lambda}_{x_b}\boldsymbol{\Phi}_b\boldsymbol{\Lambda}'_{x_b} + \boldsymbol{\Theta}_{\delta_b}}_{\text{between}}, \tag{19.4}$$

where the total covariance matrix of observed variables, $\boldsymbol{\Sigma}_T$, is a function of the within- and between-groups parts of the models. That is, $\boldsymbol{\Lambda}_{x_w}$ and $\boldsymbol{\Lambda}_{x_b}$ are as previously defined; $\boldsymbol{\Phi}_w$ and $\boldsymbol{\Phi}_b$ are the matrices of within- and between-groups latent variable variances and covariances, respectively; and $\boldsymbol{\Theta}_{\delta_w}$ and $\boldsymbol{\Theta}_{\delta_b}$ are the (usually) diagonal matrices of error variances for the within- and between-groups parts of the models, respectively.

Structural and Path Model Notation

We begin with the single-level specification for the general structural model. This is given as

$$\boldsymbol{\eta} = \boldsymbol{\alpha} + B\boldsymbol{\eta} + \boldsymbol{\Gamma}\boldsymbol{\xi} + \boldsymbol{\zeta}, \tag{19.5}$$

where $\boldsymbol{\eta}$ is an $m \times 1$ vector of endogenous latent variables and $\boldsymbol{\xi}$ is an $n \times 1$ vector of exogenous latent variables. The $m \times m$ matrix of endogenous and $m \times n$ matrix of exogenous regression coefficients are given as $\boldsymbol{B}$ and $\boldsymbol{\Gamma}$, respectively. The $m \times 1$ vector of disturbance terms is given by $\boldsymbol{\zeta}$ and $\boldsymbol{\alpha}$ is an $m \times 1$ vector of structural intercepts. Note that in a general structural model, the latent variables are measured by observed variables (through an associated measurement model, as specified above). A path analytic model is a special case of this general model where the latent variables, $\boldsymbol{\eta}$ and $\boldsymbol{\xi}$, are given by vectors of observed variables, $\boldsymbol{y}$ and $\boldsymbol{x}$.

We can generalize the structural model to accommodate a multilevel structure in the following way:

$$\boldsymbol{\eta}_{ig} = \boldsymbol{\alpha}_g + \boldsymbol{B}_g\boldsymbol{\eta}_{ig} + \boldsymbol{\Gamma}_g\boldsymbol{\xi}_{ig} + \boldsymbol{\zeta}_{ig}. \tag{19.6}$$

According to this specification, the latent variables for individual i are nested within group g. The structural parameters are then free to vary randomly across groups. It is possible under this sort of framework to develop models for the randomly varying coefficients, $\boldsymbol{\alpha}_g$, $\boldsymbol{B}_g$, and $\boldsymbol{\Gamma}_g$. One such specification for a between-groups model could written as

$$\boldsymbol{\alpha}_g = \boldsymbol{\alpha}_{00} + \boldsymbol{\alpha}_{01}\boldsymbol{\eta}_g + \boldsymbol{\alpha}_{02}\boldsymbol{\xi}_g + \boldsymbol{U}_g, \tag{19.7}$$

$$\boldsymbol{B}_g = \boldsymbol{B}_{00} + \boldsymbol{B}_{01}\boldsymbol{\eta}_g + \boldsymbol{B}_{02}\boldsymbol{\xi}_g + E_g, \tag{19.8}$$

$$\boldsymbol{\Gamma}_g = \boldsymbol{\Gamma}_{00} + \boldsymbol{\Gamma}_{01}\boldsymbol{\eta}_g + \boldsymbol{\Gamma}_{02}\boldsymbol{\xi}_g + \boldsymbol{R}_g. \tag{19.9}$$

According to these models, the intercepts, endogenous, and exogenous regression coefficients are functions of between-groups endogenous latent variables, $\boldsymbol{\eta}_g$ and between-groups exogenous latent variables, $\boldsymbol{\xi}_g$. The error variances (i.e., var($\boldsymbol{U}_g$), var($\boldsymbol{E}_g$), var($\boldsymbol{R}_g$)) provide evidence in favor of or against between-group differences in the intercepts or slopes. We note that observed variables can also serve anywhere that endogenous and exogenous latent variables appear in the multilevel structural equation model. It is a notational difference where z_g and w_g are substituted for $\boldsymbol{\eta}_g$ and $\boldsymbol{\xi}_g$, and y_{ig} and x_{ig} take the place of $\boldsymbol{\eta}_{ig}$ and $\boldsymbol{\xi}_{ig}$. It should also be clear that there is no associated measurement model, since we assume that that the observed variables perfectly represent their underlying latent construct.

Estimation

A thorough treatment of SEM estimation is beyond the scope of this chapter; however, we note a particular estimation method and a few modifications of it to motivate our choice of estimators. In general, estimation is concerned with finding values of model parameters that minimize the discrepancy between the observed data and the model. In SEM, these parameters include structural coefficients, loadings, latent variable variances and covariances, exogenous observed variable variances and covariances, latent error variances, and measurement error variances. And the fundamental equation that expresses the relationship between the observed variables and the model parameters is given by

$$\boldsymbol{\Sigma} = \boldsymbol{\Sigma}(\boldsymbol{\theta}). \tag{19.10}$$

According to Equation 19.10, the population covariance matrix of observed variables, $\boldsymbol{\Sigma}$, is equal to the population covariance matrix as a function of the parameters, $\boldsymbol{\Sigma}(\boldsymbol{\theta})$, where $\boldsymbol{\theta}$ is a column vector of the model parameters. Practically speaking, the population values are unknown and we estimate them from the sample covariance matrix of the observed variables, $\boldsymbol{S}$. Values of $\boldsymbol{\theta}$ are chosen such that the distance between $\boldsymbol{S}$ and the model implied covariance matrix, $\hat{\boldsymbol{\Sigma}}$, is minimized. The most widely used method for estimating the model parameters is via the maximum likelihood fitting function, which can be written as

$$F_{\mathrm{ML}} = \log|\boldsymbol{\Sigma}(\boldsymbol{\theta})| + \mathrm{tr}\left(\boldsymbol{S}\Sigma^{-1}(\boldsymbol{\theta})\right) - \log|\boldsymbol{S}| - (p+q). \quad (19.11)$$

In Equation 19.11, p and q refer to the number of endogenous and exogenous observed variables, respectively. When the assumption of multivariate normality in the observed variables holds, the resultant parameters are asymptotically unbiased, asymptotically efficient, and consistent. This fitting function also assumes a single-level model without sampling weights (and no intercepts); however, many analysts recognize the advantages of multilevel models and realize the importance of considering the sampling design when choosing a model. To accommodate multilevel models with sampling weights, Asparouhov (2006) and Asparouhov and Muthen (2006) have shown that a *multilevel pseudo maximum likelihood* (MPML) estimator performs well in single- and multilevel models with sampling weights. Although the technical details are beyond the scope of this chapter, this estimator is likelihood based, which adjusts the asymptotic covariance matrix of the parameters by appropriately weighting the likelihood function. In Mplus, the MPML estimator is produced by calling the MLR estimator and by assigning sampling weights. We use this estimator throughout.

Issues Associated with Applying the SEM Framework to ILSA Data

As noted elsewhere in this volume, ILSA data bring great analytic possibilities and a collection of peculiarities, to which attention must be given in order to appropriately estimate parameters. In this section, we cover the approach we used for dealing with missing data, taking the study design into account, and handling plausible values.

Clustered Data Structure

Given the inherently clustered structure of PISA data (and most other ILSA data), we include in our examples multilevel path, measurement, and structural models. Multilevel SEMs allow researchers to examine relationships at and across multiple levels. This approach also explicitly takes into account the clustered structure of the data, which allows for dependencies within groups (e.g., students within schools are similar to one another). For a detailed explication of multilevel models in general, readers should consult Raudenbush and Bryk (2002) or Snijders and Bosker (2011). Kline (2011) and Kaplan (2009) are excellent references for an introduction to multilevel SEM. In the present chapter, we use compact path model symbolism typically associated with Mplus to represent the multilevel SEMs.

Missing Data

The issue of dealing with missing ILSA data is additionally covered in several chapters of the current volume (see Chapters 18, 20, and 21 of this volume); however, we discuss the particular method we use for handling missing data here. Missing data is a problem ubiquitous to applied statistics and it is well known that ignoring missing data or using ad hoc methods can have serious consequences on parameter estimates and associated inferences (e.g., Little and Rubin 2002; Rubin 1987). The means for dealing with missing data in a principled way have grown vastly in the last two decades and increasingly these methods, including multiple imputation, are available in commercially available software. In this chapter, we employ multiple imputation, which assumes that the data are missing at random (MAR). Although the methods for imputing missing data are reasonably well established, methods for handling missing multilevel data when sampling weights are involved are less developed. In the current chapter, we take an approach that preserves the within-groups correlations among the analyzed variables.

To deal with missing data and preserve features of the data (within-school correlations), we impute student-level data by school, using the Markov chain Monte Carlo (MCMC) algorithm via the MI procedure in SAS (SAS, 2003). This method assumes multivariate normality and allows for an arbitrary missing pattern. Imputing by school also allows the within-school variance to vary across clusters, a desirable property noted by van Buuren (2011). For simplicity sake, all student-level variables in our example analyses were used in the imputation model at the student level. Researchers should make decisions regarding the components of an imputation model based on a number of factors and our approach might not suit every need. At the school level, one analysis variable, discussed subsequently, is categorical. As such, we impute school data using the fully conditional specification (FCS; van Buuren 2007), which is an experimental feature in the multiple imputation procedure (PROC MI) in SAS 9.3. The FCS method for multiple imputation does not assume multivariate normality of the variables that have missing data. Instead, analysts specify the distribution and a reasonable imputation model of each variable to be imputed. As such, the FCS approach imputes data on a variable-by-variable basis. This is in contrast to the MCMC approach, which imputes data according to the joint model for all variables. Missing data at both levels were imputed five times.

Survey Weights

In the previous section, we noted that the MPML estimator is suggested for multilevel SEMs with sampling weights. As an important preliminary step, it is recommended that the informativeness (Pfeffermann 1993) of the sampling weights first be assessed (e.g., Asparouhov 2006; Snijders and Bosker 2011). When a sampling design is informative, the distribution of the variables associated with the sampled units is different than the distribution of variables sampled directly from the generating model. In other words, the design of the study (stratification, clustered sampling) leads to samples with different parameter estimates and inferences than if simple random samples had been drawn from the population of interest. An analytic approach that ignores the study design is said to be a *model-based* approach, while one that accounts for the study design is said to be a *design-based* approach. The model-based approach assumes that the data are the outcome of some probability model with parameters to be estimated. A design-based approach is one where inferences are made to a *finite* population and are based on the probability mechanism used for the sample selection (Snijders and Bosker 2011, pp. 217–218). Choosing an appropriate approach is a matter for investigation, the details of which we provide subsequently. When a sampling design is informative at both levels of a multilevel analysis, it is important to appropriately scale the sampling weights (Asparouhov 2006, 2008) and incorporate the scaled weights into the multilevel SEM to avoid biased parameter estimates (e.g., Stapleton 2002). When a sampling design is not informative, taking a design-based approach reduces the statistical efficiency of resultant parameter estimates (Kish 1992); however, when a sampling design is informative, taking a model-based approach can result in seriously biased parameter estimates (e.g., Pfeffermann et al. 1998).

To assess the informativeness of the sampling weights at both the student and school levels, we follow the recommendation of Snijders and Bosker (2011). In particular, the authors recommend against the blind application of sampling weights (a *design-based* approach) and instead advise analysts to assess whether the sampling design is *informative* (Pfeffermann, 1993). In other words, are the residuals in the model independent of the variables used in the study design? If this is the case, then the analyst can proceed with developing a model that does not take the study design into account (a *model-based* approach).

Prior to investigating the informativeness of the sampling weights, it is first necessary to derive the multilevel sampling weights. The PISA international database includes a final school weight ($w_j = \frac{1}{\pi_j}$), where π_j represents the probability of selection for school j. We use this as a level two weight. Some manipulation is necessary for the student-level weight because only a final student weight is available ($w_{i,j} = \frac{1}{\pi_{i,j}}$), where $\pi_{i,j}$ represents the joint probability of selection for the student's school and the student, given that the school was selected. To arrive at a student-level weight, ($w_{i|j} = \frac{1}{\pi_{i|j}}$), where $\pi_{i|j}$ represents the conditional probability of selection for the student given that the student's school was chosen, we divide the final student weight by the final school weight. This results in a student weight that is adjusted for nonresponse. A slight possibility exists that, under atypical situations, this procedure could results in weights less than 1. If this occurs, analysts should fix the weight as 1 (K. Rust, personal communication, March 5, 2012).

To assess the impact of applying sampling weights in our analysis, we first consider the design effects at the student and school level. The design effect is the ratio of the *effective* sample size to the actual sample size. This value gives an indication of the loss of statistical efficiency that can result from following a design-based versus a model-based analysis. To calculate the effective sample sizes at the school and student levels, respectively, we applied the following formulas:

$$N^{\text{eff}} = \frac{\left(\Sigma_j w_j\right)^2}{\Sigma_j (w_j^2)}, \tag{19.12}$$

$$n_j^{\text{eff}} = \frac{\left(\Sigma_i w_{i|j}\right)^2}{\Sigma_j (w_{i|j}^2)}. \tag{19.13}$$

Then the design effects at the school and student levels, respectively, are calculated as

$$\text{deff}_2 = N^{\text{eff}}/N \quad \text{and} \quad \text{deff}_{1j} = n_j^{\text{eff}}/n_j. \tag{19.14}$$

Of the 144 schools in the Irish PISA 2009 data, the average design effect at the student level was .993 (SD = .009), suggesting that weights are hardly variable and that ignoring weights at the student level will not lead to biased estimates. In contrast, the school-level design effect was estimated as .773, suggesting that the sample of schools has the precision of a sample that is only three-quarters as large as a simple random sample of the same size (144 vs. 111). As such, accounting for the study design at level two is advised.

A number of options exist for treating the sampling design, including adding stratification variables to models, fitting models within each level of given stratification variables, and comparing model-based and design-based estimators separately for each cluster. Given that we are illustrating a number of models for pedagogical purposes, we take a middle ground and apply weights at the school level to all multilevel models and final student weights ($w_{i,j} = \frac{1}{\pi_{i,j}}$). to all single-level models. For all multilevel models, we also examine for differences between the model-based and design-based estimators via Asparouhov's (2006) informativeness measure, given as

$$I_2 = \frac{\hat{\gamma}_h^{\text{HLM}} - \hat{\gamma}_h^{W}}{\sqrt{\hat{\Sigma}_{hh}^{\text{HLM}}}}, \tag{19.15}$$

where $\hat{\gamma}_h^{\text{HLM}}$ is the model-based fixed component for variable X_h, $\hat{\gamma}_h^{W}$ is the design-based fixed component for variable X_h, and $\hat{\Sigma}_{hh}^{\text{HLM}}$ is the squared standard error for $\hat{\gamma}_h^{\text{HLM}}$. Where $I_2 < 2$, the sampling design has little effect on the parameter estimates and a model-based analysis is preferred. Regardless of electing a model- or design-based analysis, we acknowledge the study design throughout by correcting all standard errors to account for stratified sampling. This is done by using the TYPE = COMPLEX option in Mplus, which uses a sandwich estimator in the calculation of standard errors (Huber 1967; White 1982). Finally, we note that the best approach for dealing with the survey design will vary by study (e.g., TIMSS, PIRLS, and PISA) and by population (we only treat the design for Ireland in the current chapter).

Population Achievement Estimation

As noted elsewhere in this volume and in a body of literature on the topic (see, e.g., Mislevy 1984, 1985, 1991; Mislevy et al. 1992), population achievement is estimated via a method best likened to multiple imputation for observed variables. As such, any analysis that involves achievement as an outcome or a predictor should be repeated once for each of five achievement estimates. Given that we are also imputing missing data five times, we treat both of these issues jointly by fitting all models five times and combining the parameter estimates according to Rubin (1987). Many software packages, including Mplus, have ready facilities for handling this type of analysis.

Data

As noted previously, we selected the Irish PISA 2009 student ($M = 3937$) and school ($N = 144$) data for analysis. The 2009 PISA cycle emphasized reading as the major domain; however, math and science are also included as minor domains. For the exemplar SEMs, we chose reading achievement (for models that include achievement) and a number of student- and school-level variables, briefly described in Table 19.1. Given that the variance of reading achievement was much larger than other variables in the models, we use the natural log of reading achievement as a variance stabilizing mechanism (Weisberg 2005). Descriptive statistics for these variables can be found in Table 19.2.

Models

To illustrate a few of the possibilities for analyzing ILSA data under the SEM framework, we fit three separate types of models: a path analytic model, a measurement model, and a general or structural equation model. For each model type, we fit a single-level model and a multilevel model. In each case,

TABLE 19.1

Description of Variables

Variable with Question Stem (Where Relevant)	Variable Name	Variable Label	Scale
Reading achievement	lnread1-5	Natural log of reading achievement plausible values 1 through 5	Continuous
Urbanicity (School questionnaire)			
Which of the following definitions best describes the community in which your school is located?	SC04Q01 (recoded)	City	0 = A town with less than 100,000 people; 1 = A city with at least 100,000 people
School climate (School questionnaire)			
In your school, to what extent is the learning of students hindered by the following phenomenon?	SC17Q01	a. Teachers' low expectations of students	1 = Not at all 2 = Very little 3 = To some extent 4 = A lot
	SC17Q05	e. Teachers not meeting individual students' needs	
	SC17Q06	f. Teacher absenteeism	
	SC17Q11	k. Teachers being too strict with students	
	SC17Q13	m. Students not being encouraged to achieve their full potential	
Male (Student questionnaire)			
Are you female or male?	ST04Q01 (recoded)	Male	0 = Female 1 = Male
Reading interest (Student questionnaire)			
How much do you agree or disagree with these statements about reading?	ST24Q03	c. I like talking about books with other people	1 = Strongly disagree 2 = Disagree 3 = Agree 4 = Strongly agree
	ST24Q05	e. I feel happy if I receive a book as a present	
	ST24Q07	g. I enjoy going to a bookstore or a library	
	ST24Q10	j. I like to express my opinions about books I have read	
	ST24Q11	k. I like to exchange books with my friends	

TABLE 19.1 (continued)

Description of Variables

Variable with Question Stem (Where Relevant)	Variable Name	Variable Label	Scale
Reading disinterest (Student questionnaire)			
How much do you agree or disagree with these statements about reading?	ST24Q04	d. I find it hard to finish books	1 = Strongly disagree 2 = Disagree 3 = Agree 4 = Strongly agree
	ST24Q06	f. For me, reading is a waste of time	
	ST24Q08	h. I read only to get information that I need	
	ST24Q09	i. I cannot sit still and read for more than a few minutes	

TABLE 19.2

Descriptive Statistics

Variable	*N*	Minimum	Maximum	Mean	Standard Deviation
Student					
Male	3937	0	1	0.50	0.50
I like talking about books with other people	3823	1	4	2	1
I feel happy if I receive a book as a present	3818	1	4	2	1
I enjoy going to a bookstore or a library	3822	1	4	2	1
I like to express my opinions about books I have read	3832	1	4	2	1
I like to exchange books with my friends	3828	1	4	2	1
I find it hard to finish books	3822	1	4	2	1
For me, reading is a waste of time	3820	1	4	2	1
I read only to get information that I need	3820	1	4	2	1
I cannot sit still and read for more than a few minutes	3824	1	4	2	1
Ln(1st plausible value—reading)	3937	4.53	6.70	6.19	0.22
School					
City	127	0	1	0.28	0.45
Teachers' low expectations of students	125	1	3	1.82	0.76
Teachers not meeting individual students' needs	124	1	4	2.11	0.63
Teacher absenteeism	125	1	3	1.84	0.61
Teachers being too strict with students	125	1	4	1.74	0.66
Students not being encouraged to achieve their full potential	125	1	4	1.72	0.75

we acknowledge the study design by applying sampling weights at the school level, estimating and reporting Asparouhov's (2006) informativeness index, and adjusting standard errors for nonrandom sampling by using a robust estimator. It is notable that the models represented in this chapter are just a very small sample of what is possible in the single- and multilevel SEM framework. With respect to assessing the model-data consistency—in other words, the adequacy of the model fit to the data—we report the chi-square test of model fit and several standard fit indices including the comparative fit index (CFI), the Tucker–Lewis index (TLI), and the root mean-squared error of approximation (RMSEA). For technical and interpretive details on the chi-square test and the fit indices, interested readers are directed to Bollen (1989), Kaplan (2009), or Kline (2011).

Path Models

We illustrate the application of single- and multilevel path models to Ireland's PISA 2009 data with path model specifications illustrated in Figures 19.1 and 19.2. The single-level path model specifies that gender predicts three reading attitude variables (*talk with others*; *reading for information*; and *exchange books*). In turn, these variables are hypothesized to predict reading achievement. The multilevel path analysis adds a between-groups model that allows average reading achievement and the effect of *talk with others* on *reading achievement* to vary randomly across schools. Further, the group varying terms (represented as latent variables in the between-groups models) are predicted by *teachers' low expectation of students*. The additional random terms are represented by filled black dots in the within-groups part of the model.

Measurement Models

The single- and multilevel measurement model specifications can be found in Figures 19.3 and 19.4. In Figure 19.3, the single-level measurement model is specified as a two-factor model of *reading interest* and *reading disinterest*. Reading interest is measured by five variables (*talk with others*; *book present*;

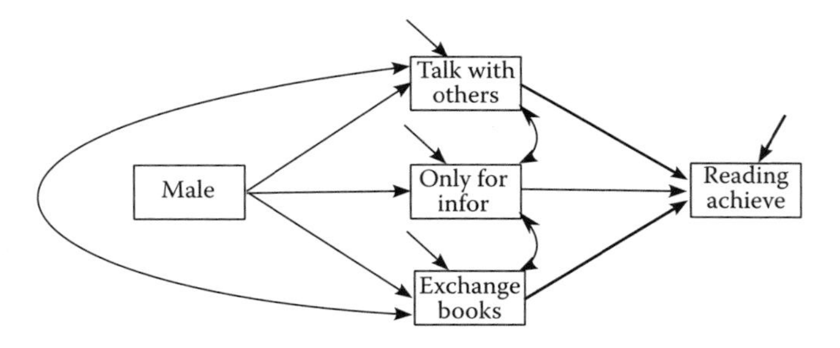

FIGURE 19.1
Single-level path model of student gender and reading attitude on achievement.

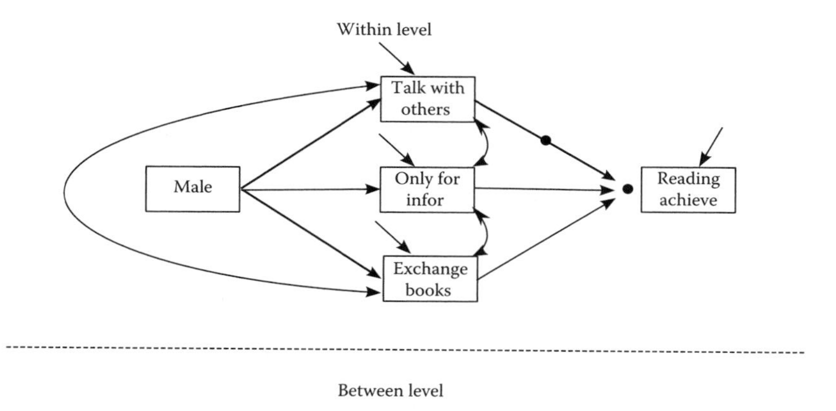

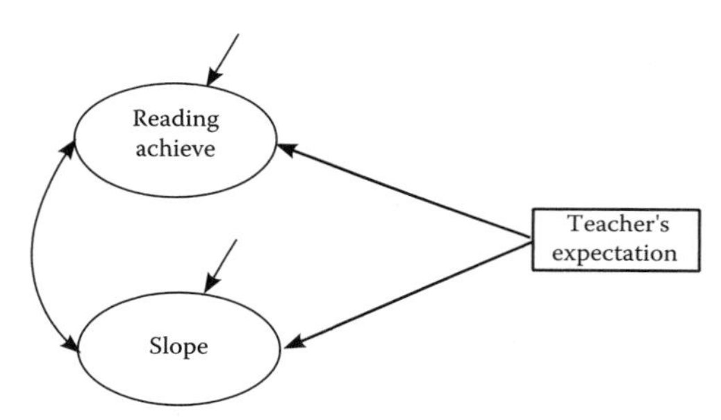

FIGURE 19.2
Two level path model with a random intercept for achievement and a random slope for the effect of talking with others; teachers expectations predict both the intercept and slope.

bookstore or library; *express opinions*; and *exchange books*). Reading disinterest is measured by four observed variables (*hard to finish*; *waste of time*; *only for information*; and *not sit still*). A clear advantage of a multilevel approach to SEM is that the between-groups model can vary substantially from the within-groups model. We illustrate this idea in Figure 19.4. Specifically, in the multilevel measurement model, the within-groups model is identical

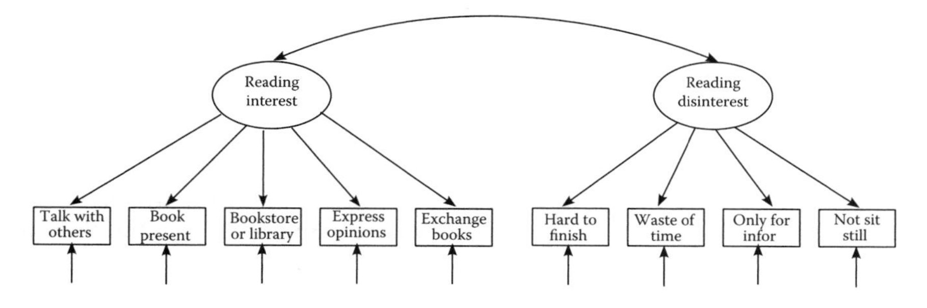

FIGURE 19.3
Single-level, two-factor measurement model of overall reading attitude.

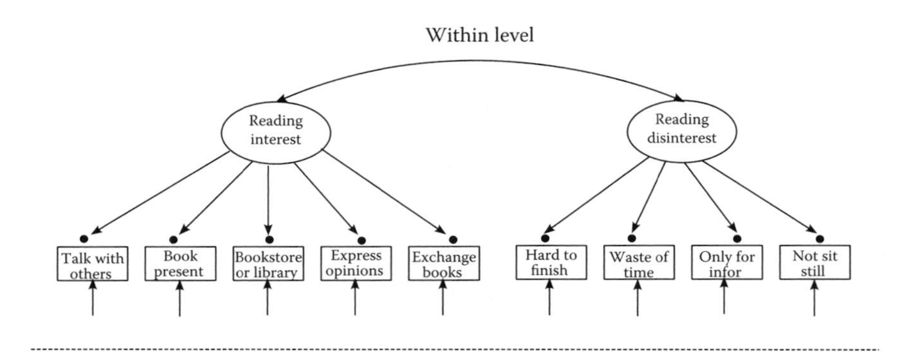

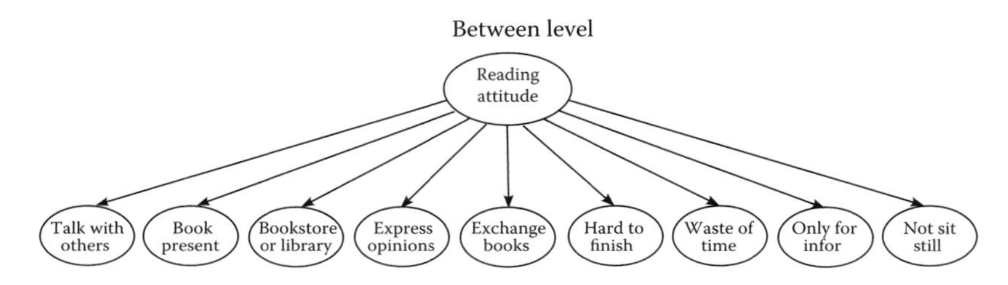

FIGURE 19.4
Two level, two-factor measurement model of overall reading attitude with random intercepts for all observed variables.

to Figure 19.3; however, the between-groups model is specified as a single-factor measurement model with all nine variables measuring a school-level factor that we describe as *reading attitude*. Note that the multilevel model specifies each of the observed variables as having random intercepts (indicated by a filled black dot). And these random intercepts are represented at the school-level as latent variables. Although the latent variables are endogenous variables and typically include a disturbance term, this portion of the model is omitted because between group residual variance is usually close to zero and the added terms add significantly to the integration burden of the estimation algorithm (L. Muthen, personal communication, July 25, 2011).

Structural Models

The single- and multilevel structural model specifications are located in Figures 19.5 and 19.6. In both examples, the latent variables, *reading interest* and *reading disinterest*, are measured by the same observed variables as the measurement model example. In the single-level specification, gender has direct effects on *reading interest*, *reading disinterest*, and *reading achievement*. Further, *reading disinterest* and *reading interest* have direct effects on *reading achievement*. The multilevel structural model has a within-groups

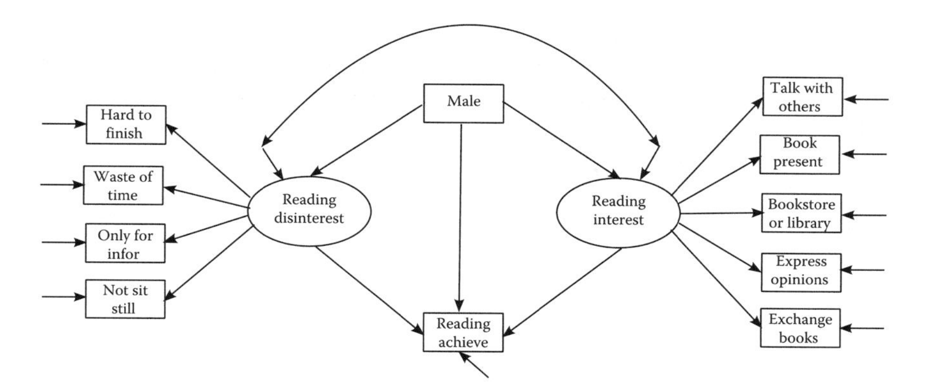

FIGURE 19.5
Single-level structural mediation model for the effect of gender and reading interest on reading achievement.

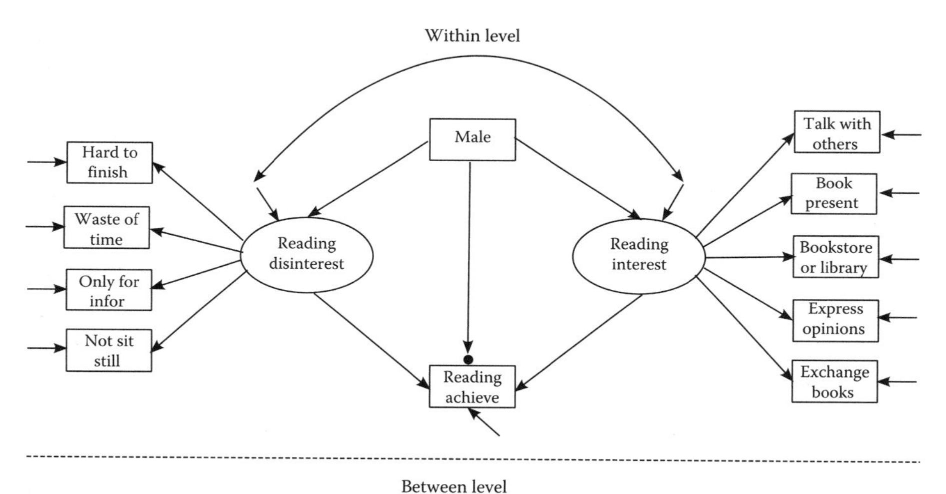

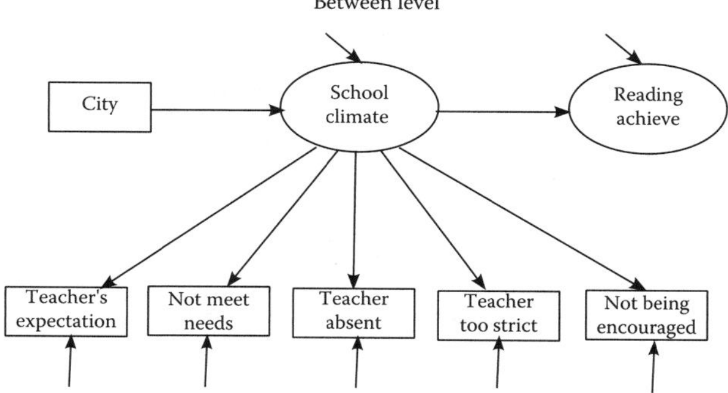

FIGURE 19.6
Two level structural mediation model for the effect of gender and reading interest on reading achievement, with an indirect effect of urbanicity on average achievement at level two.

model that is nearly identical to the single-level example with the exception of a random term for average reading achievement across schools. This random term allows average achievement to vary randomly across schools. At the between-level, group-varying achievement is modeled as a function of *school climate* (a school-level latent variable measured by *teachers' expectations; not met needs; teacher absent; teacher too strict;* and *not being encouraged*). Differences in school climate are in turn explained by the urbanicity variable *city*.

Results

Single- and Multilevel Path Analysis

The findings from the path analysis are located in Table 19.3. According to the single-level path analysis, the mediating effects of talking about books and exchanging books are positively associated with reading achievement ($\beta_{read\_talk} = 0.044$, 95% CI [0.032, 0.056], $\beta_{read\_exchange} = 0.014$, 95% CI [0.004, 0.024]), while reading for information only is negatively associated with achievement ($\beta_{read\_infor} = -0.062$, 95% CI [−0.074, −0.050]). Further, male students tend to talk about books and exchange books less than female students ($\gamma_{talk\_male} = -0.416$, 95% CI [−0.473, −0.359], $\gamma_{exchange\_male} = -0.590$, 95% CI [−0.647, −0.533]), while they read for information only more than female students ($\beta_{infor\_male} = 0.342$, 95% CI [0.289, 0.395]). It is also notable that Asparouhov's informativeness index (I_2) is quite small (<2) for all parameters. This indicates that there are only small differences between the reported model that includes sampling weights and the model that does not include sampling weights. According to the chi-square test of model fit, this is a fairly poor-fitting model and some misspecification has likely occurred. In addition, based on other indices of model fit test (not shown), CFI = 0.982, TLI = 0.816, RMSEA = 0.119, this model fits poorly.

The findings from the within-groups part of the multilevel path analysis are consistent with the single-level analysis; however, the parameter estimates are generally smaller in the multilevel model and the effect of exchanging books with friends is not significant. The standard errors are also consistently larger in the multilevel results. It is important to note here that the absence of a reported effect for the regression of reading achievement on talking about books reflects the fact that this was specified as a random slope. The overall estimate of this relationship is located in the between-school part of the model as *random slope with reading achievement.* This suggests that allowing the relationship between talking about books and achievement to vary randomly across schools results in a reversed relationship between these two variables. That said, the actual magnitude of this

TABLE 19.3

Single- and Multilevel Path Analysis Results

	Single-Level PA			**Multilevel PA**		
	Estimate	**SE**	I_2	**Estimate**	**SE**	I_2
				Within-School Level		
Reading achievement on						
Talk about books	0.044*	0.006	−0.400	—	—	—
Only for information	−0.062*	0.006	−0.333	−0.056*	0.008	−0.500
Exchange books	0.014*	0.005	0.200	0.006	0.008	0.800
Talk about books on male	−0.416*	0.029	0.250	−0.389*	0.038	−0.629
Only for information on male	0.342*	0.027	−0.519	0.310*	0.044	0.553
Exchange books on male	−0.590*	0.029	0.036	−0.553*	0.044	−0.825
Talk about books with						
Only for information	−0.314*	0.012	−0.167	−0.294*	0.016	−0.571
Exchange books	0.474*	0.014	−0.308	0.463*	0.020	−0.313
Only for information with exchange books	−0.323*	0.013	−0.333	−0.306*	0.016	−0.357
				Between-School Level		
Random slope on teacher's expectation				−0.014	0.009	1.167
Reading achievement on teacher's expectation				−0.003	0.027	−0.684
Random slope with reading achievement				−0.003	0.001	1.000
Model Fit Indices						
χ^2	54.456 (1 df)		—	—		—
AIC	29,417.107		—	20,981.756		—
BIC	29,523.153		—	21,134.472		—

Note: SE, standard error. I_2 = Asparouhov's (2006) informativeness index.
* Values are statistically significant at $p < .05$.

effect overall is quite small, suggesting few actual differences in achievement between someone who never talks with others about books and someone who often or always talks with others about books.

The findings from the between-groups part of the model indicate that between-school differences in achievement are unrelated to differences in teacher's expectations ($\gamma^B_{read\_expect} = -0.003$, 95% CI [−0.056, 0.050]). And the relationship between the intercept for reading and the slope for talking about books is negatively related ($\beta^B_{avg.read\_slope} = -0.003$, 95% CI [−0.005, −0.001]), although not statistically significant. In other words, we expect that the school-level effect for talking about books will be weaker in schools with higher average achievement. We also note that I_2 remains low for all parameters at both the within-groups and the between-groups part of the model.

Finally, it is important to note that the chi-square test of model fit is not available for multilevel SEMs that are specified with random slopes. Instead, we can rely on model selection indices, including Akaike's information criteria (AIC, e.g., Akaike 1987) and the Bayesian information criteria (BIC, Schwarz 1978) for comparing results across models. Both indices provide a penalty for model complexity while the BIC has a penalty that increases with increased sample size. In both cases, smaller values of these indices are preferred. In the example, the multilevel model exhibits better fit since both indices are smaller for this model.

Single- and Multilevel Measurement Model

The results for the single- and multilevel measurement models are presented in Table 19.4. As expected, the *reading interest* items load highly and significantly on the intended construct. The same is true of the *reading disinterest* items. It is also notable that the correlation between the two latent variables is negative, as could be expected. As was true in the path analysis example, the informativeness index for each parameter estimate is quite small, indicating the results between a sample weighted and unweighted model are not very different. According to the chi-square test, this model does not fit the data well. However, according to other indices of model fit the model fits data well, CFI = 0.970, TLI = 0.959, RMSEA = 0.057.

The findings for the within-groups part of the multilevel measurement model are consistent with the single-level analysis; however, in line with the path analysis results, the parameter estimates are smaller and the standard errors are larger in the multilevel measurement model. With respect to the between-groups part of the model, we note that the factor loadings are generally smaller than the within-groups model. This suggests that the strength of the relationship between *reading attitude* and the school-level items intended to measure this construct is weaker than at the student level (which, notably, is specified as a two-factor model). The multilevel measurement model also fits worse, according to the AIC and BIC, than does the single-level measurement model. Poor fit taken in conjunction with smaller between-groups factor loadings provides some evidence that the between-groups part of the model is likely misspecified. One plausible explanation could be that the between-groups model should also be specified as a two-factor model.

Single- and Multilevel Structural Equation Model

The results for both the single- and multilevel SEM are shown in Table 19.5. Comparison of the single-level measurement model in Table 19.4 to the single-level measurement models in Table 19.5 suggests very similar loadings across these models. Further, reading interest tends to be lower and reading

TABLE 19.4

Single- and Multilevel Measurement Model Results

	Single-Level Measurement Model			Multilevel Measurement Model		
	Estimate	SE	I_2	Estimate	SE	I_2
				Within-School Level		
Reading interest by						
Talk about books	1.000	0.000	—	1.000	0.000	—
Receive book presents	0.938*	0.021	0.095	0.952*	0.028	0.957
Go to library	0.957*	0.021	0.381	0.983*	0.026	0.667
Express opinions about books	0.908*	0.018	0.111	0.902*	0.025	0.800
Exchange books	1.019*	0.019	0.056	0.998*	0.021	−0.056
Reading disinterest by						
Hard to finish books	1.000	0.000	—	1.000	0.000	—
Waste of time	1.235*	0.038	0.000	1.205*	0.048	0.077
Only for information	1.025*	0.031	0.033	0.980*	0.037	0.258
Can't sit still and read	1.279*	0.034	−0.364	1.237*	0.033	0.167
Factor Covariances						
Reading interest with reading disinterest	−0.337*	0.012	0.000	−0.314*	0.013	−0.250
				Between-School Level		
Reading attitude by						
Talk about books				1.000	0.000	
Receive book presents				0.652*	0.102	−0.531
Go to library				0.613*	0.104	0.242
Express opinions about books				0.809*	0.072	0.147
Exchange books				1.346*	0.128	−0.462
Hard to finish books				−0.334*	0.122	−0.068
Waste of time				−0.747*	0.128	0.550
Only for information				−0.757*	0.111	0.087
Can't sit still and read				−0.746*	0.126	0.606
Model Fit Indices						
χ^2	341.288 (df 26)		—	468.994 (df 62)		—
AIC	74,553.292		—	74,758.248		—
BIC	74,727.957		—	74,989.054		—

Note: SE, standard error. I_2 = Asparouhov's (2006) informativeness index.
* Values are statistically significant at $p < .05$.

TABLE 19.5

Single- and Multilevel Structural Equation Model

	Single-Level SEM			Multilevel SEM		
	Estimate	SE	I_2	Estimate	SE	I_2
				Within-School Level		
Reading interest by						
Talk about books	1.000	0.000	—	1.000	0.000	—
Receive book presents	0.931*	0.021	0.100	0.928*	0.026	0.700
Go to library	0.953*	0.021	0.381	0.957*	0.028	0.583
Express opinions about books	0.902*	0.018	0.111	0.896*	0.026	0.650
Exchange books	1.025*	0.019	0.105	1.022*	0.024	−0.053
Reading disinterest by						
Hard to finish books	1.000	0.000	—	1.000	0.000	—
Waste of time	1.235*	0.038	−0.056	1.246*	0.050	0.025
Only for information	1.024*	0.031	0.000	0.990*	0.039	0.333
Cannot sit still and read	1.279*	0.034	−0.394	1.255*	0.038	−0.088
Reading interest on male	−0.402*	0.025	0.280	−0.376*	0.034	−0.576
Reading disinterest on male	0.214*	0.021	−0.476	0.191*	0.034	0.419
Reading achievement on						
Reading interest	0.000	0.015	−0.267	−0.004	0.026	−0.125
Reading disinterest	−0.167*	0.018	−0.278	−0.152*	0.030	−0.263
Male	−0.055*	0.007	0.000	−0.052*	0.012	0.600
Factor Covariances						
Reading interest with reading disinterest	−0.316*	0.011	−0.091	−0.313*	0.013	−0.417
				Between-School Level		
School climate by						
Teacher's expectation				1.000	0.000	
Not meet student's needs				0.985*	0.264	0.082
Teacher absenteeism				0.737*	0.211	0.510
Teachers being too strict				0.759*	0.256	0.522
Students not being encouraged				1.085*	0.233	0.641
School climate on city				−0.184	0.124	−0.327
Reading achievement on School climate				−0.048	0.032	0.048
Model Fit Indices						
χ^2	548.307 (df 40)		—	413.579 (df 54)		—
AIC	77,784.853		—	65,180.757		—
BIC	78,003.184		—	65,504.514		—

Note: SE, standard error. I_2 = Asparouhov's (2006) informativeness index.
* Values are statistically significant at $p < .05$.

disinterest tends to be higher for males ($\gamma_{interest\_male} = -0.402$, 95% CI [−0.451, −0.353], $\gamma_{disinterest\_male} = 0.214$, 95% CI [0.173, 0.255]), according to our results. And males and students with higher levels of disinterest tend to have lower average achievement ($\gamma_{read\_male} = -0.055$, [−0.069, −0.041], $\beta_{read\_disinterest} = -0.167$, 95% CI [−0.202, −0.132]). We also found evidence of a negative relationship between the two latent variables ($\phi_{disinterest\_interest} = -0.316$, 95% CI [−0.338, −0.294]). As with the previous examples, the informativeness indices for all parameters are low in the single-level case. The chi-square test suggests that the specified model does not fit the data well; however, the other indices indicate a reasonable fit of the model to the data (CFI = 0.958, TLI = 0.942, RMSEA = 0.058).

The multilevel results are consistent for the within-groups model (e.g., similar loadings and similar structural coefficients). As with prior examples, the parameter estimates for the within-groups part of the multilevel model are generally smaller while the parameter estimate errors are reliably larger. For this example, we specified only random intercepts for reading achievement. The model for school-level reading achievement was specified such that a mediating relationship exists between urbanicity and average achievement. Our hypothesized mediating variable is school climate, which is measured by five observed, school-level variables (defined previously). Our findings suggest that urbanicity is unrelated to school climate $\gamma^{B}_{climate\_city} = -0.184$, 95% CI [−0.427, 0.059]) and that school climate is unrelated to reading achievement ($\beta^{B}_{read\_climate} = -0.048$, 95% CI [−0.111, 0.015]). As with all previous examples, the informativeness index is small, suggesting that the unweighted solution is not markedly different than the weighted solution. According to the chi-square values and the values of the AIC and BIC, the multilevel model exhibits better fit than the single-level model, indicating that between-school processes contribute to the model fit.

Conclusion

Structural equation models represent a broad, flexible framework to specify and fit numerous classes of models. The general nature of this analytic approach makes it a useful and sensible means for understanding complex, multivariate relationships such as those found in educational settings. Further, ILSA data offer a rich data source for examining relationships and testing hypotheses about many behavioral phenomena including, but certainly not limited to, educational achievement, attitude toward learning, perceptions about the environment, and a litany of other topics. Although applying SEM methods to ILSA data offers a powerful means for scientific investigation, the complex nature of either in isolation can prove daunting for the uninitiated.

In this chapter, we intended to provide a few worked examples of fitting single- and multilevel SEMs (path models, measurement models, and general models) to the PISA 2009 data for Ireland. As part of this process, we demonstrated a straightforward means for dealing with sampling weights, missing data, and conducting analyses with plausible values. We also provided basic interpretations of the results to aid researchers who are interested in applying these methods to ILSA data in their own research. This discussion is considerably enhanced for the reader who consults with the supporting web page (http://www.indiana.edu/~ilsahb/), where SAS and Mplus syntax for replicating all of the results found herein. As previously noted, several applications of the methods described were specific to the data we used or the models that we fit to the data. We encourage readers to consult the other chapters in the current volume, the references noted throughout the current chapter, and the specific user manual and technical report for the database to be analyzed.

References

Akaike, H. 1987. Factor analysis and AIC. *Psychometrika*, *52*, 317–332.

Asparouhov, T. 2006. General multilevel modeling with sampling weights. *Communications in Statistics: Theory and Methods*, *35*, 439–460.

Asparouhov, T. 2008. Scaling of sampling weights for two level models in Mplus 4.2. Unpublished manuscript.

Asparouhov, T. and Muthen, B. 2006. Multilevel modeling of complex survey data. *Proceedings of the Joint Statistical Meeting in Seattle*, August 2006. ASA section on Survey Research Methods, 2718–2726.

Bauer, D. 2003. Estimating multilevel linear models as structural equation models. *Journal of Educational and Behavioral Statistics*, *28*(2), 135–167.

Bollen, K., 1989. *Structural Equations with Latent Variables*. Hoboken, NJ: Wiley & Sons, Inc.

Bollen, K. Bauer, D., Christ, S., and Edwards, M. 2010. Overview of structural equation models and recent extensions. In S. Kolenikov, D. Steinley, and L. Thombs (Eds.), *Statistics in the Social Sciences: Current Methodological Developments* (pp. 37–79). Hoboken, NJ: Wiley & Sons, Inc.

Huber, P. J. 1967. The behavior of maximum likelihood estimation under nonstandard conditions. *Proceedings of the Fifth Berkeley Symposium on Mathematical Statistics and Probability*, Berkeley, CA, pp. 221–233.

Jöreskog, K. G. 1973. A general method for estimating a linear structural equation system. In A. S. Goldberger and O. D. Duncan (Eds.), *Structural Equation Models in the Social Sciences* (pp. 85–112). New York: Seminar Press.

Jöreskog, K. G. and Sörbom, D. (1996–2001). *LISREL 8: User's Reference Guide* (2nd Edition). Lincolnwood, IL: Scientific Software International, Inc.

Kaplan, D. 2009. *Structural Equation Modeling: Foundations and Extensions* (2nd Edition). Newbury Park, CA: Sage Publications.

Kish, L. 1992. Weighting for unequal P_i. *Journal of Official Statistics*, *8*, 183–200.

Kline, R. 2011. *Principals and Practice of Structural Equation Modeling* (3rd Edition). New York: The Guilford Press.

Little, R. and Rubin, D. 2002. *Statistical Analysis of Missing Data* (2nd Edition). New York: Wiley.

MacCallum, R. and Austin, J. 2000. Applications of structural equations modeling to psychological research. *Annual Review of Psychology, 51*, 201–226.

Martens, M. 2005. The use of structural equation modeling in counseling psychology research. *The Counseling Psychologist, 33*(3), 269–298.

Mislevy, R. 1984. Estimating latent distributions. *Psychometrika, 49*(3), 359–381.

Mislevy, R. J. 1985. Estimation of latent group effects. *Journal of the American Statistical Association, 80*(392), 993–997.

Mislevy, R. 1991. Randomization-based inference about latent variables from complex samples. *Psychometrika, 56*(2), 177–196.

Mislevy, R. J., Beaton, A. E., Kaplan, B., and Sheehan, K. M. 1992. Estimating population characteristics from sparse matrix samples of item responses. *Journal of Educational Measurement, 29*(2), 133–161.

Muthén, B. 1991. Multilevel factor analysis of class and student achievement components. *Journal of Educational Measurement, 28*, 338–354.

Muthén, B. 1994. Multilevel covariance structure analysis. *Sociological Methods & Research, 22*(3), 376–398.

Muthén, B. 2002. Beyond SEM: General latent variable modeling. *Behaviormetrika, 29*, 81–117.

Muthén, L. K. and Muthén, B. O. (1998–2010). *Mplus User's Guide*. 6th Edition. Los Angeles, CA: Muthén & Muthén.

Pfeffermann, D. 1993. The role of sampling weights when modeling survey data. *International Statistical Review, 61*(2), 317–337.

Pfeffermann, D., Skinner, C. J., Holmes, D. J., Goldstein, H., and Rasbash, J. 1998. Weighting for unequal selection probabilities in multilevel models. *Journal of the Royal Statistical Society. Series B (Statistical Methodology), 60*(1), 23–40.

Raudenbush, S. W. and Bryk, A. 2002. *Hierarchical Linear Models: Applications and Data Analysis Methods* (2nd Edition). Newbury Park, CA: Sage.

Rosseel, Y. 2012. Lavaan: An R package for structural equation modeling. *Journal of Statistical Software, 48*(2), 1–36.

Rubin, D. 1987. *Multiple Imputation for Nonresponse in Sample Surveys*. New York: Wiley.

SAS (Version 9) [Computer software and manual]. 2003. Cary, NC: SAS Institute.

Schwarz, G. 1978. Estimating the dimension of a model. *The Annals of Statistics, 6*(2):461–464. doi:10.1214/aos/1176344136

Skrondal, A. and Rabe-Hesketh, S. 2004. *Generalized Latent Variable Modeling: Multilevel, Longitudinal, and Structural Equation Models* (pp. 49–63). New York: Chapman & Hall/CRC.

Snijders, T. and Bosker, R. 2011. *Multilevel Analysis: An Introduction to Basic and Advanced Multilevel Modeling*. Thousand Oaks, CA: Sage.

Stapleton, L. 2002. The incorporation of sample weights into multilevel structural equation models. *Structural Equation Modeling, 9*, 475–502.

Van Buuren, S. 2007. Multiple imputation of discrete and continuous data by fully conditional specification. *Statistical Methods in Medical Research, 16*, 219–242.

Van Buuren, S. 2011. Multiple imputation of multilevel data. In J. Hox and J.K. Roberts (Eds.), *Handbook of Advanced Multilevel Analysis*. New York: Routledge.

Weisberg, S. 2005. *Applied Linear Regression*. New York: John Wiley & Sons.

White, H. 1982. Maximum likelihood estimation of misspecified models. *Econometrica, 50*, 1–25.

20

Efficient Handling of Predictors and Outcomes Having Missing Values

Yongyun Shin
Virginia Commonwealth University

CONTENTS

Hierarchical organization of schooling in all nations ensures that international large-scale assessment data are multilevel where students are nested within schools and schools are nested within nations. Longitudinal follow-up of these students adds an additional level. Hierarchical or multilevel models are appropriate to analyze such data (Raudenbush and Bryk 2002; Goldstein 2003). A ubiquitous problem, however, is that explanatory as well as outcome variables may be subject to missingness at any of the levels, posing the data analyst with a challenge.

This chapter explains how to efficiently analyze a two-level hierarchical linear model given incompletely observed data where students at level 1 are nested within schools at level 2. This social setting may also apply to occasions nested within individuals, students nested within nations, and schools nested within nations. The efficient missing data method we use in this chapter aims to analyze all available data (Shin and Raudenbush 2007). The "all available data" include children with item as well as unit nonresponse, as they belong to a school and a nation having observed data and thus add information to strengthen inferences at higher levels (Shin and Raudenbush 2011; Shin 2012).

The section "Assumptions about Missing Data" clarifies the assumptions we make about missing data for efficient analysis of multilevel incomplete data. The section "Missing Data Methods" summarizes currently available methods for analysis of multilevel incomplete data. The section "Efficient Handling of Missing Data" introduces the missing data method we use in this chapter and explains how it efficiently estimates a hierarchical linear model given incomplete data. The section "Data Analysis" illustrates efficient analysis of a hierarchical linear model given the incompletely observed U.S. data from the Programme for International Student Assessment (PISA, OECD 2007). The section "Data Analysis with Plausible Values" illustrates an analysis strategy with plausible values given the PISA data where each missing value of the outcome variable is filled in or imputed with five plausible values, but predictors may be subject to missingness. The section "Extensions and Limitations" discusses the extensions and limitations of the efficient missing data method.

Assumptions about Missing Data

In this chapter, we consider analysis of incompletely observed two-level data with the most common missing data pattern in education, a general missing pattern. That is, the missing data method to be introduced in the section "Efficient Handling of Missing Data" efficiently handles explanatory as well as outcome variables that are subject to missingness with any missing patterns at a single level or multiple levels. Consequently, we do not distinguish different types of missing patterns produced by item or unit nonresponse.

Nearly all educational data sets are multilevel and have missing data. Until quite recently, researchers facing multilevel incomplete-data analysis have dropped cases with missing values. The complete-case analysis is more problematic in multilevel analysis than it is in single-level analysis. A missing national characteristic, for example, implies deletion of not only the nation but all nested schools and students within the nation. Such analysis lowers sample sizes at multiple levels to produce inefficient inferences. Consequently, the standard errors of parameter estimates will be larger than they should, resulting in conservative hypothesis tests and excessively wide confidence intervals. In addition, the missing data patterns may be associated with the deleted data to produce biased inferences (Little and Rubin 2002). Unbiased analysis is achieved when missing data are a random sample of the complete data, so that the missing data patterns are not associated with complete data, that is, when data are missing completely at random (MCAR, Rubin 1976). However, data MCAR is seldom a reasonable assumption.

Missing values may be imputed or filled in by ad hoc imputation methods such as a sample-mean substitution and a regression model-based prediction. The substituted or predicted values, however, underrepresent the true uncertainty in the missing values to produce underestimated standard errors of parameter estimates. Consequently, the resulting hypothesis tests will be too liberal and the confidence intervals too narrow. Missing values may also be filled in by other imputation methods such as a last-observation-carry-forward method (Krueger 1999) and hot deck imputation (Little and Rubin 2002) where missing values are replaced with observed values of similar units. These single imputation methods take each imputed value as if it is the true value for subsequent complete-data analysis. The estimation, however, does not take into account uncertainty due to missing data to yield understated standard errors of parameter estimates. In general, these ad hoc imputation methods are not recommended unless missing data consist of a small fraction of complete data.

In this chapter, we shall employ two comparatively mild assumptions in many applications that data are missing at random (MAR) and that the parameters, θ, of the desired hierarchical model are distinct from the nuisance parameters, ϕ, of the missing data generating mechanism or the model for missing patterns (Rubin 1976). The MAR assumption means that missing data patterns are conditionally independent of missing data given observed data. That is, the association between missing data patterns and complete data is explained by observed data. When variables subject to missingness are highly correlated, for example, the observed data are likely to explain the association between missing data patterns and complete data to make the MAR assumption plausible (Shin and Raudenbush 2011; Shin 2012). The MAR assumption requires that we analyze all observed data for efficient analysis. The distinct parameter assumption is reasonable if there is little reason to believe that knowing the nuisance parameters ϕ provides extra information on the desired parameters θ (Schafer 1997). In regression analysis for the effect of socioeconomic status on a math achievement outcome, for example, a student may not take the exam because of illness or because of moving to a different school due to relocation of the family. It is not reasonable to believe that knowing such a mechanism would provide more information about the desired effect. In that case, the distinct parameter assumption is reasonable. On the other hand, if low performers are more likely to miss the exam than high performers such that knowing ϕ of the missing data generating mechanism is informative about the desired effect, then the distinct parameter assumption is not reasonable. Data missing under these two assumptions are called ignorable (Little and Rubin 2002). The ignorable missing data assumption is much weaker than the MCAR assumption (Rubin 1976; Schafer 1997; Little and Rubin 2002). Note that the MCAR implies the distinct parameter assumption.

Missing data—neither MAR nor MCAR—are said to be not missing at random (NMAR, Rubin 1976; Little and Rubin 2002). Under this assumption, missing patterns are associated with observed as well as missing data. A longitudinal study, for example, produces informative dropouts where the dropout patterns are associated with unobserved as well as observed outcomes (Diggle and Kenward 1994; Little 1995; Muthén et al. 2011). Consequently, both θ and ϕ have to be estimated from the joint distribution of complete data and missing patterns. This amounts to estimating, in addition to the desired hierarchical model, the model for missing patterns. Because the joint model involves missing data, the model assumptions yield parameters that are not uniquely identifiable, or parameter estimates that are not supported by observed data (Little 2009). Such a parameter may be constrained for identification or assumed to take a value in estimation. In general, little evidence exists in observed data to support such a parameter estimate. Consequently, sensitivity analysis should follow estimation of the joint distribution over the range of plausible values of the parameter (Little 1995, 2009). Therefore, analysis given data NMAR is more challenging than that given data MAR or MCAR.

In this chapter, we employ the ignorable missing data assumption that is quite plausible in many applications (Schafer 1997; Little and Rubin 2002). It is also the weakest condition under which we produce valid inferences by analyzing the desired hierarchical linear model only, that is, by ignoring the missing data generating mechanism (Rubin 1976). The next section reviews currently available methods for analysis of ignorable multilevel missing data.

Missing Data Methods

A wide array of methods exist for efficient analysis of single-level ignorable missing data (Rubin 1976, 1987, 1996; Dempster et al. 1977; Meng 1994; Schafer 1997, 2003; Little and Rubin 2002). In particular, model-based multiple imputation (Rubin 1987, 1996) is now routinely available based on widely used software packages such as NORM (Schafer 1999) and SAS (PROC MI, Yuan 2000). These single-level methods, however, cannot be applied validly to hierarchical missing data, and their extension to multilevel data entails challenges (Dempster et al. 1981; Schafer and Yucel 2002; Goldstein and Browne 2002, 2005; Yucel 2008; Shin and Raudenbush 2007, 2010, 2011). If methods developed for the multiple imputation of single-level data are applied to multilevel data, the variance–covariance structure of the imputed data sets will not accurately represent the multilevel educational processes that generated the data, nor will the structural relations at each level be captured correctly. When multilevel data are analyzed by a single-level method or under the misspecified number of levels, the resulting inferences may be considerably biased or inefficient (Shin 2003; Shin and Raudenbush 2011; Van Buuren 2011).

Current widely available methods for efficient analysis of ignorable multilevel missing data are quite limited. A two-level multivariate hierarchical linear model, where level-1 outcomes are subject to missingness given completely observed covariates, may be efficiently estimated via software packages such as Mplus (Muthén and Muthén 2010) and MLwiN (Rasbash et al. 2009; Browne 2012). With a univariate outcome in the model, this approach amounts to the complete-case analysis. When outcomes and covariates have missing values in the hierarchical model, however, a joint distribution of the variables subject to missingness has to be formulated and estimated to efficiently handle the missing data; and given the estimated distribution, multiple imputation of the complete data may be generated for subsequent analysis of the desired hierarchical model (Rubin 1987). Software packages such as WinBUGS (Spiegelhalter et al. 2000; Lunn et al. 2009), Mplus (Muthén and Muthén 2010; Asparouhov and Muthén 2010), MLwiN (Browne 2012), and R (Yucel 2008) provide Bayesian methods that enable formulation and estimation of such a joint distribution, and generation of the multiple imputation for subsequent analysis of the hierarchical model. However, these software packages provide little guidance as to how to explicitly formulate the joint distribution corresponding to the hierarchical model. For example, formulation of the joint distribution given a level-2 covariate subject to missingness in the hierarchical model is neither automated nor clearly described by any of the software packages. In general, the transformation between the joint distribution and the hierarchical model is nontrivial, involving an identification problem, and great care should be taken in formulation of the joint distribution that will identify the hierarchical model (Meng 1994; Shin and Raudenbush 2007). Otherwise, the estimation may produce biased point and uncertainty estimates of the hierarchical model or the formulated joint distribution may be extremely high-dimensional to estimate well (Shin and Raudenbush 2007, 2013).

Multilevel ignorable missing data may be multiply imputed by univariate sequential regression models (Raghunathan et al. 2001), also known as multiple imputation by fully conditional specification (Van Buuren et al. 2006). Software packages such as IVEware (Raghunathan et al. 2001) and multivariate imputation by chained equations (MICE, van Buuren and Groothuis-Oudshoorn 2011) use a Bayesian method to produce multiple imputation. This approach specifies a univariate regression model for each variable subject to missingness conditional on all other variables and generates multiple imputation based on the fitted model. While flexible in dealing with a mixture of continuous and discrete variables subject to missingness, the chained univariate models may not be compatible with a joint model (Horton and

Kleinman 2007; van Buuren and Groothuis-Oudshoorn 2011). The implied joint model by the series of univariate regression models may not exist (Rubin 2003; Van Buuren et al. 2006). This approach has not been extended to outcomes and covariates subject to missingness at multiple levels of a hierarchical model (Van Buuren 2011).

The next section introduces an efficient missing data method via multiple imputation and its software for unbiased and efficient estimation of a two-level hierarchical linear model given ignorable missing data. A key feature is that the data analyst need only know the desired hierarchical model. This approach removes or substantially reduces the burden of the incomplete-data analysis from the data analyst as intended by the method of multiple imputation (Rubin 1987, 1996; Meng 1994). Consequently, with the software in hand, the incomplete hierarchical data analysis will introduce little more challenge than the complete-data counterpart to the data analyst.

Efficient Handling of Missing Data

This section explains how to efficiently estimate a two-level hierarchical linear model (HLM2) given incomplete data according to the missing data method of Shin and Raudenbush (2007). The method employs a six-step analysis procedure to (1) specify a desired hierarchical linear model given incompletely observed hierarchical data; (2) reparametrize as the joint distribution of variables, including the outcome, that are subject to missingness conditional on all of the covariates that are completely observed under multivariate normality; (3) efficiently estimate the joint distribution using maximum likelihood (ML); (4) generate multiple imputation of complete data based on the ML estimates of the joint model; (5) analyze the desired hierarchical model by complete-data analysis given the multiple imputation; and finally (6) combine the multiple hierarchical model estimates (Rubin 1987). These steps have been implemented in a software package *HLM 7* (Raudenbush et al. 2011). Given the hierarchical linear model that a data analyst specifies at the first step, *HLM 7* automates the rest of the analysis procedure to produce efficient analysis of the hierarchical model. Consequently, the data analyst need only know the desired hierarchical linear model, which is no different from the complete-data analysis.

In this section, we introduce two comparatively simple examples of hierarchical linear models with incomplete data to describe the problem that researchers confront in the conventional incomplete-data analysis and how the missing data method resolves the problem by enabling efficient analysis via multiple imputation. One is a random-intercept model, and the other a random-coefficients model. Then, we present a reasonably general HLM2 given incomplete data, which may be efficiently analyzed by the method. Finally, we describe how to estimate the desired parameters and make inferences given multiple imputation (Rubin 1987).

Random-Intercept Model

To see how to handle multilevel incomplete data efficiently, it is instructive to consider a simple random-intercept model (Raudenbush and Bryk 2002). Let child i attend school j for $i = 1,\ldots,n_j$ and $j = 1,\ldots,M$. We consider a simple child-level or level-1 model

$$Y_{ij} = \beta_{0j} + \beta_{1j}X_{ij} + \epsilon_{ij} \tag{20.1}$$

where Y_{ij} is a univariate outcome variable, β_{0j} is the level-1 intercept, β_{1j} is the effect of a level-1 covariate X_{ij}, and child-specific random effect ϵ_{ij} is normally distributed with mean zero and variance σ^2, that is, $\epsilon_{ij} \sim N(0,\sigma^2)$. Note that the model (Equation 20.1) is single-level within school j. Each coefficient in the level-1 model (Equation 20.1) becomes an outcome variable that may vary across schools in the school-level or level-2 model. We consider level-2 models

$$\begin{aligned} \beta_{0j} &= \gamma_{00} + \gamma_{01}Z_j + u_{0j}, \\ \beta_{1j} &= \gamma_{10} \end{aligned} \tag{20.2}$$

where γ_{00}, γ_{01}, and γ_{10} are level-2 coefficients, Z_j is a level-2 covariate, and school-specific random effect $u_{0j} \sim N(0,\tau)$ is independent of child-specific ϵ_{ij}. By replacing β_{0j} and β_{1j} in the level-1 model (Equation 20.1) with $\gamma_{00} + \gamma_{01}Z_j + u_{0j}$ and γ_{10} on the right-hand side of the level-2 models, respectively, we obtain a random-intercept model or HLM2

$$Y_{ij} = \gamma_{00} + \gamma_{01}Z_j + \gamma_{10}X_{ij} + u_{0j} + \epsilon_{ij}. \tag{20.3}$$

With data completely observed, this model may be analyzed by standard multilevel software such as SAS, *HLM 7*, and MLwiN (Rasbash et al. 2009).

Difficulties arise given incompletely observed data. We consider (Y_{ij}, X_{ij}, Z_j) all subject to missingness with a general missing pattern in the desired model (Equation 20.3). Missing data may occur under seven different patterns for child i attending school j: One, two, or all three values of (Y_{ij}, X_{ij}, Z_j) may be missing. In general, p variables subject to missingness may produce up to $2^p - 1$ different missing patterns. Complete-case analysis drops children or observations that belong to any one of the missing patterns. It also deletes school j with missing Z_j, which entails deletion of all students attending the school. The resulting inferences will be inefficient and subject to bias. Ad hoc single-imputation methods fill in missing values for subsequent complete-data analysis. The imputed data underrepresent uncertainty due to missing data in estimation. In general, hypothesis tests will be liberal, rejecting the null hypothesis too often. These methods are not recommended unless children and schools with missing values consist of a small fraction of all children and schools.

Efficient analysis of the model (Equation 20.3) has to analyze all available data. That is, rather than dropping observations that belong to any one of the seven missing patterns in the complete-case analysis, we drop child i in school j if and only if the child belongs to one missing pattern: all three values of (Y_{ij}, X_{ij}, Z_j) missing. If one of (Y_{ij}, X_{ij}, Z_j) is missing for the child, the other two values available are analyzed; and if two values out of (Y_{ij}, X_{ij}, Z_j) are missing, the one value observed is analyzed. Consequently, children with unit nonresponse are also analyzed as long as they attend schools having observed Z_j to strengthen inferences at school level. Consider, for example, school j having a single child sampled ($n_j = 1$) who misses both Y_{ij} and X_{ij}, but school j has Z_j observed. Note that school j is dropped if and only if all school mates miss both Y_{ij} and X_{ij} and the school misses Z_j.

Efficient analysis of the HLM2 (Equation 20.3) using all available data may be formalized in the joint distribution of (Y_{ij}, X_{ij}, Z_j) subject to missingness

$$\begin{bmatrix} Y_{ij} \\ X_{ij} \\ Z_j \end{bmatrix} = \begin{bmatrix} \alpha_1 \\ \alpha_2 \\ \alpha_3 \end{bmatrix} + \begin{bmatrix} b_{1j} \\ b_{2j} \\ b_{3j} \end{bmatrix} + \begin{bmatrix} e_{1ij} \\ e_{2ij} \\ 0 \end{bmatrix} \tag{20.4}$$

for the means $(\alpha_1,\alpha_2,\alpha_3)$ of (Y_{ij}, X_{ij}, Z_j), and school-specific random effects $\begin{bmatrix} b_{1j} \\ b_{2j} \\ b_{3j} \end{bmatrix} \sim N\left(\begin{bmatrix} 0 \\ 0 \\ 0 \end{bmatrix}, \begin{bmatrix} \psi_{11} & \psi_{12} & \psi_{13} \\ \psi_{12} & \psi_{22} & \psi_{23} \\ \psi_{13} & \psi_{23} & \psi_{33} \end{bmatrix}\right)$ of (Y_{ij}, X_{ij}, Z_j) independent of child-specific random effects $\begin{bmatrix} e_{1ij} \\ e_{2ij} \end{bmatrix} \sim N\left(\begin{bmatrix} 0 \\ 0 \end{bmatrix}, \begin{bmatrix} \sigma_{11} & \sigma_{12} \\ \sigma_{12} & \sigma_{22} \end{bmatrix}\right)$ of (Y_{ij}, X_{ij}), where $\psi_{11} = \text{var}(b_{1j})$, $\psi_{12} = \text{cov}(b_{1j}, b_{2j})$, $\psi_{13} = \text{cov}(b_{1j}, b_{3j})$, $\psi_{22} = \text{var}(b_{2j})$, $\psi_{23} = \text{cov}(b_{2j}, b_{3j})$, $\psi_{33} = \text{var}(b_{3j})$, $\sigma_{11} = \text{var}(e_{1ij})$, $\sigma_{12} = \text{cov}(e_{1ij}, e_{2ij})$, and $\sigma_{22} = \text{var}(e_{2ij})$. Note that Z_j stays the same among schoolmates with no level-1 random effect. The missing data method for the desired hierarchical model (Equation 20.3) via efficient estimation of the joint model (Equation 20.4) produces efficient analysis of the hierarchical model as the conditional distribution of Y_{ij} given X_{ij} and Z_j (Shin and Raudenbush 2007).

To explicitly show how to analyze all available data, we first consider children with a single value missing. If a single value Y_{ij} is missing for child i attending school j, the two observed values (X_{ij}, Z_j) of the child enable estimation of

$$\begin{bmatrix} X_{ij} \\ Z_j \end{bmatrix} = \begin{bmatrix} \alpha_2 \\ \alpha_3 \end{bmatrix} + \begin{bmatrix} b_{2j} \\ b_{3j} \end{bmatrix} + \begin{bmatrix} e_{2ij} \\ 0 \end{bmatrix} \sim N\left(\begin{bmatrix} \alpha_2 \\ \alpha_3 \end{bmatrix}, \begin{bmatrix} \psi_{22} + \sigma_{22} & \psi_{23} \\ \psi_{23} & \psi_{33} \end{bmatrix} \right) \tag{20.5}$$

which adds information to estimation for $(\alpha_2, \alpha_3, \psi_{22}, \psi_{23}, \psi_{33}, \sigma_{22})$; if a single value X_{ij} is missing, the other two observed values enable estimation of a bivariate model (Y_{ij}, Z_j) to strengthen inferences involving $(\alpha_1, \alpha_3, \psi_{11}, \psi_{13}, \psi_{33}, \sigma_{11})$; and with Z_j missing, child i with observed (Y_{ij}, X_{ij}) adds information to estimation for $(\alpha_1, \alpha_2, \psi_{11}, \psi_{12}, \psi_{22}, \sigma_{11}, \sigma_{12}, \sigma_{22})$ in a bivariate model (Y_{ij}, X_{ij}).

Let us now consider children with two values missing. Child i missing (Y_{ij}, X_{ij}) adds information to estimation of a univariate model $Z_j \sim N(\alpha_3, \psi_{33})$ at school level. Take, for example, school j having a single child ($n_j = 1$) with unit nonresponse, but the school has observed Z_j. If the child misses (Y_{ij}, Z_j), she contributes to estimation of $X_{ij} \sim N(\alpha_2, \psi_{22} + \sigma_{22})$; and if she misses (X_{ij}, Z_j), she adds information to estimation of $Y_{ij} \sim N(\alpha_1, \psi_{11} + \sigma_{11})$.

Consequently, all partially observed cases contribute to estimation of the joint model (Equation 20.4), and thus of the desired model (Equation 20.3). The only case when school j is dropped from analysis happens if and only if school j misses Z_j and all n_j schoolmates miss both Y_{ij} and X_{ij}. Therefore, the method analyzes all available data to achieve efficient analysis of the desired hierarchical model (Equation 20.3).

Completely observed covariates: We now consider completely observed covariates U_{ij} and W_j at levels 1 and 2, respectively, in addition to (Y_{ij}, X_{ij}, Z_j) subject to missingness. The desired level-1 model is

$$Y_{ij} = \beta_{0j} + \beta_{1j} X_{ij} + \beta_{2j} U_{ij} + \epsilon_{ij} \tag{20.6}$$

where β_{2j} is the effect of the level-1 covariate U_{ij} and everything else is defined in the same way as that of the model (Equation 20.1). We consider level-2 models

$$\begin{aligned} \beta_{0j} &= \gamma_{00} + \gamma_{01} Z_j + \gamma_{02} W_j + u_{0j}, \\ \beta_{1j} &= \gamma_{10}, \\ \beta_{2j} &= \gamma_{20} \end{aligned} \tag{20.7}$$

where γ_{00}, γ_{01}, γ_{02}, γ_{10}, and γ_{20} are level-2 coefficients, Z_j and W_j are level-2 covariates, and school-specific random effect $u_{0j} \sim N(0, \tau)$ is independent of child-specific $\epsilon_{ij} \sim N(0, \sigma^2)$. The desired random-intercept model or HLM2 is

$$Y_{ij} = \gamma_{00} + \gamma_{01}Z_j + \gamma_{02}W_j + \gamma_{10}X_{ij} + \gamma_{20}U_{ij} + u_{0j} + \varepsilon_{ij}. \tag{20.8}$$

To efficiently handle missing data, we formulate the joint distribution of (Y_{ij}, X_{ij}, Z_j) subject to missingness conditional on (U_{ij}, W_j) completely observed. That is, we formulate the joint model as Equation 20.4, where α_1, α_2, and α_3 are replaced with $\alpha_{10} + \alpha_{11}W_j + \alpha_{12}U_{ij}$, $\alpha_{20} + \alpha_{21}W_j + \alpha_{22}U_{ij}$, and $\alpha_{30} + \alpha_{31}W_j$, respectively, and every other component is the same as it appears in the model (Equation 20.4). Note that the level-2 covariate W_j has its effects on level-1 as well as level-2 responses (Y_{ij}, X_{ij}, Z_j), while the level-1 covariate U_{ij} affects level-1 responses (Y_{ij}, X_{ij}) only. The efficient handling of missing data for the joint model (Equation 20.4) also applies here for the joint model corresponding to the HLM2 (Equation 20.8).

Random-Coefficients Model

This section explains the strategy for efficient analysis of a random-coefficients model given incomplete data. We consider the level-1 model (Equation 20.6) where the intercept as well as the coefficient of U_{ij} vary randomly across schools at level 2. Thus, we consider level-2 models

$$\begin{aligned} \beta_{0j} &= \gamma_{00} + \gamma_{01}Z_j + \gamma_{02}W_j + u_{0j}, \\ \beta_{1j} &= \gamma_{10}, \\ \beta_{2j} &= \gamma_{20} + u_{2j} \end{aligned} \tag{20.9}$$

where school-specific random effects $u_j \sim N(0, \tau)$ are independent of child-specific $\varepsilon_{ij} \sim N(0, \sigma^2)$ for $u_j = \begin{bmatrix} u_{0j} \\ u_{2j} \end{bmatrix}$ and $\tau = \begin{bmatrix} \tau_{00} & \tau_{02} \\ \tau_{02} & \tau_{22} \end{bmatrix}$ and everything else is defined identically as the counterpart of the level-2 models (Equation 20.7). The desired random-coefficients model or HLM2 is

$$Y_{ij} = \gamma_{00} + \gamma_{01}Z_j + \gamma_{02}W_j + \gamma_{10}X_{ij} + \gamma_{20}U_{ij} + u_{0j} + u_{2j}U_{ij} + \varepsilon_{ij} \tag{20.10}$$

for (Y_{ij}, X_{ij}, Z_j) subject to missingness, and (U_{ij}, W_j) completely observed. Conventional analysis of the HLM2 (Equation 20.10) confronts the same problems with the seven missing data patterns as does that of the random-intercept model (Equation 20.3). Efficient handling of missing data for the hierarchical model (Equation 20.3) also applies here for efficient analysis of the hierarchical model (Equation 20.10). We consider the joint distribution of (Y_{ij}, X_{ij}, Z_j) subject to missingness conditional on completely observed U_{ij} and W_j as

$$\begin{bmatrix} Y_{ij} \\ X_{ij} \\ Z_j \end{bmatrix} = \begin{bmatrix} \alpha_{10} + \alpha_{11}W_j + \alpha_{12}U_{ij} \\ \alpha_{20} + \alpha_{21}W_j + \alpha_{22}U_{ij} \\ \alpha_{30} + \alpha_{31}W_j \end{bmatrix} + \begin{bmatrix} b_{0j} + b_{1j}U_{ij} \\ b_{2j} \\ b_{3j} \end{bmatrix} + \begin{bmatrix} e_{1ij} \\ e_{2ij} \\ 0 \end{bmatrix} \tag{20.11}$$

where $(\alpha_{10}, \alpha_{20}, \alpha_{30})$ are the intercepts, $(\alpha_{11}, \alpha_{21}, \alpha_{31})$ are the effects of W_j on (Y_{ij}, X_{ij}, Z_j), respectively, $(\alpha_{12}, \alpha_{22})$ are the effects of U_{ij} on (Y_{ij}, X_j), respectively, and school-specific

$$\begin{bmatrix} b_{0j} \\ b_{1j} \\ b_{2j} \\ b_{3j} \end{bmatrix} \sim N\left(\begin{bmatrix} 0 \\ 0 \\ 0 \\ 0 \end{bmatrix}, \begin{bmatrix} \psi & \psi & \psi & \psi \\ \psi & \psi & \psi & \psi \\ \psi & \psi & \psi & \psi \\ \psi & \psi & \psi & \psi \end{bmatrix} \right)$$

are independent of child-specific.

$$\begin{bmatrix} e_{1ij} \\ e_{2ij} \end{bmatrix} \sim N\left(\begin{bmatrix} 0 \\ 0 \end{bmatrix}, \begin{bmatrix} \sigma_{11} & \sigma_{12} \\ \sigma_{12} & \sigma_{22} \end{bmatrix} \right)$$

Covariates subject to missingness having random effects: Efficient estimation of the random-coefficients model (Equation 20.10) requires that covariate U_{ij} having random coefficient u_{2j} be completely observed. Difficulty arises when U_{ij} is subject to missingness in the hierarchical model. The joint model for $(Y_{ij}, X_{ij}, U_{ij}, Z_j)$ subject to missingness has to be formulated for efficient handling of missing data while U_{ij} needs to be given on the right-hand side of the model for estimation of its random coefficient. Such a joint model cannot be expressed as a multivariate normal distribution so that the normal factorization of the joint model that leads to the hierarchical model as the conditional distribution of Y_{ij} given covariates does not apply. Consequently, it is difficult to efficiently handle missing data in the hierarchical model via ML estimation of the multivariate normal joint model. We assume that covariates having random effects are completely observed, which is a limitation of the method.

General HLM2

We now express a general HLM2 given incomplete data, which can be efficiently analyzed by the missing data method. The model is

$$Y_{ij} = X_{ij}^T \gamma_x + Z_j^T \gamma_z + U_{ij}^T \gamma_u + W_j^T \gamma_\omega + D_{ij}^T u_j + \varepsilon_{ij} \tag{20.12}$$

where Y_{ij} is a univariate outcome variable, X_{ij} and Z_j are vectors of p_1 level-1 and p_2 level-2 covariates subject to missingness having fixed effects γ_x and γ_z, respectively, U_{ij} and W_j are vectors of p_3 level-1 and p_4 level-2 covariates completely observed having fixed effects γ_u and γ_ω, respectively, and D_{ij} is another vector of p_5 level-1 covariates completely observed having level-2 unit-specific random effects $u_j \sim N(0,\tau)$ independent of level-1 unit-specific random errors $\epsilon_{ij} \sim N(0,\sigma^2)$ for a p_5-by-p_5 matrix τ and scalar σ^2. The desired parameters are $\theta = (\gamma_x, \gamma_z, \gamma_u, \gamma_\omega, \tau, \sigma^2)$.

The hierarchical models considered so far are special cases of the HLM2 (Equation 20.12). For example, the random-intercept model (Equation 20.3) is a special case of the HLM2 (Equation 20.12), where X_{ij} and Z_j are scalar, $U_{ij} = 0$, $W_j = D_{ij} = 1$, $\gamma_x = \gamma_{10}$, $\gamma_z = \gamma_{01}$, $\gamma_u = 0$, $\gamma_\omega = \gamma_{00}$, and $u_j = u_{0j}$. Although the general HLM2 (Equation 20.12) is not required to represent an intercept, many applications do where the first elements of W_j and D_{ij} are equal to one with the corresponding first elements of γ_ω and u_j representing the mean intercept and the random deviation of the intercept from the mean, respectively. We require that D_{ij} be completely observed. Note that U_{ij} and D_{ij} may share common covariates. For example, D_{ij} may be a subset of covariates in U_{ij}.

The p variables (Y_{ij}, X_{ij}, Z_j) subject to missingness in the HLM2 (Equation 20.12) may produce up to $2^p - 1$ different missing patterns for $p = 1 + p_1 + p_2$. Complete-case analysis drops children who belong to any one of the missing patterns. As the number of p variables subject to missingness increases, a number of children and schools may have to be dropped from the analysis, which results in inefficient inferences that may also be substantially biased.

To describe the joint model that efficiently handles missing data in the HLM2 (Equation 20.12), let $A \otimes B$ be a Kronecker product that multiplies a b-by-b matrix B to each element of an a-by-a matrix A (Magus and Neudecker 1988), and let I_n denote an n-by-n identity matrix for a positive integer n. For example, $I_3 \otimes B$ is a $(3 \times b)$-by-$(3 \times b)$ diagonal matrix with diagonal submatrices (B, B, B) and all other elements equal to zero. Given the HLM2 (Equation 20.12) with missing data, we formulate a joint distribution of (Y_{ij}, X_{ij}, Z_j) subject to missingness conditional on (U_{ij}, W_j, D_{ij}) completely observed as

$$\begin{bmatrix} Y_{ij} \\ X_{ij} \\ Z_j \end{bmatrix} = \begin{bmatrix} U_{ij}^{*T}\alpha_1 \\ (I_{p_1} \otimes U_{ij}^{*T})\alpha_2 \\ (I_{p_2} \otimes W_j^T)\alpha_3 \end{bmatrix} \begin{bmatrix} D_{ij}^T b_{1j} \\ b_{2j} \\ b_{3j} \end{bmatrix} + \begin{bmatrix} e_{1ij} \\ e_{2ij} \\ 0 \end{bmatrix} \tag{20.13}$$

for $U_{ij}^{*} = [W_j^T \, U_{ij}^T]^T$, vectors α_1 and α_2 of the fixed effects of U_{ij}^{*} on Y_{ij} and X_{ij}, respectively, a vector α_3 of the fixed effects of W_j on Z_j, and school-specific random effects $\begin{bmatrix} b_{1j} \\ b_{2j} \\ b_{3j} \end{bmatrix} \sim N\left(\begin{bmatrix} 0 \\ 0 \\ 0 \end{bmatrix}, \begin{bmatrix} \psi_{11} & \psi_{12} & \psi_{13} \\ \psi_{12} & \psi_{22} & \psi_{23} \\ \psi_{13} & \psi_{23} & \psi_{33} \end{bmatrix}\right)$ and child-specific random effects $\begin{bmatrix} e_{1ij} \\ e_{2ij} \end{bmatrix} \sim N\left(\begin{bmatrix} 0 \\ 0 \end{bmatrix}, \begin{bmatrix} \sigma_{11} & \sigma_{12} \\ \sigma_{12} & \sigma_{22} \end{bmatrix}\right)$ independent.

The missing data method for the HLM2 (Equation 20.12) via efficient estimation of the joint model produces efficient analysis of the hierarchical model as the conditional distribution of Y_{ij} given covariates (Shin and Raudenbush 2007). Note that given complete data, the HLM2 (Equation 20.12) is $Y_{ij} = U_{ij}^T \gamma_u + W_j^T \gamma_\omega + D_{ij}^T u_j + \varepsilon_{ij}$ equal to the joint model (Equation 20.13). The same strategy described above is used for efficient handling of missing data in (Y_{ij}, X_{ij}, Z_j), where child i with at least a single value observed contributes to estimation of the joint model. See Shin and Raudenbush (2007) for ML estimation of the joint model and multiple imputation given the ML estimates. Note that the variables subject to missingness, including the outcome, appear on the left-hand side given those completely observed on the right-hand side, which is the required form of the joint model (Equation 20.13) for efficient handling of missing data and efficient computation.

Combining Estimates from Multiple Imputation

Analysis of each of m imputed or completed data sets according to the desired HLM2 (Equation 20.12) produces m sets of θ estimates and their associated variances. Following Rubin (1987) and Schafer (1997), let Q be a parameter or a function of parameters in θ. Analysis of the tth completed data set produces the ML estimate $\hat{Q}_t$ and the associated variance U_t for $t = 1, 2, \ldots, m$. The combined parameter estimate is simply the average

$$\bar{Q} = \frac{1}{m} \sum_{t=1}^{m} \hat{Q}_t. \tag{20.14}$$

The variance associated with the combined estimate is

$$T = \bar{U} + \left(1 + \frac{1}{m}\right) B \tag{20.15}$$

that consists of the average within-imputation variance

$$\bar{U} = \frac{1}{m} \sum_{t=1}^{m} U_t \tag{20.16}$$

and the between-imputation variance

$$B = \frac{1}{m-1}\sum_{t=1}^{m}(\hat{Q}_t - \bar{Q})^2. \tag{20.17}$$

The within-imputation variance U_t reflects uncertainty in estimation of Q given the tth imputed data set (as if the missing values imputed were the true values), while the between-imputation variance B conveys uncertainty across the m estimates of Q due to missing data. No missing data implies $B = 0$ so that $T = \bar{U}$. With the infinite number of imputations, the variance (Equation 20.15) associated with $\bar{Q}$ is $T = \bar{U} + B$. The term $(1 + 1/m)$ in Equation 20.15 adds extra uncertainty due to the finite number of m imputations (Rubin 1987; Schafer 1997; Little and Rubin 2002).

For inferences on a column vector Q of k elements, Equations 20.14 through 20.16 are of the same form, but Equation 20.17 becomes $B = \left(\frac{1}{m-1}\right)\Sigma_{t=1}^{m}(\hat{Q}_t - \bar{Q})(\hat{Q}_t - \bar{Q})^T$, where $(\hat{Q}_t - \bar{Q})^T$ denotes the vector $(\hat{Q}_t - \bar{Q})$ transposed.

Hypothesis Tests

Let Q be a fixed effect or a linear function of fixed effects. We make inferences about Q based on

$$\frac{\bar{Q} - Q}{\sqrt{T}} \sim t_v \tag{20.18}$$

where t_v is the t distribution with the degrees of freedom

$$v = (m-1)\left(1 + \frac{1}{r}\right)^2 \tag{20.19}$$

for

$$r = \left(1 + \frac{1}{m}\right)\frac{B}{\bar{U}} \tag{20.20}$$

estimating “the relative increase in variance due to nonresponse” (Rubin 1987). Consequently, a $(1 - \alpha) \times 100\%$ confidence interval for Q is

$$\bar{Q} \pm t_{v,1-\alpha/2}\sqrt{T} \tag{20.21}$$

where $t_{v,1-\alpha/2}$ is the $(1-\alpha/2)\times 100$th percentile from t_v. The p-value for testing a null hypothesis H_0: $Q = Q_0$ against an alternative hypothesis H_a: $Q \neq Q_0$ at a significance level α is

$$2 \times P\left(\mathcal{T} > \frac{|\bar{Q} - Q_0|}{\sqrt{T}}\right) \tag{20.22}$$

where $\mathcal{T}$ is a t_v random variable (Rubin 1987; Schafer 1997).

When the between-imputation variance B is low relative to $\bar{U}$ to yield a low r such that the degrees of freedom in Equation 20.19 are high, $t_{v,1-\alpha/2}$ in the interval (Equation 20.21) and $\mathcal{T}$ in the p-value (Equation 20.22) can be replaced with the corresponding percentile $z_{1-\alpha/2}$ and a random variable Z from the standard normal distribution, respectively. When the relative increase r in variance due to missing data is high to yield low degrees of freedom, Equation 20.20 implies that increasing the number of m imputations decreases r to raise the degrees of freedom. *HLM 7* prints the degrees of freedom v in Equation 20.19 that can be solved for

$$r = \left(\sqrt{v/(m-1)} - 1\right)^{-1}. \tag{20.23}$$

Equation 20.20 with $\bar{U}$ replaced with T estimates "the fraction of information about Q missing due to nonresponse" (Little and Rubin 2002)

$$s = \left(1 + \frac{1}{m}\right)\frac{B}{T}. \tag{20.24}$$

Both r and s are positively associated with the between-imputation variance B, but negatively associated with the number of m imputations. Frequently, researchers claim that a few imputations are enough to handle missing data reasonably well based on the fraction of missing information estimated by Equation 20.24. The size of the standard error of the Q estimate based on m imputations relative to the ideal one based on infinitely many imputations is approximately $\sqrt{1+s/m}$ (Rubin 1987, p. 114; Schafer 1997, p. 107). When the fraction of missing information is high at $s = 0.5$, for example, the standard error $\sqrt{T}$ of $\bar{Q}$ based on $m = 5$ imputations will be about $\sqrt{1+0.5/5} = 1.05$ times as high as the counterpart based on infinitely many imputations. With $s = 0.3$, as few as $m = 3$ imputations will achieve about the same efficiency in terms of the relative size of standard errors.

To compare model fits based on the likelihood ratio tests, let Q be k parameters in θ of HLM2 (Equation 20.12) or the full model. We want to test a null hypothesis H_0: $Q = Q_0$ versus an alternative one H_a: $Q \neq Q_0$. Let θ_0 be the parameters of the reduced model under $Q = Q_0$. Consider, for example, a two-dimensional $Q = \begin{bmatrix} \tau_{01} \\ \tau_{11} \end{bmatrix}$ in $\theta = (\gamma_x, \gamma_z, \gamma_u, \gamma_\omega, \tau, \sigma^2)$ for $\tau = \begin{bmatrix} \tau_{00} & \tau_{01} \\ \tau_{01} & \tau_{11} \end{bmatrix}$ and $Q_0 = \begin{bmatrix} 0 \\ 0 \end{bmatrix}$. Then, θ_0 has $k = 2$ parameters less than θ. The Q may also include a combination of fixed effects, variances, and covariances.

Let $\hat{\theta}^t$ and $\hat{\theta}_0^t$ be the ML estimates of θ and θ_0, respectively, given the tth completed data set for $t = 1,..., m$. The log likelihoods $l(\hat{\theta}^t)$ and $l(\hat{\theta}_0^t)$ evaluated at $\hat{\theta}^t$ and $\hat{\theta}_0^t$, respectively, yield the likelihood ratio statistic $d_t = 2[l(\hat{\theta}^t) - l(\hat{\theta}_0^t)]$. The test statistic proposed by Li et al. (1991) is

$$D_1 = \frac{\bar{d}/k - (m+1)(m-1)^{-1} r_1}{1 + r_1} \tag{20.25}$$

for $\bar{d} = \Sigma_{t=1}^m d_t / m$, where

$$r_1 = \left(1 + \frac{1}{m}\right)\left[\frac{1}{m-1}\sum_{t=1}^{m}\left(\sqrt{d_t} - \overline{\sqrt{d}}\right)^2\right] \tag{20.26}$$

is $(1 + 1/m)$ times the sample variance of $\sqrt{d_1}, \ldots, \sqrt{d_m}$ for $\overline{\sqrt{d}} = \Sigma_{t=1}^m \sqrt{d_t}/m$ (Little and Rubin 2002). The r_1 estimates the average relative increase in variance due to missing data across the k parameters Q (Schafer 1997). Let F_{k,v_1} denote a random variable from the F distribution with k numerator and v_1 denominator degrees of freedom for

$$v_1 = k^{-3/m}(m-1)(1 + 1/r_1)^2. \tag{20.27}$$

The p-value is given by

$$P(F_{k,v_1} > D_1). \tag{20.28}$$

With r_1 close to zero, v_1 is large so that kD_1 has the chi-square distribution with k degrees of freedom for $D_1 \approx \bar{d}/k$. Then the p-value (Equation 20.28) may also be obtained by

$$P(\chi_k^2 > kD_1) \tag{20.29}$$

for a chi-square random variable χ_k^2 with k degrees of freedom. Given multiple imputation, the likelihood ratio statistic d_t may be obtained from the tth completed data set to yield r_1, D_1, and v_1 for the hypothesis test. The test statistic D_1 yields an approximate range of p-values between one half and twice the computed p-value (Li et al. 1991).

To obtain a more accurate p-value, let $\bar{\theta} = \Sigma_{t=1}^{m} \hat{\theta}^t / m$ and $\bar{\theta}_0 = \Sigma_{t=1}^{m} \hat{\theta}_0^t / m$ so that $d_t' = 2[l_t(\bar{\theta}) - l_t(\bar{\theta}_0)]$ is the likelihood ratio test statistic evaluated at the average ML estimates $\bar{\theta}$ and $\bar{\theta}_0$ given the tth completed data set. The test statistic proposed by Meng and Rubin (1992) is

$$D_2 = \frac{\bar{d}'}{k(1+r_2)} \tag{20.30}$$

for $\bar{d}' = \Sigma_{t=1}^{m} d_t' / m$, where

$$r_2 = \frac{(m+1)}{k(m-1)}(\bar{d} - \bar{d}') \tag{20.31}$$

estimates the average relative increase in variance due to missing data across the k parameters Q (Schafer 1997). The p-value is given by

$$P(F_{k,v_2} > D_2) \tag{20.32}$$

where the denominator degrees of freedom is

$$v_2 = \begin{cases} 4 + (u-4)[1 + (1-2/u)/r_2]^2, & \text{if } u = k(m-1) > 4 \\ (m-1)(k+1)(1+1/r_2)^2/2, & \text{otherwise.} \end{cases} \tag{20.33}$$

Unlike D_1, computation of D_2 requires log likelihoods $l_t(\bar{\theta})$ and $l_t(\bar{\theta}_0)$ evaluated at the average ML estimates $\bar{\theta}$ and $\bar{\theta}_0$ given the tth completed data set that *HLM7* does not provide at the time of my writing this chapter.

The two approaches to testing H_0: $Q = Q_0$ against H_a: $Q \neq Q_0$ (Li et al. 1991; Meng and Rubin 1992) are based on the likelihood ratio statistics. When Q_0 involves variance components equal to zero given complete data, the likelihood ratio test is known to produce a conservative p-value based on the chi-square distribution with k degrees of freedom (Pinheiro and Bates 2000). Stram and Lee (1994) suggested use of a mixture of chi-square distributions to improve the accuracy of the p-value (Pinheiro and Bates 2000; Verbeke and Molenberghs 2000; Snijders and Bosker 2012). Given incomplete data, the test statistic D_1 yields an approximate range of p-values between one half and twice the observed p-value (Li et al. 1991), and the test statistic D_2 produces the p-value (Equation 20.32) that is more accurate than the corresponding p-value (Equation 20.28) (Meng and Rubin 1992; Schafer 1997; Little and Rubin 2002). The next two sections show how to analyze a hierarchical linear model given ignorable missing data by *HLM 7* according to the method explained in this section.

Data Analysis

This section illustrates how to efficiently analyze hierarchical linear model (Equation 20.12) given incompletely observed data from PISA (OECD 2007). PISA has been collecting hierarchical data about 15-year-old students attending schools nested within nations every 3 years since the year 2000. The data for analysis consists of 5611 students attending 166 schools in the United States from the PISA 2006 data collection. Table 20.1 summarizes the data. The outcome variable of interest is the mathematics achievement score (MATH). PISA imputes each missing score five times to provide five sets of plausible mathematics scores. In this section, we analyze the first set of plausible mathematics scores summarized in Table 20.1 as if they were completely observed. The next section illustrates an analysis strategy with all plausible values.

To summarize the data for analysis, at level 1, mathematics score (MATH) and age (AGE) are completely observed while the highest parental occupation status (HISEI), the highest education level of parents in the number of years of schooling (PARENTED), family wealth (WEALTH), and first-generation immigrant status (IMMIG1) are missing for 390, 61, 34, and 189 students, respectively. The 5611 students score 475 points in mathematics and are 190 months old on average; the highest occupation status and education level of parents are 52.46 units and 13.61 years on average, respectively; the average family wealth is 0.15 units; and 6% of the students are first-generation immigrants. At level 2, the student-to-teacher ratio (STRATIO) is missing for 28 schools, or 17% of the 166 schools, and the private school indicator (PRIVATE) is missing for three schools. The schools have 15.46 students per teacher on average, and 9% of the schools are private (cf. OECD 2007).

Summary statistics reveal that first-generation immigrants scored 36.08 points lower than did other students in mathematics achievement on average.

TABLE 20.1

US Data for Analysis from PISA Data Collection in 2006

Level	Variables	Description	Mean (SD[a])	Missing (%)
I	MATH	Mathematics scores	475.18 (89.87)	0 (0)
	AGE	Age in months	189.79 (3.54)	0 (0)
	HISEI	Highest parental occupation status	52.46 (16.78)	390 (7)
	PARENTED	Highest education level of parents in the number of years of schooling	13.61 (2.49)	61 (1)
	WEALTH	Family wealth	0.15 (0.80)	34 (1)
	IMMIG1	1 if first-generation immigrant	0.06 (0.24)	189 (3)
II	STRATIO	Student-to-teacher ratio	15.46 (4.64)	28 (17)
	PRIVATE	1 if private	0.09 (0.29)	3 (2)

[a] Standard deviation.

In this section, we ask how much of the difference is attributable to the individual and school characteristics summarized in Table 20.1; and, controlling for the individual and school characteristics, how first-generation immigrants compare with other students in mathematics achievement. The complete-case analysis drops 1405 students and 28 schools to produce inefficient inferences that may also be substantially biased. We compare the complete-case analysis with the efficient missing data analysis given incomplete data.

Complete-Case Analysis

Preliminary analysis reveals that the school means of the highest parent education (PARENTED), indicative of school quality, vary substantially across schools with a 95% confidence interval (11.38, 15.82). The effect of the highest parent education may vary randomly across schools of different quality. A random-coefficients model to test such a hypothesis is

$$Y_{ij} = \gamma_{00} + \gamma_{01}\text{STRATIO} + \gamma_{02}\text{PRIVATE} + \gamma_{10}\text{AGE} + \gamma_{20}\text{HISEI} + \gamma_{30}\text{PARENTED} + \lambda_{40}\text{WEALTH} + \gamma_{50}\text{IMMIG1} + u_{0j} + u_{3j}\text{PARENTED} + \epsilon_{ij} \tag{20.34}$$

a special case of the HLM2 (Equation 20.12), where

$$Y_{ij} = \text{MATH}, X_{ij} = [\text{HISEI PARENTED WEALTH IMMIG1}],$$

$$Z_j = [\text{STRATIO PRIVATE}], U_{ij} = AGE, W_j = 1, D_{ij} = [1 \text{ PARENTED}],$$

$$\gamma_x = [\gamma_{20}\,\gamma_{30}\,\gamma_{40}\,\gamma_{50}]^T, \gamma_z = [\gamma_{01}\,\gamma_{02}]^T, \gamma_u = \gamma_{10}, \gamma_w = \gamma_{00}, u_j = [u_{0j}\,u_{3j}]^T$$

for $\tau = \begin{bmatrix} \tau_{00} & \tau_{03} \\ \tau_{03} & \tau_{33} \end{bmatrix}$. We center HISEI, PARENTED, WEALTH, AGE, and STRATIO around their respective sample means, and carry out the complete-case analysis by *HLM 7* to produce the ML estimates under the heading "CC" in Table 20.2. The CC analysis analyzed 4206 students attending 138 schools. The τ_{33} estimate is 19 with the associated variance estimate 6.45^2, not shown in Table 20.2. The hypothesis test of interest is

$$H_0 : \tau_{33} = 0 \text{ against } H_a : \tau_{33} > 0.$$

Approximate normality of the ML estimator $\ln(\hat{\tau}_{33})$ for the natural logorithm $\ln(\cdot)$ produces an approximate 95% confidence interval for $\ln(\tau_{33})$, which is transformed to an approximate 95% confidence interval (9.70, 36.90) for τ_{33}. The interval far away from zero provides some evidence in support of the

TABLE 20.2

Analysis of the Random-Coefficients Model (Equation 20.34)

	CC[a]	Efficient[b]			Efficient PV[c]		
Covariate	**Coef. (se)[d]**	**Coef. (se)**	**df[e]**	***p*-value**	**Coef. (se)**	**df**	***p*-value**
Intercept	478 (3.15)[f]	471 (3.11)[f]	163	<0.001	471 (3.14)[f]	163	<0.001
STRATIO	−0.11 (0.60)	−0.18 (0.68)	99	0.792	−0.24 (0.70)	49	0.729
PRIVATE	35.85 (10.19)[f]	34.51 (11.36)[f]	163	0.003	36.52 (11.44)[f]	163	0.002
AGE	0.64 (0.32)[f]	0.90 (0.29)[f]	5214	0.002	0.97 (0.30)[f]	494	0.001
HISEI	0.82 (0.08)[f]	0.86 (0.08)[f]	124	<0.001	0.84 (0.09)[f]	26	<0.001
PARENTED	4.93 (0.70)[f]	4.18 (0.61)[f]	165	<0.001	4.24 (0.66)[f]	61	<0.001
WEALTH	6.48 (1.62)[f]	9.39 (1.42)[f]	5214	<0.001	9.25 (1.43)[f]	2058	<0.001
IMMIG1	−10.04 (5.27)	−10.81 (4.61)[f]	1191	0.019	−10.37 (5.08)[f]	66	0.045
τ	$\begin{bmatrix} 1027 & 103 \\ 103 & 19 \end{bmatrix}$	$\begin{bmatrix} 1280 & 117 \\ 117 & 19 \end{bmatrix}$			$\begin{bmatrix} 1288 & 111 \\ 111 & 14 \end{bmatrix}$		
σ^2	5229	5360			5252		

[a] Complete-case analysis.
[b] Efficient analysis of the first set of plausible outcome values.
[c] Efficient analysis of all plausible outcome values.
[d] Coefficient (standard error).
[e] Degrees of freedom.
[f] Statistically significant at the significance level $\alpha = 0.05$.

alternative hypothesis. For the hypothesis test, *HLM 7* produces a χ^2 test statistic 196.47 with 136 degrees of freedom based on 137 schools with enough data (Raudenbush and Bryk 2002, Chapter 3). The *p*-value is less than 0.001 to reject the null hypothesis. Therefore, the CC analysis shows that the effects of the highest parent education vary randomly across schools, and that attending a private school, age, the highest parental occupation status, the highest parent education, and family wealth are all positively associated with math achievement while student-to-teacher ratio and first-generation immigrant status are not significantly associated with the outcome.

Efficient Analysis

Now, we reanalyze the random-coefficients model (Equation 20.34) given incomplete data by *HLM 7* according to the efficient missing data method explained in the section "Efficient Handling of Missing Data." The ML estimates based on $m = 5$ imputations are displayed under the heading "Efficient" in Table 20.2. The efficient analysis considered 5550 students attending 166 schools after dropping 61 students with the highest parent education missing because the method requires that the covariate having a random coefficient be completely observed. The τ_{33} estimate is 19 with the associated variance estimate 5.38^2, not shown in Table 20.2, that is less than the CC counterpart 6.45^2 above. Consequently, an approximate 95% confidence interval for τ_{33} is (10.93, 33.11) narrower and farther away from zero than the corresponding CC interval (9.70, 36.90). To test H_0: $\tau_{33} = 0$ against H_a: $\tau_{33} > 0$, we first note that the null hypothesis $\tau_{33} = 0$ implies $\tau_{03} = 0$ so that $k = 2$. *HLM 7* provides multiply imputed data sets. Given the tth completed data set based on the full model, we fit both the full and reduced models to obtain the likelihood ratio test statistic d_t and the average $\bar{d}$ to compute D_1 in Equation 20.25. The average relative increase in variance due to missing data in $Q = [\tau_{03}\ \tau_{33}]^T$ is $r_1 = 0.003 \approx 0$ to yield the test statistic $D_1 \approx \bar{d}/2$. Consequently, $2D_1 \approx 50.75$ gives the p-value $P(x_2^2 > 50.75) < 0.00001$ based on Equation 20.29. This method provides the range of the p-value between one-half and twice the computed one (Li et al. 1991). This precision gives enough evidence to reject the null hypothesis in support of the alternative hypothesis that the effects of the highest parent education vary randomly across schools.

Parameter estimates with the associated standard errors, degrees of freedom and p-values are shown under "Efficient" in Table 20.2. The efficient analysis shows that attending a private school, age, the highest parental occupation status, the highest parent education and family wealth are all positively associated with mathematics achievement as the CC analysis revealed. Controlling for the individual and school characteristics, however, the first-generation immigrant status is negatively associated with the outcome while the association is not statistically significant according to the CC analysis. A main reason for the different inferences is the lower standard error 4.61 of the efficient analysis than the CC counterpart 5.27. Based on $m = 5$ imputations, the relative increase in variance due to missing data in the effect estimate is $r = (\sqrt{1191/4} - 1)^{-1} = 0.06$ based on Equation 20.23. The fraction of missing information s in Equation 20.24 is lower than $r = 0.06$ so that the standard error 4.61 is at most $\sqrt{1 + 0.06/5} = 1.006$ times as high as the ideal one based on infinitely many imputations (Rubin 1987, p. 114; Schafer 1997, p. 107). Consequently, the effect estimate based on $m = 5$ imputations loses little precision, relative to the counterpart based on infinitely many imputations. That is, five imputations provide enough precision for estimation of the effect. Overall, the standard errors associated with the effect estimates of level-1 covariates under the efficient analysis are up to 14% lower than the CC counterparts. In addition, the effect estimates 0.90 and 9.39 of age and family wealth under the efficient analysis are considerably higher than their CC counterparts 0.64 and 6.48, respectively. Furthermore, the CC analysis seems to exaggerate the goodness of fit by producing smaller variance estimates than those of the efficient analysis. At level 2, the CC analysis produces a lower standard error associated with the effect estimate of the private school indicator than does the efficient analysis. The relatively understated CC standard error reflects its positive association with the comparatively underestimated CC variance components.

On the basis of the efficient analysis, a typical nonimmigrant student attending a public school with average age, highest parental occupation status, highest parent education, and family wealth scores 471 points in mathematics achievement on average. Students attending a private school score 34.51 points higher than do those attending a public school on average, controlling for the effects of other covariates in the model. One month older in age, a unit increase in the highest parental occupation status, 1 year increment in the highest parent education, and a unit increase in family wealth are expected to raise mathematics scores by 0.90, 0.86, 4.18, and 9.39 points, respectively, *ceteris paribus*. Controlling for the individual and school characteristics in the model (Equation 20.34), the average difference in mathematics achievement between first-generation immigrants and other students reduces to 10.81 points or 30% of the initial gap, 36.08 points. Consequently, the individual and school characteristics considered in the model (Equation 20.34) explain 70% of the initial gap in mathematics achievement between first-generation immigrants and other students.

Data Analysis with Plausible Values

The efficient analysis of HLM2 (Equation 20.34) in the previous section considers the first set of plausible mathematics scores as if they were completely observed. Consequently, the $m(= 5)$ imputed or completed data sets reflect uncertainty due to the missing values of covariates, but do not take into account uncertainty from missing outcome values to produce understated standard errors of estimates. This section illustrates an efficient analysis strategy for the model (Equation 20.34) using all five sets of plausible mathematics scores. The first set has mean (standard deviation) equal to 475.18 (89.87) shown in Table 20.1. The second to the fifth sets have means (standard deviations) equal to 474.44 (89.09), 474.46 (88.95), 474.97 (88.66), and 474.54 (89.41), respectively.

In the previous section, we produced m imputations to efficiently analyze the desired HLM2 (Equation 20.34) with covariates subject to missingness given the first set of plausible outcome values. In this section, we repeat the same analysis with identical covariates given each set of plausible outcome values. With the number of q sets of plausible outcome values fixed at 5, this strategy produces $5m(= q \times m)$ completed data sets. Unlike the efficient analysis of the previous section based on the first set of plausible outcome values, the $5m$ imputations reflect uncertainty in parameter estimates due to missing values of both outcome and covariates. Note that we may obtain more imputations by increasing the number of m imputations per set of plausible outcome values. It is important flexibility to be able to increase m

that will decrease the relative increase in variance due to missing data in Equation 20.20 and, thus, increase the degrees of freedom of an estimate in Equation 20.19, in particular when the missing values of covariates account for a considerable amount of uncertainty in estimation. The degrees of freedom of an estimate is negatively associated with the p-value. This flexibility is not available to us for the outcome variable because the number of q sets of plausible outcome values is fixed at 5 by the imputer. Because the efficient analysis in the previous section reveals that uncertainty in estimation due to the missing values of covariates is not substantial, we generate $m = 1$ imputation per set of plausible outcome values to analyze 5 imputations in this section. Then we use the "Multiple Imputation" option of *HLM 7* that automates the complete-data analysis of the hierarchical model (Equation 20.34) given the multiple imputation to produce the combined estimates (Rubin 1987).

The ML estimates are displayed under the heading "Efficient PV" in Table 20.2. We compare the efficient PV analysis with the efficient analysis based on the first set of plausible values in the previous section. Again based on 5550 students attending 166 schools, the estimated τ_{33} is 14, lower than 19 produced under the efficient analysis. The associated variance estimate is 6.09^2, higher than 5.38^2 under the efficient analysis, to reflect added uncertainty due to missing outcome values. An approximate 95% confidence interval for τ_{33} is (5.97, 32.85) wider and closer to zero than the efficient analysis counterpart (10.93, 33.11). For testing H_0: $\tau_{33} = 0$ versus H_a: $\tau_{33} > 0$, the average relative increase in variance due to missing data in $Q = [\tau_{03}\ \tau_{33}]^T$ is $r_1 = 0.61$ based on Equation 20.26 to yield the test statistic $D_1 = 13.73$. Higher than the corresponding $r_1 = 0.003$ based on the first set of plausible values in the previous section, the $r_1 = 0.61$ implies that missing outcome values add a considerable amount of uncertainty to the Q estimates. The p-value (Equation 20.28) is $P(F_{2,18} > 13.73) = 0.0002$ to reject the null hypothesis in support of the alternative hypothesis that $\tau_{33} > 0$.

Both efficient PV analysis and efficient analysis produce comparable effect estimates and the same statistical inferences. However, the efficient PV analysis yields comparatively low degrees of freedom overall to reveal added uncertainty in the estimates due to the missing values of the outcome variable. In particular, the degrees of freedom for the effect estimate of the first-generation immigrant status reduce from 1191 under the efficient analysis to 66 under the efficient PV analysis. The 66 degrees of freedom translate into the relative increase in variance due to missing data $r = (\sqrt{66/4} - 1)^{-1} \approx 0.33$ based on Equation 20.23, which is a substantial increase from the corresponding $r = 0.06$ under the efficient analysis. The relative increase in r implies that added uncertainty due to missing outcome values is considerable, thereby inflating the standard error 4.61 and the p-value 0.019 under the efficient analysis to 5.08 and 0.045 under the efficient PV analysis, respectively.

Extensions and Limitations

This chapter explained how to efficiently analyze a two-level hierarchical linear model where explanatory as well as outcome variables may be subject to missingness with a general missing pattern at any of the levels (Shin and Raudenbush 2007). The key idea is to reexpress the desired hierarchical model as the joint distribution of variables, including the outcome, that are subject to missingness conditional on all of the covariates that are completely observed under multivariate normality; estimate the joint distribution by ML; generate multiple imputation given the ML estimates of the joint distribution; analyze the desired hierarchical model by complete-data analysis given the multiple imputation; and then combine the multiple hierarchical model estimates (Rubin 1987). Given the desired hierarchical model specified by a data analyst, the rest of the analysis steps can be automated for efficient estimation of the hierarchical model. The automation has been implemented in a software package *HLM* 7 that is yet to be released to the public at the time of writing this chapter. With such a software package in hand, multilevel incomplete-data analysis is no different from complete-data analysis from the data analyst's perspective.

This chapter illustrated two examples for efficient analysis of incompletely observed PISA data with *HLM* 7. The outcome variable was a mathematics achievement score subject to missingness. PISA imputed five sets of plausible values for each missing outcome value. Assuming the first set of plausible mathematics scores completely observed in the first example, we efficiently analyzed hierarchical linear model (Equation 20.34) given covariates subject to missingness at multiple levels. We compared the efficient analysis with the complete-case analysis. Overall, the complete-case analysis produced higher standard errors than did the efficient analysis, and some estimates of the complete-case analysis were considerably different from the counterparts of the efficient analysis. Consequently, the two analyses produced different statistical inferences for the effect of a key covariate. The second example was analysis of the same model (Equation 20.34) with covariates subject to missingness and all plausible outcome values. We repeated the efficient analysis of the first example with identical covariates given each set of plausible outcome values, each time imputing a single completed data set to eventually produce as many completed data sets as the five sets of plausible outcome values for subsequent complete-data analysis. The combined estimates were comparable to and produced the same statistical inferences as those of the efficient analysis in the first example. On the other hand, the degrees of freedom for estimates were considerably lower than those of the first example, overall. That is, the estimates exhibited higher relative increase in variance due to missing data than did those of the first example based on the first set of plausible values. Consequently, a substantial amount of uncertainty due to missing data in estimation was from missing outcome values.

The second example illustrates one of the difficulties in incomplete-data analysis when the imputer of the plausible outcome values is different from the data analyst of the desired hierarchical model (Equation 20.34) (Meng 1994; Rubin 1996). With $q = 5$ sets of plausible outcome values and covariates subject to missingness at multiple levels, the analysis based on 5 imputations ($m = 1$ imputation per set of plausible outcome values) reveals that a considerable amount of uncertainty due to missing data in estimation is from missing outcome values. Consequently, the data analyst may want to increase the number of q imputations of plausible outcome values, which may increase the degrees of freedom and reduce the computed p-value because the degrees of freedom (Equation 20.19) of an estimate is positively associated with the number of imputations. However, with access to 5 sets of plausible outcome values only, she is unable to reduce the relative increase in variance due to missing outcome values. Increasing the number of m imputations per set of plausible outcome values may help reduce the relative increase in variance due to the missing values of covariates, not due to missing outcome values. When both outcome and covariates contain missing values, the efficient missing data method and its software introduced in this chapter provide flexibility to manipulate the number of imputations. In that case, the data analyst is able to set the number of imputations and produce multiple imputations tailored to the analyst's own analysis with the software. Because the analyst is also the imputer, the analyst will not face the difficulty that arises when he or she is not the imputer (Meng 1994; Rubin 1996).

The two-level missing data approach in this chapter has been extended to efficient analysis of a two-level contextual-effects model (Raudenbush and Bryk 2002; Shin and Raudenbush 2010); of a three-level hierarchical linear model (Shin and Raudenbush 2011; Shin 2012); and of an arbitrary Q-level hierarchical linear model (Shin and Raudenbush 2013) given incomplete data. Three-level user-friendly software based on the missing data methods of Shin and Raudenbush (2011) and Shin (2012) is under development at the time of writing this chapter. These advances guide us with continuous variables subject to missingness.

The efficient analysis of HLM2 (Equation 20.34) in Table 20.2 involves discrete first-generation immigrant status and private school indicator subject to missingness at levels 1 and 2, respectively. Although it is not appropriate to handle the discrete missing data under the corresponding multivariate joint normal distribution (Equation 20.13) of variables subject to missingness, the implied conditional model is the desired hierarchical model (Equation 20.34). Furthermore, the joint model assumption (Equation 20.13) to handle missing data affects only imputed data, not observed data. The advantage is that the hierarchical model is analyzed by the efficient missing data method (Schafer 1997; Shin and Raudenbush 2007). However, with the missing rate high, the normal joint model assumption becomes nontrivial. A useful future extension of this approach is to efficient and robust handling of normal and nonnormal multilevel missing data.

It entails challenges to extend the missing data method introduced in this chapter to analysis of multilevel incomplete data from multistage sampling where different selection probabilities of units are used (OECD 2007; Tourangeau el al. 2009). The extent to which complicated sampling weights affect the missing data analysis is not well known. To minimize possible adverse impact such as biased inferences, the sampling weights may be applied at the final stage of complete-data analysis given multiple imputation (Graubard and Korn 1996; Pfeffermann et al. 1998; Korn and Graubard 2003). An important future research topic is extension of the efficient missing data method to analysis of multilevel incomplete data generated from multistage sampling with different selection probabilities of units.

Another limitation of the missing data method is that the covariates having random coefficients must be completely observed. With such covariates subject to missingness, the joint distribution of variables subject to missingness may not be expressed as a multivariate normal distribution. Consequently, subsequent analysis given the estimated normal joint model by ML does not apply. Research is under way to relax the assumption, which will broaden the applicability of this method.

Acknowledgments

This work was supported by the Institute of Education Sciences, U.S. Department of Education, through Grants R305D090022 to NORC and R305D130033 to VCU. The opinions expressed are those of the author and do not represent views of the Institute or the U.S. Department of Education.

References

Asparouhov, T. and Muthén, B. 2010. Multiple Imputation with Mplus. Technical Report. www.statmodel.com.

Browne, W.J. 2012. MCMC estimation in MLwiN Version 2.25. Centre for Multilevel Modelling, University of Bristol.

Dempster, A.P., Laird, N.M., and Rubin, D.B. 1977. Maximum likelihood from incomplete data via the EM algorithm. *Journal of the Royal Statistical Society B*, **76**, 1–38.

Dempster, A.P., Rubin, D.B., and Tsutakawa, R.K. 1981. Estimation in covariance components models. *Journal of the American Statistical Association*, **76**, 341–353.

Diggle, P. and Kenward, M.G. 1994. Informative drop-out in longitudinal data analysis. *Journal of the Royal Statistical Society C*, **43**, 49–93.

Goldstein, H. and Browne, W. 2002. Multilevel factor analysis modelling using Markov Chain Monte Carlo estimation. In G. Marcoulides and I. Moustaki (eds.), *Latent Variable and Latent Structure Models*, London: Lawrence Erlbaum.

Goldstein, H. 2003. *Multilevel Statistical Models*, London: Edward Arnold.

Goldstein, H. and Browne, W. 2005. Multilevel factor analysis models for continuous and discrete data. In A. Olivares and J.J. McArdle (eds.), *Contemporary Psychometrics. A Festschrift to Roderick P. McDonald*. Mahwah, NJ: Lawrence Erlbaum.

Graubard, B.I. and Korn, E.L. 1996. Modelling the sampling design in the analysis of health surveys. *Statistical Methods in Medical Research*, 5, 263–281.

Horton, N.J. and Kleinman, K.P. 2007. Much ado about nothing: A comparison of missing data methods and software to fit incomplete data regression models. *The American Statistician*, **61**, 79–90.

Korn, E.L. and Graubard, B.I. 2003. Estimating variance components by using survey data. *Journal of the Royal Statistical Society B*, **65**, 175–190.

Krueger, A.B. 1999. Experimental estimates of education production functions. *The Quarterly Journal of Economics*, **114**, 497–532.

Li, K., Meng, X., Raghunathan, T.E., and Rubin D.B. 1991. Significance levels from repeated p-values with multiply-imputed data. *Statistica Sinica*, **1**, 65–92.

Little, R.J.A. 1995. Modeling the drop-out mechanism in repeated-measures studies. *Journal of the American Statistical Association*, **90**, 1112–1121.

Little, R.J.A. and Rubin, D.B. 2002. *Statistical Analysis with Missing Data*, New York: Wiley.

Little, R.J.A. 2009. Selection and pattern-mixture models. In G. Fitzmaurice, M. Davidian, G. Verbeke, and G. Molenberghs (eds.), *Longitudinal Data Analysis* (pp. 409–431), Boca Raton, FL: Chapman & Hall/CRC.

Lunn, D., Spiegelhalter, D., Thomas, A., and Best, N. 2009. The BUGS project: Evolution, critique and future directions. *Statistics in Medicine*, **28**, 3049–3082.

Magnus, J.R. and Neudecker, H. 1988. *Matrix Differential Calculus with Applications in Statistics and Econometrics*, New York: Wiley.

Meng, X. and Rubin, D.B. 1992. Performing likelihood ratio tests with multiply-imputed data sets. *Biometrika*, **79**, 103–111.

Meng, X. 1994. Multiple-imputation inferences with uncongenial sources of input. *Statistical Science*, **9**, 538–558.

Muthén, L. and Muthén, B. 2010. *Mplus Users Guide* (6th ed.). Los Angeles, CA: Muthén and Muthén.

Muthén, B., Asparouhov, T., Hunter, A.M., and Leuchter, A.F. 2011. Growth modeling with nonignorable dropout: alternative analyses of the STAR*D antidepressant trial. *Psychological Methods*, **16**, 17–33.

OECD. 2007. *PISA 2006: Science Competencies for Tomorrow's World*, Paris: OECD.

Pfeffermann, D., Skinner, C.J., Holmes, D.J., Goldstein, H., and Rasbash, J. 1998. Weighting for unequal selection probabilities in multilevel models. *Journal of the Royal Statistical Society B*, **60**, 23–40.

Pinheiro, J.C. and Bates, D.M. 2000. *Mixed-Effects Models in S and S-plus*, New York: Springer.

Raghunathan, T., Lepkowski, J., Van Hoewyk, J., and Solenberger, P. 2001. A multivariate technique for multiply imputing missing values using a sequence of regression models. *Survey Methodology*, **27**, 85–95.

Rasbash, J., Steele, F., Browne, W., and Goldstein, H. 2009. A Users Guide to MLwiN Version 2.10, Centre for Multilevel Modelling, University of Bristol.

Raudenbush, S.W. and Bryk, A.S. 2002. *Hierarchical Linear Models*, Newbury Park, CA: Sage.

Raudenbush, S.W., Bryk, A.S., Cheong, Y., Congdon, R.T., and Mathilda du Toit. 2011. *HLM 7: Hierarchical Linear and Nonlinear Modeling*. Lincolnwood, IL: Scientific Software International.

Rubin, D.B. 1976. Inference and missing data. *Biometrika*, **63**, 581–592.

Rubin, D.B. 1987. *Multiple Imputation for Nonresponse in Surveys*, New York: J. Wiley & Sons.

Rubin, D.B. 1996. Multiple imputation after 18+ years. *JASA*, **91**, 473–489.

Rubin, D.B. 2003. Nested multiple imputation of NMES via partially incompatible MCMC. *Statistica Neerlandica*, **57**, 3–18.

Schafer, J.L. 1997. *Analysis of Incomplete Multivariate Data*, London: Chapman & Hall.

Schafer, J.L. 1999. *NORM: Multiple Imputation of Incomplete Multivariate Data under a Normal Model [Computer Software]*. University Park: Pennsylvania State University, Department of Statistics.

Schafer, J.L. and Yucel, R.M. 2002. Computational strategies for multivariate linear mixed-effects models with missing values. *Journal of Computational and Graphical Statistics*, **11**, 437–457.

Schafer, J.L. 2003. Multiple imputation in multivariate problems when imputation and analysis models differ. *Statistica Neerlandica*, **57**, 19–35.

Shin, Y. 2003. *Inference and Applications in Hierarchical Linear Models with Missing Data.* Unpublished doctoral dissertation, University of Michigan.

Shin, Y. 2012. Do black children benefit more from small classes? Multivariate instrumental variable estimators with ignorable missing data. *Journal of Educational and Behavioral Statistics*, **37**, 543–574.

Shin, Y. and Raudenbush, S.W. 2007. Just-identified versus over-identified two-level hierarchical linear models with missing data. *Biometrics*, **63**, 1262–1268.

Shin, Y. and Raudenbush, S.W. 2010. A latent cluster mean approach to the contextual effects model with missing data. *Journal of Educational and Behavioral Statistics*, **35**, 26–53.

Shin, Y. and Raudenbush, S.W. 2011. The causal effect of class size on academic achievement: Multivariate instrumental variable estimators with data missing at random. *Journal of Educational and Behavioral Statistics*, **36**, 154–185.

Shin, Y. and Raudenbush, S.W. 2013. Efficient analysis of *Q*-level nested hierarchical general linear models given ignorable missing data. *International Journal of Biostatistics* (to appear).

Snijders, T.A.B. and Bosker, R.J. 2012. *Multilevel Analysis: An Introduction to Basic and Advanced Multilevel Modeling*, London: Sage.

Spiegelhalter, D.J., Thomas, A., and Best, N.G. 2000. WinBUGS version 1.3: User manual. Medical Research Council Biostatistics Unit, Cambridge.

Stram, D.O. and Lee, J.W. 1994. Variance components testing in the longitudinal mixed effects model. *Biometrics*, **50**, 1171–1177.

Tourangeau, K., Nord, C., Le, T., Sorongon, A.G., and Najarian, M. 2009. *Early Childhood Longitudinal Study, Kindergarten Class of 1998-99 (ECLS-K), Combined User's Manual for the ECLS-K Eighth-Grade and K-8 Full Sample Data Files and Electronic Codebooks* (NCES 2009-004). Washington, DC: NCES, IES, DOE.

Van Buuren, S., Brand, J., Groothuis-Oudshoorn, C., and Rubin, D. 2006. Fully conditional specification in multivariate imputation. *Journal of Statistical Computation and Simulation*, **76**, 1049–1064.

Van Buuren, S. and Groothuis-Oudshoorn, K. 2011. MICE: Multivariate imputation by chained equations in R. *Journal of Statistical Software*, **45**, 1–67.

Van Buuren, S. 2011. Multiple imputation of multilevel data. In J.J. Hoxand and J.K. Roberts. (eds.), *The Handbook of Advanced Multilevel Analysis*, Chapter 10, pp. 173–196. Milton Park, UK: Routledge.

Verbeke, G. and Molenberghs, G. 2000. *Linear Mixed Models for Longitudinal Data*, New York: Springer.

Yuan, Y.C. 2000. Multiple imputation for missing data: Concepts and new development. In *Proceedings of the Twenty-Fifth Annual SAS Users Group International Conference (Paper No. 267)*. Cary, NC: SAS Institute.

Yucel, R.M. 2008. Multiple imputation inference for multivariate multilevel continuous data with ignorable non-response. *Philosophical Transactions of the Royal Society A*, **366**, 2389–2403.

21

Multilevel Modeling of Categorical Response Variables

Carolyn J. Anderson
University of Illinois at Urbana-Champaign

Jee-Seon Kim
University of Wisconsin-Madison

Bryan Keller
Teachers College, Columbia University

CONTENTS

The most common type of item found on large-scale surveys has response options that are categorical. Models for binary, ordinal, and nominal variables are relatively common and well developed (e.g., Agresti, 2002, 2007; Fahrmeir and Tutz, 2001; McCullagh and Nelder, 1989; Powers and Xie, 1999); however, three major complications arise when modeling responses from large-scale surveys that employ a complex sampling design. These challenges result from the fact that data are typically clustered, the probability of selection is not equal for all units, and missing data is the norm rather than the exception. To deal with the clustered or nested structure, extensions of standard models for categorical data to clustered categorical data are presented. The particular random effects models presented in this chapter are logistic regression models for dichotomous responses, multinomial logistic regression models for nominal responses, and continuation ratios, adjacent categories, partial adjacent categories, proportional odds, and partially proportional odds models for ordinal data. The development of multilevel versions of these models and the software to fit them has progressed to the point that the models can be fit to data using a number of common software programs.

The second modification to the routine use of standard models for categorical data is dealing with unequal probability sampling of primary sampling units (e.g., schools) and secondary units (e.g., students). Ignoring the sample design can lead to biased results. Weighting of data may be necessary at each level of the model. The theory for adding weights to multilevel models is relatively straightforward (Asparouhov and Muthén 2006; Grilli and Pratesi 2004; Pfeffermann et al. 1998; Rabe-Hesketh and Skronkal 2006); however, the availability of software for discrete data that implements the theory is much less common. Design weights are incorporated during estimation and the estimated parameters are based on maximizing the pseudolikelihood rather than maximizing the likelihood. The estimating equations for discrete models are given in this chapter because they open up alternative software options with minimal programming in statistical software programs.

The third issue when modeling data from large-scale surveys is the problem of missing data. Deleting cases that have missing data can lead to biased parameter estimates and is also very wasteful. Although numerous methods have been developed and are regularly used to deal with missing data, the options for missing data when data are hierarchically structured are limited. Multiple imputation for continuous (normal) data has been developed by Shin and Raudenbush (2007, 2010) and is described in Chapter 20. Van Buuren (2011, 2012) presents a method based on fully conditionally specified models that can be used to impute missing values for both continuous and discrete variables. Unfortunately, this method has only been developed for imputing values at the lowest level of hierarchically structured data. At the possible cost of reduced efficiency, our approach to missing data imputes data at Level 2 and for each Level 1 unit, thus preserving the multilevel structure.

In the first section of this chapter, the data used to illustrate the models and methodology are described along with how weights are computed and how missing data were imputed. In the sections "Dichotomous Response Variables," "Nominal Response Variables," and "Ordinal Response Variables," models for dichotomous responses, nominal responses, and ordinal responses, respectively, are presented along with example applications to the data described in the section "Data." Current software options are described briefly in the section "Software." Example input code for fitting models to data using Mplus and SAS and a short document explaining the input code can be downloaded from http://faculty.ed.uiuc.edu/cja/homepage. Lastly, we conclude in the section "Discussion" with a summary of challenges in applications of models presented in this chapter.

Data

The models presented in this chapter are illustrated using the U.S. data from the 2006 Progress in International Reading Literacy Study (PIRLS). The data are from a complex two-stage cluster sampling design with unequal probability weighting. Students are nested within classrooms (teachers) and classes are nested within schools. Although some schools have students from multiple classrooms, most schools have only one classroom in the study. Therefore, the data are treated as consisting of two levels: students within schools. There are $M = 182$ schools with a total of $N = 5128$ students.* The number of students per school n_j ranges from 7 to 66 with an average of 27. Data from the fourth grade student questionnaire and the school questionnaire are used here.

Response and Explanatory Variables

Given the increasing role of technology and the Internet in society, our goal is to study how various student and school factors are related to Internet usage, in particular the use of the Internet as a source of information for school-related work. The response or "dependent" variable is an item asking students how often they use the Internet to "Look up information for school" (Schnet). The response options are "Every day or almost every day," "Once or twice a week," "Once or twice a month," and "Never or almost never." The distribution of the responses is given in Table 21.1. The response options

* These totals reflect the fact that one school was excluded that had no responses on the school questionnaire, 9 students were deleted who has no responses on the student questionnaire, and 3 students were deleted who has missing data but the imputation method approach used failed due to too few students from their schools.

TABLE 21.1

Distribution of Response Options to the Item Regarding Use of the Internet to Look Up Information for School

Response	Frequency	Percent	Cumulative Percent
Every day or almost every day	746	14.94	14.94
Once or twice a week	1240	24.83	39.77
Once or twice a month	1377	27.57	67.34
Never or almost never	1631	32.66	100.00

are discrete with a natural ordering. All four categories are used to illustrate the nominal and ordinal models, and to illustrate models for binary logistic regression the responses are dichotomized as

$$Y_{ij} = \begin{cases} 1 & \text{If student } i \text{ in school } j \text{ uses the Internet at least once a week} \\ 0 & \text{If student } i \text{ in school } j \text{ uses the Internet at most twice a month} \end{cases}.$$

The within-school or Level 1 explanatory variables include the students' gender (Girl), how much time per day a student spends reading for homework (TimeRdg), and the amount of time per day a student spends watching TV, videos, playing computer games, and so on (ScreenTime). The between-school or Level 2 explanatory variables include the number of fourth grade students per computer designated for fourth grade student use (NperComp), shortages of materials and staff at the school (Shortages), the location of the school (Urban, Suburban, Rural), and whether some, all, or no students at a school receive free or reduced price lunches (AllFree, SomeFree, NoneFree). To ensure the proper interpretation of model results, more information about these variables and basic summary statistics are given in Table 21.2.

Weights

The recommendations for weights given by Rutkowski et al. (2010) are used here. The weights for students (Level 1 or secondary units) will be computed as the product of student and class weights:

$$w_{1|ij\ell} = \underbrace{(WF_{ij\ell} \times WA_{ij\ell})}_{\text{student } i} \underbrace{(WF_{j\ell} \times WA_{j\ell})}_{\text{class } \ell}, \tag{21.1}$$

where $WF_{ij\ell}$ is the inverse of the probability of selection of student i from class ℓ in school j, $WF_{j\ell}$ is the inverse of the probability of selection of class ℓ from school j, $WA_{ij\ell}$ is the weight adjustment for nonresponse by student i in class ℓ from school j, and $WA_{j\ell}$ is the weight adjustment for nonresponse for class ℓ from school j. The weight adjustments are for those students and

TABLE 21.2

Basic Summary Statistics and Information about Explanatory Variables Used the Examples

Variable	Coding	Mean or Percent	Std Dev	Min	Max
Within or Level 1 Variables					
Girl	1 girl 0 boy	50.00% 50.00%			
TimeRdg	How much time per day spent reading for homework (ordinal but treated as quantitative)	2.34	0.75	1	4
ScreenTime	Mean of 2 items regarding how much time per day spent watching TV, videos, electronic games (high score means more time)	2.34	1.13	0	4
Between or Level 2 Variables					
NperComp	The number of fourth grade students per computer (number)	3.05	3.30	0.44	25.25
Shortages	Mean of 14 items dealing with various shortages (high score means more shortages)	0.59	0.54	0	3.0
Urban	1 Urban school 0 Suburban or rural	29.44% 70.65%			
Suburban	1 Suburban school 0 Urban or rural	42.78% 57.22%			
Rural	1 Rural school 0 Urban or suburban	27.78% 72.22%			
AllFree	1 All students free/reduced price lunch 0 Some or none	11.67% 83.33%			
SomeFree	1 Some students free/reduced price lunch 0 All or none	79.44% 20.56%			
NoneFree	1 Few students free/reduced price lunch 0 All or some	8.89% 91.11%			

Note: Values are based on nonmissing data.

classes that were selected but chose not to participate (Heeringa et al. 2010; Rutkowski et al. 2010). The school weights used are

$$w_{2|j} = WF_j \times WA_j, \tag{21.2}$$

where WF_j is the inverse of the probability of selecting school j, and WA_j is the weight adjustment for school j for nonresponse.

There are two views on the use of sampling or design weights in an analysis. One approach is design-based and advocates using sampling weights when analyzing data. The other approach is model-based and advocates not using the sampling weights. Snijders and Bosker (2012) discuss this issue in the context of multilevel models. The main argument in favor of a model-based approach is that the sample design is irrelevant when the model is the "true" one and the sampling procedure is independent of the probability model. If this is the case, taking into account sampling weights results in a loss of efficiency and less precise estimates (Heeringa et al. 2010). The main argument for a design-based approach is that parameter estimates could be seriously biased, especially if the sampling weights vary widely over units in the population (i.e., the weights are informative). Since there is a trade-off between efficiency (model-based has better efficiency) and bias (design-based is unbiased), both model- and design-based results are reported.

To determine whether the design may influence the results, one can examine the variability of the weights, the effective sample size, and the design effect. For example, if the schools had the same probability of being selected and every school selected responds, then the weights for the schools would all be equal. When weights are all the same, their variance equals 0 and they could simply be set to 1. With equal weights, the design would be simple random sampling. The same is true for students. If each student has the same probability of selection and each student responds, then the variance of the weights would be 0. For the PIRLS data, the mean and standard deviation of the weights for schools equal 305.42 and 218.64, respectively, which suggest the school weights are informative. For the students, the means of the Level 1 weights of students within schools mostly equal 1. Furthermore, for 74% of the schools, the standard deviations of the student weights equal 0, for 21% of the schools, they are less than 0.05, and for 5% of the schools, they are less than 0.13. The relatively large value of the standard deviation of the school weights suggests that the school weights will have an impact on the results; however, the small standard deviations for the student weights suggest these will have a negligible impact on the results.

Another way to assess the potential impact of weights is to examine the effective sample sizes. The effective sample sizes for the Level 2 (primary) and Level 1 (secondary) units are defined as

$$N_{\text{effective}} = \frac{\left(\sum_j w_{2|j}\right)^2}{\sum_j \left(w_{2|j}^2\right)} \quad \text{and} \quad n_{\text{effective},j} = \frac{\left(\sum_i \sum_\ell w_{1|ij\ell}\right)^2}{\sum_i \sum_\ell \left(w_{1|ij\ell}^2\right)}, \tag{21.3}$$

respectively (Heeringa et al. 2010; Snijders and Bosker 2012). These formulas define an effective sample size such that the information in a weighted sample equals the effective sample size from simple random sampling (Snijders and Bosker 2012). If the weights are all equal, then the effective sample sizes would equal the actual sample size. If the weights are informative, then the effective sample size will be less than the actual sample size. For the PIRLS data, the effective sample size for the schools is only 120, which is considerably less than the number of schools (i.e., $M = 182$). The effective sample sizes for students within schools mostly equal n_j, the number of students from school j.

A third measure of impact of the sampling design is based on the effective sample size. The design effects for Level 2 and Level 1 are defined as the ratios of the effective sample size over the actual sample size:

$$\text{Design} = \frac{N_{\text{effective}}}{M} \quad \text{and} \quad \text{Design}_j = \frac{n_{\text{effective},\, j}}{n_j}, \tag{21.4}$$

where M equals the number of clusters (Snijders and Bosker 2012). If weights are noninformative, then effective sample size equals the actual sample size and the ratio of the effective over the actual sample size equals 1. For the PIRLS data, the Level 2 design effect only equals 0.66; however, the Level 1 design effects are all mostly equal to one (i.e., $\text{Design}_j = 1$). In the analysis of our data, the Level 1 weights will have a negligible effect on the results, but the Level 2 weights will likely have a noticeable impact.

The last issue related to weights is how to scale them. In the context of multilevel models, two major suggestions are given for scaling of the weights (Pfeffermann et al. 1998): sums of weights equal effective sample size or sums of weights equal sample size. In the example analyses using the PIRLS data, we use the more common approach and scale weights such that their sums equal the sample sizes (i.e., $\sum_j w_{2|j} = 182$ and $\sum_i \sum_\ell w_{1|ij\ell} = n_j$).

The weights are incorporated when estimating a model's parameters. How weights are incorporated is described in detail in the section "Estimation" for the binary logistic regression model and is subsequently modified for the nominal and ordinal models in the sections "Nominal Response Variables" and "Ordinal Response Variables." The capability to include weights for the models discussed in this chapter has not been implemented in many of the common programs. For the examples presented in this chapter, SAS (version 9.3) was used to fit all models to the PIRLS data and Mplus (Muthén and Muthén 2010) was also used for the models that Mplus is capable of fitting to data (i.e., random effects binary logistic regression and proportional odds models). Software options are discussed in more detail in the section "Software."

Missing Data

A vexing problem for large, complex surveys (and longitudinal data) is missing data. Dealing with missing data is particularly difficult when data are clustered. If students or schools with missing data on the variables listed in Table 21.2 were excluded, there would only be $M = 158$ schools and a total of 3994 students. Simply removing students or whole schools due to missing data is not only a waste of data, but also can result in biased parameter estimates (Allison 2002; Enders 2010; Schafer 1997; Van Buuren 2011, 2012). The program Mplus can fit models using maximum likelihood estimation (MLE) with missing values on the response variable, but cannot handle missing predictors. Multiple imputation is an alternative approach to impute missing response and predictor variables.

A few procedures have been developed for multiply imputing missing clustered data. Chapters 18 and 20 of this book deal with missing data in multilevel models; however, this only pertains to normally distributed variables (Shin and Raudenbush 2007, 2010). Another proposal for multilevel data that is described by Van Buuren (2011) (see also Van Buuren, 2012) only deals with simple cases without missing Level 2 variables. Neither of these two solutions works for our data. Yet another proposal has been put forth that includes dummy variables for each cluster (Reiter et al. 2006); however, this presupposes that clusters only differ in terms of their intercepts. In our example, we do not want to make this assumption and using dummy variables is impractical due to the large number of clusters; therefore, this method is not pursued here. Nearly nonexistent are proposals for incorporating sampling weights into the imputation model for missing data. The one exception is Amer (2009), but this only deals with two clusters.

Ignoring the nested structure of the data when multiple imputing missing values can lead to severely biased results (Reiter et al. 2006; Van Buuren 2011, 2012); therefore, we took a practical approach. Since the PIRLS data are clustered, the imputations were carried out separately by imputing school-level missing values (i.e., one analysis on the 182 schools) and imputing missing values for student level. The latter was done by carrying out 182 analyses, one for each school. Although this method is not optimally efficient (leading to perhaps overly conservative inference), this approach retains associations among school (Level 2) variables, and preserves the heterogeneity between schools (i.e., random differences between schools in terms of intercept and the effects of variables). Furthermore, since the Level 1 weights are noninformative, these have minimal impact on the quality of the Level 1 imputed data.

For the PIRLS data, a reasonable assumption is that the data are missing at random, and we included additional variables in the imputation model to increase the likelihood of meeting this assumption. The patterns of missing data in the current study for the student variables are given in Table 21.3 and those for the school variables are given in Table 21.4. Each row represents a pattern of missing data where an "X" indicates that the column variable is not missing and a "." indicates that the variable is missing. The last two columns give the frequency and percent at which the row patterns occur, and the last two rows give the frequency and percent of missing values for each variable. The percent of students that have no missing data is fairly high (i.e., 92%), and the percentages of missing values for each variable are relatively small (i.e., less than 5.5%). In Table 21.4, two school-level variables that comprise StdComp are given (i.e., Num4th, the number of fourth grade students, and SchComp, the number of computers available to fourth grade students). From Table 21.4, we find that 84% of the schools' data are completely

TABLE 21.3

Missing Data on Student Questionnaire

Pattern	Girl	Schnet	TimeRdg	ScreenTime	Frequency	Percent
1	X	X	X	X	4715	91.95
2	X	X	X	.	219	4.27
3	X	.	X	X	69	1.35
4	X	.	X	.	51	0.99
5	X	X	.	X	48	0.94
6	X	X	.	.	7	0.14
7	X	.	.	X	12	0.23
8	X	.	.	.	4	0.08
9	.	X	X	X	3	0.06
Frequency	3	136	71	281	5128	
Percent	0.06	2.6548	1.38	5.48		

observed and most school-level variables have less than 5% missing values. The one exception is SchComp that has 7.7% of the values missing.

Multiple imputation using fully conditionally specified models as implemented in SAS (versions 9.3) was used to impute missing school and student-level variables (Van Buuren 2011, 2012). Fifteen imputed data sets were created. The relative efficiency was examined for each imputed variable to determine whether 15 data sets were sufficient. The relative efficiency measures the multiple imputed standard error relative to its theoretical minimum value, more specifically it equals (Enders 2010)

$$\left(1-\frac{\text{Fraction of missing information}}{\text{Number of imputed data sets}}\right)^{-1}.$$

The imputation models included additional auxiliary variables from the student and school questionnaires and imputations were done at the scale

TABLE 21.4

Missing Data on School Questionnaire

Group	Lunch	Location	Num4th	Short	Schcomp	Frequency	Percent
1	X	X	X	X	X	152	83.52
2	X	X	X	X	.	11	6.04
3	X	X	X	.	X	6	3.30
4	X	X	X	.	.	2	1.10
5	X	X	.	X	X	6	3.30
6	X	X	.	.	.	1	0.55
7	X	.	X	X	X	2	1.10
8	.	X	X	X	X	2	1.10
Frequency	2	2	7	9	14	182	
Percent	1.10	1.10	3.85	4.95	7.69		

level (rather than at the item level). For the school questionnaire data, nine variables were used in the imputation model, four of which were auxiliary variables. The relative efficiencies were all larger than .99. For the student-level data, the imputation model was kept relatively simple (i.e., only four variables, one of which was an auxiliary variable) due to small sample sizes in some schools. When imputing the student-level data, only 2 of the 182 schools were problematic and these had small sample numbers of students. The three students from these two schools (one from one school and two from the other) were dropped. The mean relative efficiencies for the within-school imputations were all greater than .99. The final number of students in the data set equaled 5128.

After the data were imputed, the school and student data sets were combined, composite and dummy variables that are used in the analyses were created, and the design weights were computed and scaled.

Dichotomous Response Variables

When a response variable is dichotomous, natural and common choices to model the responses are logistic and probit regression models, and when respondents are nested or clustered within larger units, there are multilevel random effects versions of these models. In this chapter, we focus on the multilevel logistic regression model.

In the section "Multilevel Binary Logistic Regression," the multilevel random effects logistic regression model is presented as a statistical model for clustered data, and in the section "A Latent Variable Approach," the model is presented in terms of a latent variable, including a random utility formulation. In the section "Estimation," estimation that incorporates design weights is discussed. In the section "Example for Binary Response Variable," an analysis of the PIRLS data is presented to illustrate the model.

Multilevel Binary Logistic Regression

A single-level logistic regression model can be generalized to a multilevel model by allowing the regression coefficients to differ randomly over clusters. Let the response variable for individual i in group j be coded as $Y_{ij} = 1$ for responses in a target category and $Y_{ij} = 0$ for responses in the other category. The distribution of Y_{ij} is assumed to be Bernoulli (or binomial if data are tabulated). A cluster-specific or Level 1 logistic regression model for the probability that individual i in cluster j has a response in the target category (i.e., $P(Y_{ij} = 1)$) is

$$P(Y_{ij}=1)=\frac{\exp(\beta_{0j}+\beta_{1j}x_{1ij}+\cdots+\beta_{Pj}x_{Pij})}{1+\exp(\beta_{0j}+\beta_{1j}x_{1ij}+\cdots+\beta_{Pj}x_{Pij})}, \tag{21.5}$$

where x_{pij} is the value of the pth Level 1 predictor variable for individual i in cluster j, β_{0j} is the intercept for cluster j, and β_{pj} is the regression coefficient for x_{pij} in cluster j. Note that for $P(Y_{ij} = 0)$, the β_{pj}s in the numerator all equal zero. The term in the denominator ensures that $P(Y_{ij} = 1) + P(Y_{ij} = 0) = 1$.

The between-cluster or Level 2 models are the same as those in the multilevel models for normal response variables discussed in Chapter 18 (i.e., hierarchical linear models or HLMs). Suppose that there are Q possible between-cluster predictors, $z_{1j}, \ldots, z_{Qj}$. The Level 2 models are linear models for each Level 1 regression coefficient:

$$\begin{aligned}
\beta_{0j} &= \sum_{q=0}^{Q} \gamma_{0q} z_{qj} + U_{0j} \\
\beta_{1j} &= \sum_{q=0}^{Q} \gamma_{1q} z_{qj} + U_{1j} \\
&\vdots \\
\beta_{Pj} &= \sum_{q=0}^{Q} \gamma_{Pq} z_{qj} + U_{Pj},
\end{aligned} \tag{21.6}$$

where $z_{0j} = 1$ for an intercept. The z_{qj}s are predictors or explanatory variables that model systematic differences between clusters, the γ_{pq}s are fixed regression coefficients for the z_{qj}s, and the U_{pj}s are the cluster-specific random effects. The first subscript on γ_{pq} corresponds to the effect in the Level 1 model and the second subscript corresponds to the predictor in the Level 2 model. For example, γ_{pq} is the regression coefficient for z_{qj} in the model for β_{pj}.

The predictors in the models for the β_{pj}s (i.e., the z_{qj}s) need not be the same over the models in Equation 21.6. For example, in the PIRLS data, shortages of supplies may predict differences between schools' intercepts, but not predict the differences between schools in terms of the effect of the amount of time that a student spends using electronics (i.e., ScreenTime).

The unexplained, random, or stochastic differences between clusters are modeled by the random effects. The distributional assumption for the U_{pj}s is

$$\boldsymbol{U}_j = \begin{pmatrix} U_{0j} \\ U_{1j} \\ \vdots \\ U_{Pj} \end{pmatrix} \sim MVN\left(\begin{pmatrix} 0 \\ 0 \\ \vdots \\ 0 \end{pmatrix}, \begin{pmatrix} \tau_{00} & \tau_{10} & \cdots & \tau_{P0} \\ \tau_{10} & \tau_{11} & \cdots & \tau_{P1} \\ \vdots & \vdots & \ddots & \vdots \\ \tau_{P0} & \tau_{P1} & \cdots & \tau_{PP} \end{pmatrix} \right) \quad i.i.d., \tag{21.7}$$

where *MVN* stands for multivariate normal and *i.i.d.* stands for independent and identically distributed. For short, the assumption in Equation 21.7 can be written as $\boldsymbol{U}_j \sim MVN(\boldsymbol{0}, \boldsymbol{T})$ *i.i.d.*

Substituting the Level 2 models into the regression coefficients of the Level 1 model yields our combined or cluster-specific model for the probabilities

$$P(Y_{ij}=1 \mid \boldsymbol{x}_{ij}, \boldsymbol{z}_j, \boldsymbol{U}_j) = \frac{\exp\left(\Sigma_{p=0}^{P}\left(\Sigma_{q=0}^{Q} \gamma_{pq} z_{qj} + U_{pj}\right) x_{pij}\right)}{1+\exp\left(\Sigma_{p=0}^{P}\left(\Sigma_{q=0}^{Q} \gamma_{pq} z_{qj} + U_{pj}\right) x_{pij}\right)}, \quad (21.8)$$

where $x_{0ij} = z_{0j} = 1$ for the intercept. To emphasize that these are conditional models for probabilities that depend on the observed and unobserved variables, the conditioning is explicitly indicated here where $\boldsymbol{x}_{ij}$ consists of within-cluster predictors, $\boldsymbol{z}_j$ the observed between-cluster predictors, and $\boldsymbol{U}_j$ the unobserved cluster random effects.

Similar to single-level logistic regression models, $\exp(\gamma_{pq})$ equals an odds ratio. More specifically, based on the model for probabilities in Equation 21.8, the odds that $Y_{ij} = 1$ versus $Y_{ij} = 0$ equals

$$\frac{P(Y_{ij}=1 \mid \boldsymbol{x}_{ij}, \boldsymbol{z}_j, \boldsymbol{U}_j)}{P(Y_{ij}=0 \mid \boldsymbol{x}_{ij}, \boldsymbol{z}_j, \boldsymbol{U}_j)} = \exp\left[\sum_{p=0}^{P}\left(\sum_{q=0}^{Q} \gamma_{pq} z_{qj} + U_{pj}\right) x_{pij}\right]. \quad (21.9)$$

To illustrate the interpretation of the γ_{pq}s, consider a model with two Level 1 (within-cluster) predictors, x_{1ij} and x_{2ij}, one Level 2 (between-group) predictor z_{1j} of the intercept and the slope of x_{2ij}, and a random intercept U_{0j}. The odds for this model are

$$\frac{P(Y_{ij}=1 \mid x_{1ij}, x_{2ij}, z_{1j}, U_{0j})}{P(Y_{ij}=0 \mid x_{1ij}, x_{2ij}, z_{1j}, U_{0j})} = \exp(\gamma_{00} + \gamma_{01} z_{1ij} + \gamma_{10} x_{1ij} + \gamma_{20} x_{2ij} + \gamma_{21} z_{1j} x_{2ij} + U_{0j}). \quad (21.10)$$

To interpret the parameter of a "main effect," for example, the effect of x_{1ij}, holding *all other variables constant*, the odds that $Y_{ij} = 1$ for a one unit increase in x_{1ij} is $\exp(\gamma_{10})$ times the odds for x_{1ij}; that is, the odds ratio equals $\exp(\gamma_{10})$. If $\gamma_{10} > 0$, then the odds increase; if $\gamma_{10} < 0$, then the odds decrease; and if $\gamma_{10} = 0$, then the odds do not change (i.e., the odds are equal). This interpretation of the effect of x_{1ij} requires that the values of all other variables be constant; however, it does not depend on the particular values of the other predictor variables or the particular value of x_{1ij}. To interpret a (cross-level) interaction when the focus is on the effect of x_{2ij} on Y_{ij}, we would consider the odds for a one unit increase in x_{2ij}. This odds ratio is $\exp(\gamma_{20} + \gamma_{21} z_{1j})$ and it depends on the value of z_{1j}. In reporting and explaining the effect of an interaction, representative values of z_{1j} could be used (e.g., 25th, 50th, and 75th percentiles). Alternatively, when the focus is on the effect of z_{1j} on Y_{ij}, we could report and explain the odds ratio for a one unit increase in z_{ij}, which is $\exp(\gamma_{01} + \gamma_{21} x_{2ij})$ and depends on x_{2ij}.

The interpretation of γ_{pq} is always qualified by "all other variables constant," including the random effects U_{0j} and other observed predictors. The interpretation is cluster-specific or within clusters because the random school effect U_{0j} has to be held constant. This is different from the interpretation of γ_{pq}s in HLM. The γ_{pq}s in HLM are interpretable as cluster specific and as marginal or population average effects. This difference stems from the fact that marginal distributions of multivariate normal random variables (as we have in HLM) are normally distributed. This is not true for a multilevel random effects logistic regression model or any of the models covered in this chapter. When one collapses over the unobserved random effects of the cluster-specific logistic regression model to get the marginal distribution, the result is not a logistic regression model (Demidenko 2004; Raudenbush and Bryk 2002; Snijders and Bosker 2012). The marginal effects in multilevel random effects logistic regression models are smaller than the cluster-specific ones.

A Latent Variable Approach

An alternative approach to random effects models for dichotomous responses (and later multicategory responses) is to propose a latent continuous variable that underlies the observed data (Muthén 2004; Skrondal and Rabe-Hesketh 2000). This approach can also be framed as a random utility model or a discrete choice model (McFadden, 1974). The former hypothesizes that observed data are related to the latent variable Y_{ij}^* as follows:

$$Y_{ij} = \begin{cases} 1 & \text{if } Y_{ij}^* > 0 \\ 0 & \text{if } Y_{ij}^* \le 0 \end{cases}, \tag{21.11}$$

where 0 is the threshold. A linear random effects model is then proposed for the latent variable Y_{ij}^*:

$$Y_{ij}^* = \sum_{p=0}^{P} \left(\sum_{q=0}^{Q} \gamma_{pq} z_{qj} + U_{pj} \right) x_{pij} + \varepsilon_{ij}, \tag{21.12}$$

where $z_{0j} = x_{0ij} = 1$ (for intercepts), γ_{pq} are the fixed effects parameters, U_{pj} are the random effects for clusters, and ε_{ij} is a random residual for individual i within cluster j. This linear model can also be arrived at using a multilevel perspective as was done in the previous section.

The distribution assumed for ε_{ij} determines the model for data. If $\varepsilon_{ij} \sim N(0, \sigma^2)$, the model for Y_{ij} is a probit model, and if ε_{ij} follows a logistic distribution, the model for Y_{ij} is a logistic regression.*

Instead of 0 as the cutoff or threshold relating the observed and latent variables in Equation 21.11, some authors and programs (e.g., Muthén and Muthén in Mplus) estimate a nonzero threshold, say ξ. This is equivalent to using a 0 cutoff by noting that for $Y_{ij} = 1$

$$Y_{ij}^* = \gamma_{00} + \sum_{q=1}^{Q} \gamma_{0q} z_{0qk} + U_{0j} + \sum_{p=1}^{P} \left(\sum_{q=0}^{Q} \gamma_{pq} z_{pqk} + U_{pj} \right) x_{pij} + \varepsilon_{ij} > 0,$$

which is equivalent to

$$\sum_{q=1}^{Q} \gamma_{0q} z_{0qk} + U_{0j} + \sum_{p=1}^{P} \left(\sum_{q=0}^{Q} \gamma_{pq} z_{pqk} + U_{pj} \right) x_{pij} + \varepsilon_{ij} > -\gamma_{00}.$$

The cutoff or threshold is $\xi = -\gamma_{00}$.

The latent variable approach facilitates the extension of the model to multicategory data. In the case of ordered response variables, assuming the existence of a latent variable is especially appealing because there may be some underlying continuum that gives rise to ordered observed responses. Many psychometric models are based on just such an assumption (e.g., random utility models, discrete choice models, Guttman scale, item response theory models, and others).

Similar to HLM for normal response variables, an intraclass correlation (ICC) can be computed for random intercept models. In the case of a random intercept logistic regression model with no predictors (i.e., the only random effect in the model is U_{0j}), the ICC equals

$$\text{ICC} = \frac{\tau_{00}}{\tau_{00} + \pi^2/3}.$$

The ICC is a measure of within-cluster homogeneity and equals the proportion of variance due to between-cluster differences. A *residual* ICC, computed

* In the random utility or choice model formulation, a linear model is given for the latent variable (utility) for each category (i.e., Y_{ijk}^* for category k, where $k = 0,\ldots,K$), and the category with the largest value of the latent variable/utililty is selected. For identification, the random and fixed effects for the utility associated with the reference category (i.e., $Y_{ij0}^* = 0$) equal zero; therefore, Y_{ijk}^* in Equation 21.11 represents the difference between utilities. The distribution of the difference between ε_{ij}s determines the choice option selected. When the residuals in the model for Y_{ijk}^* are assumed to follow a Gumbel (extreme value) distribution, the difference between the ε_{ij}s follows a logistic distribution leading to the logistic regression model.

when there are fixed effect predictors and a random intercept, measures within-cluster homogeneity and variance due to between-cluster differences given the predictors.

Estimation

Typically, MLE is used to estimate single-level and multilevel random effects logistic regression model parameters; however, with weights, the method used is a version of pseudolikelihood estimation. Following the general approach of Pfeffermann et al. (1998) (see also, Asparouhov and Muthén 2006; Grilli and Pratesi 2004; Rabe-Hesketh and Skronkal 2006), the weights are incorporated into the likelihood equations and the parameters are found by maximizing the modified equations or pseudolikelihood equations. The basic procedure described below is also used for other models covered in this chapter with slight variations (i.e., the distribution of the cluster-specific response and the model for probabilities). Given the multilevel structure, first we show how the Level 1 weights are included to yield cluster-specific pseudolikelihood equations and then show how the Level 2 weights are incorporated.

Let $\mathcal{L}(y_{ij}|\boldsymbol{x}_{ij},\boldsymbol{z}_j,\boldsymbol{U}_j)$ equal the logarithm of the likelihood conditional on the observed predictor variables and random effects $\boldsymbol{U}_j$ for cluster j. To incorporate the Level 1 weights $w_{1|ij}$, each of the likelihoods for individuals within a cluster are weighted by their values of $w_{1|ij}$ and then the random effects are collapsed over to yield the cluster-specific log-pseudolikelihood $\mathcal{L}(\boldsymbol{y}_j)$ as follows:

$$\mathcal{L}(\boldsymbol{y}_j)=\log\left[\int_u \exp\left\{\sum_{i=0}^{n_j} w_{1|ij}\mathcal{L}(y_{ij}\mid \boldsymbol{x}_{ij},\boldsymbol{z}_j,\boldsymbol{U}_j)\right\} f(\boldsymbol{U}_j)d\boldsymbol{U}_j\right], \qquad (21.13)$$

where integration is over all random effects in $\boldsymbol{U}_j$, and $f(\boldsymbol{U}_j)$ is the multivariate normal distribution with mean $\boldsymbol{0}$ and covariance matrix $\boldsymbol{T}$ (Asparouhov and Muthén 2006; Grilli and Pratesi 2004; Rabe-Hesketh and Skronkal 2006). In Equation 21.13, the unobserved predictors are integrated out to yield cluster-specific log-pseudolikelihoods. For logistic regression, the individual likelihood equals the Bernouli distribution function and $\mathcal{L}(y_{ij}|\boldsymbol{x}_{ij},\boldsymbol{z}_j,\boldsymbol{U}_j)$ is

$$\mathcal{L}(y_{ij}\mid \boldsymbol{x}_{ij},\boldsymbol{z}_j,\boldsymbol{U}_j)=\log\left\{P(Y_{ij}=1\mid \boldsymbol{x}_{ij},\boldsymbol{z}_j,\boldsymbol{U}_j)^{y_{ij}}(1-P(Y_{ij}-1\mid \boldsymbol{x}_{ij},\boldsymbol{z}_j,\boldsymbol{U}_j))^{(1-y_{ij})}\right\},$$

where $P(Y_{ij}=1|\boldsymbol{x}_{ij},\boldsymbol{z}_j,\boldsymbol{U}_j)$ is given by Equation 21.8, and y_{ij} is an indicator variable that equals 1 if the response by individual i in cluster j is the target category and 0 otherwise.

The group-specific log-pseudolikelihoods in Equation 21.13 must be combined to yield the log-pseudolikelihood using all the data. The Level 2 weights enter at this point. Assuming independence between clusters, the log-pseudolikelihood for all the responses equals (Asparouhov and Muthén 2006; Grilli and Pratesi 2004; Rabe-Hesketh and Skronkal 2006)

$$\mathcal{L}(\boldsymbol{y})=\sum_{j=1}^{M} w_{2|j}\mathcal{L}(\boldsymbol{y}_j). \tag{21.14}$$

The parameters (i.e., γ_{pq} and τ'_{pp} for all p, p', and q) that maximize Equation 21.14 are the maximum pseudolikelihood estimates. Although the log-pseudolikelihood equations have only been given here for two-level models, Rabe-Hesketh and Skronkal (2006) give a general set of formulas for higher-level models and illustrate their use for a three-level model fit to Program International Student Assessment (PISA) data.

A number of estimation algorithms exist that attempt to find the parameters that maximize either the log-likelihood (i.e., $w_{2|j} = w_{1|ij} = 1$) or the log-pseudolikelihood. Two algorithms, marginal quasi-likelihood and penalized quasi-likelihood, attempt to approximate the model by linearizing the model using a Taylor series expansion and then using an algorithm designed for a linear mixed model (SAS Institute Inc., 2011b). Unfortunately, these two strategies yield parameters estimates that are severely biased. A better approach is to find the parameters that maximize the function in Equation 21.14. The "gold standard" is adaptive Gaussian quadrature (i.e., numerical integration); however, this becomes computationally very difficult and time consuming for multiple (correlated) random effects. A third common alternative is Bayesian methods. For the examples presented in this chapter, adaptive quadrature is used to estimate model parameters.

When MLE is used to obtain parameter estimates, the standard errors of parameters can be estimated based on the model and are valid provided that the correct model is specified. Under pseudolikelihood estimation, the model-based standard errors will be biased; therefore, robust (sandwich or empirical) estimators of the standard errors are recommended. The sandwich estimates are based on the data.* The robust standard errors can be used to compute test statistics for the fixed effects, in particular, the test statistic equals $\hat{\gamma}_{pq}/\widehat{\text{se}}$, where $\hat{\gamma}_{pq}$ is the pseudolikelihood estimate of γ_{pq} and $\widehat{\text{se}}$ is the sandwich estimate.

Example for Binary Response Variable

In this example, we model how much time a student spends looking up information for school on the Internet where the response variable was coded $Y_{ij} = 1$ for at least once a week and $Y_{ij} = 0$ for at most twice a month. The predictor variables that were considered are given in Table 21.2. Each model

* The sandwich estimators are also used in MLE with a misspecified model.

was fit to each of the 15 imputed data sets and the results combined using Rubin's method (Rubin 1987), which is given in Chapter 18 of this book. As a measure of the impact of missing data on the analysis, the missing fraction of information for estimating a parameter was computed. This measures how much information about a parameter is lost due to missingness (Snijders and Bosker, 2012). This fraction equals

$$\text{Missing fraction} = \frac{(1+1/15)\,\text{var}\left(\hat{\gamma}_{pq}\right)}{\widehat{\text{var}}\left(\hat{\gamma}_{pq}\right) + (1+1/15)\,\text{var}\left(\hat{\gamma}_{pq}\right)}, \tag{21.15}$$

where $\widehat{\text{var}}(\hat{\gamma}_{pq})$ is the estimated sandwich variance from combining over imputations, and $\text{var}(\hat{\gamma}_{pq})$ is the variance of the parameters over the 15 imputed data sets. Small values indicate little loss of information.

The models were fit using both Mplus (version 6) and SAS PROC NLMIXED (version 9.3) and empirical (sandwich) standard errors were computed by each program. We started with a relatively simple random intercept model with few fixed effects and increased the complexity by adding more fixed effects at Level 1 and Level 2, including cross-level interactions. Some of the Level 1 predictor variables showed a fair amount of variability between schools; therefore, Level 1 variables were centered around their school means and the means used as predictors of the intercept. The first such model had school-mean centered ScreenTime (i.e., $\text{ScreenTime}_{ij} - \overline{\text{ScreenTime}}_j$) and school-mean centered TimeRdg (i.e., $\text{TimeRdg}_{ij} - \overline{\text{TimeRdg}}_j$) as within or Level 1 predictors, and the means $\overline{\text{ScreenTime}}_j$ and $\overline{\text{TimeRdg}}_j$ were entered into the model as predictors of school intercepts. Once a relatively complex fixed effects model was successfully fit to the data, fixed effects that were nonsignificant were successively dropped; however, at some steps, some school effects were put back in the model and retested. After arriving at a fixed effects structure that seemed to be best, we fit three more models, each of which had a random slope for each of the Level 1 variables. The final model chosen was

$$\text{Level 1: } \log\left(\frac{P(Y_{ij}=1)}{P(Y_{ij}=0)}\right) = \beta_{0j} + \beta_{1j}\text{Girl}_{ij} + \beta_{2j}\text{ScreenTime}_{ij} + \beta_{3ij}(\text{TimeRdg}_{ij} - \overline{\text{TimeRdg}}_j), \tag{21.16}$$

and

$$\begin{aligned}\text{Level 2: } \beta_{0j} &= \gamma_{00} + \gamma_{01}\overline{\text{TimeRdg}}_j + \gamma_{02}\text{Shortages}_j + \gamma_{03}\text{AllFree}_j + U_{0j}\\ \beta_{1j} &= \gamma_{10}\\ \beta_{2j} &= \gamma_{20}\\ \beta_{3j} &= \gamma_{30},\end{aligned}$$

where $U_{0j} \sim N(0, \tau_{00})$ *i.i.d.* Although we arrived at this model including weights, we report estimated parameters, standard errors, and various statistics in Table 21.5 for this model with and without weights. Models were also fit to data using only Level 2 weights; however, since these yielded essentially identical results as those using both Level 1 and Level 2 weights, only the results with both weights included are reported. This is true here and in later sections. Note that the missing fractions are for the most part small indicating that missing data had a small impact of the results. There are some differences between the results depending on whether weights are used or not. In particular, note that the standard errors are larger when weights are used. The effect of Girl is significant at the .05 level when weights are not used but is not significant when weights are used in the estimation.

Before settling on this model and interpreting the parameters, we also performed some model diagnostics. The U_js were estimated after the model was fit to the data using empirical Bayes estimation. One of the assumptions of the multilevel model was that the U_{pj}s are normal. The estimated U_{0j}s were plotted and found to be roughly normal, which is consistent with the normality assumption but is not conclusive. The normality of estimated U_{0j} only reveals if the assumption is tenable (Verbeke and Molenberghs 2000).

We adapted a method, receiver operating characteristic (ROC)* curve analysis, used in single-level logistic regression (Agresti 2002) to multilevel logistic regression. This analysis gives us a sense of how much is gained from the model and a way to compare models. After a model was fit to data, predicted probabilities setting $U_j = 0$ (i.e., "unconditional probabilities") were computed and probabilities conditional on the schools (i.e., using estimates of U_{0j}s) were also computed. Predicted responses were determined using multiple cut-points from 0 to 1 incremented by .025. The probabilities of true positives (i.e., $P(\hat{Y}_{ij} = 1 | y_{ij} = 1)$) and false positives (i.e., $\hat{Y}_{ij} = 1 | y_{ij} = 0$) were plotted against each other (one point for each cut-point). As an example, the ROC curves for the final model are plotted in Figure 21.1 with the cut-points indicated. The ROC curves for different imputations were nearly indistinguishable so what is plotted in Figure 21.1 is the average of the 15 curves. Chance (random prediction) is indicated by the straight line. The further a model's ROC curve is above the straight line, the better the fit of the model to data. The area under the ROC curves equals the concordance index. Note that including estimates $\hat{U}_{0j}$ for the schools yields better predictions than simply setting $U_{0j} = 0$, which is also evidence for the importance of a random effect for the intercept.

Likelihood ratio tests and information criteria can be used to compare models estimated by MLE. Under MLE, the hypothesis for random effects (e.g., H_0: $\tau_{00} = 0$ for a random intercept model) is a nonstandard test because the conditions for standard test statistics fail (e.g., $\tau_{00} = 0$ is on the

* ROC analysis was originally used in signal detection analyses where decisions are made under uncertainty, such as whether a blip on a radar is a signal or a noise.

TABLE 21.5

Estimated Parameters, Robust Standard Errors, and Various Statistics for Final Model Fit to Binary Response Variable Where Weights Are Not Used and Both Level 1 and Level 2 Weights Are Used

	Maximum Likelihood Estimation (No Weights)						Pseudolikelihood Estimation (With Weights)					
Effect	Estimate	s.e.	Est/s.e.	p	Missing Fraction	Odds Ratio	Estimate	s.e.	Est/s.e.	p	Missing Fraction	Odds ratio
Within or Level 1												
Intercept	−2.617	0.503	5.20	<.01	.047	0.07	−3.326	0.684	4.855	<.01	.022	0.04
Girl	0.150	0.060	2.50	.01	.042	1.16	0.133	0.071	1.889	.06	.051	1.14
ScreenTime	0.077	0.030	2.56	.01	.084	1.08	0.105	0.038	2.726	.01	.142	1.11
TimeRdg − $\overline{\text{TimeRdg}}$	0.246	0.041	5.95	<.01	.041	1.28	0.287	0.057	5.035	<.01	.019	1.33
Between or Level 2												
$\overline{\text{TimeRdg}}$	0.860	0.211	5.96	<.01	.038	2.36	1.112	0.278	3.998	<.01	.018	3.04
Shortages	−0.232	0.081	4.08	<.01	.043	0.79	−0.209	0.101	−2.084	.04	.052	0.81
AllFree	0.353	0.118	2.86	<.01	.047	1.42	0.394	0.153	2.580	.01	.088	1.48
Random intercept												
Variance	0.250	0.055			.017		0.262	0.058	4.483		.021	
			Mean (std error)									
−2lnlike			6698.53 (2.95)									
AIC			6714.53 (2.95)									
BIC			6740.16 (2.95)									

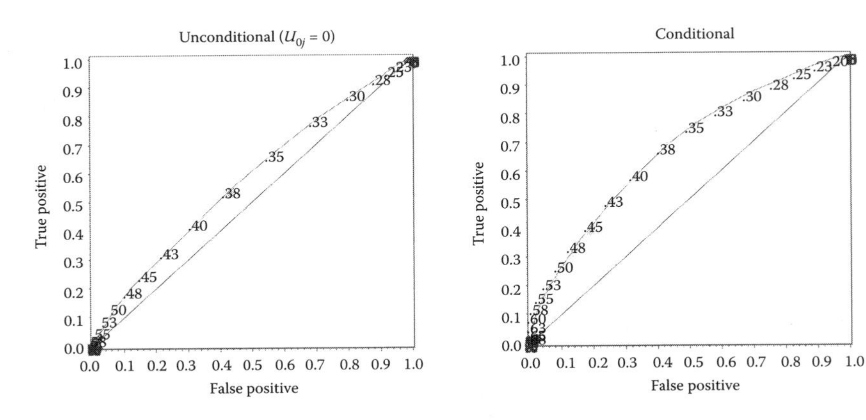

FIGURE 21.1
ROC curves for unconditional (i.e., $U_{0j} = 0$) and conditional predictions (i.e., $\hat{U}_{0j}$) based on the final logistic regression model from pseudolikelihood estimation. The numbers are the cut-points used to determine predicted responses from predicted probabilities.

boundary of the parameter space). However, variances can be tested by computing a likelihood ratio test statistic and compare it to a mixture of chi-square distributions (for more details see Chapter 18 or Verbeke and Molenberghs 2000; Molenberghs and Verbeke 2005; Self and Liang 1987; Stram and Lee 1994). Under pseudolikelihood, simple likelihood ratio tests of fixed effects and information criteria for model comparisons should not be used. A correction exists for likelihood ratio tests (Asparouhov and Muthén 2006); however, it requires extra computations to obtain the correction factor. The corrected likelihood ratio test is implemented in Mplus and was studied by Asparouhov and Muthén (2006). They found the adjusted likelihood ratio test to be superior to the uncorrected one in terms of rejection rates. It should be noted that the corrected test's performance depends on cluster sample size (i.e., larger cluster sizes lead to better results).

Based on the pseudolikelihood parameter estimates, holding all other variables constant (including the random effect), the odds that a girl will use the Internet for school at least weekly are 1.14 times that for boys. In other words, within a given school, girls are more likely than boys to use the Internet to look up information for school. Additionally, students who spend more time in front of a screen (watching TV, DVDs, playing computer or video games) are more likely to use the Internet, and students who spend more time reading for school relative to their classmates are more likely to use the Internet for school.

Student's use of the Internet for school work depends on the school they attend. First note that the residual *ICC* for the final model equals .262/$(0.262 + \pi^2/3) = .07$. For the sake of comparison when only the Level 1 variables are in the model, $\hat{\tau}_{00} = 0.335(\widehat{se} = 0.068)$ and $ICC = 0.335/(0.335 + \pi^2/3) = .09$. The between-level variables accounted for $(.335 - .262)/.335 \times 100 = 21\%$ of variance of the intercept (differences between schools). Holding observed and unobserved variables constant, the odds ratio that a student will use the Internet for school for a one unit increase in school-mean time spent reading equals 3.04; that is, higher values of $\overline{\text{TimeRdg}_j}$ are associated with larger odds of Internet use. Greater shortages in a school are associated with decreased odds of Internet use; however, larger odds are associated with schools where all students have free or reduced priced lunch.

Nominal Response Variables

The binary logistic regression model is a special case of the baseline multinomial logistic regression model. After presenting the multinomial model, estimation is briefly discussed followed by an example using the PIRLS data.

Baseline Multinomial Model

Let k index the K response options such that $Y_{ij} = k$ for $k = 1,\ldots,K$. The Level 1 or cluster-specific model is

$$P(Y_{ij} = k) = \frac{\exp\left(\beta_{0jk} + \sum_{p=1}^{P} \beta_{pjk} x_{pij}\right)}{\sum_{k=1}^{K} \exp\left(\beta_{0jk} + \sum_{p=1}^{P} \beta_{pjk} x_{pjk}\right)}. \tag{21.17}$$

The sum in the denominator ensures that the sum of the probabilities over response options equals 1. Note that there is a different intercept and slope for each response option.* This model can become very complex rather quickly. Typically, one response option is chosen as a baseline and the parameters for this response option are set equal to 0. There may be a natural baseline (e.g., "none") or an arbitrary response option can be used as the baseline. We use K here as the baseline or reference category and the β_{pjK}s are set equal to 0.

The Level 2 model is just like the Level 2 model for the binary logistic model, except that there is now a Level 2 model for each Level 1 predictor and each $(K-1)$ response option. The regression coefficients are assumed to be (multivariate) normal; that is

$$\beta_{0jk} = \sum_{q=0}^{Q} \gamma_{0qk} z_{qi} + U_{0jk} \tag{21.18}$$

$$\vdots$$

* This model can be reparameterized as a conditional multinomial logistic regression model such that the predictors can be attributes of *response options*, as well as of individuals (i.e., x_{pijk}). In the conditional model, there is a single β_j for each school and for each response variable x_{ijk}, but there are many more response variables (Agresti, 2002, 2007; Anderson and Rutkowski, 2008).

$$\beta_{Pjk} = \sum_{q=0}^{Q} \gamma_{Pqk} z_{qi} + U_{pjk} \tag{21.19}$$

for $k = 1, \ldots, (K-1)$. The random terms may depend on the response categories as well as the cluster.

The most natural interpretation of the fixed effects parameters is in terms of odds ratios. Given that response K was chosen as the reference category, the cluster-specific model is

$$\log\left(\frac{P(Y_{ij}=k)}{P(Y_{ij}=K)}\right) = \exp\left(\sum_{p=0}^{P}\left(\sum_{q=0}^{Q} \gamma_{pqk} z_{qj} + U_{pjk}\right) x_{pij}\right). \tag{21.20}$$

The interpretation of the main effect and interaction effects are the same as those for the binary logistic regression model; that is, exponentials of γ_{pqk}s are estimates of odds ratios.

There are a number of ways to simplify the model. One way is to assume that the random effects do not depend on the category, in which case the k subscript will be dropped (e.g., U_{pj}). Another way to simplify the model is to set some fixed effects to be equal for different response categories (e.g., $\gamma_{pqk} = \gamma'_{pqk'}$ where $k \neq k'$). These simplifications are illustrated in the section "Multinomial Example."

Estimation

Estimation of the multinomial model with weights is basically the same as it was for the binary logistic regression model, except that the binomial distribution is replaced by the more general multinomial distribution and the cluster-specific multilevel nominal logistic regression model replaces the multilevel logistic regression model. With these changes, the cluster-specific log-likelihood is

$$\begin{aligned}\mathcal{L}(y_{ijk} \mid \boldsymbol{x}_{ij}, \boldsymbol{z}_j, \boldsymbol{U}_{jk}) = \log\Big[& P(Y_{ij}=1 \mid \boldsymbol{x}_{ij}, \boldsymbol{z}_j, \boldsymbol{U}_{jk})^{y_{ij1}}\, P(Y_{ij}=2 \mid \boldsymbol{x}_{ij}, \boldsymbol{z}_j, \boldsymbol{U}_{jk})^{y_{ij2}} \\ & \ldots P(Y_{ij}=K \mid \boldsymbol{x}_{ij}, \boldsymbol{z}_j, \boldsymbol{U}_{jk})^{y_{ijK}}\Big],\end{aligned} \tag{21.21}$$

where y_{ijk} is an indicator of the observed response and it is coded as $y_{ijk} = 1$ if the response from individual i in cluster j is k, and zero otherwise. The cluster-specific multinomial logistic regression model is

$$P(Y_{ij}=k \mid \boldsymbol{x}_{ij}, \boldsymbol{z}_j, \boldsymbol{U}_{jk}) = \frac{\exp\left(\sum_{p=0}^{P}\left(\sum_{q=0}^{Q} \gamma_{pqk} z_{qj} + U_{pjk}\right) x_{pij}\right)}{\sum_{k=1}^{K}\left(\exp\left(\sum_{p=0}^{P}\left(\sum_{q=0}^{Q} \gamma_{pqk} z_{qj} + U_{pjk}\right) x_{pij}\right)\right)}. \tag{21.22}$$

The cluster-specific log-pseudolikelihoods are combined as in Equation 21.14 to yield a log-pseudolikelihood for the entire data set.

Multinomial Example

The parameters of models reported in this section were fit using SAS/NLMIXED (version 9.3) using the EMPIRICAL option to compute the sandwich standard errors. The sandwich standard errors were used in the statistical tests for the fixed effects. In the models, the baseline or reference category was "None or almost never."

Our modeling process started with fitting models with one fixed effect predictor, various combinations of fixed effects, and a simple random structure (i.e., only a single random intercept for each school, $U_{0jk} = U_{0j}$ for all $k = 1,\ldots,4$). Predictors that were consistently significant were retained in the model. The fixed effects structure settled on is similar to that in the binary model example, except that free lunch was not significant. The results of this random intercept model with complex fixed effects estimated with and without weights are reported in Table 21.6.

Possible simplifications of the model are suggested by examining the parameters reported in Table 21.6. For example, the parameter estimates for Girl are similar in value for the first three response options, and only the parameters for the first response option for ScreenTime is significant. Parameters that are similar in value can be equated and those that are not significant can be set to 0 (i.e., equated to the value for the baseline category). Significance tests can be computed for these restrictions using Wald tests. We can test whether the coefficients for effects are all the same, such as those for Girl, H_0: $\gamma_{11} = \gamma_{12} = \gamma_{13}$. Let P^* equal the number of fixed effects parameters and $\boldsymbol{L}$ equal an $(r \times P^*)$ matrix whose rows define linear combinations of the P^* parameters, $\boldsymbol{\gamma}$ is a $(P^* \times 1)$ vector of parameters, and $\boldsymbol{S}_{\hat{\gamma}}$ the covariance matrix of parameter estimates. The null hypothesis for Girls is

$$H_0 : \boldsymbol{L\gamma} = \begin{pmatrix} 0 & 0 & 0 & 1 & -1 & 0 & \ldots & 0 \\ 0 & 0 & 0 & 1 & 0 & -1 & \ldots & 0 \end{pmatrix} \begin{pmatrix} \gamma_{01} \\ \gamma_{02} \\ \gamma_{03} \\ \gamma_{11} \\ \gamma_{12} \\ \gamma_{13} \\ \vdots \\ \gamma_{P3} \end{pmatrix} = \begin{pmatrix} \gamma_{11} - \gamma_{12} \\ \gamma_{11} - \gamma_{13} \end{pmatrix} = \begin{pmatrix} 0 \\ 0 \end{pmatrix}.$$

and

$$\text{Wald} = \boldsymbol{\gamma}'\boldsymbol{L}'(\boldsymbol{L}\boldsymbol{S}_{\hat{\gamma}}\boldsymbol{L}')^{-1}\boldsymbol{L}\boldsymbol{\gamma} \sim \chi_r^2.$$

TABLE 21.6

Estimated Parameters, Robust Standard Errors, p Value from Tests of Fixed Effects and Missing Fraction of Information for the Random Intercept Multinomial Logistic Regression Model without Weights (MLE) and with Level 1 and 2 Weights (Pseudolikelihood)

	Maximum Likelihood Estimation (No Weights)				Pseudolikelihood Estimation (With Weights)			
Effect	**Est**	**s.e.**	p	**Missing Fraction**	**Est**	**s.e.**	p	**Missing Fraction**
Intercept 1	−3.949	0.720	<.001	.045	−5.320	1.005	<.001	.024
Intercept 2	−3.169	0.729	<.001	.024	−3.673	1.064	<.001	.012
Intercept 3	−1.714	0.696	.014	.022	−2.195	0.986	.026	.007
Girl 1	0.411	0.101	<.001	.029	0.387	0.120	.001	.046
Girl 2	0.332	0.080	<.001	.032	0.309	0.100	.002	.037
Girl 3	0.453	0.072	<.001	.018	0.422	0.096	<.001	.019
ScreenTime 1	0.161	0.051	.002	.207	0.165	0.066	.011	.232
ScreenTime 2	0.006	0.039	.868	.084	0.026	0.047	.575	.164
ScreenTime 3	−0.047	0.037	.200	.163	−0.068	0.039	.084	.167
CTimeRdg 1	0.377	0.069	<.001	.041	0.444	0.094	<.001	.029
CTimeRdg 2	0.251	0.055	<.001	.031	0.275	0.068	<.001	.017
CTimeRdg 3	0.129	0.055	.020	.044	0.129	0.064	.043	.026
MTimeRdg 1	1.182	0.300	<.001	.029	1.717	0.409	<.001	.015
MTimeRdg 2	1.274	0.302	<.001	.022	1.471	0.435	<.001	.012
MTimeRdg 3	0.704	0.288	.015	.020	0.938	0.405	.020	.008
Shortages 1	−0.229	0.138	.098	.020	−0.177	0.141	.211	.035
Shortages 2	−0.364	0.111	.001	.033	−0.378	0.142	.007	.041
Shortages 3	−0.267	0.134	.047	.023	−0.275	0.149	.063	.034
Variance	0.315	0.063	—	.023	0.325	0.065	—	.057
		Mean (Std Error)						
−2lnlike		13,565.27 (11.20)						
AIC		13,603.27 (11.20)						
BIC		13,664.15 (11.20)						

For the above hypothesis and using the sandwich covariance matrix for $S_{\hat{\gamma}}$, Wald = 1.34 with degrees of freedom $r = 2$ and $p = .51$, which supports the conclusion that a single coefficient for Girl is sufficient. Testing whether the coefficients for CTimeRdg are the same yields Wald = 13.98, df = 2, and $p < .01$ and indicates that these should not be equated. Note that when testing a single fixed effect, such as H_0: $\gamma_{pq} = 0$, the test statistic reduces to $(\hat{\gamma}_{pq}/\widehat{\text{se}}_{\gamma pq})^2$.

Based on a number of Wald tests, in the next round of modeling, some parameters were equated and others were set equal to 0. Given a simpler fixed effects structure, the random structure of the model was developed.

Category-specific random effects were first added to the model that only had an intercept. The variances of U_{0j1} and U_{0j2} were similar in value and the estimated correlation between them equaled .79. This suggested setting $U_{0j1} = U_{0j2}$ and implies that the random effect for a school is the same when the response is either "Every day or almost every day" and "Once or twice a week." With this restriction, the model with both the fixed effects structure found previously and a semicomplex random structure was fit to the data and selected as our final model. The parameter estimates and various statistics for the final model are reported in Table 21.7. The models fit by MLE reported in Tables 21.6 and 21.7 can be compared using Akaike Information Criterion (AIC) and Bayesian Information Criterion (BIC). The model in Table 21.7 fit by MLE with the more complex random structure is better in terms of both AIC and BIC than they are in Table 21.6.

Even after simplifying the model, there are still a large number of fixed effects parameters. Comparing all pairs of response options can aid interpretation of the results. For example, suppose that we wish to look at the effect of school-mean centered reading time on the odds of "Every day or almost every day" (Daily) versus "Once or twice a week" (Weekly). Using the pseudolikelihood parameter estimates, this odds ratio equals exp(0.452 − 0.281) = 1.19. Since there were no interactions in the final model, all the odds ratios were computed using $\exp(\gamma_{pqk} - \gamma_{pqk'})$ and are reported in Table 21.8. For each effect, the odds ratios below the diagonals were computed using the pseudolikelihood estimates and those above were from MLE. An entry in the table is the odds ratio of more frequent use of the Internet versus less frequent use for unit change of the predictor variable. For example, the pseudolikelihood estimate of the odds ratio of Daily use of the Internet versus Monthly for a one unit change in school-mean centered reading time is 1.39 and the maximum likelihood estimate is 1.29.

Not all the pairwise odds ratios are necessarily significant. A test for an odds ratio can be conducted or a confidence interval can be placed on the odds ratios. In general, a test statistic for H_0: $\exp(\gamma_{pqk}) = \exp(\gamma_{pqk'})$ equals

$$\frac{\hat{\gamma}_{pqk} - \hat{\gamma}_{pqk'}}{\widehat{\text{se}}\left(\hat{\gamma}_{pqk} - \hat{\gamma}_{pqk'}\right)},$$

where that standard error of the difference equals

$$\widehat{\text{se}}\left(\hat{\gamma}_{pqk} - \hat{\gamma}_{pqk'}\right) = \sqrt{\text{var}\left(\hat{\gamma}_{pqk}\right) + \text{var}\left(\hat{\gamma}_{pqk'}\right) - 2\,\text{cov}\left(\hat{\gamma}_{pqk}, \hat{\gamma}_{pqk'}\right)},$$

and sandwich estimates are used for the variances and covariances. Given the large sample sizes, the above test statistic can be compared to a standard normal distribution. In Table 21.8, the significant odds ratios are in bold face.

TABLE 21.7

Estimated Parameters, Robust Standard Errors, p Value from Tests of Fixed Effects, Missing Fraction of Information, and Estimated Odds Ratios for the Final Multinomial Logistic Regression Model Fit Using MLE (No Weights) and Pseudolikelihood Estimation (with Weights)

	Maximum Likelihood Estimation (No Weights)					Pseudolikelihood Estimation (With Weights)				
Effect	**Est**	**s.e.**	p	**Missing Fraction**	**Odds Ratio**	**Est**	**s.e.**	p	**Missing Fraction**	**Odds Ratio**
Intercept 1	−4.259	0.662	<.001	.031	0.01	−5.274	0.989	<.001	.018	0.01
Intercept 2	−3.252	0.656	<.001	.021	0.04	−4.154	0.966	<.001	.012	0.02
Intercept 3	−1.824	0.670	.007	.019	0.16	−2.268	0.945	.016	.009	0.10
Girl 1	0.409	0.062	<.001	.026	1.52	0.383	0.080	<.001	.029	1.47
Girl 2	0.409				1.52	0.383				1.47
Girl 3	0.409				1.52	0.383				1.47
ScreenTime 1	0.166	0.044	<.001	.218	1.18	0.170	0.056	.003	.237	1.19
ScreenTime 2	0.000				1.00	0.000				1.00
ScreenTime 3	0.000				1.00	0.000				1.00
CTimeRdg 1	0.382	0.069	<.001	.042	1.47	0.452	0.095	<.001	.031	1.57
CTimeRdg 2	0.256	0.055	<.001	.032	1.29	0.281	0.070	<.001	.019	1.32
CTimeRdg 3	0.127	0.055	.021	.046	1.14	0.126	0.063	.046	.026	1.13
MTimeRdg 1	1.242	0.279	<.001	.022	3.46	1.632	0.405	<.001	.013	5.12
MTimeRdg 2	1.242				3.46	1.632				5.12
MTimeRdg 3	0.679	0.285	.017	.018	1.97	0.893	0.395	.024	.008	2.44
Shortages 1	0.000				1.00	0.000				1.00
Shortages 2	−0.180	0.084	.033	.038	0.84	−0.234	0.108	.031	.061	0.79
Shortages 3	−0.180				0.84	−0.234				0.79
$\tau_{11} = \tau_{22}$	0.436	0.081		.030		0.456	0.085		.053	
τ_{33}	0.315	0.070		.011		0.304	0.075		.029	
$\tau_{13} = \tau_{23}$	0.246	0.065	<.001	.023		0.255	0.068		.060	
	Mean (Std Error)									
−2lnlike	13,515.02 (3.21)									
AIC	13,543.10 (3.21)									
BIC	13,587.87 (3.21)									

For a $(1 - \alpha) \times 100\%$ confidence interval for an odds ratio, an interval is first computed for the γs:

$$\left(\hat{\gamma}_{pqk}\hat{\gamma}_{pqk'}\right) \pm z_{\alpha/2}\widehat{\text{se}}_{(\hat{\gamma}_{pqk}\hat{\gamma}_{pqk'})},$$

and then the exponential of the endpoints is taken for the interval of the odds ratio.

TABLE 21.8

All Possible Odds Ratios from Final Model of More Frequent versus Less Frequent Where Those below the Diagonals Are Estimated Using Pseudolikelihood (with Weights) and Those above the Diagonal from Maximum Likelihood Estimation (No Weights)

		Response Options			
Predictor Variable	**Response Option**	**Daily**	**Weekly**	**Monthly**	**Never**
Girl	Daily		1.00	1.00	**1.52**
	Weekly	1.00		1.00	**1.52**
	Monthly	1.00	1.00		**1.52**
	Never	**1.47**	**1.47**	**1.47**	
Screen	Daily		**1.18**	**1.18**	**1.18**
Time	Weekly	**1.19**		1.00	1.00
	Monthly	**1.19**	1.00		1.00
	Never	**1.19**	1.00	1.00	
Centered	Daily		**1.35**	**1.29**	**1.47**
Time	Weekly	**1.19**		**1.14**	**1.29**
Reading	Monthly	**1.39**	**1.17**		**1.14**
	Never	**1.57**	**1.32**	**1.13**	
Mean	Daily		1.00	**1.75**	**3.46**
Time	Weekly	**1.00**		**1.75**	**3.46**
Reading	Monthly	**2.09**	**2.10**		**1.97**
	Never	**5.11**	**5.12**	**2.44**	
Shortages	Daily		**1.20**	**1.20**	1.00
	Weekly	**1.26**		1.00	**0.84**
	Monthly	**1.26**	1.00		**0.84**
	Never	1.00	**0.79**	**0.79**	

Note: The odds ratios in bold are significantly different from 1 and those in boxes are adjacent categories.

The ordinal nature of the responses is apparent in Table 21.8. In general, the odds are nondecreasing as the reported amount of time using the Internet decreases (i.e., odds ratios tend to increase as one goes down the columns for the pseudolikelihood or across the rows for the maximum likelihood ones). The exception is for Shortages where the odds ratios are greater than 1 for Daily versus Weekly or Monthly, but they are less than 1 for Weekly or Monthly versus Never. The direction of the effect of Shortages is different for different pairs of response options. This illustrates one of the strengths of the multinomial model. The multinomial model permits a fine-grained analysis, including the possibility of reversals in the direction of effects.

Ordinal Response Variables

When response options have a natural ordering, as they do in our PIRLs data, the ordering can be incorporated into an analysis by selecting a model that explicitly uses the ordering of the response options. In ordinal models, the response options are dichotomized based on the ordering of the categories, which yields $(K-1)$ dichotomies. The major difference between ordinal models is how responses are dichotomized and whether restrictions are placed on parameters over the dichotomies. Three of the most common ordinal models are continuation ratios, adjacent categories, and proportional odds models. These are presented in sections "Continuation Ratios," "Adjacent Categories," and "Cumulative Probabilities." Less restrictive versions of the adjacent categories and proportional odds models are also discussed. Throughout this section, it is assumed that categories are ordered from $k=1$ to K.

Continuation Ratios

One common and simple approach for ordinal responses is to form continuation ratios

$$\frac{P(Y_{ij}=k)}{P(Y_{ij}=k+1)+\cdots+P(Y_{ij}=K)} \quad \text{for } k=1,\ldots,K-1, \tag{21.23}$$

or

$$\frac{P(Y_{ij}=k+1)}{P(Y_{ij}=1)+\cdots+P(Y_{ij}=K)} \quad \text{for } k=1,\ldots,K-1. \tag{21.24}$$

With continuation ratios, multilevel binary logistic regression models are fit separately to each of the $(K - 1)$ ratios.

For the PIRLS data, multilevel logistic regression models could be fit to each probability of a more frequent use of the Internet versus a less frequent use; that is, multilevel binary logistic regression models could be fit to each of the following ratios:

$$\text{Ratio I} \quad P(\text{Daily}) \,/\, P(\text{Weekly, Monthly or Never}) \tag{21.25}$$

$$\text{Ratio II} \quad P(\text{Weekly}) \,/\, P(\text{Monthly or Never}) \tag{21.26}$$

$$\text{Ratio III} \quad P(\text{Monthly}) \,/\, P(\text{Never}). \tag{21.27}$$

The advantages of this approach over the other ordinal models discussed in this section are that binary logistic regression models are much simpler to use and different models may be found for each way of forming odds ratios. A relative disadvantage of this model is that except for one ratio (e.g., Equation 21.25), the models for the other ratios (e.g., Equations 21.26 and 21.27) are based on a subset of the data and hence smaller sample sizes.

Adjacent Categories

A model that compares pairs of response options is the adjacent categories logit model (Hartzel et al. 2001). The cluster-specific (Level 1) model of the multilevel adjacent categories logit models is

$$\log\left(\frac{P(Y_{ij}=k)}{P(Y_{ij}=k+1)}\right)=\sum_{p=0}^{P}\beta_{pjk}x_{pij} \quad \text{for} \quad k=1,\ldots,K-1. \tag{21.28}$$

It may be reasonable that the effect of predictors is the same for each pair of adjacent responses; therefore, in the Level 2 models, the fixed effects are specified so that they do not depend on k. Only the fixed effects for the Level 2 intercept of the Level 1 intercept are allowed to depend on k. The Level 2 model for the intercept is

$$\beta_{0jk}=\gamma_{00k}+\sum_{q=1}^{Q}\gamma_{0q}z_{qj}+U_{0jk} \quad \text{for} \quad k=1,\ldots K-1. \tag{21.29}$$

From Equation 21.29, we can see that the intercept of the intercepts γ_{00k} and the random effects U_{0jk} can differ over pairs of adjacent logits, but the coefficients for the predictors of the intercept γ_{0q} are fixed over k.

The Level 2 models for the Level 1 predictors x_{pij} for $p > 0$ are

$$\beta_{pjk} = \sum_{q=0}^{Q} \gamma_{pq} z_{qj} + U_{pjk} \quad \text{for} \quad k = 1,\ldots,K-1. \tag{21.30}$$

Note that the fixed effects do not depend on k, but the random effects may depend on the specific pair of responses, k and $k + 1$.

To show that the fixed effects of the predictors are the same for neighboring categories, we replace the β_{pjk}s in the Level 1 model (Equation 21.28) by their Level 2 models (Equations 21.29 and 21.30) and obtain

$$\log\left(\frac{P(Y_{ij}=k)}{P(Y_{ij}=k+1)}\right) = \underbrace{\gamma_{00k} + \sum_{p=0}^{P} U_{pjk}}_{\text{Depends on } k} + \underbrace{\sum_{q=1}^{Q} \gamma_{0q} z_{qj} + \sum_{p=1}^{P}\sum_{q=0}^{Q} \gamma_{pq} z_{qj} x_{pij}}_{\text{Does not depend on } k}. \tag{21.31}$$

The fixed effects for x_{pij} and z_{qj} are the same for all adjacent categories. The exponentials of the γ_{pq}s are interpretable as odds ratios just as they are in binary and multinomial logistic regression. The restriction that the fixed effects of x_{pij} and z_{qj} are the same for neighboring responses is a strong assumption.

To estimate the adjacent categories model, we use the fact that the model is a special case of the baseline multinomial model. For the adjacent categories model, the log of the likelihood and the model for the $P(Y_{ij} = k|\boldsymbol{x}_{ij}, \boldsymbol{z}_j, \boldsymbol{U}_{jk})$ are the same as those for the baseline multinomial model; however, linear restrictions must be placed on the fixed effects parameters of the multinomial model. To see the connection between these models and to find the proper restrictions on the multinomial parameters, we first note that the baseline logits can be written as the sum of the adjacent category:

$$\log\left(\frac{P(Y_{ij}=k)}{P(Y_{ij}=K)}\right) = \log\left(\frac{P(Y_{ij}=k)}{P(Y_{ij}=k+1)}\right) + \cdots + \log\left(\frac{P(Y_{ij}=K-1)}{P(Y_{ij}=K)}\right), \tag{21.32}$$

for $k = 1,\ldots,K - 1$. Substituting the adjacent categories model Equation 21.31 into Equation 21.32 and simplifying yields

$$\log\left(\frac{P(Y_{ij}=k)}{P(Y_{ij}=K)}\right) = \sum_{h=k}^{K-1} \gamma_{00h} + \sum_{p=0}^{P}\left(\sum_{h=k}^{K-1} U_{pjh}\right) x_{pij} + (K-k)\left(\sum_{q=1}^{Q} \gamma_{0q} z_{qj} + \sum_{p=1}^{P}\sum_{q=0}^{Q} \gamma_{pq} z_{qj} x_{pij}\right). \tag{21.33}$$

Equation 21.33 is a restricted version of the multinomial logistic regression model. The correspondence between parameters of the two models is as follows, where the multinomial parameters have the superscript "[mlt]" and the adjacent categories parameters have no superscript:

$$\text{Multinomial model} \quad \text{Adjacent categories}$$

$$\gamma_{00k}^{[\text{mlt}]} = \sum_{h=k}^{K-1} \gamma_{00h}$$

$$\gamma_{pqk}^{[\text{mlt}]} = (K-k)\gamma_{pq} \quad \text{for } p > 0$$

$$U_{pjk}^{[\text{mlt}]} = \sum_{h=k}^{K-1} U_{pjh}.$$

Therefore, the adjacent categories intercept parameters equal $\gamma_{00k} = \gamma_{00k}^{[\text{mlt}]} - \gamma_{00,k-1}^{[\text{mlt}]}$. The fixed effects for predictors in the adjacent categories model equal $\gamma_{pq} = \gamma_{pqk}^{[\text{mlt}]} - \gamma_{pq,k-1}^{[\text{mlt}]}$; that is, the difference $\gamma_{pqk}^{[\text{mlt}]} - \gamma_{pq,k-1}^{[\text{mlt}]}$ must be restricted to equal a constant value for a given p and q. This restriction implies that the odds ratios for adjacent categories are the same regardless of which two neighboring categories are compared $\left(\text{i.e., } \exp\left[\left(\gamma_{pq,k+1}^{[\text{mlt}]} - \gamma_{pqk}^{[\text{mlt}]}\right) x_{pij}\right] = \exp\left[\gamma_{pq} x_{pij}\right]\right)$. The odds ratios do not depend on which two adjacent categories are compared, which is a property of the adjacent categories model. It follows that if a multinomial model has been fit to data, the results can provide information about the plausibility of an adjacent categories model, in particular, whether the effects of the x_{pij}s and z_{qj}s depend on the response options.

For the PIRLS example on Internet usage, we can use our results from the baseline multinomial model described in the section "Nominal Response Variables" and examine the estimated odds ratios for adjacent categories that are in boxes in Table 21.8. If the adjacent categories logit models holds, then the odds ratios for different response options for a predictor should be equal. The only odds ratios in Table 21.8 that are comparable in value are those for school-mean centered time reading (i.e., for pseudolikelihood estimation, 1.19, 1.17, and 1.13, and for MLE, 1.35, 1.14, and 1.14). The fact that some of the adjacent categories odds ratios for predictors are significantly different from 1 but others for the same predictor equal 1 implies that the adjacent categories model is not appropriate. For example, the odds ratio for Girls comparing Daily versus Weekly and Weekly versus Monthly both equal 1, but Monthly versus Never equals 1.47, which is significantly different from 1. Since the adjacent categories' models parameters equal multinomial model parameters with linear restrictions on them, Wald tests as described in the section "Multinomial Example" can be constructed to test the restrictions implied by the adjacent categories models. However, given the odds ratios in Table 21.8, it is unlikely that the assumption of equal effects of the predictors will hold, except for school-mean centered time reading. This implies the adjacent category model will not fit the data well. Although we go no further with this example, one possible model that would be interesting to investigate is a *partial adjacent categories* model where some but not all the predictors have equal effects for adjacent response options.

Cumulative Probabilities

A third common choice for ordinal response data is the proportional odds model where the cumulative probabilities, $P(Y_{ij} \le k)$ are modeled using a multilevel logit formulation. The cluster specific (Level 1) proportional odds model is

$$\log\left(\frac{P(Y_{ij} \le k)}{P(Y_{ij} > k)}\right) = \beta_{0jk} + \sum_{p=1}^{P} \beta_{pj} x_{pij} \quad \text{for} \quad k=1,\ldots,K-1, \tag{21.34}$$

where the intercepts have the same order as the response options (i.e., $\beta_{0j1} \le \beta_{0j2} \le \ldots \le \beta_{0jK}$), and the cluster-specific regression coefficients for the predictors are the same regardless of the response option.

Similar to the adjacent categories model, since there is a single regression coefficient for each x_{pij}, the effect of x_{pij} is the same regardless of which cumulative odds ratio is examined. In other words, the odds ratios for a unit change in x_{pij} (where x_{pij} only has a main effect) equals

$$\frac{P(Y_{ij} \le k \mid \boldsymbol{x}_{-p,ij}, x_{pij})}{P(Y_{ij} > k \mid \boldsymbol{x}_{-p,ij}, x^{*}_{pij})} = \exp\left[\beta_{pj}\left(x_{pij} - x^{*}_{pij}\right)\right], \tag{21.35}$$

where x_{pij} and x^{*}_{pij} are two values of predictor p, and $\boldsymbol{x}_{-p,ij}$ are the remaining predictors that are held fixed to some value. For a one unit change in a predictor (i.e., $(x_{pij} - x^{*}_{pij})=1$), the odds ratios equal $\exp(\beta_{pj})$. The odds ratios do not depend on the response options and they only depend on the difference between two values of the predictor variable and the value of β_{pj}.

The Level 2 models for the cluster-specific regression coefficients for β_{0jk} and β_{pj} are the same as those for the adjacent categories model (i.e., Equations 21.29 and 21.30, respectively), except that the random effects do not depend on k. Furthermore, the order of these fixed effects γ_{00k} reflects the ordering of the response options; that is, $\gamma_{001} \le \gamma_{002} \le \ldots \le \gamma_{00K}$.

The proportional odds model also has a latent variable interpretation. As in the section "A Latent Variable Approach," a latent variable Y^{*}_{ij} is proposed; however, since there are now multiple categories of the response variable, there are multiple ordered thresholds that determine the observed response based on the latent variable. Specifically

$$Y_{ij} = \begin{cases} 1 & Y_{ij}^* \leq \xi_{j1} \\ 2 & \xi_{j1} < Y_{ij}^* \leq \xi_{j2} \\ \vdots & \\ K & \xi_{j,K-1} < Y_{ij}^*, \end{cases}$$

and

$$Y_{ij}^* = \sum_{q=1}^{Q} \gamma_{0q} z_{qj} + \sum_{p=1}^{P} \left(\sum_{q=0}^{Q} \gamma_{pq} z_j + U_{pj} \right) x_{pij} + \varepsilon_{ij} + U_{0j}.$$

The above model for Y_{ij}^* does not have a fixed intercept γ_{00k}. The thresholds equal the negative of the fixed intercept γ_{00k} in the Level 2 model for β_{00k}; that is, $\gamma_{00k} = -\xi_{jk}$. In this latent variable formulation, Y_{ij}^* represents an individual's value along some underlying continuum. The individual's value may be random due to ε_{ij} and U_j, but the thresholds are fixed and ordered. The distribution of ε_{ij} determines the probability model for the observed responses (i.e., normal or logistic).*

Estimation and the incorporation of design weights for the proportional odds models is the same as that for the baseline multinomial logistic regression model. The only difference is that the probabilities $P(Y_{ij} = k|\boldsymbol{x}_{ij}, z_j, \boldsymbol{U}_j)$ based on the cluster-specific proportional odds model are used in Equation 21.21 rather than those based on the multinomial model. To represent the probabilities, we first note that replacing the β_{pj}s in the Level 1 model by their Level 2 models gives us the cluster-specific model for the cumulative probabilities

$$\log\left(\frac{P(Y_{ij} \leq k \mid \boldsymbol{x}_{ij}, z_j, \boldsymbol{U}_j)}{P(Y_{ij} > kk \mid \boldsymbol{x}_{ij}, z_j, \boldsymbol{U}_j)}\right) = \gamma_{00k} + \sum_{q=1}^{Q} \gamma_{0q} z_{qj} + U_{0j} + \sum_{p=1}^{P} \sum_{q=0}^{Q} (\gamma_{pq} + U_{qj}) x_{ij}$$

$$\text{for } k = 1, \ldots, K-1.$$

Using these cluster-specific cumulative probabilities, the probability of a specific response option is found by subtraction as follows:

$$P(Y_{ij} = k|\boldsymbol{x}_{ij}, z_j, \boldsymbol{U}_j) = P(Y_{ij} \leq k|\boldsymbol{x}_{ij}, z_j, \boldsymbol{U}_j) - P(Y_{ij} \leq (k-1)|\boldsymbol{x}_{ij}, z_j, \boldsymbol{U}_j) \quad \text{for } k = 2, \ldots, \text{K},$$

$$\text{and } P(Y_{ij} = 1|\boldsymbol{x}_{ij}, z_j, \boldsymbol{U}_j) = P(Y_{ij} \leq 1|\boldsymbol{x}_{ij}, z_j, \boldsymbol{U}_j).$$

* A random utility model can also be proposed where the latent variables also depend on the response option and a person chooses the response that have the largest value Y_{ij}^*. See the footnote in the section "A Latent Variable Approach" for more details.

Example

For the proportional odds model, we used the same predictors as used for the baseline multinomial model. The cumulative probabilities of more frequent to less frequent were modeled; that is

$$P(Y_{ij}=\text{Daily})/P(Y_{ij}=\text{Daily, Weekly, Monthly or Never})$$

$$P(Y_{ij}=\text{Daily or Weekly})/P(Y_{ij}=\text{Monthly or Never})$$

$$P(Y_{ij}=\text{Daily, Weekly or Monthly})/P(Y_{ij}=\text{Never})$$

The model fit to the data was

$$\log\left(\frac{P(Y_{ij}\le k)}{P(Y_{ij}>k)}\right)=\gamma_{0k}+\gamma_{10}\text{Girl}_{ij}+\gamma_{20}\text{ScreenTime}_{ij}+\gamma_{30}\text{CTimeRdg}_{ij}$$
$$+\gamma_{40}\text{MTimeRdg}_{j}+\gamma_{50}\text{Shortages}_{j}+U_{0j},$$

where $U_{0j} \sim N(0, \tau_{00})$ *i.i.d.*

The estimated parameters are reported in Table 21.9. Unlike the baseline multinomial model, the predictor ScreenTime_{ij} is no longer significant and whether design weights are incorporated or not leads to different conclusions for Shortages j (i.e., it is not significant for pseudolikelihood but is

TABLE 21.9

Parameter Estimates of Proportional Odds Models with and without Weights

	Maximum Likelihood (No Weights)					Pseudolikelihood (Weights)				
Effect	Est	s.e.	p	Odds Ratio	Missing Fraction	Est	s.e.	p	Odds Ratio	Missing Fraction
Intercept 1	–4.179	0.492	<.01	0.01	.03	–5.108	0.738	<.01	0.02	.02
Intercept 2	–2.791	0.488	<.01	0.03	.04	–3.681	0.733	<.01	0.03	.02
Intercept 3	–1.573	0.484	<.01	0.09	.03	–2.432	0.731	<.01	0.09	.01
Girl	0.252	0.055	<.01	1.26	.04	0.233	0.064	<.01	1.26	.05
ScreenT	0.071	0.028	.01	1.08	.15	0.080	0.036	.03	1.08	.20
CTimeRdg	0.257	0.042	<.01	1.34	.03	0.296	0.056	<.01	1.34	.01
MTimeRdg	0.927	0.203	<.01	3.58	.03	1.275	0.300	<.01	3.58	.01
Shortages	–0.209	0.086	.02	0.83	.03	–0.182	0.095	.06	0.83	.04
Variance	0.240	0.049			.02	0.251	0.049			.03
	Mean		(Std Error)							
–2lnlike	13608.41		(2.95)							
AIC	13626.41		(2.95)							
BIC	13626.44		(2.95)							

Note: The standard errors are empirical (sandwich) ones.

significant for maximum likelihood). One possibility is the effect of predictors are not the same over response options. Proportional odds is a strong assumption. A *partial proportional odds model* can be fit by relaxing the equal slopes assumption for some or all of the predictors (Peterson and Harrell 1990); that is, some (or all) of the β_{ij}s could depend on the response options. The equality of the predictors can be tested in the same way as described in the section "Multinomial Example." For the PIRLS example, fixed effects differing over response options were estimated for ScreenTime_{ij}, CTimeRdg_{ij}, and Shortages_j. Models with different coefficients for MTimeRdg_j failed to converge. The results from the proportional odds model that relaxes the assumption for three predictors are reported for pseudolikelihood estimation on the left side of Table 21.10. The equality of γ_{pqk}s were tested and the proportional odds assumption appeared to be valid for only school-mean centered reading time. The tests also indicated that some of the slopes for other effects could be combined and others could be set equal to 0. A final model was fit to the data incorporating the changes suggested by the tests. When estimated by MLE, this final partial proportional odds model has the smallest BIC among the models fit to cumulative odds and has essentially the same AIC as the other model in Table 21.10, which are both smaller than the AIC from the proportional odds model. The results for the final model fit to the data by pseudolikelihood estimation are on the right side of Table 21.10.

TABLE 21.10

Parameter Estimates of Partial Proportional Odds Models with Weights (Pseudolikelihood) Where Standard Errors Are Empirical (Sandwich) Ones

Effect	Est	s.e.	p	Est	s.e.	p	Odds Ratio	Fraction Missing
Intercept 1	−5.512	0.772	<.01	−5.378	0.747	<.01	0.01	0.01
Intercept 2	−3.783	0.752	<.01	−3.801	0.744	<.01	0.02	0.01
Intercept 3	−2.133	0.754	<01	−2.219	0.732	<.01	0.11	0.01
Girl	0.231	0.065	<.01	0.232	0.064	<.01	1.26	0.04
ScreenT 1	0.196	0.059	<.01	0.145	0.031	<.01	1.16	0.14
ScreenT 2	0.127	0.038	<.01	0.145	.		1.16	
ScreenT 3	−0.013	0.039	.73	0.000	.		1.00	
CTimeRdg 1	0.338	0.086	<.01	0.298	0.056	<.01	1.35	0.02
CTimeRdg 2	0.311	0.059	<.01	0.298	.		1.35	
CTimeRdg 3	0.263	0.061	<.01	0.298	.		1.35	
MTimeRdg	1.266	0.308	<.01	1.274	0.306	<.01	3.58	0.01
Shortages 1	0.042	0.122	.73	0.000	.		1.00	
Shortages 2	−0.167	0.099	.09	−0.227	0.074	<.01	0.80	0.06
Shortages 3	−0.286	0.117	.01	−0.227	.		0.79	
Variance	0.263	0.051		0.255	0.050			0.04

Note: A "." for the standard error indicates that this parameter was fixed or equated with another.

Over all the analysis on this data set, the basic story is the same. Holding other predictors constant, the odds of more regular usage of the Internet for school is higher for girls than boys, larger for students who spend more time per day using electronic entertainment (ScreenTime), larger for students who read more for homework, and larger for students in schools where the average time spent by students reading for homework is larger. The effect of shortages appears somewhat mixed. The parameter estimates from the binary logistic model and the proportional odds suggest that an increase in shortages is associated with a decrease in the odds of more regular usage of the Internet. The partial proportional odds model indicates that the odds ratio for shortages equals 1 when comparing Daily usage versus less regular usage; however, the other two cumulative odds ratios equal 0.79, which is more similar to the binary and proportional odds models. The results from the baseline multinomial model help to explain the conflicting results. The estimated odds ratios from the multinomial model in Table 21.8 for Daily versus Weekly and for Daily versus Monthly both equal 1.26 (i.e., an increase in shortages is associated with an increase in odds); however, the odds ratios for Weekly versus Never and for Monthly versus Never both equal 0.79 (i.e., an increase in shortages is associated with a decrease in the odds of usage of the Internet). Lastly, the odds ratio for Weekly versus Monthly equals 1. The direction of the effect of shortages changes depending on which response options are compared. How to form logits and which model should be used depends on the researcher's goal, hypotheses, and how the results will be used.

Software

Any section on software will quickly become out of date; however, we briefly mention those programs that are currently available and meet the following criteria: is capable of estimating models parameters using adaptive quadrature, can easily compute sandwich standard errors, and can incorporate weights. Based on the current state of the art (and our knowledge), this list includes SAS/NLMIXED (SAS Institute Inc. 2011a), STATA/GLAMM (Rabe-Hesketh and Skronkal 2012), and Mplus (Muthén and Muthén 2010).

Mplus is capable of estimating parameters for the random effects binary logistic regression and proportional odds models, but it is not capable of estimating random effects multinomial, adjacent categories, partial proportional odds models, or variations of these models. SAS/NLMIXED and STATA/GLAMM can estimate a wider range of models. The procedure NLMIXED afforded a great deal of flexibility with minimal programing effort, including imposing restrictions on model parameters. Grilli and Pratesi (2004) described how to use SAS/NLMIXED for the random effects binary logistic regression and proportional odds models with weights; however, we greatly simplified and expanded their code to fit a wider array of models. At least for the current release of SAS (version 9.3), the SAS command REPLICATE that is used to incorporate Level 2 weights permits noninteger values.* The second difficulty that Grilli and Pratesi (2004) encountered was the computation of standard errors; however, sandwich estimates may be easily obtained by using the NLMIXED option EMPIRICAL. STATA/GLAMM can fit models with more than two levels; whereas Mplus and SAS/NLMIXED can only deal with two-level models.

Discussion

A wealth of information is available from large-scale national and international survey data, much of which is open source. Although many surveys are designed to measure educational achievement, a large number of surveys and items on even those created to measure educational attainment can be used to study a variety of topics. For example, in this chapter we studied Internet usage for school work by fourth grade students. Given that questions often have categorical response options, they should be modeled using a model designed for discrete data. Which specific model a researcher uses depends on what is appropriate for the data and meets the researcher's goal and hypotheses. For example, if a researcher wants to contrast or compare response options for pairs of ordered categories, then the adjacent categories model would be useful; however, if one wants to make statements about effects above or below various points on the response scale, a (partial) proportional odds model might be the best choice. Substantive considerations are of paramount importance, but so is taking into consideration the nature and characteristics of the data at hand.

The methodology for analyzing complex survey data with multilevel models for discrete response variables is a quickly changing field. This chapter presented what we feel is the current state of affairs. Additional software options for fitting multilevel survey data with design weights are likely to become available over time. Furthermore, the methods developed for missing data in the context of multilevel models is an active area of research and we expect that more efficient methods than what was employed here will become available (e.g., Kim and Swoboda 2012; Swoboda 2011). Regardless of these shortcomings, models, methodology, and tools exist to analyze discrete response data from large-scale survey data in an appropriate manner.

* The SAS/NLMIXED documentation describing the command REPLICATE is not accurate. The documentation states that the variable must have positive integer values; however, this is not the case (Kathleen Kiernan, SAS Technical Support Statistician, personal communication, June 2012). The values of the variable may be positive real numbers. When the REPLICATE variable contains noninteger values, the number of clusters reported in the output will be incorrect.

Acknowledgment

This work was supported in part by grant R305D120005 from the Institute of Education Sciences, the U.S. Department of Education awarded to the second author.

References

Agresti, A. 2002. *Categorical Data Analysis*. Wiley, Hoboken, NJ, 2nd edition.

Agresti, A. 2007. *Introductory Categorical Data Analysis*. Wiley, Hoboken, NJ, 2nd edition.

Allison, P. D. 2002. *Missing Data*. Sage, Newbury Park, CA.

Amer, S. R. 2009. Neural network imputation in complex survey design. *International Journal of Electrical and Electronics Engineering*, 3:52–57.

Anderson, C. J. and Rutkowski, L. 2008. Multinomial logistic regression. In Osborne, J., ed., *Best Practices in Quantitative Methods*, pp. 390–409. Sage, Thousand Oaks, CA.

Asparouhov, T. and Muthén, B. 2006. Mutlilevel modeling of complex survey data. In *Proceedings of the Joint Statistical Meeting*, pp. 2718–2726. ASA section on Survey Research Methods, Seattle, WA.

Demidenko, E. 2004. *Mixed Models: Theory and Applications*. Wiley, Hoboken, NJ.

Enders, C. K. 2010. *Applied Missing Data Analysis*. Guilford Press, New York, NY.

Fahrmeir, L. and Tutz, G. 2001. *Multivariate Statistical Modelling Based on Generalized Linear Models*. Springer, New York, NY, 2nd edition.

Grilli, L. and Pratesi, M. 2004. Weighted estimation in multilevel ordinal and binary models in the presence of informative sampling designs. *Survey Methodology*, 30:93–103.

Hartzel, J., Agresti, A., and Caffo, B. 2001. Multinomial logit random effects models. *Statistical Modelling*, 1:81–102.

Heeringa, S. G., West, B. T., and Berglund, P. A. 2010. *Applied Survey Data Analysis*. Chapman Hall/CRC, Boca Raton, FL.

Kim, J.-S. and Swoboda, C. 2012. Strategies for imputing missing values in hierarchical data: Multilevel multiple imputation. Paper presented at the *Annual Meeting of American Educational Research Association*, Vancouver, CA.

McCullagh, C. E. and Nelder, J. 1989. *Generalized Linear Mixed Models*. Chapman and Hall, London, 2nd edition.

McFadden, D. 1974. Conditional logit models analysis of qualitative choice behavior. In Aarembka, P., ed., *Frontiers of Econometrics*. Academic Press, New York, NY.

Molenberghs, G. and Verbeke, G. 2005. *Models for Discrete Longitudinal Data*. Springer, New York, NY.

Muthén, B. O. 1998–2004. Mplus technical appendices. Technical report, Muthén & Muthén, Los Angeles, CA.

Muthén, L. K. and Muthén, B. 1998–2010. *Mplus User's Guide*. Muthén & Muthén, Los Angeles, CA, 6th edition.

Peterson, B. and Harrell, F. 1990. Partial proportional odds models for ordinal response variables. *Applied Statistics*, 39:205–217.

Pfeffermann, D., Skinner, C., Holmes, D., Goldstein, H., and Rasbash, J. 1998. Weighting for unequal selection probabilities in multilevel models (with discussion). *Journal of the Royal Statistical Society, Series B*, 60:23–56.

Powers, D. D. and Xie, Y. 1999. *Statistical Methods for Categorical Data Analysis*. Emerald Group, Bingley, UK.

Rabe-Hesketh, S. and Skronkal, A. 2006. Mulitlevel modeling of complex survey data. *Journal of the Royal Statistical Society, Series A*, 169:805–827.

Rabe-Hesketh, S. and Skronkal, A. 2012. *Multilevel and Longitudinal Modeling Using Stata, Volume II, 3rd Edition*. Stata Press, College Station, TX.

Raudenbush, S. W. and Bryk, A. S. 2002. *Hierarchical Linear Models*. Sage, Thousand Oaks, CA, 2nd edition.

Reiter, J., Raghunathan, T., and Kinney, S. 2006. The importance of modeling smapling design in multiple imputation for missing data. *Survey Methodology*, 32:143–149.

Rubin, D. 1987. *Multiple Imputation for Nonresponse in Surveys*. Wiley, NY.

Rutkowski, L., Gonzalez, E., Joncas, M., and von Davier, M. 2010. International large-scale assessment data: Issues in secondary analysis and reporting. *Educational Researcher*, 39:142–151.

SAS Institute Inc 2011a. *The SAS System*. SAS Institue, Cary, NC, version 9.3 edition.

SAS Institue Inc 2011b. *SAS/STAT 9.3 User's Guide: The GLIMMIX Procedure*. SAS Institue, Cary, NC, version 9.3.

Schafer, J. L. 1997. *Analysis of Incomplete Missing Data*. Chapman & Hall, London.

Self, G. and Liang, K. 1987. Asymptotic properties of maximum likelihood estimators and likelihood ratio tests under nonstandard conditions. *Journal of the American Statistical Association*, 82:605–610.

Shin, Y. and Raudenbush, S. W. 2007. Just-identified versus overidentified two-level hierarchical linear models with missing data. *Biometrics*, 63:1262–1268.

Shin, Y. and Raudenbush, S. W. 2010. A latent cluster-mean approach to the contextual effects model with missing data. *Journal of Educational and Behavioral Statistics*, 35:26–53.

Skrondal, A. and Rabe-Hesketh, S. 2000. *Generalized Latent Variable Modeling*. Chapman Hall/CRC, Boca Raton, FL.

Snijders, T. A. B. and Bosker, R. J. 2012. *Multilevel Analysis*. Sage, Thousand Oaks, CA, 2nd edition.

Stram, D. and Lee, J. 1994. Variance components testing in the longitudinal mixed effects model. *Biometrics*, 50:1171–1177.

Swoboda, C. M. 2011. *A New Method for Multilevel Multiple Imputation*. Unpublished doctoral dissertation, University of Wisconsin–Madison.

Van Buuren, S. 2011. Multiple imputation of multilevel data. In Hox, J. and Roberts, K., eds, *Handbook of Advanced Multilevel Analysis*. Taylor & Francis, New York, NY.

Van Buuren, S. 2012. *Flexible Imputation of Missing Data*. Chapman Hall/CRC, Boca Raton, FL.

Verbeke, G. and Molenberghs, G. 2000. *Linear Mixed Models for Longitudinal Data*. Springer, NY.

22

Causal Inference and Comparative Analysis with Large-Scale Assessment Data

Joseph P. Robinson
University of Illinois at Urbana-Champaign

CONTENTS

Causal Inference and Comparative Analysis with Large-Scale Assessment Data

Education researchers are increasingly interested in making causal inferences rather than simply describing correlations among variables. Drawing causal inferences from data hinges upon the design of the study, with experimental designs facilitating causal inferences most readily. However, oftentimes experimental designs are infeasible or impractical; moreover, researchers may want to make use of the vast amounts of publicly available large-scale secondary datasets. This chapter discusses the assumptions required for making causal inferences. Then, I provide an introduction to several techniques that may facilitate quasi-experimental designs when applied to secondary data, and in the process I note the specific sets of assumptions for inferring causality when using each technique. I conclude with a discussion of special considerations for researchers using international datasets, and I include examples where causal inferences may be possible. In cases where causal inferences are not possible, I discuss how researchers can use the techniques described in this chapter for careful comparative analyses.

Rubin Causal Model

The Rubin causal model* (Rubin 1974, 1977, 1978) is foundational for modern causal modeling and uses a potential outcomes framework.† The Rubin causal model begins with the premise that each individual i has a potential outcome Y for each condition. That is, if she received treatment condition t, she would have outcome Y_i^t, and if she received control condition c, she would have outcome Y_i^c. With these two potential outcomes defined, we can define the causal effect (or treatment effect) δ of treatment t versus control c for individual i as the difference in outcome Y for individual i when she receives t versus c, all else equal:

$$\delta_i = Y_i^t - Y_i^c$$

This estimand δ_i is defined in theory for each individual; however, in practice, we will not be able to observe it because any individual i will only be in

* The Rubin causal model was named so by Holland (1986), but note that this model is sometimes referred to as the Neyman–Rubin–Holland (or some other ordering) causal model, in recognition of Neyman's (e.g., 1923/1990) and Holland's (e.g., 1986) contributions to modern causal modeling in addition to Rubin's.

† The study of causation in philosophy has a long history, which is beyond the scope of this chapter. For a succinct review of the philosophical history of causation and how that history relates to modern causal modeling, see Kaplan (2009) and Holland (1986).

one condition—either t or c—and never in both at the same time. Thus, in practice, we cannot estimate the causal effect of t versus c on Y for any one *individual* because we will be missing (at least) half of the necessary data—Holland (1986) refers to this dilemma as the "fundamental problem of causal inference."

Although we are seldom interested in the causal effect for a particular individual, the missingness of relevant data is the core problem when estimating average treatment effects (ATEs) because such estimates are nothing more than the average of individual treatment effects:

$$\text{ATE} = \bar{\delta} = \sum_{i=1}^{N} \frac{\delta_i}{N} = \sum_{i=1}^{N} \frac{Y_i^t - Y_i^c}{N}$$

Hence, we are still missing (at least) half of the necessary data. Rather than giving up at this point, though, we make assumptions that will allow us to approximate the missing data—the missing counterfactuals, or what we would have observed if individuals were in the alternative condition.

Recall that in the potential outcomes framework of the Rubin causal model, each person has a potential outcome Y_i^t if she were in the treatment condition and a potential outcome Y_i^c if she were in the control condition. Because we observe an individual in (at most) one of the conditions, we cannot see that individual's outcome in the alternative condition even though she has a *potential* outcome in that condition. That is, if we observe someone in condition t, keep in mind that she could have been in condition c, and thus she has potential outcomes in both conditions t and c but an observed outcome in condition t only. For making causal inferences, we need to assume that an individual's observed condition (whether through choice or assignment) was unrelated to Y_i^t and Y_i^c—that is, for each i, we assume $\{t,c\} \perp \{Y_i^t, Y_i^c\}$. Another way of thinking of this is that we assume individuals are not placed into one condition versus the other because they were expected to have a higher or lower outcome in a given condition.

For example, when using secondary datasets, we need to assume that individuals did not self-select into the condition from which they thought they would benefit most—in many cases, this is an unrealistic assumption. Moreover, the condition that potential outcomes are unrelated to observed condition means that individuals with higher potential baseline outcomes (i.e., potential outcomes in condition c) were not overrepresented in either condition independent of their individual treatment effects (Winship and Morgan 1999). For example, even if the effect of a private tutoring program was constant across individuals and zero (i.e., $\bar{\delta} = \delta_i = 0 \,\forall\, i$), we could incorrectly infer a positive treatment effect if individuals with higher test scores in the absence of treatment self-selected into the treatment ($Y_{i \in T}^c > Y_{i \in C}^c$, where T refers to the set of individuals observed in treatment condition t, and C is the set observed in control condition c).

Experimental designs—usually regarded as the "gold standard" for causal inferences—are still prone to the "fundamental problem of causal inferences," which is that we cannot observe the same individual in each condition, all else equal (Holland 1986). Nevertheless, we regard experiments so highly because if individuals in experiments were randomly assigned to conditions, then the assumption that the observed condition is unrelated to potential outcomes is very plausible. That is, of course, if we also assume that experimenters and participants did not manipulate condition assignment and (in the case of longitudinal experiments) that any differential attrition is unrelated to potential outcomes.

Before moving on to discuss quasi-experiments, it is important to note two final details that pertain to experiments and quasi-experiments alike. First, when estimating treatment effects, condition assignment must be manipulable (Holland 1986). That is, an individual should conceivably be able to be in either condition, such as would be the case when estimating the effect of a curricular intervention or an after-school tutoring program. However, one could not estimate the causal effect of nonmanipulable dimensions such as race. Second, we will assume (and have assumed to this point) that each person has only one potential outcome for a condition; this is referred to as the "stable unit treatment value assumption" (SUTVA; Rubin 1977; see also, Cox 1958). In other words, SUTVA means that the treatment effect of t versus c for any individual i is constant and does not depend on the condition received by other individuals. For example, SUTVA would be violated if an individual's treatment effect of t versus c depended on whether her best friend was in the same condition. If SUTVA were not assumed, the number of causal effects per individual could grow beyond the number of combinations of other individuals' condition assignments (which then is exacerbated when aggregating to the full sample), rendering ATEs virtually inestimable.

Quasi-Experimental Designs

For estimating causal effects, I focus on quasi-experimental designs (rather than experiments) in the remainder of this chapter because they are likely to be of greater use when researchers use secondary datasets. Having begun with a discussion of the Rubin causal model and randomized experiments to provide a foundation in the basic principles that undergird modern causal inferences, I now introduce a subset of designs known as "quasi-experimental designs." Quasi-experiments use existing data that may or may not have originally been used in an experiment and, through careful design, can provide a causal-effect estimate. Introductions for quasi-experimental designs will be provided for (1) instrumental variables, (2) regression discontinuity, and (3) propensity score matching.

Instrumental Variables

Introduction, Theory, and Assumptions

Instrumental variables (IVs) are variables that predict treatment condition but have no direct effect on outcomes, although they can have an indirect effect on the outcome via their influence on treatment condition. By using IVs, one can carve out exogenous variation (i.e., variation not caused by the participant) in a condition that is related to outcomes, and thus estimate a causal effect (Imbens and Angrist 1994; Bound et al. 1995). This method may be best understood by analogy to an experiment. In an experiment, a fair coin toss randomly assigns a subject to a condition (treatment or control). By itself, the outcome of the coin toss has no direct effect on the end outcome of interest. That is, the coin toss only affects outcomes *indirectly*, through the condition a subject is placed in, which in turn affects outcomes. With nonexperimental data, we can approximate a coin toss if we can identify some variable that affects the likelihood of receiving the treatment but has no direct effect on outcomes—such a variable is called an "instrumental variable." For example, the coin toss in an experiment is the source of exogenous variation in the observed treatment condition for the individuals whose observed conditions were affected by the coin toss (this need not be all individuals, but it helps if the proportion influenced is large; Bound et al. 1995; Stock and Yogo 2005). When using an IV approach, we are essentially focusing on the individuals whose observed conditions were influenced by the exogenous source.

To introduce the idea of IVs, I will use an example where the IV is random assignment (as random assignment is the ideal IV), but random assignment is not the only case where IV can be used, as I will discuss later. Imagine students are randomly assigned to treatment ($z = 1$) or control ($z = 0$) in a tutoring program. Some students comply with their assignment, meaning that their treatment assignment would match their observed condition ($z = t$) regardless of which condition they were randomly assigned to—hence, we call them "compliers" (Angrist et al. 1996). However, some students do not comply with their assignment—these noncompliers come in three different varieties: (1) "always-takers," those who will always take the treatment ($t = 1$) regardless of whether they were assigned to treatment ($z = 1$) or control ($z = 0$); (2) "never-takers," those who will never take the treatment ($t = 0$) regardless of assignment; and (3) "defiers," those who will defy the experimenter by always choosing the opposite condition ($z \neq t$) regardless of which condition they were assigned to.

The presence of these four types of individuals (the "compliers" and the three types of "noncompliers") presents some challenges to estimating causal effects, so we need to make some simplifying assumptions. First, we assume that there are no defiers. Second, we assume that treatment assignment (z) affects observed condition (t). Third, we assume that the IV (e.g., random assignment to condition) has no effect on outcome Y other than through its influence on observed condition ($t = \{0,1\}$); this assumption is sometimes referred to as the "exclusion restriction" because it assumes that the instrument(s) can be excluded from the model predicting the outcome, provided the treatment condition predicted by the instrument(s) is included in the outcome model (discussed below). Imbens and Angrist (1994) demonstrate that these assumptions allow us to estimate ATEs for the compliers, which they refer to as the "local average treatment effect" (LATE). If the assumptions are valid, this estimate results because the first assumption means that defiers do not figure into our estimates because they do not exist in our data; the second assumption means that compliers do exist; and the third assumption means that always-takers and never-takers do not figure into our estimates because their assignments did not alter their observed treatment conditions.

Estimation

How do we estimate the LATE? Instrumental-variable approaches can be thought of as a two-stage regression (even though IV approaches are actually implemented in one step, which accounts for error in the prediction in the first-stage regression). In the first-stage regression, we predict the observed treatment condition by the instrument (or set of instruments); given our IV assumptions above, this first stage allows us to estimate the proportion of the sample that complies with treatment assignment:

$$t_i = \alpha_0 + \alpha_1 z_i + \nu_i$$

In the second-stage regression, we use the *predicted* value of the observed treatment (i.e., predicted from the first-stage regression) instead of the actual observed treatment variable to predict outcome Y:

$$Y_i = \beta_0 + \beta_1 \hat{t}_i + \varepsilon_i$$

The value of β_1 is the LATE because it tells us the average relationship between treatment (vs. control) and outcome Y for the compliers, as only the compliers have $cov(z,t) \neq 0$. Because $\hat{t}$ reflects the variation in t that is predicted only by the instrument z, $\beta_1 = (cov(z,Y))/(cov(z,t))$.

It is important to note that IV analyses cannot tell us the ATE (unless only compliers exist, in which case LATE = ATE) and cannot tell us the ATEs for those who always choose either the treatment or the control.

EXAMPLE

Note that I presented a simple case of random assignment (z) with imperfect compliance. I did so for illustration purposes, but IV analyses are not limited to only circumstances where one initially had an experimental design. IVs can come in all forms, provided that the IV assumptions are strongly plausible. For example, Dee (2004) was interested in the effect that education has on civic engagement (e.g., voting, staying informed of one's community). However, education is an endogenous variable, meaning that individuals themselves influence how many years of education they consume. Thus, to answer his question regarding the effect of education on civic engagement, Dee needed to isolate a portion of the years of education one obtains that are exogenous (i.e., determined by forces outside of the individual).

To do so, in one of his analyses, Dee (2004) used the IV of whether the state had restrictive child-labor laws (z) to predict exogenous variation in educational attainment (t) in the first stage, and then predict the LATE of an additional year of schooling (for those who would continue their education if their state had a restrictive child-labor law) on civic engagement (Y) in the second stage. The restrictive child-labor law in a state is arguably exogenous (and thus may be suitable as an IV) because it is imposed at the state level and individuals are subject to these laws that affect their employability, which may in turn affect their decisions about whether to stay in school longer if they cannot gain employment due to the state's child-labor law.* In his first-stage regression, Dee found that individuals in states with restrictive child-labor laws attended school longer (at least through high school), suggesting that compliers are likely to exist. In the second-stage equation predicting civic engagement, engagement is predicted by the *predicted* years of education attained (i.e., $\hat{t}$) based on the first-stage equation, rather than on the actual years of education attained (i.e., t). Using $\hat{t}$ (instead of t) in the second-stage equation, Dee's results suggest that educational attainment positively affected civic engagement for the compliers. For example, in his final IV model, an additional year of education was associated with an almost 7-percentage-point increase in the likelihood of voting in the last Presidential election.

* On the other hand, one may argue that the IV may be invalid because states with restrictive child labor laws may be more forward-thinking in a general sense, and this forward-thinking approach may be correlated with greater levels of civic participation in ways that do not necessarily relate to education. If this were true, then the IV itself has a direct correlation with the outcome of civic participation that does not operate through the education variable; this would render the IV invalid.

Other Considerations

It is important for the researcher to argue on theoretical grounds that the IV is (or set of IVs are) exogenous and, therefore, meet the exclusion restriction. However, researchers may be able to bolster their theoretical arguments with empirical evidence if the number of instruments is greater than the number of endogenous variables requiring instruments—in such instances, the model is said to be "overidentified." When the model is overidentified, overidentification tests developed by Sargan (1958) and Hansen (1982) can be used to test the null hypothesis that the IVs are uncorrelated with the second-stage residuals. Failure to reject this null hypothesis can add empirical support to the argument that the IVs are valid.

Regression Discontinuity Designs

Introduction, Theory, and Assumptions

Studies using regression discontinuity designs (RDDs) have recently been permitted by the U.S. Department of Education's What Works Clearinghouse to receive the highest category for evidence-based research ("meets evidence standards without reservations"), which speaks to the regard for such designs (Schochet et al. 2010). Why are RDDs held in such high esteem? Quite simply, when compared to the relatively stronger (and often-implausible-in-practice) assumptions of other quasi-experimental designs, such as IVs and propensity score matching (discussed next), RDDs require weaker assumptions to estimate causal effects.

In an RDD, treatment condition (t) is determined by attaining (or failing to attain) a cutscore on some dimension (r), often referred to as the rating-score variable, the forcing variable, or the running variable. For example, we might be interested in the effects of passing a high school exit exam on the likelihood of graduation (see, e.g., Reardon et al. 2010). The treatment here is passing the exit exam ($t = 1$) and the control is failing the exit exam ($t = 0$). Let us assume the exit exam score (our rating-score variable r) has a range of 0–100, and scores at or above 60 are considered passing. Among the exam-passers, there is a range of exam scores; likewise, there is a range of exam scores among the exam-failers. Thus, the average scores of exam-passers and exam-failers is not equal, which leads us to worry that other factors (e.g., motivation) differ between the groups; therefore, we are concerned that we cannot simply look at the raw mean difference in graduation rates of the two intact groups as evidence of the causal effect of exit exams. However, we could obtain a more plausible causal-effect estimate if we compared the graduation rates of those who barely passed the exam (with a score of 60) and those who barely failed (with a score just below 60).

This is the straightforward logic behind an RDD, but rather than limit our analyses to just those at 60 points and just below 60 points, we can include a larger range of observations and estimate regressions on either side of the cutscore. For simplicity, we begin by recentering our rating-score variable about the cutscore (in our example, that means the recentered rating score $r^* = r - 60$). We can write the expectation of the graduation rate Y among barely passers (i.e., students just meeting the exit-exam threshold) as a limit function among the subset of passers (i.e., those for whom $t = 1$):

$$E\left(Y|\, t=1, r^*=0\right) = \lim_{a \to 0^+} (Y \,|\, t=1, r^*=a)$$

Similarly, we can write the expectation of the graduation rate Y among barely failers (i.e., students just failing to meet the exit-exam threshold) as a limit function among the subset of failers (i.e., those for whom $t = 0$):

$$E\left(Y|\, t=0, r^*=0\right) = \lim_{a \to 0^-} \left(Y \,|\, t=0, r^*=a\right)$$

Differencing these expectations, we obtain the LATE at the cutscore (Hahn et al. 2001)—in our example, the LATE is the average effect of just barely passing the exit exam when the threshold for passing is set at 60. If the effect of passing the exit exam on graduation rate varies by where the threshold is set (e.g., the effect might be larger if the threshold were set at 90 because students close to the 90-point threshold are likely to meet additional criteria for graduation), then the LATE will likely differ from the ATE and will depend on where the precise threshold is set (see Robinson [2011] for a discussion of threshold setting in the context of an RDD-based analysis of English learner reclassification decisions).

Estimation

RDD estimation can be done either parametrically or "nonparametrically"/ semiparametrically. In parametric models, the analyst assumes a particular functional form (e.g., quadratic) relating the rating-score variable to the outcome; often, the functional form is allowed to vary on each side of the cutscore, and this can be accomplished by interacting the rating-score variable terms with the treatment indicator. For example, one might fit the following parametric model:

$$Y_i = \beta_0 + \beta_1 r_i^* + \beta_2 r_i^{*2} + \beta_3 t_i + \beta_4 \left(r_i^* \times t_i\right) + \beta_5 \left(r_i^{*2} \times t_i\right) + \varepsilon_i$$

In the above equation, β_3 is the average difference in outcomes between the treatment and control groups when $r^* = 0$—that is, at the cutscore.

If one is unsure of the correct functional form over the entire range of the rating-score variable r^*, then the parametric approach may seem unsettling. Oftentimes, multiple parametric RDDs are estimated, such as a linear model, a quadratic model, a cubic model, and so on. Greater confidence in the LATE may be achieved from a LATE that becomes stable and relatively insensitive to changes in the parametric form (e.g., as one moves from a cubic to quartic model).

Note too that one could reduce the range of r^* and estimate a lower-order polynomial (e.g., a linear function on either side of the cutscore) over this reduced range. This reduces the reliance on points far from the cutscore in estimating the LATE. Estimating RDD-based causal effects in this manner places less reliance on getting the functional form of the model correct across a wide range of r^*, but instead places emphasis on identifying the optimal bandwidth of r^* over which sufficient data is retained and a low-order polynomial model is the correct relationship (see Hahn et al. [2001] for a discussion of local linear regression for RDD). Data outside of the bandwidth will not be included in the estimation, and thus this approach to RDD is likely to result in reduced power when compared to approaches that use all of the data and fit higher-order polynomials. The size of the bandwidth can be determined by theory, but it is usually determined via a cross-validation process (Imbens and Lemieux 2008).

EXAMPLE

In the early childhood longitudinal study, kindergarten class of 1998–1999 (ECLS-K) dataset, children who came from a home where a language other than English is spoken (i.e., language-minority children) were given a test of their English language proficiency to determine the language in which subsequent tests should be administered. Language-minority children from Spanish-speaking homes who scored at least 37 points (on a scale from 0 to 60) on the English proficiency test (r) were given a mathematics test in English (i.e., $r \geq 37 \rightarrow t = 0$); those scoring fewer than 37 points were given the mathematics test in Spanish (i.e., $r < 37 \rightarrow t = 1$). I performed an RDD analysis to estimate the effect of native-language test translation on the mathematics scores of students near the proficiency cutscore. Using a series of nonparametric RDDs, I found that test translations had a substantial positive effect on the ability of students near the cutscore to demonstrate their mathematics knowledge (Robinson 2010). For example, in the preferred model, students in the spring of kindergarten who barely passed the English proficiency test (and therefore were given the mathematics test in English) scored 0.85 standard deviations (SDs) lower than students who just barely failed the English proficiency test (and therefore were given the mathematics test in Spanish). One should note, however, that the RDD-based test-translation effect estimate may not generalize to students with English proficiency scores far from the cutscore; for example, we might expect language-minority students who are highly proficient in English to actually perform better on the mathematics test in English than on the test in Spanish.

Other Considerations

Following estimation of RDD-based treatment effects, it is important to provide evidence that the results were not due to something other than the treatment contrast. This evidence usually comes in a number of forms, including demonstrating that the density of observations varies smoothly across the cutscore (McCrary 2008), other variables (e.g., gender and race) vary smoothly across the cutscore, and treatment effects are not found at "pseudo" cutscores (Imbens and Lemieux 2008).

Finally, it is important to note that RDDs can be extended beyond the simple case of a single rating score determining treatment status. First, RDD can be combined with IV to obtain the complier ATEs when attaining a cutscore *influence*, but does not fully *determine* treatment assignment (Hahn et al. 2001; see also Angrist and Lavy 1999). That is, when some individuals do not comply with the treatment assignment based on their cutscore attainment, using the RDD approach will estimate the average effect of crossing the cutscore (which was *intended* to affect observed condition) over (a) individuals whose conditions were affected as intended (i.e., compliers), (b) individuals whose conditions were not affected (i.e., always-takers and never-takers), and (c) individuals whose conditions were affected but not as intended (i.e., defiers). This average may seem uninformative because it is a composite of so many types of individuals (including noncompliers), but for policy makers, this "intent to treat (ITT) effect" is often of greater use than a treatment effect estimate because the ITT reflects the policy's total effect (as some people in the population will be influenced to take the treatment due to the policy, and some will not). If, however, one wishes to estimate the RDD LATE on the compliers, this can be accomplished by combining RDD with IV. Bear in mind, though, that combining RDD and IV means that the treatment effect being estimated is for the individuals near the cutscore whose treatment status is determined by which side of the cutscore they are on—this can be thought of as the LATE for the compliers. Second, RDDs can be applied to cases where more than one rating-score variable influences or determines treatment status (Reardon and Robinson 2012; Wong et al. 2013). Robinson (2011) presents an example that combines RDD with multiple rating-score variables with IV, where attaining passing scores of five different rating-score variables influenced English learner status.

Propensity Score Matching

Introduction, Theory, and Assumptions

When inferring causality from experiments and quasi-experiments, it is imperative that the treatment effect estimate is indeed caused by the treatment/control contrast and not due to other variables that may differ systematically between individuals in the treatment and control conditions. If confounding variables exist and are not properly accounted for in the estimation model, then the ATE estimate will be biased.

Many of us have dealt with confounding variables by including them as covariates in regression analysis (when those confounding variables are available in the dataset). Although I have not discussed regression analysis as a quasi-experimental technique, regression analysis could be viewed as a quasi-experimental design if the functional form of the regression model is correctly specified. That is, all variables (whether in the dataset or not) having a nonzero partial correlation with both the treatment and the outcome variables must be included in the regression to obtain an unbiased estimate of the average conditional treatment effect. I bring this up here because it provides a bridge from the more familiar (regression analysis) to perhaps the less familiar (matching-based treatment-effect estimators). When using matching estimators, such as propensity score matching-based estimators, we assume that we have matched treatment and control conditions on all relevant variables that are related to both treatment status and outcome, such that the only thing that varies between them is their treatment status (and any resulting difference between groups in outcomes is then attributed to treatment)—that is, we assume that there are no confounders (either observed or unobserved) after matching. This sounds similar to the correct-functional-form assumption we make in regression analysis. There are several key differences, though, between propensity score matching and regression analysis. These differences are outlined below.

When propensity score matching is executed properly, we can identify a subset of individuals in the control condition who have a similar propensity score (and thus similar profile of covariates in the vector **X**; Rosenbaum and Rubin 1983a, 1985) to individuals in the treatment condition. This matched set then overlaps in their profile of **X** (note that **X** is a vector of potential confounds, e.g., income, prior education, and language proficiency), meaning that among this matched set, there is a near-zero correlation between each covariate in **X** and treatment status. Restricting treatment-effect estimation to just this matched set means that the "correct functional form" assumption is substantially weakened in the estimation stage because the matched set of treatment and control cases have (nearly) equivalent values of each covariate in **X**; however, the correct functional form assumption remains strong in the matching phase.

This functional form assumption in the matching stage is referred to as the "conditional independence assumption" (CIA; Rubin 1977) because we assume that treatment status is independent of potential outcomes conditional on the true propensity score (and thus on the covariates **X** of which the propensity score is comprised; Rosenbaum and Rubin 1983a).* The CIA can be disaggregated into two different component assumptions. However, before discussing the two component assumptions, it will be important to think about the different types of individuals that exist in the population and the different types of treatment effects that can be obtained through matching. After we discuss the different types of treatment effects, we will return to the CIA components necessary for estimating each of these effects.

This chapter began with a discussion on the ATE for a population. However, some people in the population will always take the treatment (like we discussed in the section "Instrumental Variables"), so these individuals will always be in a subset of the set of the treated group (i.e., $i \in AT \subset T$); likewise, some individuals will always choose the control (these are the never-takers [NT]), and thus will be in a subset of the set of the treated group (i.e., $i \in NT \subset C$). Some others will sometimes choose the treatment and sometimes choose the control. Thinking about these different types of individuals helps us more clearly define the causal estimand of interest: that is, we might find that we are not in fact interested in the ATE, but rather we are interested in the ATE among those who are likely to be treated, which we refer to as the "ATE among the treated" (ATT). For example, imagine we want to estimate the effect on mortality rates of an invasive surgical procedure (vs. no procedure), and we expect this procedure will only be beneficial to individuals with a rare medical condition. In this case, we want to find out the average effect of the invasive treatment among individuals who are candidates for the treatment. Thus, the ATE (over the entire population) is not of substantive interest here. With propensity score matching, we can estimate the ATE, ATT, or ATE on the control individuals (ATC). This is distinctly different from regression analysis, which instead could estimate the

* Rosenbaum and Rubin (1983a) require the assumption that treatment assignment is a "strongly ignorable" conditional on **X**, which adds to the ignorability or conditional independence assumption (i.e., $\{Y_i^t, Y_i^c\} \perp t_i \mid \mathbf{X}_i$) the requirement that the individual could potentially be in either condition given **X** (i.e., that the individual has a nonzero probability of being in both the treatment and control conditions; $0 < \Pr(t_i = 1 \mid \mathbf{X}_i) < 1$, thus assuming overlap in probability distributions). In recent years, researchers in the social sciences (and even writers on the subject of causal inference) tend to reference only the first part of the assumption of strong ignorability, and thus use the term interchangeably with the terms conditional independence, unconfoundedness, and exogeneity (Morgan and Winship 2007). Although these terms tend to be used interchangeably in contemporary literature, one should note that the strongly ignorable treatment assignment condition is stronger than the conditional independence assumption because strong ignorability also requires the overlap assumption. Thus, when I reference the CIA in the text, I am focusing then on the first part of the strong ignorability assumption, but the second part (i.e., overlap) of strong ignorability is also assumed.

average conditional treatment effect, provided the functional form is correctly specified.

Knowing which estimand we are interested in then leads us to identify which part(s) of the CIA are required for our analysis. If we are interested in the ATT, we want to estimate

$$\text{ATT} = \bar{\delta}_{i \in T} = \sum_{\mathbf{x}} \frac{n_{T_{\mathbf{x}}}}{N_T} \left[\left(\bar{Y}_{i \in T}^{t} \mid \mathbf{X} = \mathbf{x} \right) - \left(\bar{Y}_{i \in T}^{c} \mid \mathbf{X} = \mathbf{x} \right) \right]$$

The above equation is simply the ATE of the individuals in the treatment condition. The first quantity inside the square brackets is the average outcome of individuals in the treatment condition with a specific profile of **X** (e.g., Asian, female, college graduates, and currently employed) when they receive the treatment. The second quantity inside the square brackets is the average outcome of individuals in the treatment condition with the same specific profile of **X** when they receive the control. This bracketed difference is then weighted by the proportion of individuals in the treatment condition with a specific profile of **X**, and these weighted differences are summed over all possible profiles of **X**. Note, however, the second quantity inside the square brackets is missing, so we need to approximate it. To do so, we seek to identify individuals in the control set to act as proxies for the missing counterfactuals of individuals in the treatment condition. This requires us to make the CIA for the treatment condition, which states that conditional on **X**, the outcomes of individuals from the control set can serve as proxies for the outcomes of individuals from the treatment set if they received the control:

$$\left(\bar{Y}_{i \in T}^{c} \mid \mathbf{X} = \mathbf{x} \right) = \left(\bar{Y}_{i \in C}^{c} \mid \mathbf{X} = \mathbf{x} \right)$$

Substituting the matched individuals for the missing counterfactuals, we can estimate the ATT:

$$\widehat{\text{ATT}} = \hat{\bar{\delta}}_{i \in T} = \sum_{\mathbf{x}} \frac{n_{T_{\mathbf{x}}}}{N_T} \left[\left(\bar{Y}_{i \in T}^{t} \mid \mathbf{X} = \mathbf{x} \right) - \left(\bar{Y}_{i \in C}^{c} \mid \mathbf{X} = \mathbf{x} \right) \right]$$

If instead one were interested in the ATC, one would estimate

$$\text{ATC} = \bar{\delta}_{i \in C} = \sum_{\mathbf{x}} \frac{n_{C_{\mathbf{x}}}}{N_C} \left[\left(\bar{Y}_{i \in C}^{t} \mid \mathbf{X} = \mathbf{x} \right) - \left(\bar{Y}_{i \in C}^{c} \mid \mathbf{X} = \mathbf{x} \right) \right]$$

Here, we are missing the first quantity inside the square brackets, and thus we need to make the second part of the CIA:

$$\left(\bar{Y}_{i \in C}^{t} \mid \mathbf{X} = \mathbf{x} \right) = \left(\bar{Y}_{i \in T}^{t} \mid \mathbf{X} = \mathbf{x} \right)$$

This allows us to estimate ATC:

$$\widehat{\text{ATC}} = \hat{\bar{\delta}}_{i \in C} = \sum_{\mathbf{x}} \frac{n_{C_{\mathbf{x}}}}{N_C} \left[\left(\bar{Y}_{i \in T}^{t} \mid \mathbf{X} = \mathbf{x} \right) - \left(\bar{Y}_{i \in C}^{c} \mid \mathbf{X} = \mathbf{x} \right) \right]$$

The ATT can be viewed as reweighting the average outcomes of individuals in the control condition to match the density (along **X**) of individuals in the treatment condition. Likewise, the ATC can be viewed as reweighting the average outcomes of individuals in the treatment condition to match the density of those in the control condition.

The composition of the vector **X** deserves some attention. Our goal with propensity score matching is to identify two groups of individuals who could just as easily be in one condition or the other; in essence, we want to model the selection process such that any remaining difference in treatment condition is as if random. This requires the analyst to have access to the variables that factor into the selection process, and large-scale datasets may or may not contain the required variables. For instance, if the dataset only contains demographic variables (e.g., gender, race/ethnicity, and socioeconomic status), then a propensity score matching-based analysis will likely yield biased estimates of a treatment effect unless the true selection model can be fit by using only demographic variables (Cook et al. 2008).

Before discussing estimation in practice, it is worth noting an additional benefit of matching-based estimators: these estimators allow us to identify observably similar sets of individuals in each condition without even looking at the outcome variable. That is, the outcome variable does not play a role in matching-based estimators until after the matched sets are determined, allowing the researcher to avoid "fishing expeditions." By contrast, in regression analyses, the outcome variable is always present as the left-hand side of the equation, and the analyst may be tempted to tinker with the functional form of the model until a desired result is achieved. Essentially, matching-based estimators can help keep researchers honest.

Estimation

In practice, treatment-effect estimation via propensity score matching involves a series of steps (even if one uses an all-in-one computer program to estimate these effects, the program executes a series of steps). First, treatment condition is predicted by the theorized vector **X**. This can be done with a logit or probit model with a binary treatment indicator. Second, following logit/probit estimation, each individual's propensity score is predicted by the combination of her/his values of **X** and the coefficients estimated via the logit/probit model. Third, individuals in the different conditions are matched to one another. This matching stage can be done using any of a variety of options (or even a combination of these options), including nearest-neighbor matching (finding the person with the closest propensity score in the opposite condition, either with or without replacement), caliper matching (using all individuals in the opposite condition within a small range of the propensity score as matches), stratification (dividing the propensity score into a number of quantiles, and the quantile serves as the basis for matching), kernel weighting (matching individuals to multiple others from the opposite condition, but giving greater weight to observations with more similar propensity scores), and inverse probability of treatment weighting (reweighting observations as the inverse of their likelihood to be in a condition). Caliendo and Kopeinig (2008) and Dehejia and Wahba (2002) present examples that compare estimates from different matching methods. Fourth, the quality of the matches is assessed. In the past, this "balance check" stage has been implemented simply as a series of tests of statistical significance; however, that approach alone does not reveal the extent of practical imbalance (in either the means or variances) that may remain after matching. As such, researchers suggest performing a number of balance checks, including statistical significance tests, but also reporting the mean standardized difference in each variable before and after matching and the proportion of bias explained by matching, as well as assessing the ratio of variance in the treated and matched groups (Ho et al. 2007; Morgan and Todd 2008; Rubin 2001; Stuart 2007). If adequate balance is not achieved, then the researchers should revisit their original model that generated the propensity score, perhaps adding additional covariates, including polynomials of included variables and interaction terms. Note that when the original propensity score-generating model is refit (in step one), steps two through four must also be repeated for the new model. This can be a very time-consuming task, but it is necessary before one can move on to the final step.

Once balance is achieved, the final step (i.e., effect estimation) can be performed. This final step can be implemented very simply as a regression of Y on t among the matched sample. However, when implementing a matching-based approach, you will likely note that, although adequate balance was achieved (e.g., in the sense that standardized differences in the matched sample were small and not statistically significant), small mean differences in variables will still exist between the treatment and control groups in the matched sample. These remaining (but small) differences can be statistically accounted for by adding the vector $\mathbf{X}$ to the estimation model (Ho et al. 2007).

EXAMPLE

Reardon et al. (2009) performed a propensity-score-matching analysis using ECLS-K data to estimate the effect of attending Catholic school on mathematics and reading achievement in elementary school. Prior to matching, students attending Catholic and public schools differed considerably on key variables related to achievement, such as prior achievement: for example, prior to matching, Catholic school students scored over 0.4 SDs greater than public school students on both mathematics and reading achievement tests in the fall of kindergarten. Other factors related to achievement, such as family income, mother's age at first childbirth, and poverty status, also differed significantly between Catholic school and public school students prior to matching. These and other variables (with the exception of achievement at the fall of kindergarten) were used to estimate each child's propensity score; achievement in the fall of kindergarten was not included because it may have itself been affected by the treatment because it was measured approximately 2 months into the school year. Following propensity-score estimation, a combination of caliper and nearest-neighbor matching was implemented to identify matched pairs. The matched sample demonstrated good balance: for example, students differed by 0.43 SDs on the fall kindergarten mathematics score prior to matching, but differed by 0.008 SDs after matching (in the national matched sample). Using the matched sample, Reardon et al. estimated the ATT and found that Catholic school students scored significantly lower on mathematics tests in Grades 3 and 5 than matched peers who attended public schools, suggesting a negative average effect of Catholic schooling on the mathematics achievement of the type of student who attends Catholic school.

Other Considerations

Recall that the internal validity of propensity score matching-based estimates hinges upon the plausibility of the CIA(s). One might ask how different the estimates would be if the CIA was violated because we could not include an unobserved variable related to both treatment condition and outcome(s). We can address this question somewhat by performing sensitivity analyses (Pan and Frank 2003; Rosenbaum and Rubin 1983b; Rosenbaum 1986; see also Frank 2000). The idea behind sensitivity analysis is that we want to assess how sensitive our effect estimates are to confounders; this is a simulation where we construct a variable that has a relationship of a given magnitude with both the treatment and the outcome (the relationship need not be the same for the treatment and outcome), and we calculate what the treatment effect estimate would be if such a variable were observed and included in the model.

Other issues that may arise with large-scale secondary data include multilevel analysis, the use of sampling weights, and cases of more than two conditions. Each of these topics is relatively understudied. Regarding multilevel analysis, where there is a probability of treatment for individuals (Level 1) and clusters of individuals (Level 2), one may estimate the propensity score at each level (Hong and Raudenbush 2006). An alternative approach is to estimate the individuals' propensity scores while including random or fixed effects for the Level 2 clusters, and then matching within the Level 2 clusters (see, e.g., Reardon, Cheadle, and Robinson 2009; Robinson and Espelage 2012). Regarding the use of sampling weights, weights should often be included in analyses using complex survey designs (see Winship and Radbill 1994); typically, weights are included in the model that estimates the propensity score and then matched cases are weighted by a function of the sampling weights in the estimation stage (Zanutto 2006; also see, e.g., Reardon et al. 2009). Finally, regarding cases of more than two conditions, propensity score matching can still be used, and there are several options for matching cases (Imbens 2000).

Causal Mediation Analysis

Much of the causal inference literature—as well as research using the methods just described—has focused on estimating the effect of a treatment relative to a control. However, some argue that this is a "black-box" approach that provides an estimate of a treatment effect without revealing what mechanism(s) produced the effect or what the effects are of the specific mechanisms. For example, a study may find that Catholic school attendance (vs. public school attendance) has a negative effect on math achievement (e.g., Reardon et al. 2009), but this alone does not signal what specifically about Catholic schools led to this effect. Was it that Catholic schools spend relatively less time teaching math, and thus the effect of time teaching math mediates the Catholic school effect? Or perhaps was the math assessment simply better aligned to the curriculum taught in public schools, and thus the effect of curriculum-assessment alignment mediates the Catholic school effect? Once potential mechanisms have been identified, one could perform a causal mediation analysis.

Importantly, though, if one is interested in causal mediation analysis (e.g., the effect of the mediator time spent teaching math), then the estimated effect of the mediator must be unbiased (Bullock et al. 2008, 2010; Bullock and Ha 2011; Imai et al. 2011; Judd and Kenny 1981; MacKinnon 2008; Sobel 2009), a key principle that is often overlooked in mediation analyses (Bullock et al. 2010; Imai et al. 2011). Imai et al. (2011) provide a framework for decomposing the ATE into the average causal mediation effects (ACMEs) and the average direct effects (ADEs). They first propose that an additional assumption of sequential ignorability is required. This assumption can be thought of as two components: (1) treatment assignment is exogenous, conditional on pretreatment covariates, and (2) the mediator is exogenous, conditional on pretreatment covariates and treatment assignment. Although randomization in experimental designs helps ensure the first part of the sequential ignorability assumption, it does not ensure the second part because certain pretreatment covariates may influence both the mediator and the outcome (Imai et al. 2011). If the second part of the sequential independence assumption is violated, the estimates of ACME and ADE would be biased even though the estimated ATE would be unbiased if the first part of the assumption were valid. Imai et al. (2011) propose strategies for ensuring the sequential mediation assumption is valid.

In a recent example, Robinson et al. (in press) implemented a causal mediation analysis with ECLS-K data to decompose the total difference in gender-gap growth between kindergarten and first grade into the portion that is causally mediated by teachers' perceptions of boys' and girls' math proficiency and then any remaining growth in the gender gap.* To do this, the authors first matched boys and girls using propensity score matching, but this step only ensures that the boys and girls had similar characteristics (this is related to the first part of the sequential ignorability assumption). Per the second part of the sequential ignorability assumption, teachers' perceptions had to be exogenous also. To address this part of the assumption, among the matched sample of boys and girls, the authors used an IV approach to carve out exogenous variation in perceptions by predicting current teachers' perceptions from prior teachers' perceptions—the authors have overidentified IV models, permitting them to provide empirical support for the IV assumptions. Using this IV approach to causal mediation analysis, Robinson et al. (in press) found that teachers' perceptions of girls being less mathematically proficient than observationally similar boys likely contributes substantially to the development of math gender gaps in early elementary school.

Scenarios of Causal and Comparative Analyses

When estimating causal effects, the researcher must believe that individuals in the treatment and control conditions serve as reasonable proxies for the missing counterfactuals in the opposite condition. When individuals in the treatment and control conditions differ in other ways that affect outcomes, estimates of treatment effects will be biased. In addition to the considerations mentioned above in the discussions of each quasi-experimental technique, some other considerations should be mentioned with regard to large-scale datasets: in particular, one should always bear in mind that a particular substantive question may not lend itself to causal analysis. Below, I describe three different scenarios to illustrate how quasi-experimental techniques might be applied, and whether causal inferences could be drawn. For inferring causality, two questions will be of paramount importance: (1) Is the missing counterfactual possible? (2) Can we plausibly approximate the

* Robinson et al. (in press) were not estimating the "effect" of gender on achievement growth (because gender is not manipulable), but rather were interested in decomposing the achievement growth associated with gender into two parts: the part mediated by teacher expectations (this maps onto the ACME) and the unmediated part of the association (this maps onto the ADE in Imai et al.'s terminology).

missing counterfactual? If we cannot answer these two questions affirmatively, causal inferences will not be possible for the scenario; however, it is important to note that techniques associated with causal inferences can nevertheless be used to perform careful comparative analyses.

Scenario 1. An After-School Tutoring Program

Assume the state education agency institutes an after-school tutoring program in mathematics for students who did not score above 130 points on the prior-year mathematics standardized achievement test. Further assume that no one anticipated this program going into effect, and there is no other reason to expect manipulation. The state wants to know if the tutoring program has an effect on mathematics achievement on next year's test.

This scenario does lend itself to causal inferences because each student's counterfactual is possible (i.e., any student could be in the opposite condition, and thus each child has a potential outcome in the other condition) and the missing counterfactuals could be reasonably approximated (at least, for a subset of students). The quasi-experimental design that could be used to estimate a causal effect is regression discontinuity because treatment assignment was determined by attaining a cutscore. Note, however, that RDD estimates the LATE, and the state was interested in a broader range of treatment effects. Therefore, although the analyst can apply RDD to estimate the LATE at the cutscore (and therefore have reasonably good internal validity), the analyst should clarify to the state agency that the results may not generalize to students further away from the cutscore.

Scenario 2. Comparing Achievement Scores for Students of Different Ancestries

Suppose one is interested in estimating the effect on mathematics achievement scores (Y) of being of Chinese-only ancestry (t) versus at least some non-Chinese ancestry (c) in the U.S. educational system. We first need to consider whether each person has a potential outcome in each condition: that is, do Y_i^t and Y_i^c potentially exist for all individuals? If someone is of Chinese-only ancestry ($i \in T$), by definition they cannot possibly be of at least some other non-Chinese ancestry (i.e., there is no potential outcome $Y_{i \in T}^c$); likewise, for individuals in set C. Thus, we cannot estimate the "effect" of being of Chinese-only ancestry (see Holland 1986). As such, we should also not use the terms "treatment" and "control" for this scenario.

While we cannot estimate a causal effect, we could, however, do a careful job of comparing the outcomes of students of Chinese-only ancestry with those of non-Chinese ancestry. Similar to approaches taken by Crosnoe (2005), Robinson and Espelage (2012), and Zanutto (2006), one could perform a matching-based analysis to identify groups of individuals with either ancestry type who share a common set of characteristics **X**. By doing so, the functional-form assumption in the estimation stage is weaker than the one required by regression analyses, particularly if the two groups differed substantially prior to matching (Rubin 2001). Therefore, although the original question in this scenario does not lend itself to causal inferences, we can use some of the techniques associated with causal analysis to perform more careful group comparisons.

Scenario 3. Cross-Ancestry Comparisons of After-School Program Effects

This final scenario combines the above two scenarios. Assume we want to know if the LATE at the cutscore of the after-school tutoring program is different for students of Chinese-only ancestry when compared to the LATE for students on non-Chinese-only ancestry. Based on the assumption in scenario 1, we can estimate the LATE for students from each ancestry group, thereby estimating a within-ancestry-group LATE. Moreover, we can examine whether these LATEs are different from each other; however, we cannot say that we estimated the "effect" of being of Chinese-only ancestry on the LATE. This may seem like semantics, but it is important to understand that we can only estimate the effects of manipulable causes in modern causal inference.

EXAMPLE

The literature using international large-scale assessment data provides some excellent examples of research that uses quasi-experimental designs to estimate treatment effects *within* countries, and then compares these estimated effects *across* countries. For example, Bedard and Dhuey (2006) used 1995 and 1999 TIMSS data to examine the effect of a student's relative age (i.e., relative to his classmates) on that student's achievement. As the authors note, simply regressing student achievement on age will lead to biased estimates of the effect of age if students retained in grade (and thus older when observed) were lower achieving. To address this concern, Bedard and Dhuey used an IV approach, where the endogenous variable of *actual* age was instead *predicted* by the arguably exogenous variable of the student's age relative to the school birthday cutoff for enrollment. That is, the authors used IV estimation to examine the relationship between a student's math/science achievement and their *predicted* age in 10 different counties for 9-year-olds and in 18 different countries for 13-year-olds. The IV model was estimated separately for each country/subject/age combination. Significant effects of relative age were found in nearly all country/subject/age combinations, but the effects did appear to differ across countries. For instance, a 1-month difference in relative age was associated with a 0.19-point difference in 9-year-old math achievement in Canada, but an even larger 0.48-point difference in New Zealand. For another example of research examining relative age effects on achievements within countries—but with PISA data—see Sprietsma (2010).

Conclusion

Research designs for causal inference extend beyond pure experimental designs, thus providing a range of techniques to potentially apply to secondary datasets. However, as with experimental studies, careful design and the plausibility of key assumptions are critical for inferring causality from quasi-experimental designs. When causal inferences are possible, the researcher should provide strong evidence to support the assertion that the estimate is indeed a causal one, and the researcher should note any limitations to generalizability. When causal inferences are not possible, quasi-experimental techniques may still be helpful to researchers seeking to provide careful comparative analyses.

Finally, please note that the goal of this chapter was to provide an introduction to causal inference and quasi-experimental designs. The methods and approaches reviewed here are by no means exhaustive. For further reading on causal inference and a broad range of quasi-experimental designs, see Angrist and Pischke (2009), Morgan and Winship (2007), Murnane and Willett (2011), Pearl (2009), and Shadish et al. (2002). Also, Imbens and Lemieux (2008) and Bloom (2009) provide excellent guides to implementation for RDD, and Caliendo and Kopeinig (2008) and Guo and Fraser (2010) do likewise for propensity score matching.

References

Angrist, J. D., Imbens, G. W., and Rubin, D. B. 1996. Identification of causal effects using instrumental variables. *Journal of the American Statistical Association, 91*(434), 444–455.

Angrist, J. D. and Lavy, V. 1999. Using Maimonides' rule to estimate the effect of class size on scholastic achievement. *The Quarterly Journal of Economics, 114*(2), 533–575.

Angrist, J. D. and Pischke, J.-S. 2009. *Mostly Harmless Econometrics: An Empiricist's Companion.* Princeton, NJ: Princeton University Press.

Bedard, K. and Dhuey, E. 2006. The persistence of early childhood maturity: International evidence of long-run age effects. *The Quarterly Journal of Economics, 121*(4), 1437–1472.

Bloom, H. S. 2009. *Modern Regression Discontinuity Analysis.* New York: MDRC.

Bound, J., Jaeger, D. A., and Baker, R. M. 1995. Problems with instrumental variables estimation when the correlation between the instruments and the endogenous explanatory variable is weak. *Journal of the American Statistical Association, 90*(430), 443–450.

Bullock, J. G. Green, D. P., and Ha, S. E. 2008. Experimental approaches to mediation: A new guide for assessing causal pathways. Retrieved from http://isps.research.yale.edu/conferences/isps40/downloads/BullockGreenHa.pdf

Bullock, J. G. Green, D. P., and Ha, S. E. 2010. Yes, but what's the mechanism? (Don't expect an easy answer). *Journal of Personality and Social Psychology, 98,* 550–558.
Bullock, J. G. and Ha, S. E. 2011. Mediation analysis is harder than it looks. In J. N. Druckman, D. P. Green, J. H. Kuklinski, and A. Lupia (Eds.), *Cambridge Handbook of Experimental Political Science* (pp. 508–521) New York: Cambridge University Press.
Caliendo, M. and Kopeinig, S. 2008. Some practical guidance for the implementation of propensity score matching. *Journal of Economic Surveys, 22*(1), 31–72.
Cook, T. D., Shadish, W. R., and Wong, V. C. 2008. Three conditions under which experiments and observational studies produce comparable causal estimates: New findings from within-study comparisons. *Journal of Policy Analysis and Management, 27*(4): 724–750.
Cox, D. R. 1958. *The Planning of Experiments*. New York: Wiley.
Crosnoe, R. 2005. Double disadvantage or signs of resilience? The elementary school contexts of children from Mexican immigrant families. *American Educational Research Journal, 42*(2), 269–303.
Dee, T. S. 2004. Are there civic returns to education? *Journal of Public Economics, 88*(9–10), 1697–1720.
Dehejia, R. H. and Wahba, S. 2002. Propensity score matching methods for non-experimental causal studies. *Review of Economics and Statistics, 84*(1), 151–161.
Frank, K. A. 2000. The impact of a confounding variable on a regression coefficient. *Sociological Methods and Research, 29*(2), 147–194.
Guo, S. and Fraser, M. W. 2010. *Propensity Score Analysis: Statistical Methods and Applications*. Thousand Oaks: Sage.
Hahn, J., Todd, P., and van der Klaauw, W. 2001. Identification and estimation of treatment effects with a regression-discontinuity design. *Econometrica, 69*(1), 201–209.
Hansen, L. P. 1982. Large sample properties of generalized method of moments estimators. *Econometrica, 50*(4), 1029–1054.
Ho, D. E., Imai, K., King, G., and Stuart, E. A. 2007. Matching as nonparametric preprocessing for reducing model dependence in parametric causal inference. *Political Analysis, 15*(3), 199–236.
Holland, P. W. 1986. Statistics and causal inference. *Journal of the American Statistical Association, 81*(396), 945–960.
Hong, G. and Raudenbush, S. W. 2006. Evaluating kindergarten retention policy: A case study of causal inference for multilevel observational data. *Journal of the American Statistical Association, 101*(475), 901–910.
Imai, K., Keele, L., Tingley, D., and Yamamoto, T. 2011. Unpacking the black box of causality: Learning about causal mechanisms from experimental and observational studies. *American Political Science Review, 105*(4), 765–789.
Imbens, G. W. and Angrist, J. D. 1994. Identification and estimation of local average treatment effects. *Econometrica, 62*(2), 467–475.
Imbens, G. W. and Lemieux, T. 2008. Regression discontinuity designs: A guide to practice. *Journal of Econometrics, 142,* 615–635.
Imbens, G. W. 2000. The role of the propensity score in estimating dose-response functions. *Biometrika, 87*(3), 706–710.
Judd, C. M. and Kenny, D. A. 1981. Process analysis: Estimating mediation in treatment evaluations. *Evaluation Review, 5,* 602–619.
Kaplan, D. 2009. Causal inference in non-experimental educational policy research. In G. Sykes, B. Schneider, and D. N. Plank (eds.), *Handbook on Education Policy Research*. New York: Taylor & Francis.

MacKinnon, D. P. 2008. *Introduction to Statistical Mediation Analysis*. Mahwah, NJ: Erlbaum.

McCrary, J. 2008. Manipulation of the running variable in the regression discontinuity design: A density test. *Journal of Econometrics, 142*, 698–714.

Morgan, S. L. and Todd, J. J. 2008. A diagnostic routine for the detection of consequential heterogeneity of causal effects. *Sociological Methodology, 38*(1), 231–281.

Morgan, S. L. and Winship, C. 2007. *Counterfactuals and Causal Inference: Methods and Principles for Social Research*. New York, NY: Cambridge University Press.

Murnane, R. J. and Willett, J. B. 2011. *Methods Matter: Improving Causal Inference in Educational and Social Science Research*. New York, NY: Oxford University Press.

Neyman, J. 1923/1990. On the application of probability theory to agricultural experiments. Essays on principles. Section 9. *Statistical Science, 5*(4), 465–480.

Pan, W. and Frank, K. A. 2003. A probability estimate of the robustness of a causal inference. *Journal of Behavioral and Educational Statistics, 28*(4), 315–337.

Pearl, J. 2009. *Causality: Models, Reasoning, and Inference* (2nd Ed.). Cambridge, UK: Cambridge University Press.

Reardon, S. F., Arshan, N., Atteberry, A., and Kurlaender, M. 2010. Effects of failing a high school exit exam on course taking, achievement, persistence, and graduation. *Educational Evaluation and Policy Analysis, 32*(4), 498–520.

Reardon, S. F., Cheadle, J. C., and Robinson, J. P. 2009. The effects of Catholic school attendance on reading and math achievement in kindergarten through fifth grade. *Journal of Research on Educational Effectiveness, 2*(1), 45–87.

Reardon, S. F., and Robinson, J. P. 2012. Regression discontinuity designs with multiple rating-score variables. *Journal of Research on Educational Effectiveness, 5*(1), 83–104.

Robinson, J. P. 2010. The effects of test translation on young English learners' mathematics performance. *Educational Researcher, 39*(8), 582–590.

Robinson, J. P. 2011. Evaluating criteria for English learner reclassification: A causal-effects approach using a binding-score regression discontinuity design with instrumental variables. *Educational Evaluation and Policy Analysis, 33*(3), 267–292.

Robinson, J. P. and Espelage, D. L. 2012. Bullying explains only part of LGBTQ–heterosexual risk disparities: Implications for policy and practice. *Educational Researcher, 41*(8), 309–319.

Robinson, J. P., Lubienski, S. T., Ganley, C. M., and Copur-Gencturk, Y. (in press). Teachers' perceptions of students' mathematics proficiency may exacerbate early gender gaps in achievement. *Developmental Psychology.*

Rosenbaum, P. R. 1986. Dropping out of high school in the United States: An observational study. *Journal of Educational Statistics, 11*(3), 207–224.

Rosenbaum, P. R. and Rubin, D. B. 1983a. The central role of the propensity score in observational studies for causal effects. *Biometrika, 70*(1), 41–55.

Rosenbaum, P. R. and Rubin, D. B. 1983b. Assessing sensitivity to an unobserved binary covariate in an observational study with binary outcome. *Journal of the Royal Statistical Society Series B, 45*(2), 212–218.

Rosenbaum, P. R. and Rubin, D. B. 1985. Constructing a control group using multivariate matched sampling methods that incorporate the propensity score. *The American Statistician, 39*, 33–38.

Rubin, D. B. 1974. Estimating causal effects of treatments in randomized and nonrandomized studies. *Journal of Educational Psychology, 66*(5), 688–701.

Rubin, D. B. 1977. Assignment to treatment group on the basis of a covariate. *Journal of Educational Statistics, 2*(1), 1–26.

Rubin, D. B. 1978. Bayesian inference for causal effects: The role of randomization. *The Annals of Statistics, 6*(1), 34–58.

Rubin, D. B. 2001. Using propensity scores to help design observational studies: Application to the tobacco litigation. *Health Services & Outcomes Research Methodology, 2,* 169–188.

Sargan, J. D. 1958. The estimation of economic relationships using instrumental variables. *Econometrica, 26,* 393–415.

Schochet, P., Cook, T., Deke, J., Imbens, G., Lockwood, J. R., Porter, J., and Smith, J. 2010. *Standards for Regression Discontinuity Designs*. Retrieved from What Works Clearinghouse website: http://ies.ed.gov/ncee/wwc/pdf/wwc—rd.pdf.

Shadish, W. R., Cook, T. D., and Campbell, D. T. 2002. *Experimental and Quasi-Experimental Designs for Generalized Causal Inference.* Boston, MA: Houghton Mifflin.

Sobel, M. 2009. Causal inference in randomized and non-randomized studies: The definition, identification, and estimation of causal parameters. *The Sage Handbook of Quantitative Methods in Psychology* (pp. 3–22). Thousand Oaks: Sage.

Sprietsma, M. 2010. Effect of relative age in the first grade of primary school on long-term scholastic results: International comparative evidence using PISA 2003. *Education Economics, 18*(1), 1–32.

Stock, J. H. and Yogo, M. 2005. Testing for weak instruments in linear IV regression. In D. W. K. Andrews and J. H. Stock (Eds.), *Identification and Inference for Econometric Models: Essays in Honor of Thomas Rothenberg*, pp. 80–108. Cambridge, UK: Cambridge University Press.

Stuart, E. A. 2007. Estimating causal effects using school-level data. *Educational Researcher, 36*(4), 187–198.

Winship, C. and Morgan, S. L. 1999. The estimation of causal effects from observational data. *Annual Review of Sociology, 25,* 659–706.

Winship, C. and Radbill, L. 1994. Sampling weights and regression analysis. *Sociological Methods and Research, 23,* 230–257.

Wong, V. C., Steiner, P. M., and Cook, T. D. 2013. Analyzing regression-discontinuity designs with multiple assignment variables: A comparative study of four estimation methods. *Journal of Educational and Behavioral Statistics, 38*(2), 107–141.

Zanutto, E. L. 2006. A comparison of propensity score and linear regression analysis of complex survey data. *Journal of Data Science, 4,* 67–91.

23

Analyzing International Large-Scale Assessment Data within a Bayesian Framework

David Kaplan
University of Wisconsin-Madison

Soojin Park
University of Wisconsin-Madison

CONTENTS

Analyzing International Large-Scale Assessment Data within a Bayesian Framework

> . . .it is clear that it is not possible to think about learning from experience and acting on it without coming to terms with Bayes' theorem.
>
> —**Jerome Cornfield**

When listening to the news or reading newspapers, it not uncommon to find periodic surges of interest regarding which countries are performing the best in major academic subject domains such as reading, mathematics, and science. Information on the comparability of educational outcomes across countries is provided by international large-scale assessments (ILSAs) such as the Organization for Economic Cooperation and Development (OECD)-sponsored Program for International Student Assessment (PISA; OECD 2012), the International Association for the Evaluation of Educational Achievement (IEA)-sponsored Trends in International Mathematics and Science Study (TIMSS; Martin et al. 2008), and the IEA-sponsored Progress in International Reading Literacy Study (PIRLS; Mullis et al. 2008).

The policy consequences of these cross-country comparisons can be quite profound. For example, according to the results of PISA 2000, the average score of German students was below the OECD average, and it was found to be lower than other countries that had similar levels of per capita gross domestic product (GDP). Not only did Germany exhibit relatively low performance in reading, but PISA 2000 results also showed inequities in schooling outcomes in terms of the socioeconomic background of students. Subsequent PISA surveys pointed out that the early tracking policies of the German educational system were not associated with better overall performance, but in fact were associated with low equity (OECD 2003b, 2010). The results of PISA 2000 triggered a unilateral decision by the German federal government to agree to educational reforms and national standards, which moved the German curricula toward a more practical focus (Wiseman 2010). Of course, the policy changes enacted by Germany on the basis of PISA 2000 are not guaranteed to produce the same positive results in another country or economy.

In addition to the use of ILSAs for policy analysis, a great deal of basic research has been conducted that utilizes a variety of ILSAs covering different domains of interest as well as different age or grade levels. A review of the extant literature shows that the vast majority of research conducted with PISA, TIMSS, and PIRLS utilizes a range of statistical methods, including simple regression analysis, logistic regression modeling, multilevel modeling, factor analysis, item response theory, and structural equation modeling. In almost every instance, these methods have been conducted within the Fisherian and Neyman/Pearson schools of statistics. These schools constitute the "classical" school of statistics that rest on the foundations of the frequentist view of probability.

In contrast to the frequentist school of statistics, the Bayesian school presents a coherent, internally consistent, and (arguably) more powerful alternative to the classical school. However, Bayesian statistics has long been ignored in the quantitative methods training of social scientists. Typically, the only introduction that a student might have to Bayesian methodology is a brief overview of Bayes' theorem while studying probability in an introductory statistics class. Two reasons can be given for lack exposure to Bayesian methods. First, until recently, it was not possible to conduct statistical modeling from a Bayesian perspective because of its complexity and lack of available software. Second, Bayesian statistics represents a compelling alternative to frequentist statistics, and is, therefore, controversial. However, in recent years, there has been renewed interest in the Bayesian alternative along with extraordinary developments in the extension and application of Bayesian statistical methods to the social and behavioral sciences. This growth has been attributed mostly to developments in powerful computational tools that render the specification and estimation of complex models such as those used in the analysis of ILSA data feasible from a Bayesian perspective. However, and in addition, growing interest in Bayesian inference has also stemmed from an overall dissatisfaction with the internal inconsistencies of the classical approach (Howson and Urbach 2006).

The orientation of this chapter will be toward those who conduct research using ILSA data and who are well trained in statistical modeling within the frequentist paradigm. The goal is to introduce concepts of model specification, estimation, and evaluation from the Bayesian perspective. Nevertheless, the scope of this chapter is, by necessity, limited, because the field of Bayesian statistics is remarkably wide ranging and space limitations preclude a full development of Bayesian theory. Moreover, not all concepts covered in this chapter will be demonstrated in our examples, owing to present software limitations. Thus, the goal of this chapter will be to lay out the fundamental ideas of Bayesian statistics as they pertain specifically to the analysis if ILSA data.

The organization of this chapter is as follows. The first section provides an overview of Bayesian statistical inference. This is followed by a section describing how Bayesian models are evaluated and compared. Next, we briefly discuss the elements of Bayesian computation. We then present three examples of common analyses conducted on ILSA data from a Bayesian perspective. Our examples will utilize the PISA study, but the issues and implications outlined are applicable to TIMSS, PIRLS, and other ILSA endeavors. We will specify models using data from PISA 2009, comparing the case where priors are not used to the case where priors based on results from PISA 2000 are used. First, we will specify a simple country-level regression to examine the influence of priors in a small sample size case. Next, we will estimate a hierarchical linear model using U.S. PISA data. Finally, we will estimate a school-level confirmatory factor analysis model of school administrator perceptions of school climate, again using U.S. PISA data. All analyses utilize the Mplus software program (Muthén and Muthén 2012).

Overview of Bayesian Statistical Inference

This section provides an overview of Bayesian inference and follows closely the recent overview by Kaplan and Depaoli (2012a) here discussed within the context of ILSA data generally, and, in particular, PISA.

To begin, denote by Y a random variable that takes on a realized value y. In the context of PISA, Y could be the PISA index of economic, social, and cultural status (ESCS).* In the context of more advanced methods, Y could be vector-valued, such as items on the PISA school climate survey. Once the student responds to the survey items, Y becomes realized as y. In a sense, Y is unobserved—it is the probability distribution of Y that we wish to understand from the actual data values y.

Next, we denote by θ a parameter that we believe characterizes the probability model of interest. The parameter θ can be a scalar, such as the mean or the variance of the ESCS distribution, or it can be vector valued, such as the set of all parameters of a factor analysis of the PISA school climate survey.

We are concerned with determining the probability of observing y given the unknown parameters θ, which we write as $p(y \mid \theta)$. In statistical inference, the goal is to obtain estimates of the unknown parameters given the data. This is expressed as the likelihood of the parameters given the data, denoted as $L(\theta \mid y)$. Often, we work with the log-likelihood written as $l(\theta \mid y)$. In accordance with the likelihood principle (see, e.g., Royall 1997), the likelihood function summarizes all of the statistical information in the data.

The key difference between Bayesian statistical inference and frequentist statistical inference concerns the nature of the unknown parameters θ. In the frequentist framework, θ is assumed to be unknown, but fixed. In Bayesian statistical inference, any quantity that is unknown, such as θ, is considered to be random, possessing a probability distribution that reflects our uncertainty about the true value of θ. Because both the observed data y and the parameters θ are considered to be random, we can model the joint probability of the parameters and the data as a function of the conditional probability distribution of the data given the parameters, and the prior probability distribution of the parameters. More formally

$$p(\theta, y) = p(y|\theta)p(\theta). \tag{23.1}$$

Because of the symmetry of joint probabilities

$$p(y|\theta)p(\theta) = p(\theta|y)p(y). \tag{23.2}$$

* From the OECD glossary of statistical terms, ESCS "was created on the basis of the following variables: the International Socio-Economic Index of Occupational Status (ISEI); the highest level of education of the students parents, converted into years of schooling; the PISA index of family wealth; the PISA index of home educational resources; and the PISA index of possessions related to 'classical' culture in the family home."

Therefore

$$p(\theta|y) = \frac{p(\theta, y)}{p(y)} = \frac{p(y|\theta)p(\theta)}{p(y)} \tag{23.3}$$

where $p(\theta|y)$ is referred to as the *posterior distribution* of the parameters θ given the observed data y. Thus, from Equation 23.3, the posterior distribution of θ given y is equal to the data distribution $p(y|\theta)$ times the prior distribution of the parameters $p(\theta)$ normalized by $p(y)$ so that the distribution integrates to one. Equation 23.3 is *Bayes' theorem*. For discrete variables

$$p(y) = \sum_{\theta} p(y|\theta)p(\theta), \tag{23.4}$$

and for continuous variables

$$p(y) = \int_{\theta} p(y|\theta)p(\theta)\, d\theta. \tag{23.5}$$

As above, the denominator in Equation 23.3 does not involve model parameters so we can omit the term and obtain the *unnormalized posterior distribution*

$$p(\theta \mid y) \propto p(y \mid \theta)p(\theta) \tag{23.6}$$

or equivalently

$$p(\theta \mid y) \propto L(\theta \mid y)p(\theta). \tag{23.7}$$

Equation 23.6 represents the core of Bayesian statistical inference and it is what separates Bayesian statistics from frequentist statistics. Specifically, Equation 23.6 states that our uncertainty regarding the parameters of our model, as expressed by the prior distribution $p(\theta)$, is *moderated* by the actual data $p(y|\theta)$, or equivalently, $L(\theta|y)$, yielding an updated estimate of the model parameters as expressed by the posterior distribution $p(\theta|y)$. Again, in the context of the ESCS index, Equation 23.6 states that the posterior distribution of the parameters underlying ESCS (e.g., the mean and/or the variance) is proportional to the prior information based, perhaps, on previous research, moderated by the actual sample data on ESCS as summarized by the likelihood function. The Bayesian framework thus encodes our prior knowledge of the parameters via the prior distribution. Updated knowledge of the parameters of our model is obtained from a summary of the posterior distribution. Summaries of the parameters of the posterior distribution (described later) can (and should) be used as new priors in a subsequent study. In essence, this is how the Bayesian framework supports evolutionary knowledge development and it is what separates it from the frequentist school of statistics.

Important Assumption: Exchangeability

In most discussions of likelihood, and indeed, in the specification of most statistical models, it is common to invoke the assumption that the data $y_1, y_2, \ldots, y_n$ are independently and identically distributed—often referred to as the *i.i.d.* assumption. Bayesians, however, invoke the deeper notion of *exchangeability* to produce likelihoods and address the issue of independence.

Exchangeability arises from de Finetti's representation theorem (de Finetti, 1974) and implies that the subscripts of a vector of data, for example, $y_i, y_2, \ldots, y_n$, do not carry information that is relevant to describing the probability distribution of the data. In other words, the joint distribution of the data, $f(y_i, y_2, \ldots, y_n)$ is invariant to permutations of the subscripts.*

As a simple example of exchangeability, consider the response that student i would have to the question appearing in PISA 2009, "How much do you agree or disagree with each of the following statements about teachers at your school?"—"Most teachers are interested in my well-being," where for simplicity we recode the responses as

$$yi = \begin{cases} 1, & \text{if student } i \text{ agrees} \\ 0, & \text{if student } i \text{ disagrees.} \end{cases} \quad (23.8)$$

Next, consider three possible responses to 10 randomly selected students

$$p(1,0,1,1,0,1,0,1,0,0)$$
$$p(1,1,0,0,1,1,1,0,0,0)$$
$$p(1,0,0,0,0,0,1,1,1,1)$$

Exchangeability implies that only the total number of agreements matter, not the location of those agreements in the vector. This is a subtle assumption insofar as it means that we believe that there is a parameter θ that generates the observed data via a stochastic model and that we can describe that parameter without reference to the particular data at hand (Jackman 2009).

* Technically, according to de Finetti (1974), this refers to *finite* exchangeability. Infinite exchangeability is obtained by adding the provision that every finite subset of an infinite sequence is exchangeable.

As Jackman (2009) points out, the fact that we can describe θ without reference to a particular set of data is, in fact, what is implied by the idea of a prior distribution. In fact, as Jackman notes, "the existence of a prior distribution over a parameter is a *result* of de Finetti's Representation Theorem (de Finetti 1974), rather than an assumption" (Jackman 2009, p. 40).

It is important to note that exchangeability is weaker than the statistical assumption of independence. In the case of two events—say A and B—independence implies that $p(A \mid B) = p(A)$. If these two events are independent, then they are exchangeable—however, exchangeability does not imply independence.

A generalization of de Finetti's representation theorem that is of relevance to the analysis of ILSA data relates to the problem of *conditional exchangeability*. In considering the PISA 2009 example, a more realistic situation arises because students are nested in schools. Thus, exchangeability at the student level would not be expected to hold, because in considering the entire sequence of responses, school subscripts ($g = 1,2,\ldots,G$) on the individual response (e.g., y_{ig}) are not exchangeable. However, within a given school, students might be exchangeable, and the schools themselves (within a country) might be exchangeable. This issue also leads to the more general idea of Bayesian hierarchical models, in which case exchangeability applies not just to data but also to parameters (Jackman 2009). In our multilevel modeling example below, we will assume at least conditional exchangeability.

Types of Priors

The distinguishing feature of Bayesian inference is the specification of the prior distribution for the model parameters. The difficulty arises in how a researcher goes about choosing prior distributions for the model parameters. We can distinguish between two types of priors: (1) *noninformative* and (2) *informative priors* based on how much information we believe we have prior to data collection and how accurate we believe that information to be.

Noninformative Priors

The argument being made throughout this chapter is that prior cycles of PISA, and indeed information gleaned from other cognate surveys, can provide a knowledge base to be used in subsequent studies. However, in some cases, a researcher may lack, or be unwilling to specify, prior information to aid in drawing posterior inferences. From a Bayesian perspective, this lack of information is still important to consider and incorporate into our statistical models. In other words, it is equally important to quantify our ignorance as it is to quantify our cumulative understanding of a problem at hand.

The standard approach to quantifying our ignorance is to incorporate a noninformative prior distribution into our specification. Noninformative prior distributions are also referred to as *objective*, *vague*, or *diffuse* priors. Arguably, the most common noninformative prior distribution is the uniform distribution over some sensible range of values. Care must be taken in the choice of the range of values over the uniform distribution. Specifically, a uniform $[-\infty,\infty]$ distribution would be an *improper* prior distribution because it does not integrate to 1.0 as required of probability distributions. Another type of noninformative prior is the so-called *Jeffreys' prior*, which handles some of the problems associated with uniform priors. An important treatment of noninformative priors can be found in Press (2003), and a discussion of "objective" Bayesian inference can be found in Berger (2006).

Informative Priors

In many practical situations, there may be sufficient prior information on the shape and scale of the distribution of a model parameter that it can be systematically incorporated into the prior distribution. Such priors are referred to as *informative*. One type of informative prior is based on the notion of a *conjugate prior* distribution. A conjugate prior distribution is one that, when combined with the likelihood function, yields a posterior that is in the same distributional family as the prior distribution. This is a very important and convenient feature because if a prior is not conjugate, the resulting posterior distribution may not have a form that is analytically simple to solve. Arguably, the existence of numerical simulation methods for Bayesian inference, such as Markov chain Monte Carlo sampling, may render nonconjugacy less of a problem.

Point Estimates of the Posterior Distribution

Bayes' theorem shows that the posterior distribution is composed of encoded prior information weighted by the data. With the posterior distribution in hand, it is relatively straightforward to fully describe its components—such as the mean, mode, and variance. In addition, interval summaries of the posterior distribution can be obtained. Summarizing the posterior distribution provides the necessary ingredients for Bayesian hypothesis testing.

In the general case, the expressions for the mean and variance of the posterior distribution come from expressions for the mean and variance of conditional distributions generally. Specifically, for the continuous case, the mean of the posterior distribution can be written as

$$E(\theta|y) = \int_{-\infty}^{+\infty} \theta p(\theta|y)\, d\theta, \tag{23.9}$$

and is referred to as the *expected a posteriori* or EAP estimate. Thus, the conditional expectation of θ is obtained by averaging over the marginal distribution of y. Similarly, the conditional variance of θ can be obtained as (see, Gill 2002)

$$\begin{aligned} \text{var}(\theta|y) &= E\left[(\theta - E\left[\left(\theta|y\right]\right)^2 |y\right), \\ &= E(\theta^2|y) - E\left(\theta|y\right)^2. \end{aligned} \tag{23.10}$$

The conditional expectation and variance of the posterior distribution provide two simple summary values of the distribution. Another summary measure would be the mode of the posterior distribution; the so-called *maximum a posteriori* (MAP) estimate. Those measures, along with the quantiles of the posterior distribution, provide a complete description of the distribution.

Posterior Probability Intervals

One important consequence of viewing parameters probabilistically concerns the interpretation of *confidence intervals*. Recall that the frequentist confidence interval is based on the assumption of a very large number of repeated samples from the population characterized by a fixed and unknown parameter μ. For any given sample, we obtain the sample mean $\bar{y}$ and form, for example, a 95% confidence interval. The correct frequentist interpretation is that 95% of the confidence intervals formed this way capture the true parameter μ under the null hypothesis. Notice that from this perspective, the probability that the parameter is in the interval is either zero or one.

In contrast, the Bayesian perspective forms a *posterior probability interval* (PPI; also known as a *credible interval*). Again, because we assume that an unknown parameter can be described by a probability distribution, when we sample from the posterior distribution of the model parameters, we can obtain its quantiles. From the quantiles, we can directly obtain the probability that a parameter lies within a particular interval.

Formally, a $100(1-\alpha)\%$ PPI for a particular subset of the parameter space θ is defined as

$$1 - \alpha = \int_C p\left(\theta|y\right) d\theta. \tag{23.11}$$

So, for example, a 95% PPI means that the probability that the parameter lies in the interval is 0.95. Notice that the interpretation of the PPI is entirely different from the frequentist confidence interval.

Bayesian Model Evaluation and Comparison

Posterior Predictive Model Checking

An important aspect of Bayesian model evaluation that sets it apart from its frequentist counterpart is its focus on posterior prediction. The general idea behind posterior predictive model checking is that there should be little, if any, discrepancy between data generated by the model and the actual data itself. In essence, posterior predictive model checking is a method for assessing the specification quality of the model from the viewpoint of predictive accuracy. Any deviation between the model-generated data and the actual data suggests possible model misspecification.

Posterior predictive model checking utilizes the posterior predictive distribution of replicated data. Following Gelman et al. (2003), let y^{rep} be data replicated from our current model. That is

$$p\,(y^{rep}|y) = \int p(y^{rep}|\theta)\,p(\theta|y)\;d\theta$$
$$= \int p(y^{rep}|\theta)p(y|\theta)p(\theta)\;d\theta. \tag{23.12}$$

Notice that the second term, $p(\theta\,|\,y)$, on the right-hand side of Equation 23.12 is simply the posterior distribution of the model parameters. In the context of PISA, Equation 23.12 states that given current data y, the distribution of future observations on, say, a model predicting student reading scores from background characteristics, denoted as $p(y^{rep}\,|\,y)$, is equal to the probability distribution of the future observations based on the model given the parameters, $p(y^{rep}\,|\,\theta)$, weighted by the posterior distribution of the model parameters, $p(y\,|\,\theta)p(\theta)$. Thus, posterior predictive checking accounts for uncertainty in both the parameters underlying the model and uncertainty in the data itself.

As a means of assessing the fit of the model, posterior predictive checking implies that the replicated data should match the observed data quite closely if we are to conclude that the model fits the data. One statistic that can be used to measure the discrepancy between the observed data and replicated data is the likelihood ratio chi-square statistic. In Mplus (Muthén and Muthén 2012), each draw of the posterior estimates is used to generate replicated data. Then, the likelihood ratio chi-square is computed comparing the observed data to the replicated data for each draw. A scatterplot can be drawn that displays the likelihood ratio for the replicated data against the likelihood ratio for the observed data.

An approach to summarizing posterior predictive checking incorporates the notion of Bayesian p-values. Denote by $T(y)$ the likelihood ratio test statistic based on the data and the model parameters estimated at the tth MCMC iteration. Further, let $T(y^{rep})$ be the same test statistic but defined for the replicated data based on the model parameters estimated at the tth MCMC iteration (described below). Then, the Bayesian p-value is defined to be

$$p\text{-value} = p(T((y) < T(y^{\text{rep}})y). \qquad (23.13)$$

Lower values of Equation 23.13 suggest poor model fit insofar as the test statistic based on the data does not equal or exceed the test statistic based on replicated data generated from the model parameters themselves. Mplus will produce a 95% confidence interval around the difference between $T(y)$ and $T(y^{rep})$. If the lower limit of the confidence interval is positive, it suggests poor model fit. Good model fit is considered to be a Bayesian p-value of approximately 0.50 (Muthén and Asparouhov, 2012a). Mplus will also produce a posterior predictive checking scatterplot, where the number of points above the 45-degree line corresponds to the Bayesian p-value. We will demonstrate posterior predictive checking in our examples.

Deviance Information Criterion

As suggested earlier in this chapter, the Bayesian framework does not adopt the frequentist orientation to null hypothesis significance testing. Instead, as with posterior predictive checking, a key component of Bayesian statistical modeling is a framework for model choice, with the idea that the chosen model will be used for prediction. For this chapter, we will focus on the *deviance information criterion* (DIC; Spiegelhalter et al. 2002) as a method for choosing among a set of competing models.

The DIC is one of many different types of *information criteria* for model selection. Arguably, the most popular method for model selection is the *Bayesian information criterion.* The BIC is derived from so-called *Bayes factors* (Kass and Raftery 1995). In essence, a Bayes factor provides a way to quantify the odds that the data favor one hypothesis over another, where the hypotheses do not need to be nested. When the prior odds of favoring one hypothesis over another are equal, the Bayes factor reduces to the ratio of two integrated likelihoods. The BIC can then be derived from this ratio (see Kass and Raftery 1995; Raftery 1995).

Although the BIC is derived from a fundamentally Bayesian perspective, it is often productively used for model comparison in the frequentist domain. However, the DIC is an explicitly Bayesian approach to model comparison that was developed based on the notion of *Bayesian deviance.* Consider a particular model proposed for a set of data, denoted as $p(y|\theta)$. Then, we begin by defining *Bayesian deviance* as

$$D(\theta) = -2\log\left[p(y|\theta)\right] + 2\log\left[h(y)\right]. \qquad (23.14)$$

where the term $h(y)$ is a standardizing factor that does not involve model parameters and thus is not involved in model selection. Note that although Equation 23.14 is similar to the BIC, it is not, as currently defined, an explicitly Bayesian measure of model fit. To accomplish this, we use Equation 23.14 to obtain a posterior mean over θ by defining

$$\overline{D(\theta)} = E_\theta\left[-2\log\left[p(y \mid \theta) \mid y\right] + 2\log\left[h(y)\right]\right]. \tag{23.15}$$

Next, let $D(\bar{\theta})$ be a posterior estimate of θ. From here, we can define the *effective dimension* of the model as

$$q_D = \overline{D(\theta)} - D(\bar{\theta}), \tag{23.16}$$

which is the mean deviance minus the deviance of the means. Notice that q_D is a Bayesian measure of model complexity. With q_D in hand, we simply add the model fit term $\overline{D(\theta)}$ to obtain the DIC—namely

$$\text{DIC} = \overline{D(\theta)} + q_D, \tag{23.17}$$

$$= 2\overline{D(\theta)} - D(\bar{\theta}) \tag{23.18}$$

Similar to the BIC, the model with the smallest DIC among a set of competing models is preferred. The DIC is available in Mplus (Muthén and Muthén 2012) and will be demonstrated in the examples below.

Brief Overview of MCMC Estimation

As stated in the introduction, the key reason for the increased popularity of Bayesian methods in the social and behavioral sciences has been the advent of powerful computational algorithms now available in proprietary as well as open-source software. The most common algorithm for Bayesian estimation is based on MCMC sampling. In the interest of space, we will not discuss the details of MCMC sampling and instead refer the reader to number of very important papers and books that have been written about MCMC sampling (see, e.g., Gilks et al. 1996). Suffice to say, the general idea of MCMC is that instead of attempting to analytically solve for the moments and quantiles of the posterior distribution, MCMC instead draws specially constructed samples from the posterior distribution $p(\theta|y)$ of the model parameters.

For the purposes of this chapter, we will use the Gibbs sampler (Geman and Geman 1984) as implemented in Mplus (Muthén and Muthén 2012). Informally, the Gibbs sampler proceeds as follows. Consider that the goal is to obtain the joint posterior distribution of two model parameters—say, θ_1 and θ_2, given some data y, written as $f(\theta_1,\theta_2 | y)$. These two model parameters can be regression coefficients from a simple multiple regression model. Dropping the conditioning on y for simplicity, what is required is to sample from $f(\theta_1 | \theta_2)$ and $f(\theta_2 | \theta_1)$. In the first step, an arbitrary value for θ_2 is chosen, say θ_2^0. We next obtain a sample from $f(\theta_1 | \theta_2^0)$. Denote this value as θ_1^1. With this new value, we then obtain a sample θ_2^1, from $f(\theta_2 | \theta_1^1)$. The Gibbs algorithm continues to draw samples using previously obtained values until two long chains of values for both θ_1 and θ_2 are formed. It is common that the first m of the total set of samples is dropped. These are referred to as the *burn-in* samples. The remaining samples are then considered to be draws from the marginal posterior distributions of $f(\theta_1)$ and $f(\theta_2)$.

An important part of MCMC estimation is assessing the convergence of the algorithm. Here too, a number of approaches exist to determine if the algorithm has converged (see, e.g., Sinharay 2004). A variety of these diagnostics are reviewed and demonstrated in Kaplan and Depaoli (2012a), including the Geweke convergence diagnostic (Geweke 1992), the Heidelberger and Welch convergence diagnostic (Heidelberger and Welch 1983), the Raftery and Lewis convergence diagnostic (Raftery and Lewis 1992), and the Brooks, Gelman, and Rubin diagnostic (Gelman and Rubin 1992a,b; Gelman 1996).

Visual diagnostics of convergence include the trace plot and the autocorrelation plot. The trace plot shows the value of the estimate at the tth iteration of the algorithm. Convergence is indicated by a tight "caterpillar-like" band centered around the modal value of the estimate. The autocorrelation plot shows the degree to which the current value of the parameter is dependent on the immediate value of the parameter. High autocorrelation suggests poor convergence and that the MCMC algorithm did not do a good job of exploring the posterior distribution (see Kim and Bolt 2007). We will present both plots in the examples below.

When implementing the Gibbs sampler with multiple chains, one of the most common diagnostics is the Brooks, Gelman, and Rubin diagnostic (see, e.g., Gelman and Rubin 1992a,b; Gelman 1996). This diagnostic is based on analysis of variance and is intended to assess convergence among several parallel chains with varying starting values. Specifically, Gelman and Rubin (1992a) proposed a method where an overestimate and an underestimate of the variance of the target distribution is formed. The overestimate of variance is represented by the between-chain variance and the underestimate is the within-chain variance (Gelman 1996). The theory is that these two estimates should be approximately equal at the point of convergence. The comparison of between and within variances is referred to as the *potential scale reduction factor* (PSRF) and larger values typically indicate that the chains have not fully explored the target distribution. Specifically, a variance ratio that is computed with values approximately equal to 1.0 indicates convergence. Brooks and Gelman (1998) added an adjustment for sampling variability in the variance estimates and also proposed a multivariate extension, which does not include the sampling variability correction. The changes by Brooks and Gelman reflect the diagnostic as implemented in Mplus (Muthén and Muthén 2012). Once it has been determined that the algorithm has converged, summary statistics, including the posterior mean, mode, standard deviation, and PPI, can be obtained.

EXAMPLE 23.1: BAYESIAN REGRESSION ANALYSIS

In this section, we provide an example of Bayesian regression analysis applied to a country-level analysis of reading performance using data from PISA 2000 and 2009. Recall that the year 2000 was the first cycle of PISA and the major domain was reading. PISA 2009 represented the first complete domain cycle of PISA concentrating again on reading. The goal in presenting this example is twofold. First, we wish to demonstrate Bayesian extensions of a commonly used method applied to a sensible question of policy and research relevance. Second, we wish to compare the results of analyses using PISA 2009 when we have no prior information (the noninformative prior case) to the case where we use information gleaned from PISA 2000 to provide informative priors (the informative prior case).

We begin by discussing the basic model with noninformative and informative priors. We then turn to the results, which provide a comparison of choice of priors in the context of a relatively small sample size problem.

Model

Consider a very simple model regressing country-level reading proficiency on country-level background predictors. To begin, let $\mathbf{y}$ be an n-dimensional vector $(y_1,y_2,...,y_n)'$ $(i = 1,2,...,n)$ of scores from n countries on the PISA reading assessment, and let $\mathbf{X}$ be an $n \times k$ matrix containing k background measures, such as GDP, country average teacher salaries, and so on. Then, the normal linear regression model can be written as

$$\mathbf{y} = \mathbf{X}\boldsymbol{\beta} + \mathbf{u}, \tag{23.19}$$

where $\boldsymbol{\beta}$ is a $k \times 1$ vector of regression coefficients and where the first column of $\boldsymbol{\beta}$ contains an n-dimensional unit vector to capture the intercept term. We assume that country-level PISA reading scores are generated from a normal distribution—specifically

$$\mathbf{y} \sim N(\mathbf{X}\boldsymbol{\beta}, \sigma^2\mathbf{I}), \tag{23.20}$$

where **I** is an identity matrix. Moreover, we assume that the n-dimensional vector **u** of disturbance terms is assumed to be independently, identically, and normally distributed—specifically

$$\mathbf{u} \sim N\left(0, \sigma^2 \mathbf{I}\right). \tag{23.21}$$

From standard linear regression theory, the likelihood of the model parameters $\boldsymbol{\beta}$ and σ^2 can be written as

$$L\left(\boldsymbol{\beta}, \sigma^2 \middle| \mathbf{X}, \mathbf{y}\right) = \left(2\pi\sigma^2\right)^{-n/2} \exp\left\{-\frac{1}{2\sigma^2}(\mathbf{y} - \mathbf{X}\boldsymbol{\beta})'(\mathbf{y} - \mathbf{X}\boldsymbol{\beta})\right\}. \tag{23.22}$$

Noninformative Priors

In the context of the normal linear regression model, the uniform distribution is typically used as a noninformative prior. That is, we assign an improper uniform prior to the regression coefficient β that allows β to take on values over the support $[-\infty, \infty]$.* This can be written as $p(\beta) \propto$ c, where c is a constant.

Next, we assign a uniform prior to $\log(\sigma^2)$ because this transformation also allows values over the support $[0, \infty]$. From here, the joint posterior distribution of the model parameters is obtained by multiplying the prior distributions of $\boldsymbol{\beta}$ and σ^2 by the likelihood give in Equation 23.22. Assuming that $\boldsymbol{\beta}$ and σ^2 are independent, we obtain

$$\begin{aligned} p\left(\boldsymbol{\beta}, \sigma^2 \middle| \mathbf{y}, \mathbf{X}\right) &\propto L\left(\boldsymbol{\beta}, \sigma^2 \middle| \mathbf{y}, \mathbf{X}\right) p(\boldsymbol{\beta}) p\left(\sigma^2\right), \\ &\propto \left(\sigma^2\right)^{-n/2} \exp\left\{-\frac{1}{2\sigma^2}(\mathrm{y} - \mathbf{X}\boldsymbol{\beta})'(\mathrm{y} - \mathbf{X}\boldsymbol{\beta})\right\} \times \mathrm{c} \times \sigma^{-2}. \end{aligned}$$

Noting that c does not contain model parameters, and so drops out with the proportionality, we obtain

$$p\left(\boldsymbol{\beta}, \hat{\sigma}^2 \middle| \mathbf{y}, \mathbf{X}\right) \propto \left(\sigma^2\right)^{(-n/2+1)} \exp\left\{-\frac{1}{2\sigma^2}(\mathbf{y} - \mathbf{X}\boldsymbol{\beta})'(\mathbf{y} - \mathbf{X}\boldsymbol{\beta})\right\} \tag{23.23}$$

As pointed out by Lynch (2007), the posterior distribution of the model parameters in Equation 23.23 differs from the likelihood only in the leading exponent $(n/2 + 1)$, which although is of no consequence in large samples, may be of consequence in the subsequent example. Here, we see how the Bayesian approach and the frequentist

* The *support* of a distribution refers to the smallest closed interval (or "set" if the distribution is multivariate) whose elements are actually members of the distribution. The complement to this set has elements with probabilities of zero.

approach align when samples are large and priors are noninformative. When samples are small, however, priors can dominate the likelihood and have much greater influence on summaries of the posterior distribution.

Informative Conjugate Priors

Turning to conjugate priors, the most sensible conjugate prior distribution for the vector of regression coefficients $\boldsymbol{\beta}$ of the linear regression model is the multivariate normal prior. The argument for using the multivariate normal distribution as the prior for $\boldsymbol{\beta}$ lies in the fact that the asymptotic distribution of the regression coefficients is normal (Fox 2008).

The conditional prior distribution of the vector $\boldsymbol{\beta}$ given σ^2 can be written as

$$p(\boldsymbol{\beta}|\sigma^2) = (2\pi)^{k/2}|\Sigma|^{1/2}\exp\left[-\frac{1}{2}(\boldsymbol{\beta}-\mathbf{B})'\Sigma^{-1}(\boldsymbol{\beta}-\mathbf{B})\right], \tag{23.24}$$

where k is the number of variables, $\mathbf{B}$ is the vector of mean hyperparameters assigned to $\boldsymbol{\beta}$, and $\Sigma = \sigma^2\mathbf{I}$ is the diagonal matrix of constant disturbance variances.

The conjugate prior for the variance of the disturbance term σ^2 is the inverse-gamma distribution, with hyperparameters a and b. We write the conjugate prior distribution for σ^2 as

$$p(\sigma^2) \propto (\sigma^2)^{-(a+1)} e^{-b/\sigma^2} \tag{23.25}$$

With the likelihood $L(\boldsymbol{\beta},\sigma^2|\mathbf{X},\mathbf{y})$ defined in Equation 23.22 as well as the prior distributions $p(\boldsymbol{\beta}|\sigma^2)$ and $p(\sigma^2)$, we have the necessary components to obtain the joint posterior distribution of the model parameters given the data. Specifically, the joint posterior distribution of the parameters $\boldsymbol{\beta}$ and σ^2 is given as

$$p(\boldsymbol{\beta},\sigma^2|y,\mathbf{X}) \propto L(\boldsymbol{\beta},\sigma^2|\mathbf{X},\mathbf{y}) \times p(\boldsymbol{\beta}|\sigma^2) \times p(\sigma^2), \tag{23.26}$$

which, after some algebra, yields

$$p(\boldsymbol{\beta},\sigma^2|\mathbf{y},\mathbf{X}) \propto \sigma^{-n-a}\exp\left[-\frac{1}{2\sigma^2}\begin{pmatrix}\hat{\sigma}^2(n-k)+(\boldsymbol{\beta}-\hat{\beta})'\mathbf{X}'\mathbf{X}(\boldsymbol{\beta}-\hat{\beta}) \\ +2b+(\boldsymbol{\beta}-\mathbf{B})'(\boldsymbol{\beta}-\mathbf{B})\end{pmatrix}\right] \tag{23.27}$$

which has the form of a multivariate normal distribution.

Data

This example presents a small sample size Bayesian regression analysis. Data from PISA (OECD 2003b, 2010) and *Education at a Glance* (EAG; OECD 2003a, 2009) were obtained from 25 countries both in 2000 and 2009. Variables used in the regression were teacher salary, class size, GDP per capita, and aggregated reading performance. Teacher salary and per capita GDP were retrieved from EAG, and class size and reading performance were obtained from PISA 2000 and 2009 results (OECD 2003a,b, 2010). Per capita GDP was measured in equivalent U.S. dollars converted using a purchasing power parity formula. Teacher salary was measured relative to national income, that is, a ratio of the teacher salary after 15 years of experience (minimum training) to per capita GDP.

For this example, we use the Mplus software program (Muthén and Muthén 2012). Our focus on Mplus is based on the fact that it has a very general framework that allows for the specification of Bayesian models.

Results

The top two panels of Figure 23.1 show the trace plots and autocorrelation plots for the regression coefficient relating country-aggregated teacher salary to country-level reading competency. Remaining plots are available upon request. An inspection of the trace plots and autocorrelation plots show evidence of convergence. Identical results were found for remaining model parameters. Moreover, the scale reduction factor is approximately 1.0 for both cases, indicating excellent convergence of the two chains.

Table 23.1 shows the results of the Bayesian regression analysis using noninformative and informative priors as described above. Specifically, for the noninformative priors case, a normal prior was chosen for the regression coefficients, with a mean of zero and variance of 10^{10}, and a noninformative inverse-gamma prior was chosen for the residual variance. For the informative case, a normal prior was again chosen for the regression coefficients with means based on the results of a conventional regression using the PISA 2000 data. The prior variances of the regression coefficients were obtained by squaring the standard errors obtained from the conventional regression. This example demonstrates the use of prior information based on a previous ILSA cycle.

An inspection of Table 23.1 reveals, as expected, that the posterior standard deviations and PPIs are wider for the noninformative case than the informative case, reflecting our uncertainty regarding the model parameters. The top panel of Figure 23.2 shows the posterior density plots for slope of teacher salary on reading, where we can clearly see the differences between the noninformative and informative cases.

The bottom panel of Figure 23.2 shows the posterior predictive checking scatterplot under the noninformative and informative analyses. Recall that this plot is used to aid in evaluating the model's goodness-of-fit, with the proportion of observations above the 45-degree line

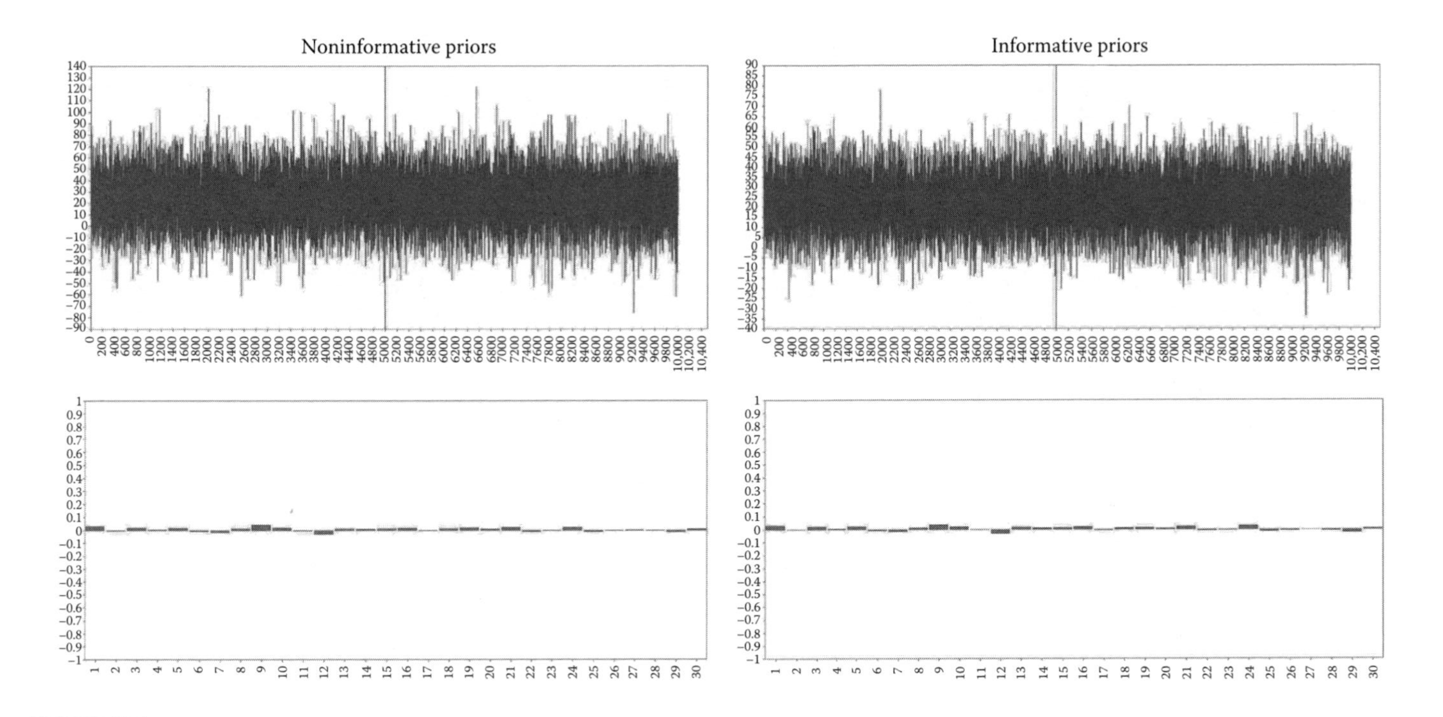

FIGURE 23.1
Trace and autocorrelation plots for teacher salary slope: Bayesian regression analysis.

TABLE 23.1

Between-Country Bayesian Regression Estimates Using PISA 2009 and EAG Data

Parameter	MAP	SD	*p*-value	95% PPI
Noninformative Priors				
READING on GDP Per cap	0.003	0.001	0.001	0.001, 0.004
READING on teacher salary	23.94	20.34	0.12	−17.44, 63.32
READING on class size	−0.84	1.44	0.50	−2.84, 2.88
Informative Priors (PISA 2000)				
READING on GDP per cap	0.003	0.001	0.000	0.002, 0.004
READING on teacher salary	22.32	12.27	0.04	−2.67, 46.10
READING on class size	−0.58	0.97	0.58	−1.68, 2.14

Note: MAP, maximum *a posteriori*; SD, posterior standard deviation; *p*-value is one-tailed.

corresponding to the Bayesian *p*-value. We find that the regression model with informative priors shows slightly better fit than the model with noninformative priors. Finally, the DIC values for the noninformative and informative priors cases is 243.64 and 241.448, respectively, favoring the model with informative priors.

We conclude this example by noting that in the case of small sample sizes (here, 25 countries) the influence of the priors is fairly noticeable. Our results regarding the precision of the estimates based on using information from the PISA 2000 cycle are considerably different than if we had not utilized this information.

EXAMPLE 23.2: BAYESIAN MULTILEVEL MODELING

A common feature of ILSA data collection is that students are nested in higher organizational units such as classrooms and/or schools. Indeed, in many instances, the substantive problem concerns specifically an understanding of the role that classrooms or school characteristics play in predicting an outcome of interest. For example, the PISA structure deliberately samples schools (within a country) and then takes an age-based sample of 15-year-olds within sampled schools. Such data collection plans are generically referred to as *clustered sampling designs*. Data from such clustered sampling designs are then collected at both levels for the purposes of understanding each level separately, but also to understand the inputs and processes of student- and school-level variables as they predict both school- and student-level outputs.

It is probably without exaggeration to say that one of the most important contributions to the empirical analysis of data arising from clustered sampling designs such as PISA has been the development of multilevel models. Important contributions to the theory of multilevel modeling can be found in Raudenbush and Bryk (2002) and references therein. In this example, we present Bayesian multilevel modeling.

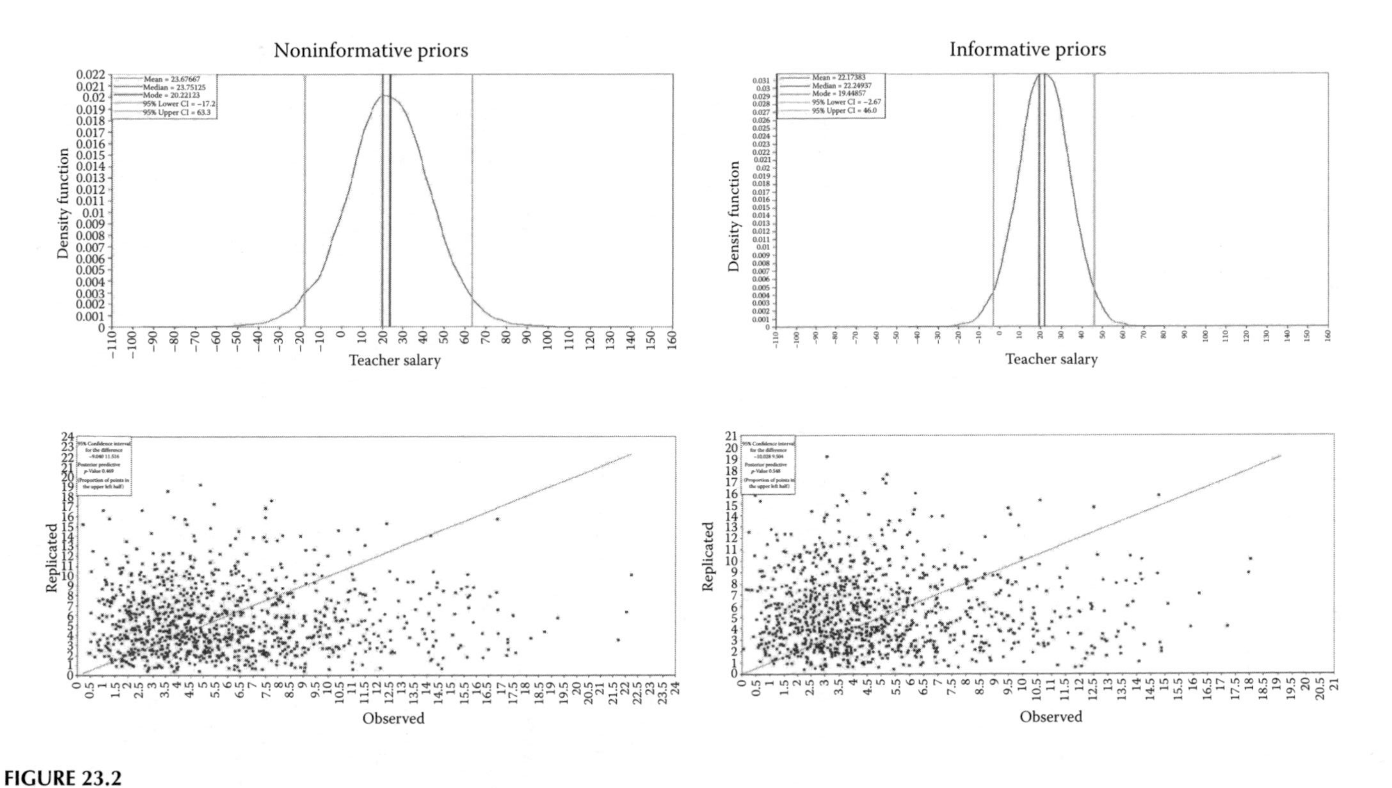

FIGURE 23.2
Posterior density and posterior probability scatter plots for teacher salary slope: Bayesian regression analysis.

Model

Perhaps the most basic multilevel model is the random effects analysis of variance model. As a simple example consider whether there are differences among G schools ($g = 1,2,\ldots,G$) on the outcome of student achievement y obtained from n students ($i = 1,2,\ldots,n$). In this example, it is assumed that the G schools are a random sample from a population of schools.* The model can be written as a two-level, random effects ANOVA model as follows. Let

$$y_{ig} = \beta_g + \epsilon_{ij}, \tag{23.28}$$

where y_{ig} is an achievement score for student i and school g, β_g is the school random effect, and ϵ_{ig} is an error term with homoskedastic variance σ^2. The model for the school random effect can be written as

$$\beta_g = \mu + \delta_g, \tag{23.29}$$

where μ is a grand mean and δ_g is an error term with homoskedastic variance ω^2 that picks up the school effect over and above the grand mean. Inserting Equation 23.29 into Equation 23.28 yields

$$y_{ig} = \mu + \delta_g + \varepsilon_{ig}, \tag{23.30}$$

which expresses the outcome y_{ig} in terms of an overall grand mean μ, a between-school effect δ_g, and a within-school effect ε_{ig}.

Recall that a fully Bayesian perspective requires specifying prior distributions on all model parameters. For the model in Equation 23.30, we first specify the distribution of the achievement outcome y_{ig} given the school effect δ_g and the within-school variance σ^2. Specifically

$$y_{ig} \left| \delta_g, \sigma^2 \sim N\left(\delta_g, \sigma^2\right)\right. \tag{23.31}$$

Given that parameters are assumed to be random in the Bayesian context, we next specify the prior distribution on the remaining model parameters. For this model, we specify normal conjugate priors for the school effects δ_g and the overall grand mean μ—viz.

$$\delta_g \left| \mu, \omega^2 \sim N(\mu, \omega^2)\right., \tag{23.32}$$

$$\mu \sim N\left(b_0, B_0\right), \tag{23.33}$$

* In many large-scale studies of schooling, the schools themselves may be obtained from a complex sampling scheme. However, we will stay with the simple example of random sampling.

where b_0 and B_0 are the mean and variance hyperparameters on μ that are assumed to be *fixed* and *known*. For the within-school and between-school variances, we specify the conjugate inverse-gamma priors—viz.

$$\sigma^2 \sim \text{inverse-gamma}(v_0/2, v_0\sigma_0^2/2), \tag{23.34}$$

$$\omega^2 \sim \text{inverse-gamma}(k_0/2, k_0\omega_0^2/2), \tag{23.35}$$

where v and k are degrees of freedom and σ_0^2 and ω_0^2 are hyperparameter values (Gelman et al. 2003).

To see how this specification fits into a Bayesian hierarchical model, note that we can arrange all of the parameters of the random-effects ANOVA model into a vector $\boldsymbol{\theta}$ and write the prior density as

$$p(\boldsymbol{\theta}) = p(\delta_1, \delta_2, \ldots, \delta_G, \mu, \sigma^2, \omega^2), \tag{23.36}$$

where under the assumption of exchangeability of the school effects δ_g we obtain (see, e.g., Jackman 2009)

$$p(\boldsymbol{\theta}) = \prod_{g=1}^{G} p\left(\delta_g \middle| \mu, \omega^2\right) p(\mu) p\left(\sigma^2\right) p\left(\omega^2\right) \tag{23.37}$$

Slopes and Intercepts as Outcomes Model

In the simple, random-effects ANOVA model, exchangeability warrants the existence of prior distributions on the school means β_g. We noted that a condition where exchangeability might not hold is if we are in possession of some knowledge about the schools, for example, if some are public schools and others are private schools. In this case, exchangeability across the entire set of schools is not likely to hold, and instead we must invoke *conditional exchangeability*. That is, we might be willing to accept exchangeability within school types. Our knowledge of school type, therefore, warrants the specification of a more general multilevel model that specifies the school means as a function of school-level characteristics. The addition of covariates at the student and school levels was first discussed in the educational context by Burstein (1980) and later developed by Raudenbush and Bryk (2002).*

* In this section, we have made the distinction between random-effects ANOVA and multilevel models. This is simply a matter of nomenclature. One could consider all of these models as a special case of hierarchical Bayesian models.

Data

This example presents a Bayesian multilevel regression analysis based on an unweighted sample of 5000 15-year-old students in the United States who were administered PISA 2009 (OECD 2012). The first plausible value of reading performance served as a dependent variable and was regressed on a set of student-level and school-level predictors. Student-level predictors included student background variables—specifically, gender (gender), immigrant status (native), language that they use (slang; coded 1 if test language is the same as language at home, 0 otherwise), and a measure of the student's ESCS. In addition, measures of student engagement and strategies in reading were included as predictors; specifically, enjoyment of reading (joyread), diversity in reading (divread), memorization strategies (memor), elaboration strategies (elab), control strategies (cstrat), student relationship with teachers (studrel), disciplinary climate (disclima), and class size (clsiz). A random slope is specified for the regression of reading performance on ESCS.

School-level predictors included school background variables; that is, school average socioeconomic background (xescs); school size (schsize) and square of school size (schsize2); city (coded 1 for both a small city and large city; 0 otherwise); and rural (coded 1 for a village, hamlet, rural area, or a small town; 0 otherwise). In addition, measures of school climate and policies were included; that is, school average student relationship with teachers (xstudrel), school average disciplinary climate (xdisclim), student behavior (studbeha), teacher behavior (teacbeha), student selection policies (selsch), transferring policy (transfer), school autonomy (respires, respcur), private school (private), school policies on assessment (stdtest, assmon, asscomp), school average language-learning time (xlmins), school average science-learning time (xsmins), school average mathematics-learning time (xmmins), shortage in staff (tcshort), and educational material (scmatedu). The slope of reading performance on ESCS is regressed on school average student relations with teachers (xstudrel) school disciplinary climate (xdisclim).

Method

Two multilevel regressions were conducted in a manner similar to Example 23.1. First, a Bayesian multilevel regression was conducted with weakly informative priors. This analysis assumes a normal distribution for the regression coefficients with a mean of zero and variance of 10^{10}. Weakly informative inverse-gamma priors were chosen for the residual variances. Thus, although the prior has a mode, the precision is so small as to be effectively noninformative. Second, a Bayesian multilevel regression was conducted with informative priors obtained from a conventional multilevel regression analysis of the PISA 2000 data, a manner similar to the regression analysis in Example 1 except that weakly informative inverse-gamma priors were used for the residual variances.

Results

The analysis used the Gibbs sampler as implemented in Mplus with two chains, 100,000 iterations with 50,000 burn-in and a thinning interval of

50. The default in Mplus is to discard half of the total number of iterations as burn-in. Thus, summary statistics on the model parameters are based on 1,000 draws from the posterior distribution generated via the Gibbs sampler. Figure 23.3 shows the trace plots and autocorrelation plots for both the noninformative and informative priors cases focusing on the random slope of reading performance on joyread. An inspection of these plots shows evidence of convergence. Moreover, in each case, the PSRF is very close to 1.0, indicating that the two chains have converged. Plots for all remaining parameters also indicate convergence. These plots are available upon request.

Tables 23.2 and 23.3 present selected results for the multilevel model based on noninformative and informative priors, respectively. We concentrate on predictors that are not part of the sampling design, thus these estimates are conditioned not only on the predictors in Table 23.2, but also on design variables not shown. The influence of priors can be clearly seen when examining the random slope regression. Recall the SLOPE refers to the regression coefficient relating reading performance to parental social and cultural status. For the noninformative priors case, the MAP estimate of SLOPE regressed on XSTUDREL is 0.42 (s.d. = 6.86) with a one-tailed p-value of 0.36. The 95% PPI ranges from –11.22 to 16.04. By contrast, the results of the informative case show a MAP estimate of –1.41 (s.d. = 5.15) with a one-tailed p-value of 0.75 and a 95% PPI ranging from –6.77 to 13.53. PPC plots and DIC values are not presently available in Mplus for Bayesian multilevel models.

EXAMPLE 23.3: BAYESIAN CONFIRMATORY FACTOR ANALYSIS

Model

Recent discussions of Bayesian confirmatory factor analysis and its extension to Bayesian structural equation modeling can be found in Kaplan and Depaoli (2012b), Lee (2007), and Muthén and Asparouhov (2012a). Following the general notation originally provided by Jöreskog (1969), write the confirmatory factor analysis model as

$$\mathbf{y} = \boldsymbol{\alpha} + \boldsymbol{\Lambda}\boldsymbol{\eta} + \boldsymbol{\varepsilon}, \tag{23.38}$$

where $\boldsymbol{y}$ is a vector of manifest variables, $\boldsymbol{\alpha}$ is a vector of measurement intercepts, $\boldsymbol{\Lambda}$ is a factor loading matrix, $\boldsymbol{\eta}$ is a vector of latent variables, and $\boldsymbol{\varepsilon}$ is a vector of uniquenesses with covariance matrix $\boldsymbol{\Psi}$, typically specified to be diagonal. Under conventional assumptions (see, e.g., Kaplan 2009), we obtain the model expressed in terms of the population covariance matrix $\boldsymbol{\Sigma}$ as

$$\boldsymbol{\Sigma} = \boldsymbol{\Lambda}\boldsymbol{\Phi}\boldsymbol{\Lambda}' + \boldsymbol{\Psi}, \tag{23.39}$$

where $\boldsymbol{\Phi}$ is the covariance matrix of the common factors. The distinction between the confirmatory factor analysis (CFA) model in Equation 23.38

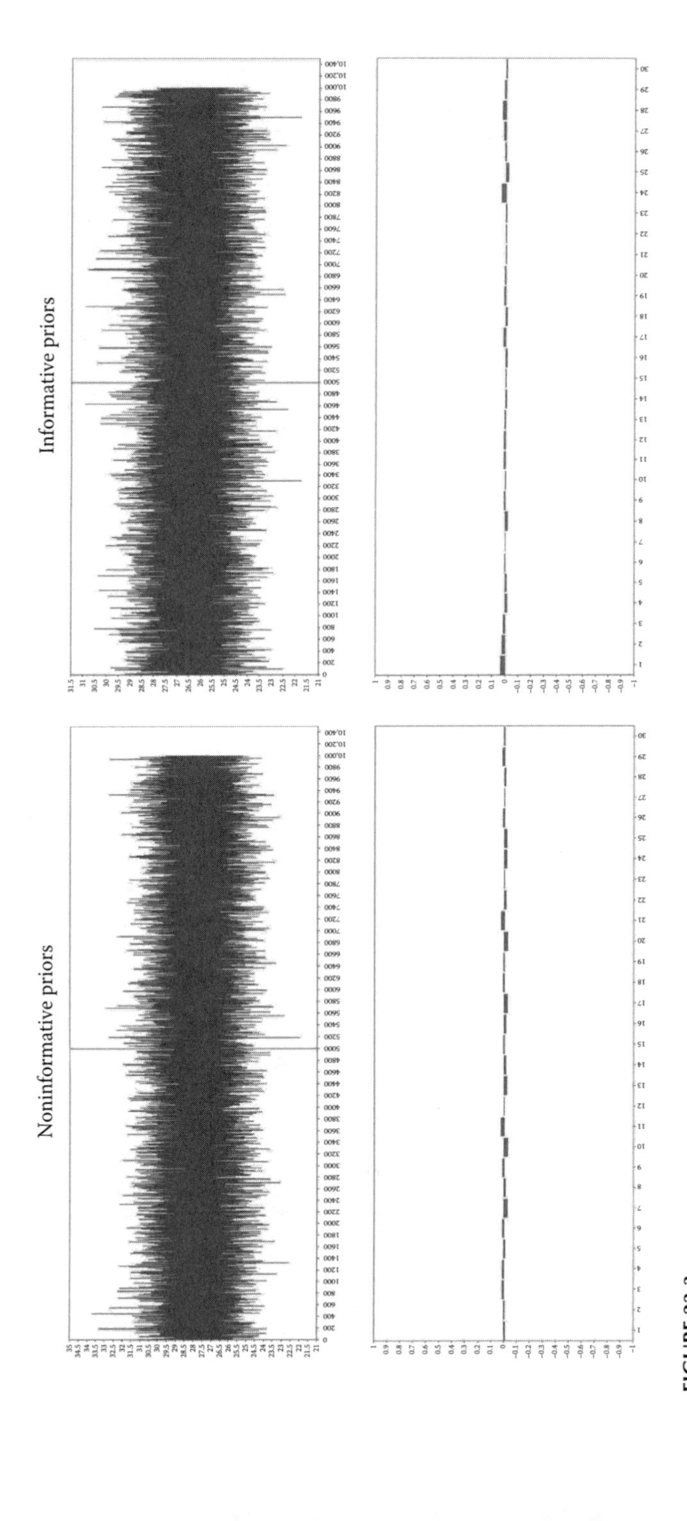

FIGURE 23.3
Trace and autocorrelation plots for joy-of-reading slope: Bayesian hierarchical linear model.

TABLE 23.2

Multilevel Bayesian Regression Estimates with Noninformative Priors

Parameter	MAP	SD	p-value	95% PPI
Within Level				
READING ON JOYREAD	27.28	1.36	0.000	25.10, 30.45
READING ON DIVREAD	−3.43	1.34	0.004	−6.16, −0.93
READING ON MEMOR	−13.08	1.46	0.000	−18.28, −12.55
READING ON ELAB	−12.38	1.39	0.000	−14.60, −9.17
READING ON CSTRAT	21.70	1.61	0.000	18.56, 24.83
READING ON STUDREL	1.31	1.24	0.11	−0.94, 3.89
READING ON DISCLIMA	6.40	1.31	0.000	4.87, 10.01
READING ON CLSIZ	0.71	0.19	0.001	0.24, 0.97
Between Level				
SLOPE ON XSTUDREL	0.42	6.90	0.36	−11.22, 16.04
SLOPE ON XDISCLIM	5.69	6.30	0.09	−3.95, 20.67
READING ON XSTUDREL	−13.40	10.50	0.11	−33.71, 7.60
READING ON XDISCLIM	24.24	9.41	0.01	12.34, 49.59
READING ON STUDBEHA	0.83	3.91	0.05	−1.21, 14.19
READING ON TEACBEHA	0.28	3.77	0.54	−7.57, 7.20
READING ON SELSCH	3.88	5.28	0.23	−6.45, 14.28
READING ON TRANSFER	1.03	6.89	0.42	−12.03, 14.91
READING ON RESPRES	−6.02	3.20	0.34	−7.73, 5.13
READING ON RESPCUR	5.55	2.24	0.30	−3.15, 5.54
READING ON PRIVATE	36.74	24.57	0.06	−10.00, 86.16
READING ON STDTEST	−73.08	33.01	0.04	−124.72, 5.26
READING ON ASSMON	−4.15	18.53	0.43	−39.61, 33.58
READING ON ASSCOMP	12.67	8.26	0.18	−8.74, 23.27
READING ON XLMINS	−0.03	0.09	0.76	−0.11, 0.24
READING ON XSMINS	−0.14	0.08	0.09	−0.19, 0.04
READING ON XMMINS	0.15	0.09	0.49	−0.18, 0.18
READING ON TCSHORT	−2.39	2.98	0.23	−7.90, 3.65
READING ON SCMATEDU	2.38	2.32	0.78	−6.36, 2.76

Note: MAP, maximum *a posteriori*; SD, posterior standard deviation; p-value is one-tailed.

and exploratory factor analysis typically lies in the number and location of restrictions placed in the factor loading matrix $\mathbf{\Lambda}$ (see, e.g., Kaplan 2009).

Conjugate Priors for SEM Parameters

To specify the prior distributions, it is notationally convenient to arrange the model parameters as sets of common conjugate distributions. For this model, let $\boldsymbol{\theta}_{\text{norm}} = \{\boldsymbol{\alpha}, \mathbf{\Lambda}\}$ be the set of free model parameters that are

TABLE 23.3

Multilevel Bayesian Regression Estimates with Informative Priors Based on PISA 2000

Parameter	MAP	SD	*p*-value	95% PPI
Within Level				
READING ON JOYREAD	28.44	1.13	0.000	24.21, 28.65
READING ON DIVREAD	−2.48	1.13	0.007	−3.87, 0.53
READING ON MEMOR	−16.47	1.26	0.000	−18.94, −13.99
READING ON ELAB	−10.25	1.19	0.000	−14.07, −9.44
READING ON CSTRAT	20.36	1.35	0.000	19.74, 25.01
READING ON STUDREL	1.02	1.06	0.04	−0.18, 3.97
READING ON DISCLIMA	5.44	1.10	0.000	2.29, 6.57
READING ON CLSIZ	0.65	0.18	0.000	0.36, 1.07
Between Level				
SLOPE ON XSTUDREL	−1.41	5.15	0.75	−6.77, 13.53
SLOPE ON XDISCLIM	9.12	4.80	0.23	−5.85, 12.78
READING ON XSTUDREL	−2.28	7.00	0.54	−12.82, 14.29
READING ON XDISCLIM	16.37	6.75	0.01	2.34, 28.71
READING ON STUDBEHA	2.14	3.03	0.14	−2.69, 9.13
READING ON TEACBEHA	5.34	2.88	0.38	−4.74, 6.47
READING ON SELSCH	−4.03	3.70	0.21	−10.32, 4.21
READING ON TRANSFER	−5.37	5.56	0.43	−12.05, 9.81
READING ON RESPRES	−2.43	2.13	0.39	−4.84, 3.57
READING ON RESPCUR	−1.10	1.72	0.33	−4.07, 2.70
READING ON PRIVATE	20.18	13.85	0.02	1.62, 56.47
READING ON STDTEST	−16.46	10.55	0.03	−40.12, 1.12
READING ON ASSMON	−0.58	9.68	0.72	−13.56, 24.43
READING ON ASSCOMP	6.90	6.66	0.35	−10.51, 15.50
READING ON XLMINS	0.07	0.06	0.20	−0.06, 0.17
READING ON XSMINS	−0.02	0.05	0.39	−0.11, 0.08
READING ON XMMINS	−0.05	0.06	0.15	−0.18, 0.06
READING ON TCSHORT	−0.16	1.99	0.32	−4.85, 2.98
READING ON SCMATEDU	−4.67	1.95	0.15	−5.93, 1.86

Note: MAP, maximum *a posteriori*; SD, posterior standard deviation; *p*-value is one-tailed.

assumed to follow a normal distribution and let $\boldsymbol{\theta}_{IW} = \{\boldsymbol{\Phi}, \boldsymbol{\Psi}\}$ be the set of free model parameters that are assumed to follow an inverse-Wishart distribution. Thus

$$\boldsymbol{\theta}_{\text{norm}} \sim N(\boldsymbol{\mu}, \boldsymbol{\Omega}), \tag{23.40}$$

where $\boldsymbol{\mu}$ and $\boldsymbol{\Omega}$ are the mean and variance hyperparameters, respectively, of the normal prior. The uniqueness covariance matrix $\boldsymbol{\Psi}$ is assumed to follow an inverse-Wishart distribution. Specifically

$$\boldsymbol{\theta}_{IW} \sim \mathrm{IW}(\mathbf{R},\delta), \tag{23.41}$$

where $\mathbf{R}$ is a positive definite matrix, and $\delta > q - 1$, where q is the number of observed variables. Different choices for $\mathbf{R}$ and δ will yield different degrees of "informativeness" for the inverse-Wishart distribution.

Data

This example is based on a reanalysis of a confirmatory factor analysis described in the OECD technical report (OECD 2012). In the report, the confirmatory factor analysis was employed to construct two indices indicating teacher and student behavioral problems (TEACBEHA and STUDBEHA), using a weighted sample of students from the OECD countries. For this example, we used an unweighted sample of 165 school principals in the United States who participated in PISA 2009. The principals were administered a questionnaire asking to what extent student learning is hindered by student or teacher behavioral problems. Each item has the following four categories: not at all, very little, to some extent, and a lot.

The CFA model in this example was specified to have two factors, which are teacher and student behavioral problems. The factor related to teacher behavioral problems contains the following seven items: teachers' low expectation of students (SC17Q01), poor student–teacher relations (SC17Q03), teachers not meeting individual students' needs (SC17Q05), teacher absenteeism (SC17Q06), staff resisting change (SC17Q09), teachers being too strict with students (SC17Q11), and students not being encouraged to achieve their full potential (SC17Q13). The second factor relating to student behavioral problems contains the following six items: student absenteeism (SC17Q02), disruption of classes by students (SC17Q04), students skipping classes (SC17Q07), students lacking respect for teachers (SC17Q08), student use of alcohol or illegal drugs (SC17Q10), and students intimidating or bullying other students (SC17Q12).

Results

The analysis used the Gibbs sampler as implemented in Mplus with two chains, 100,000 iterations with 50,000 burn-in and a thinning interval of 50. Thus, summary statistics on the model parameters are based on 1000 draws from the posterior distribution generated via the Gibbs sampler. Figure 23.4 presents the trace plots and autocorrelation plots for both the noninformative and informative cases. The plots show evidence of convergence, and the PSRF (not shown) is very close to 1.0.

Selected results of the CFA model for the noninformative (upper panel) and informative (lower panel) cases is displayed in Table 23.4. For the noninformative case, a normal prior was chosen for the factor loadings, with a mean of zero and variance of 10^{10}, and a noninformative inverse-gamma prior was chosen for the factor variances and unique variances. For the informative priors case, priors on the factor loadings were based

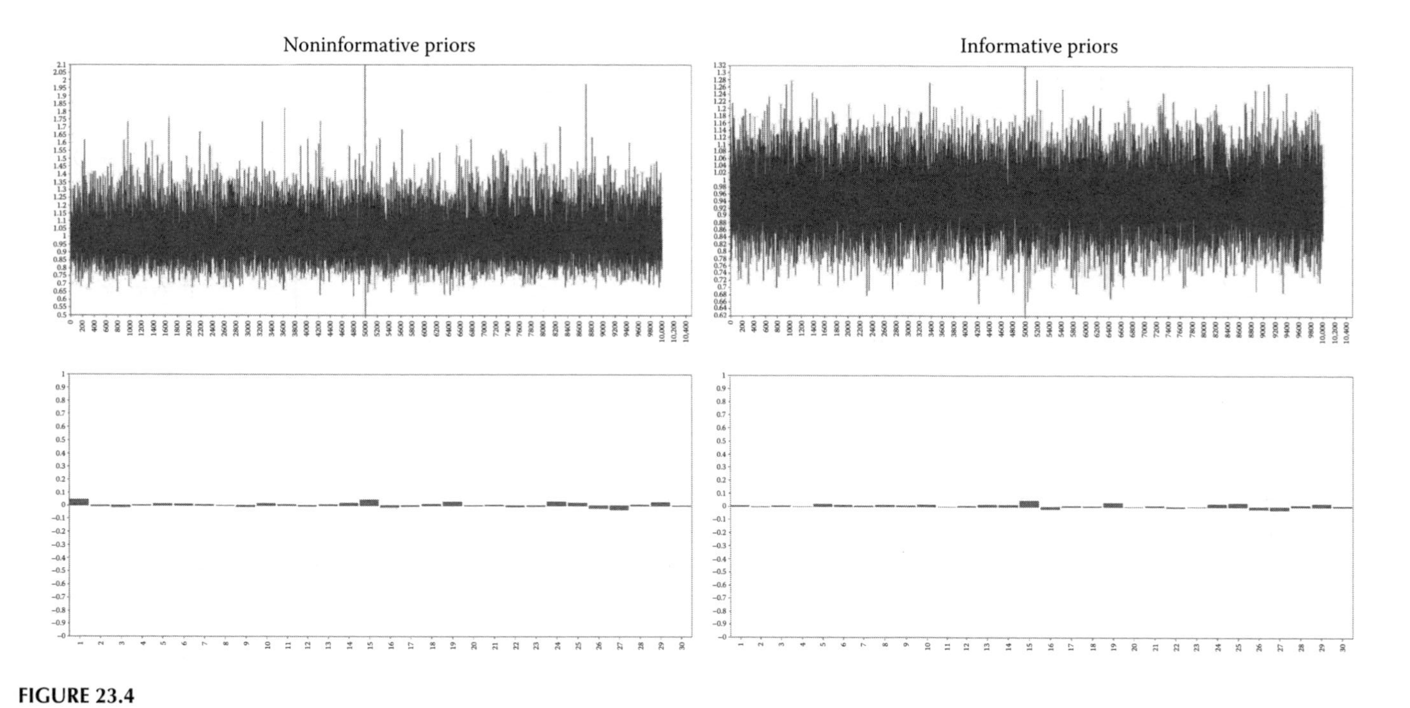

FIGURE 23.4
Trace and autocorrelation plots for factor loading three: Bayesian confirmatory factor analysis.

TABLE 23.4

Selected Bayesian CFA Estimates with Noninformative and Informative Priors

Parameter	MAP	SD	*p*-value	95% PPI
	Noninformative Priors			
Loadings: TEABEHA by				
SC17Q03	0.99	0.13	0.00	0.78, 1.31
SC17Q05	0.94	0.13	0.00	0.70, 1.20
SC17Q06	0.68	0.12	0.00	0.48, 0.95
SC17Q09	0.87	0.14	0.00	0.73, 1.28
SC17Q11	0.56	0.11	0.00	0.31, 0.73
SC17Q13	0.96	0.14	0.00	0.74, 1.27
Loadings: STUDBEHA by				
SC17Q04	0.85	0.13	0.00	0.69, 1.18
SC17Q07	0.98	0.15	0.00	0.82, 1.41
SC17Q08	0.93	0.14	0.00	0.78, 1.30
SC17Q10	0.59	0.11	0.00	0.34, 0.78
SC17Q12	0.56	0.09	0.00	0.42, 0.78
TEABEHA with				
STUDBEHA	0.18	0.04	0.00	0.12, 0.27
	Informative Priors			
Loadings: TEABEHA by				
SC17Q03	1.00	0.08	0.00	0.80, 1.11
SC17Q05	1.07	0.09	0.00	0.82, 1.17
SC17Q06	0.77	0.09	0.00	0.58, 0.94
SC17Q09	0.97	0.10	0.00	0.82, 1.19
SC17Q11	0.64	0.08	0.00	0.40, 0.72
SC17Q13	1.06	0.10	0.00	0.84, 1.21
Loadings: STUDBEHA by				
SC17Q04	0.92	0.09	0.00	0.75, 1.11
SC17Q07	1.12	0.12	0.00	0.95, 1.42
SC17Q08	1.02	0.10	0.00	0.86, 1.24
SC17Q10	0.73	0.10	0.00	0.48, 0.88
SC17Q12	0.67	0.08	0.00	0.51, 0.83
TEABEHA with				
STUDBEHA	0.15	0.03	0.00	0.12, 0.23

Note: MAP, maximum *a posteriori*; SD, posterior standard deviation.

on a previous factor analysis of the PISA 2000 data in a manner similar to Example 23.1. Weakly informative inverse-gamma priors were chosen for the factor variances and unique variances.

As expected, including informative priors based on a conventional CFA of the PISA 2000 data yields smaller posterior standard deviations and narrower 95% PPIs when compared to the noninformative priors case. An inspection of the posterior density plot for one of the loadings (TEABEHA by SC17Q06) in the upper panel of Figure 23.5 shows the

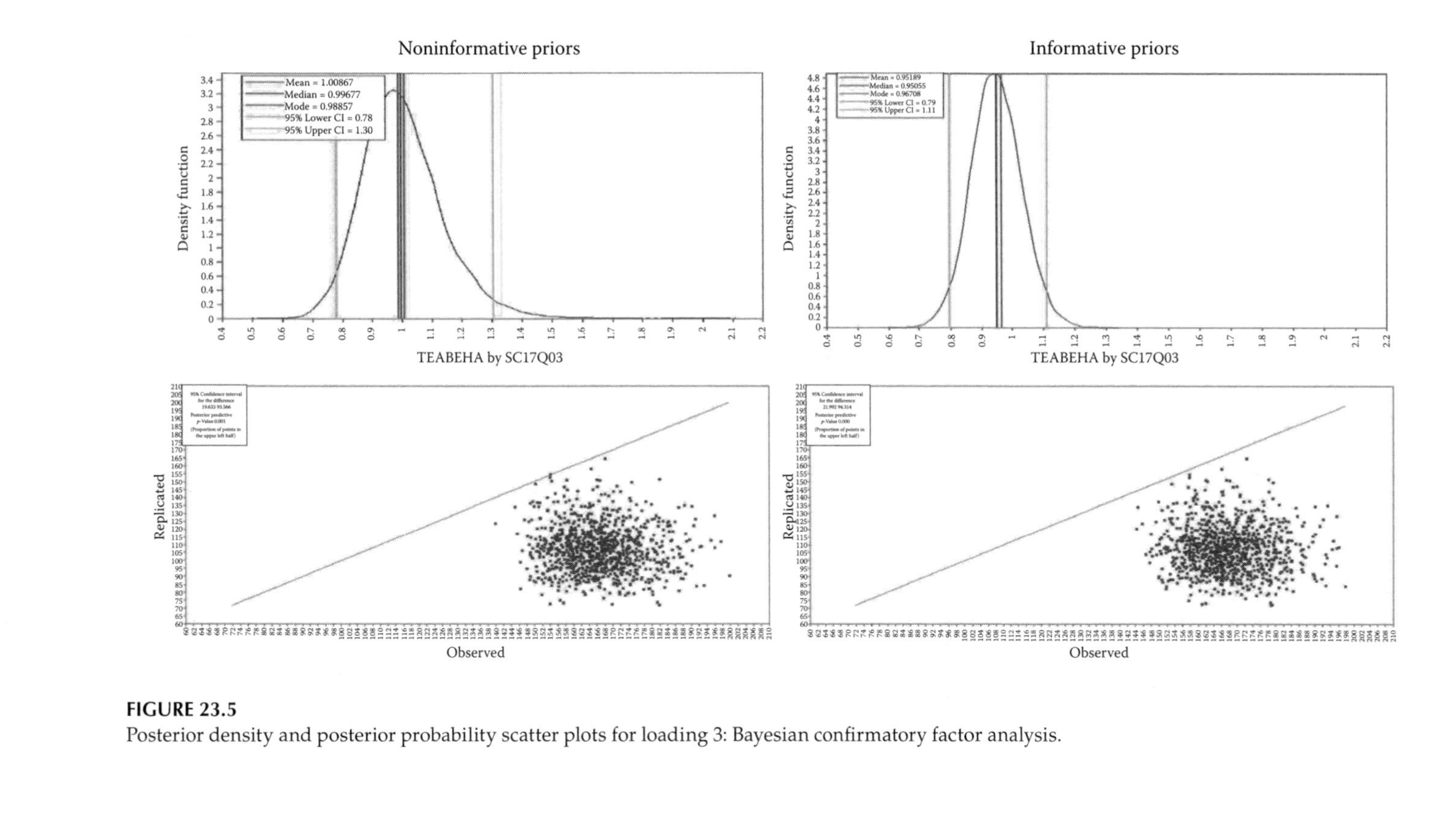

FIGURE 23.5
Posterior density and posterior probability scatter plots for loading 3: Bayesian confirmatory factor analysis.

difference between the noninformative case and informative case with respect to the shape of the posterior density.

An inspection of the lower panel of Figure 23.5 shows the PPC scatterplot for noninformative priors and informative priors cases. We see that virtually all of the likelihood ratio chi-square values fall below the 45-degree line, indicating poor model fit to the posterior replicated data. As in conventional confirmatory factor analysis, lack of model fit may be due to the restrictions placed on the factor loading matrix in line with the theory that there are two factors underlying these data. In the interest of space, we do not modify this model; however, see Muthén and Asparouhov (2012a) for an example of model modification in the Bayesian CFA context. Finally, the DIC values for the noninformative and informative priors cases for the CFA model are 3505.32 and 3503.50, slightly favoring the CFA model with informative priors.

Conclusion

The purpose of this chapter was to discuss and illustrate the Bayesian approach to the analysis of ILSA data. The chapter provided a brief overview of the elements of Bayesian inference along with an example illustrating the implementation of a multilevel model from a Bayesian perspective.

It is worth asking why one would choose to adopt the Bayesian framework for the analysis of ILSA data—particularly when, in large samples, it can often provide results that are very close to that of frequentist approaches such as maximum likelihood. The answer lies in the major distinction between the Bayesian approach and the frequentist approach; that is, in the elicitation, specification, and incorporation of prior distributions on the model parameters. It must be noted that despite the similarities in the results, the interpretations are completely different. First, from the Bayesian perspective, parameters are viewed as random and unknown, reflecting our uncertainty about unknown quantities, with probability serving as the language of uncertainty. This is in contrast to the frequentist approach, which views parameters as fixed and unknown. Second, the Bayesian perspective evaluates the quality of a substantive model in terms of posterior prediction, with competing models judged in terms of their support within the data. This is in contrast to conventional null hypothesis testing with its focus on assessing a hypothesis that is known *a priori* not to be true. Finally, the summary of the posterior distribution of the model parameters reflects our current or "updated" knowledge about the parameters of interest, and this updated knowledge should be incorporated in future studies in the form of new priors. No such notion of "updating" knowledge exists in the frequentist framework, and each analysis is treated as though nothing was learned from previous studies.

Clearly, then, the critical difference relates to reflecting uncertainty via the specification of the prior distribution.

A fair question to ask of the Bayesian approach centers on how priors should be obtained. What Bayesian theory forces us to recognize is that it is possible to bring in prior information on the distribution of model parameters, but that this requires a deeper understanding of the "elicitation problem" (O'Hagan et al. 2006; Abbas et al. 2008, 2010). In some cases, elicitation of prior knowledge can be obtained from experts and/or key stakeholders (however, see Muthén and Asparouhov [2012b] for a discussion of the dangers of using informative priors favored by a researcher). In the context of ILSAs, however, we have demonstrated how informative prior information can be gleaned directly from previous waves of the same ILSA—in our case, PISA 2000—and incorporated into a Bayesian model specification. Alternative elicitations from different cycles of the same ILSA and even different ILSAs can be directly compared via Bayesian model selection measures, such as use of the DIC or Bayes factors.

To summarize, we believe that conventional frequentist statistical modeling cannot exploit all that can be learned from ILSAs such as PISA. In contrast, we believe that Bayesian inference, with its focus on formally combining current data with previous research, can provide a methodological framework for the *evolutionary* development of knowledge about the inputs, processes, and outcomes of schooling. The practical benefits of the Bayesian approach for international educational research will be realized in terms of how it provides insights into important substantive problems.

Acknowledgments

The research reported in this chapter was supported by the Institute of Education Sciences, U.S. Department of Education, through Grant R305D110001 to the University of Wisconsin—Madison. The opinions expressed are those of the authors and do not necessarily represent views of the institute or the U.S. Department of Education.

References

Abbas, A. E., Budescu, D. V., and Gu, Y. 2010. Assessing joint distributions with isoprobability contours. *Management Science*, 56, 997–1011.

Abbas, A. E., Budescu, D. V., Yu, H.-T., and Haggerty, R. 2008. A comparison of two probability encoding methods: Fixed probability vs. fixed variable values. *Decision Analysis*, 5, 190–202.

Berger, J. 2006. The case for objective Bayesian analysis. *Bayesian Analysis*, 3, 385–402.
Brooks, S. P. and Gelman, A. 1998. General methods for monitoring convergence of iterative simulations. *Journal of Computational and Graphical Statistics*, 7, 434–455.
Burstein, L. 1980. The analysis of multilevel data in educational research and evaluation. *Review of Research in Education*, 8, 158–233.
de Finetti, B. 1974. *Theory of Probability*, vols. 1 and 2. New York: John Wiley and Sons.
Fox, J. 2008. *Applied Regression Analysis and Generalized Linear Models*, 2nd edition. Newbury Park , CA: Sage.
Gelman, A. 1996. Inference and monitoring convergence. In: W. R. Gilks, S. Richardson, and D. J. Spiegelhalter (Eds.), *Markov Chain Monte Carlo in Practice*. New York: Chapman & Hall, pp. 131–143.
Gelman, A., Carlin, J. B., Stern, H. S., and Rubin, D. B. 2003. *Bayesian Data Analysis*. 2nd edition. London: Chapman and Hall.
Gelman, A. and Rubin, D. B. 1992a. Inference from iterative simulation using multiple sequences. *Statistical Science*, 7, 457–511.
Gelman, A. and Rubin, D. B. 1992b. A single series from the Gibbs sampler provides a false sense of security. In: J. M. Bernardo, J. O. Berger, A. P. Dawid, and A. F. M. Smith (Eds.), *Bayesian Statistics 4*. Oxford: Oxford University Press, pp. 625–631.
Geman, S. and Geman, D. 1984. Stochastic relaxation, Gibbs distributions and the Bayesian restoration of images. *IEEE Transactions on Pattern Analysis and Machine Intelligence*, 6, 721–741.
Geweke, J. 1992. Evaluating the accuracy of sampling-based approaches to calculating posterior moments. In J. M. Bernardo, J. O. Berger, A. P. Dawid, and A. F. M. Smith (Eds.), *Bayesian Statistics 4*. Oxford: Oxford University Press.
Gilks, W. R., Richardson, S., and Spiegelhalter, D. J. (Eds.). 1996. *Markov Chain Monte Carlo in Practice*. London: Chapman and Hall.
Gill, J. 2002. *Bayesian Methods: A Social and Behavioral Sciences Approach*. London: Chapman and Hall/CRC.
Heidelberger, P. and Welch, P. 1983. Simulation run length control in the presence of an initial transient. *Operations Research*, 31, 1109–1144.
Howson, C. and Urbach, P. 2006. *Scientific Reasoning: The Bayesian Approach*. 3rd edition. Chicago: Open Court.
Jackman, S. 2009. *Bayesian Analysis for the Social Sciences*. New York: John Wiley.
Jöreskog, K. G. 1969. A general approach to confirmatory maximum likelihood factor analysis. *Psychometrika*, 34, 183–202.
Kaplan, D. 2009. *Structural Equation Modeling: Foundations and Extensions*. 2nd edition. Newbury Park , CA: Sage Publications.
Kaplan, D. and Depaoli, S. 2012a. Bayesian statistical methods. In T. D. Little (Ed.), *Oxford Handbook of Quantitative Methods*. Oxford: Oxford University Press.
Kaplan, D. and Depaoli, S. 2012b. Bayesian structural equation modeling. In: R. Hoyle (Ed.), *Handbook of Structural Equation Modeling*. Guilford Publishing, Inc, pp. 650–673.
Kass, R. E. and Raftery, A. E. 1995. Bayes factors. *Journal of the American Statistical Association*, 90, 773–795.
Kim, J.-S. and Bolt, D. M. 2007. Estimating item response theory models using Markov chain Monte Carlo methods. *Educational Measurement: Issues and Practice*, 26, 38–51.
Lee, S.-Y. 2007. *Structural Equation Modeling: A Bayesian Approach*. New York: Wiley.
Lynch, S. M. 2007. *Introduction to Applied Bayesian Statistics and Estimation for Social Scientists*. New York: Springer.

Martin, M. O., Mullis, I. V. S., and Foy, P. 2008. *TIMSS 2007 International Science Report: Findings from IEA's Trends in International Mathematics and Science Study at the Fourth and Eighth Grades*. Chestnut Hill, MA: Boston College.

Mullis, I. V. S., Martin, M. O., Kennedy, A. M., and Foy, P. 2008. *PIRLS 2006 International Science Report: IEA's Progress In International Reading Literacy Study in Primary Schools in 40 Countries*. Chestnut Hill, MA: Boston College.

Muthén, B. and Asparouhov, T. 2012a. Bayesian SEM: A more flexible representation of substantive theory. *Psychological Methods*, 17, 313–335.

Muthén, B. and Asparouhov, T. 2012b. Rejoinder: Mastering a new method. *Psychological Methods*, 17(3), 346–353. doi: 10.1037/a0029214.

Muthén, L. K. and Muthén, B. 2012. *Mplus: Statistical Analysis with Latent Variables*. Los Angeles: Muthén & Muthén.

OECD. 2003a. *Education at a Glance 2003: OECD Indicators*. Paris: OECD.

OECD. 2003b. *Literacy Skills for the World of Tomorrow: Further Results from PISA 2000*. Paris: Author.

OECD. 2009. *Education at a Glance 2009: OECD Indicators*. Paris: OECD.

OECD. 2010. *Results: What Makes a School Successful? Resources, Policies and Practices (Volume IV)*. Paris: Author.

OECD. 2012. *PISA 2009 Technical Report*. Paris: Author.

O'Hagan, A., Buck, C. E., Daneshkhah, A., Eiser, J. R., Garthwaite, P. H., Jenkinson, D. J., Oakley, J. E., and Rakow, T. 2006. *Uncertain Judgements: Eliciting Experts' Probabilities*. West Sussex, England: Wiley.

Press, S. J. 2003. *Subjective and Objective Bayesian Statistics: Principles, Models, and Applications*. 2nd edition. New York: Wiley.

Raftery, A. E. 1995. Bayesian model selection in social research (with discussion). In: P. V. Marsden (Ed.), *Sociological Methodology*. Vol. 25. New York: Blackwell, pp. 111–196.

Raftery, A. E. and Lewis, S. M. 1992. How many iterations in the Gibbs sampler? In: J. M. Bernardo, J. O. Berger, A. P. Dawid, and A. F. M. Smith (Eds.), *Bayesian Statistics 4*. Oxford: Oxford University Press, pp. 763–773.

Raudenbush, S. W. and Bryk, A. S. 2002. *Hierarchical Linear Models: Applications and Data Analysis Methods*. 2nd edition. Thousands Oaks, CA: Sage Publications.

Royall, R. 1997. *Statistical Evidence: A Likelihood Paradigm*. New York: Chapman and Hall.

Sinharay, S. 2004. Experiences with Markov chain Monte Carlo convergence assessment in two psychometric examples. *Journal of Educational and Behavioral Statistics*, 29, 461–488.

Spiegelhalter, D. J., Best, N. G., Carlin, B. P., and Linde, A. van der. 2002. Bayesian measures of model complexity and fit (with discussion). *Journal of the Royal Statistical Society, Series B (Statistical Methodology)*, 64, 583–639.

Wiseman, A. W. 2010. *The Impact of International Achievement Studies on National Education Policy Making*. Bingley, UK: Emerald, Publishing.

24

A General Psychometric Approach for Educational Survey Assessments: Flexible Statistical Models and Efficient Estimation Methods

Frank Rijmen
CTB/McGraw-Hill

Minjeong Jeon
University of California, Berkeley

Matthias von Davier
Educational Testing Service

Sophia Rabe-Hesketh
University of California, Berkeley

CONTENTS

General Background: Changes in the Assessment Frameworks and Designs

The National Assessment of Educational Progress (NAEP) has been "the nation's report card" for more than four decades. NAEP includes multiple survey assessments and special studies that yield a snapshot of what students in the United States know and can do (nces.ed.gov/nationsreportcard). On the international scene, similar purposes are served by large-scale student survey assessments such as the Program for International Student Assessment (PISA), the Trends in Mathematics and Science Study (TIMSS), and the Progress in International Reading Literacy Study (PIRLS). For adults, there are international assessments such as the International Adult Literacy Study (IALS) and the Programme for the International Assessment for Adult Competencies (PIAAC).

At the basis of each large-scale survey assessment is a *framework*. The framework provides the theoretical basis for what is measured in the assessment. It forms a blueprint that determines the content to be assessed and guides item development for the different content domains. Over the years, assessment frameworks have become more explicit about the assessment design and content. As an example, the current NAEP framework for reading was established for the 2009 assessment (Winick et al. 2009). It distinguishes between text types and processes of reading. There are two broad types of text: literary versus informational texts. Items are developed for each text type. At the same time, items are indicators of one of three reading processes: locate and recall, integrate and interpret, and critique and evaluate. Hence, a cross-classification exists for the items of the reading assessment: One classification is related to content (text types), and the other is related to cognitive processes. A similar cross-classification scheme is also in place for NAEP science and mathematics, and for the aforementioned international survey assessments.

Another evolution is that the advance of information technology is starting to affect the assessment *design*s in a profound way. The use of computers is not merely a change in the mode of administration (paper-and-pencil vs. computer-delivered). Information technology allows for the development and delivery of interactive and integrative tasks. Such tasks present the student with a complex simulated environment that potentially taps into multiple skills at the same time. The features of the environment can change in response to specific student actions. Solving the task typically involves multiple actions, and there can be more than one way of solving it.

A direct consequence of the use of complex integrated tasks for the assessment design is that item clustering is likely to become more prominent. This is because the stimulus material will be rich enough to allow for the assessment of multiple aspects of student performance within each task. Each of these aspects can be thought of as a separate item that is clustered within an integrated task. Also, integrated tasks tend to be more complex, and, thus, more time consuming, so the number of integrated tasks that can be administered to a single student will be limited. Note that item clustering is already present in the current assessments. For example, in NAEP, PISA, and PIRLS reading assessments, students typically have to respond to several items upon reading a text passage. However, item cluster effects are expected to be more pronounced in integrated tasks because each of these tasks is specifically designed to be embedded within a realistic scenario.

Current Psychometric Approach

The psychometric procedures currently in operation for large-scale survey assessments adhere to a divide-and-conquer approach (von Davier et al. 2006; see also Chapter 7 in this volume). This approach divides the analysis into three stages. In the first stage, item parameters are estimated assuming that a unidimensional item response theory model holds for each content domain. A separate model is estimated for each content domain. For example, for NAEP reading, two unidimensional models are estimated: one for items pertaining to informational text types, and one for items pertaining to literary text types. No background variables are included at this stage. Next, a multidimensional item response theory model is estimated that includes person background variables. In this model, the item parameters are considered to be known and equal to the item parameters estimated in the first stage. Each of the ability dimensions is modeled as a function of background variables through a multivariate (latent) linear regression model. The item response model is a between-item multidimensional model: Every item is allowed to load on a single dimension only. Third, a set of plausible ability values is drawn from the estimated posterior distribution of each student derived from the model estimated in the second stage. These plausible values are used to obtain statistics on subgroups of interest, for example, differences in average ability between girls and boys.

The dimensions of the multidimensional model represent different constructs across and within assessments. In NAEP, a separate multidimensional model is estimated for reading, science, and mathematics. The dimensions of the models for each subject represent content domains, for example literary and informational reading. In PISA, several multidimensional models are fit. For the 2006 cycle, two five-dimensional models were estimated. In the first model, the dimensions corresponded to reading, science, mathematics, and two attitudinal scales. In the second model, there was one dimension for reading, one for mathematics, and one dimension for each of the three identified science competencies (identifying scientific issues, explaining scientific phenomena, and using scientific evidence). The psychometric approach as followed in PISA seems to be driven primarily by technical limitations of the current operational procedures. It is hard to justify theoretically that different sets of plausible values are generated from models that incorporate conflicting assumptions about the dimensionality of, in the case of the 2006 cycle, science. Conceptually, it would make more sense to fit a higher-order model in which the competency dimensions are nested within an overall science dimension, and generate all plausible values from the same model.

The effects of item clustering are not taken into account by the psychometric models that are currently in use for large-scale assessments. However, the use of item clusters may result in conditional dependencies between items belonging to the same cluster. Not taking into account these sources of dependencies may result in biased estimates of parameters and, to an even larger degree, standard errors (Bradlow et al. 1999). One way of taking item clusters into account is to incorporate an additional dimension for each item cluster (Bradlow et al. 1999). However, there can be a substantial number of item clusters in any given assessment. For example, in NAEP reading, item clusters correspond to reading passages and associated items. Incorporating a separate dimension for each item cluster would result in a model that far exceeds the computational capabilities of the estimation methods currently in use.

The psychometric approach that will be outlined in the next section overcomes in many cases the technical challenges that are associated with high-dimensional models. For those cases, the computational load can be dramatically reduced by exploiting the confirmatory multidimensional structure. This will be explained in more detail in the section on graphical models. For now, it suffices to say that the new approach can accommodate both a higher-order model in which competencies are nested within subject areas, and a model that includes dimensions for item clustering.

Finally, an aspect not mentioned so far but one that is consequential for the psychometric analysis is that the students who participate in NAEP or other large-scale educational survey assessment studies are sampled according to a complex multistage sampling design (nces.ed.gov/nationsreportcard/tdw/sample_design/). As a result, not all students of the population have the same probability of being included in the study. The current psychometric model takes the unequal sampling probabilities into account through the inclusion of sampling weights. However, the use of sampling weights does not take into account the fact that the responses of students belonging to the same cluster (classroom, school, etc.) can be expected to be more similar than the responses of students pertaining to different clusters. Especially in the first stage, when item parameters are estimated, failing to take into account the full dependence structure present in the data may lead to biased estimates of the item parameters and their corresponding standard errors (Molenberghs and Verbeke 2005). To the extent that within-cluster dependencies can be explained by the background variables included in the study, this may be less of a concern in the second and third stages of the analysis. The proposed psychometric approach can also take the effects of the sampling design into account.

Proposed Psychometric Approach

Our aim is to develop psychometric models that reflect assessment designs and data collection designs of survey assessments better than current methods do. This way, the results of the psychometric modeling effort will provide information that more accurately reflects each of those aspects. For example, by incorporating dimensions corresponding to cognitive processes as well as content domains in a comprehensive model, information will be available as to whether and to what degree the processes and/or the content domains constitute distinguishable sources of individual differences.

Compared to the current psychometric models in operational use, the proposed approach will take into account more potential sources of dependencies between items through a richer dimensional model structure. Additional sources of dependencies between persons can be taken into account through the incorporation of a multilevel model structure. Therefore, it can be expected that the resulting estimates will more accurately reflect differences between policy-relevant subpopulations.

The proposed approach borrows its flexibility from its specification within a generalized linear and nonlinear mixed-model framework. The computational burden that has hitherto hampered the use of multidimensional item response models can be overcome in many cases through the use of graphical models. Our unique contribution lies in bringing together a graphical model framework and a generalized (and nonlinear) mixed-model framework. An additional advantage of our proposed approach is concurrent estimation of all model parameters, as opposed to the divide-and-conquer approach currently in use.

Generalized Linear and Nonlinear Mixed-Model Framework

It is generally acknowledged that item response theory models are closely related to item factor analysis models. In turn, both can be formulated as generalized linear and nonlinear mixed models (Rijmen et al. 2003; Skrondal and Rabe-Hesketh 2004; von Davier 2005). Ability dimensions in item response theory models are conceptualized as random effects in the mixed-model framework, and the responses to items correspond to repeated measurements of the same individual. Random effects are unobserved or latent variables that correspond to sources of individual differences. They account for the dependencies that are typically observed among responses clustered within the same person.

The advantages of working within this overarching framework are substantial. First, the common framework helps to understand the commonalities and differences between various statistical models that have been developed in different research areas such as educational measurement, econometrics, and biostatistics. Well-known models such as item response theory and discrete latent structure models developed under different theoretical frameworks are indeed special cases of generalized linear and nonlinear mixed models. Second, models can be extended—at least conceptually—in a straightforward way. For example, item response theory models can be extended by including random effects at multiple levels when persons are nested in classrooms nested in schools (e.g., Raudenbush and Sampson 1999; Fox and Glas 2001; Rabe-Hesketh et al., 2004). Third, theoretical and empirical findings can be more easily communicated through the use of a common terminology. Fourth, general-purpose software becomes available that allows for the estimation of a broad range of models. This avoids the need to get acquainted with a different specialized software package for every different model (or the need to program an estimation routine oneself). Also, it makes it more straightforward to compare the performance of different models because they are all estimated within the same software environment.

Item parameters and the regression parameters of both item characteristics and person background variables are conceptualized as fixed parameters. Because item responses are categorical, linear mixed models do not apply. Instead, item response probabilities are related to a predictor through a link function. For models of the one-parameter family, the predictor is linear in the fixed and random effects, and hence those models are generalized linear mixed models. For models of the two-parameter family, the predictor is no longer linear in the fixed and random effects, and therefore models are nonlinear mixed models. The generalized linear and nonlinear mixed-model framework is already increasingly used as a modeling framework for NAEP and other large-scale assessments (Li et al. 2009; Xu and von Davier 2006). The results reported in this chapter are further continuations and extensions of this line of research. However, the high dimensionality of the models that we are proposing poses technical challenges for the currently available methods within the mixed-model framework.

Model Specification

Let y_{ij} denote the categorical response variable of person i on the jth item, $i = 1,\ldots, I$; $j = 1,\ldots, J$. The vector of all responses of person i is denoted by $\mathbf{y}_i$. Conditional on a D-dimensional vector of latent variables $\boldsymbol{\theta}_i = (\theta_{i1},\ldots,\theta_{id},\ldots,\theta_{iD})'$, the responses are assumed to be statistically independent

$$\Pr(\mathbf{y}_i|\boldsymbol{\theta}_i) = \prod_{j=1}^{J} \Pr(y_{ij}|\boldsymbol{\theta}_i) \tag{24.1}$$

The latent variables $\boldsymbol{\theta}_i$ are treated as random effects, with a distribution $p(\boldsymbol{\theta}_i; \xi)$ over persons that is characterized by the parameter vector ξ. The latent variables can be either continuous and normally distributed variables, or discrete variables with locations and weights to be estimated from the data.

Furthermore, $\pi_{ij} = \Pr(y_{ij} = 1|\boldsymbol{\theta}_i)$ is related to a linear function of the latent variables through a link function $g(\cdot)$

$$g(\pi_{ij}) = \boldsymbol{\alpha}_j'\boldsymbol{\theta}_i + \beta_j \tag{24.2}$$

where $g(\cdot)$ is typically the probit or logit link function for binary data, and a generalization thereof for polytomous data.

The parameter β_j is the intercept parameter for item j. It can be further structured as a weighted sum of item properties (De Boeck and Wilson 2004; Fischer 1973; Rijmen et al. 2003):

$$\beta_j = \mathbf{q}_j'\boldsymbol{\delta}, \tag{24.3}$$

where $\mathbf{q}_j$ is a P-dimensional vector of item properties and δ the corresponding vector of fixed parameters. Item properties are denoted by the letter "q" since the design matrix (q_{jp}) is often called the Q-matrix in the psychometric literature. For the model to be identified, Q has to be of full column rank (and hence $P \leq J$). If Q is the identity matrix, no constraints are imposed on the item intercept parameters β_j.

As $\boldsymbol{\alpha}_j$ is the vector of item loadings, each element a_{jd} indicates to what degree item j is associated with latent variable θ_{id}. In the "two-parameter" family of models, the elements of $\boldsymbol{\alpha}_j$ are parameters that are estimated from the data. When the elements of $\boldsymbol{\alpha}_j$ are assumed to be known, a "one-parameter" model is obtained. Additionally, an item-guessing parameter can also be incorporated into the expressions for the π_{ij}s, resulting in a "three-parameter" model. The two-parameter model is equivalent to item factor analysis.

In this chapter, we focus on confirmatory models. In confirmatory models, only some of the elements of $\boldsymbol{\alpha}_j$ are estimated for each item j, and the other elements are set to zero. This way, the assessment framework and design can be translated into the dimensional structure of the model. For example, the effects of item clustering can be taken into account through the specification of a bifactor model (Gibbons and Hedeker 1992). In the bifactor model, every item has only two nonzero loadings: Every item loads on a general dimension and on a dimension that is specific to the item cluster to which the item belongs. Equation 24.2 then becomes

$$g(\pi_{ij}) = \alpha_{jg}\theta_{ig} + \alpha_{jk[j]}\theta_{ik[j]} + \beta_j, \tag{24.4}$$

where the subscript g refers to the general dimension and $k = 1,\ldots, K$ refers to each of the specific dimensions (hence, $D = K + 1$). The notation $k[j]$ is used to indicate explicitly that specific dimension k has item j nested within it. Sometimes we will omit the square brackets to simplify the notation. Furthermore, all of the dimensions are assumed to be independent. To identify the model, the location and scale of all dimensions have to be fixed. More complex confirmatory models can be specified to take into account the assessment frameworks and designs. This will be illustrated in the real data application of this chapter.

Furthermore, each of the latent variables can be modeled as a linear function of person characteristics through a so-called latent regression model (Adams et al. 1997; Mislevy 1987; Mislevy et al. 1992; Zwinderman 1991):

$$\theta_{id} = \mathbf{x}'_{id}\boldsymbol{\lambda}_d + \zeta_{id}, \tag{24.5}$$

where $\mathbf{x}_{id}$ is a vector of person characteristics, $\boldsymbol{\lambda}_d$ a vector of fixed latent regression parameters, and ζ_{id} the residual for person i on dimension d. Because $\boldsymbol{\theta}_i$ is considered to be a vector of random effects with a distribution defined over persons, so is ζ_i.

The clustering effects at the person side (of persons within classrooms, classrooms within schools, etc.) can be taken into account through the addition of a multilevel component in the latent regression part of the model (Rabe-Hesketh et al. 2004; Li et al. 2009). For example, when persons are clustered within schools, an intercept can be included in Equation 24.5 that is random over schools:

$$\theta_{id} = \gamma_{ds[i]} + \mathbf{x}'_{id}\boldsymbol{\lambda}_d + \zeta_{id}. \tag{24.6}$$

The subscript $s[i]$ indicates that person i is clustered within school s, $s = 1,\ldots,S$. γ_{ds} is the random intercept for school s, and $\boldsymbol{\gamma}_s = (\gamma_{1s},\ldots,\gamma_{ds},\ldots,\gamma_{Ds})'$ follows a distribution $p(\boldsymbol{\gamma}_s;\boldsymbol{\psi})$ that is defined over schools and is characterized by the parameter vector $\boldsymbol{\psi}$. The distribution of the random person effects is now a defined conditional on the random effect for school s, $\boldsymbol{\theta}_i \sim p\left(\boldsymbol{\theta}_i \middle| \boldsymbol{\gamma}_{s[i]};\boldsymbol{\xi}\right)$.

Technical Challenges

A High-Dimensional Latent Space

Maximum likelihood estimation of model parameters in nonlinear mixed models involves integration over the space of all random effects. In general, the integrals have no closed-form solution. Numerical integration over the joint space of all latent variables becomes computationally very demanding as the number of dimensions grows (e.g., von Davier and Sinharay 2007). Specifically, when the integral over the random effects distribution is evaluated using Gaussian quadrature (e.g., Bock and Aitkin 1981), the number of calculations involved increases exponentially with the number of random effects. Even though the number of quadrature points per dimension can be reduced when using adaptive Gaussian quadrature (Pinheiro and Bates 1995; Rabe-Hesketh et al. 2005; Schilling and Bock 2005), the total number of points again increases exponentially with the number of dimensions. Furthermore, adaptive Gaussian quadrature involves the computation of the mode of the posterior of the random effects and the Hessian of the log-posterior at the mode for each response pattern. The use of Monte Carlo techniques (e.g., stochastic EM, the Gibbs sampler in a Bayesian framework) has increased the number of dimensions that can be incorporated into a model, but for the high-dimensional models that are proposed, these techniques remain very computationally intensive (von Davier and Sinharay 2007, 2010).

As an alternative, so-called limited-information techniques have been developed in the field of structural equation modeling to deal with ordered categorical observed (indicator) variables (Jöreskog 1994; Muthén 1984). Unlike maximum likelihood estimation methods, the limited-information techniques do not take into account the complete joint contingency table of all items, but only marginal tables up to the fourth order (Mislevy 1985). In this way, parameter estimation can be carried out using weighted least squares estimation and is reasonably fast, even for high-dimensional models. However, the number of elements in the optimal weight matrix, which has to be invertible, grows with the fourth power of the number of items (Mislevy 1985), so that the sample size needed to estimate an item response theory model with many items, which is the typical situation in educational measurement, becomes prohibitive in many practical applications. Alternatively, Muthén et al. (1997) proposed a robust weighted least squares approach where the optimal weight matrix is replaced by a diagonal matrix, having as elements the diagonal elements of the optimal weight matrix. By relying on marginal item frequency tables, limited-information techniques are not suited for incomplete data collection designs. Since all large-scale survey assessments employ an incomplete data collection design (each participant receives a booklet containing a subset of items), limited-information techniques are of limited value in this context.

Item response theory models that do not incorporate item discrimination parameters, such as the Rasch (1960) or the partial credit model (Masters 1982), can be formulated as generalized linear mixed models (Rijmen et al. 2003). For these models, quasi-likelihood methods (Breslow and Clayton 1993) have been developed as relatively fast alternatives to MML estimation methods. Research showing that the early versions of these methods resulted in inconsistent estimates has led to attempts to improve upon these methods (Goldstein and Rasbash 1996; Lin and Breslow 1996; Raudenbush et al. 2000).

Notwithstanding the widespread use of limited-information and quasi-likelihood estimation techniques, and ongoing efforts for further improvements in these methods, one can safely assume that many researchers would prefer or at least consider using maximum likelihood estimation methods if their computational burden could be overcome.

Traditionally, the number of latent variables in a model has been considered to be the key determinant of the computational cost of maximum likelihood parameter estimation. However, depending on the conditional dependency relations that are assumed under the model, the actual computational cost can be far lower by exploiting these conditional relations during model estimation. In particular, the set of conditional independence relations implied by a model can be used to partition the joint space of all latent variables into smaller subsets that are conditionally independent (Rijmen et al. 2008). As a consequence, brute force numerical integration over the joint latent space can be replaced by a sequence of integrations over smaller subsets of latent variables. In the context of (Bayesian) Monte Carlo techniques, sampling schemes can be constructed in an analogous way, which will be more efficient than their naïve counterparts (Chib 1996; Scott 2002). The gain in efficiency may be dramatic in some cases.

Though possible in principle, the algebraic manipulations of the likelihood of a specific model that are involved in turning the conditional independence relations into efficient estimation schemes may get quite complicated, and will be different from model to model. This may explain why, with a few notable exceptions (Cai 2010; Gibbons and Hedeker 1992; Gibbons et al. 2007; Vermunt 2003), brute force numerical integration (or sampling) has remained the norm. Fortunately, graphical model theory offers a general procedure for exploiting conditional independence relations during parameter estimation that can be applied in a general way.

Use of Graphical Models

The use of graphical models is central to our approach. Graphical models have roots in statistical physics and genetics (Lauritzen 1996), and are being used in diverse fields such as image analysis (Besag and Green 1993), speech recognition (Rabiner 1989), and artificial intelligence (Jensen 1996). A common underlying realization for the use of graphical models in these areas of research is that brute force calculations become technically too cumbersome for high-dimensional statistical models. Hence, the need to break down complex models into smaller manageable pieces. Graphs turn out to be extremely useful in such a modular approach (Cowell et al. 1999).

A first step is to represent the statistical model in a directed acyclic graph in which the nodes correspond to random variables, and the directed edges represent conditional dependence relations. Graphs have been used extensively in the literature to visualize statistical models. They offer a convenient way of representing and communicating the structure of a statistical model. As an illustration, the directed acyclic graph for the bifactor model is presented in Figure 24.1 for the case of four-item clusters. Note that each $\mathbf{y}_{ik}$ represents a vector of item responses pertaining to the same item cluster k.

The reason that directed acyclic graphs form a convenient way of representing a statistical model is that the joint probability function of all (latent and observed) variables always can be factorized into a set of conditional probability functions according to the directed acyclic graph. For the bifactor model, the joint density function for the observed responses and latent variables can be factorized as

$$p(\mathbf{y}_i,\boldsymbol{\theta}_i) = p(\theta_{ig})\prod_{k=1}^{K}\Pr(\mathbf{y}_{ik}|\theta_{ig},\theta_{ik})p(\theta_{ik}). \tag{24.7}$$

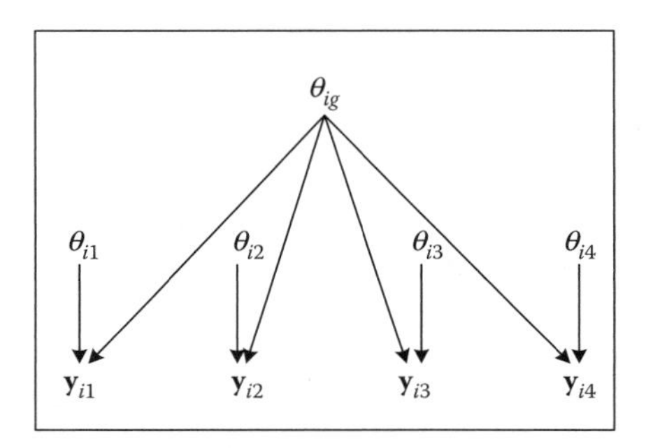

FIGURE 24.1
Directed acyclic graph of a bifactor model.

In the behavioral and social sciences, graphs are commonly used for model visualization purposes. In particular, most researchers using structural equation and path models routinely rely on graphs to represent statistical models. However, it was largely overlooked how graphical models can be used to construct efficient maximum likelihood estimation methods. This will be illustrated in the following for the bifactor model.

Maximum likelihood estimation involves the computation of the score function, which is obtained by integrating the gradient of the logarithm of the expression in Equation 24.7 over the posterior distribution of the latent variables $p(\boldsymbol{\theta}_i|\mathbf{y}_i)$. The efficiency of an estimation method depends on whether this posterior distribution can be factorized into subsets of smaller dimensionality, avoiding numerical integration over the joint (posterior) latent space of the model.

The core of the construction of efficient computational schemes relies on the transformation of the directed acyclic graph into a triangulated moral graph, and the subsequent construction of a junction tree. In a junction tree, the nodes correspond to subsets of variables of the statistical model. A result of utmost importance is that those subsets of variables are conditionally independent of each other. Therefore, the junction tree can be used to partition the high-dimensional space of all latent variables into subsets of lower dimensionality, and numerical integration over the joint (posterior) latent space can be carried out through a sequence of computations in these lower-dimensional subspaces. Consequently, the number of function evaluations involved in maximum likelihood estimation is not determined by the number of latent variables per se, but by the dimensionality of the latent spaces of the subsets of variables that are conditionally independent. Because one can rely on algorithms defined on a graphical representation of the statistical model, the sets of conditionally independent variables can be obtained in an automatic way and for a whole family of statistical models. There is no need for tedious algebraic manipulations of the likelihood for each specific model.

As an illustration for the use of graphical models to obtain efficient maximum likelihood estimation methods, we shortly discuss the case of the bifactor model (see also Rijmen 2009). A first step is to transform the directed acyclic graph into an undirected, triangulated, moral graph. The moral graph for the bifactor model is given in Figure 24.2. The moral graph is obtained by adding an undirected edge between nodes that have a common "child" (e.g., θ_{ig} and θ_{i1} have $\mathbf{y}_{i1}$ as a common child, see Figure 24.1), and by making all directed edges undirected. Note that in Figure 24.2 the latent variables θ_{ik} are replaced by the latent variables z_{ik}. z_{ik} is a discrete approximation to θ_{ik} in case the latter is a continuous variable. Otherwise, $z_{ik} = \theta_{ik}$. The reason is that although graphical models have been proposed for mixed sets of both continuous and discrete variables, some of the useful results break down when continuous parents (θ_{ik}) have discrete children ($\mathbf{y}_{ik}$). Therefore, continuous θ_{ik} are approximated by discrete variables z_{ik}.

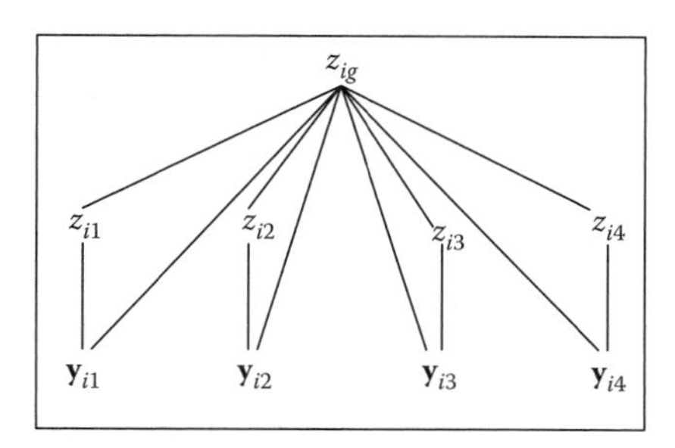

FIGURE 24.2
Moral graph for the bifactor model.

As a matter of fact, this is tantamount to what is done when evaluating the integral over $\boldsymbol{\theta}_i$ using numerical integration techniques.

Next, a junction tree is formed. In the junction tree, nodes correspond to cliques, which are the maximal subsets of nodes that are all interconnected in the moral graph. A junction tree possesses the running intersection property: Any variable that is in the intersection of two nodes of the junction tree also appears in any other node of the junction tree on the path between those two nodes. A junction tree for the bifactor model is given in Figure 24.3.

Since the joint posterior distribution $p(\boldsymbol{\theta}_i|\mathbf{y}_i)$ can be factorized according to the junction tree, and the number of latent variables in any clique is two (see Figure 24.3), maximum likelihood estimation can be carried out through a sequence of computations in two-dimensional spaces. The importance of this result can hardly be overstated. Full-information, maximum likelihood estimation for multidimensional models has been hampered because the number of computations involved in naïve numerical integration methods over the joint latent space increases exponentially with the number of dimensions. Using a graphical model framework, the very same maximum likelihood estimates can be obtained with a method whose complexity is only linear in the number of dimensions. This brings a whole family of models within reach that hitherto were deemed to be too complex to be fitted within a realistic time period. As an example, let us consider a bifactor model for the 2009 NAEP reading assessment for Grade 8. The model incorporates one general dimension, and one specific dimension for each of the 13 "new regular" item clusters. Therefore, the number of computations using naïve numerical integrations methods is of the order q^{14}, with q the number of quadrature points used for any given dimension. Even with q as low as

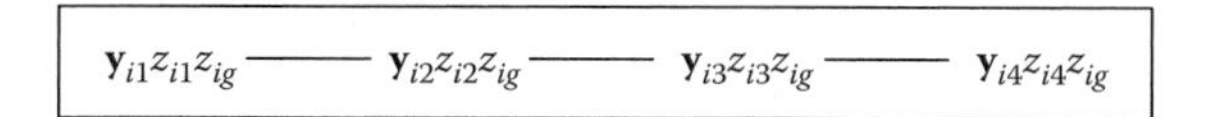

FIGURE 24.3
Junction tree for the bifactor model.

five, the total number of quadrature points exceeds one billion. Obviously, naïve numerical integration is not feasible in reasonable time. In contrast, the number of computations involved using a graphical model approach for the same bifactor model is of the order $2 \times 13 \times q^2$, which equals 650 when $q = 5$, and 10,400 when $q = 20$. Even in the latter case, this number poses no computational difficulties.

A major advantage of the graphical modeling framework is its generality. It can be used to determine the computational complexity of any model that can be represented in a directed acyclic or undirected graph. Through the transformation of the graph into a junction tree, efficient estimation algorithms can be constructed in an automatic way by exploiting the conditional independence relations implied by the model.

Efficient parameter estimation through maximum likelihood will not be attainable for every model that can be defined through Equations 24.1 through 24.6. For example, a multidimensional model in which all the correlations between dimensions are free parameters will not lend itself to a reduction of the computational load, because all latent variables will end up in the same clique. However, the use of graphical models will enable us to determine whether or not a reduction of the computational load is possible, and, if so, what the reduction looks like. As said before, we do not depend on our ability (or perseverance) to obtain efficient estimation methods through algebraic manipulations of the likelihood, and we do not have to implement these methods separately for each model.

Application

In the remainder of this chapter, we will apply our proposed psychometric framework to the 2006 PIRLS assessment. We will present the results of a variety of multidimensional item response theory models to showcase the versatility of the approach. We focus on models that have a conditional independence structure that allows for efficient maximum likelihood parameter estimation.

PIRLS is an internationally comparative reading assessment that has been carried out every 5 years since its inception in 2001. In 2006, 40 countries participated, with a total sample size of 215,137. A balanced incomplete booklet design was used, where each booklet consisted of two item blocks. The 2006 PIRLS assessment contained 10 item blocks. In every item block, a text material was presented, followed by a set of questions. The number of items within an item block varied from 11 to 14 (Martin et al. 2007), with a total of 126 items. Both constructed-response and multiple-choice items were included. All multiple-choice items were binary items, whereas some of the constructed-response items were partial credit items. In all analyses reported below, the logit link function was used for binary items, $g(\pi_{ij}) = \log(\pi_{ij}/(1-\pi_{ij}))$. For polytomous items with $C_j + 1$ response categories, cumulative link functions were used, $g(\pi_{ij}^{c+}) = \log(\pi_{ij}^{c+}/(1-\pi_{ij}^{c+}))$, with c denoting the response category, and $\pi_{ij}^{c+} = P(y_{ijc} > c|\theta_g, \theta_k)$ for $c = 0, \ldots, C_j - 1$.

Two overarching purposes of reading are assessed in PIRLS: reading to acquire and use information, and reading for literary experience (Mullis et al. 2006). Each of both purposes is assessed through a set of questions that are clustered within text materials. In addition, four processes of comprehension are distinguished (Mullis et al. 2006): focus on and retrieve explicitly stated information; make straightforward inferences; interpret and integrate ideas and information; and examine and evaluate content, language, and textual elements. Within each item block accompanying a text material, each of the four comprehension processes is assessed by one or more items. Taken together, items are both clustered within reading purposes and within comprehension processes. Reading purposes are crossed with comprehension processes. Furthermore, item blocks corresponding to test materials are nested within reading purposes, and crossed with comprehension processes. Finally, it should be noted that the cross-classification of items is not completely balanced: the proportions of items measuring the four comprehension processes is not constant across item blocks or reading purposes.

Like other large-scale assessments, participants in PIRLS are sampled according to a complex two-stage clustered sampling design. The sampling design calls for the use of sampling weights during model estimation, as recently discussed by Rutkowski et al. (2010). Within a country, sampling weights are computed as the inverse of the selection probability, and their sum approximates the size of the population (Foy and Kennedy 2008). When combining data from several countries, using these "total" weights would lead to results that are heavily influenced by the large countries. Therefore, in the following analyses, the so-called senate weights were used. The senate weights are a renormalization of the total weights within each country so that they add up to the same constant (often 500 is chosen) for each country, giving equal weight to each country in the analyses (Rutkowski et al. 2010).

In the following, we present a set of multidimensional item response theory models that do take into account the effects of item clustering within item blocks and reading purposes on the one hand, and within comprehension processes on the other hand. The models should be considered as first steps toward the fully integrated approach sketched in the first part of this chapter. For example, they do not yet take into account person covariates or the multilevel structure on the person side. Nevertheless, the presented models already exhibit a richer dimensional structure than the models currently in operational use for PIRLS, and all model parameters are estimated concurrently.

In a first stage, models were fit that incorporated separate dimensions for either comprehension processes or item blocks in addition to a dimension for overall reading ability. Both bifactor and second-order models were formulated. The bifactor model was already discussed above (Equation 24.4). The second-order multidimensional item response theory model that takes into account the effects of item clusters incorporates a specific dimension for each of them, just like the bifactor model. It also contains a general dimension, but unlike the bifactor model, items do not directly depend on this general dimension. Rather, items only directly depend on their respective specific dimensions, which in turn depend on the general dimension. It is assumed that the specific dimensions are conditionally independent. That is, all associations between the specific dimensions are assumed to be taken into account by the general dimension. The directed acyclic graph for the second-order model is displayed in Figure 24.4.

The model equations for the second-order model defined for binary data look as follows:

$$g(\pi_{ij}) = \alpha_{jk[j]}\theta_{ik[j]} + \beta_j, \tag{24.8}$$

$$\theta_{ik} = \alpha_{kg}\theta_{ig} + \xi_{ik}, \tag{24.9}$$

where α_{kg} indicates to what extent the specific dimension θ_{ik} is explained by the general dimension θ_{ig}, and ξ_{ik} is the part of θ_{ik} that is unique. Because it is assumed that all the dependencies between the specific dimensions are accounted for through the general dimension, all ξ_{ik} are assumed to be statistically independent from each other and from θ_{ig}. Combining Equations 24.8 and 24.9 yields

$$g(\pi_{ij}) = \alpha_{jk[j]}\alpha_{kg}\theta_{ig} + \alpha_{jk[j]}\xi_{ik[j]} + \beta_j \tag{24.10}$$

From Equation 24.10, it is easily verified that model identification requires that the location and scale of all the unique dimensions and of the general dimension have to be fixed (see Rijmen 2010). Assuming a normal distribution

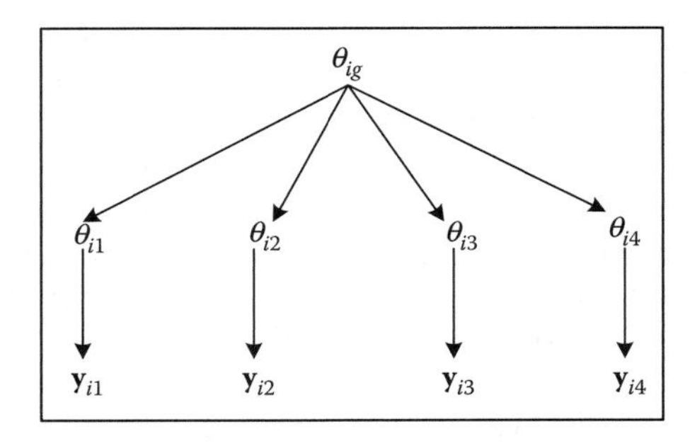

FIGURE 24.4
Directed acyclic graph of a second-order model.

for the dimensions, as we will assume for all fitted models in this section, the location and scales can be fixed by assuming univariate standard normal distribution for the latent variables θ_{ig}, $\xi_{i1},\ldots,\xi_{iK}$.

Comparing Equation 24.10 to Equation 24.4, it follows that the second-order model is a restricted bifactor model, where within each item cluster, the loadings on the specific dimensions are proportional to the loadings on the general dimension. In general, it has been shown that a higher-order model can always be reformulated as a hierarchical model with proportionality constraints on the loadings (Yung et al. 1999). The second-order model for discrete observed variables was introduced in the literature on item response theory under the name of the testlet model (Bradlow et al. 1999; Wainer et al. 2007). The fact that these authors used a slightly different notation may have contributed to the fact that the formal equivalence between the testlet model and a second-order model has been largely ignored.

The top six rows of Table 24.1 present the number of parameters, the number of dimensions, the deviance, and the Akaike information criterion (AIC; defined as the deviance plus twice the number of parameters; Akaike 1973) for the unidimensional model, the bifactor models, and the second-order models. The unidimensional model was estimated with both 10 and 20 quadrature points, whereas 10 quadrature points were used for all multidimensional models. The individual contributions of the tested students to the log likelihood were weighted by their sampling weights in computing

TABLE 24.1

Number of Parameters, Number of Dimensions, Deviance, and Aikaike Information Criterion for the Estimated Models

	#Par	#Dim	Deviance	AIC
2PL_10	290	1	706,956	707,386
2PL_20	290	1	706,431	707,011
BF_IB	415	11	704,925	**705,755**
2O_IB	300	11	705,603	706,203
BF_CP	415	5	705,965	706,795
2O_CP	294	5	706,743	707,331
2D2PL	291	2	706,161	706,743
2DBF_IB	416	12	704,737	**705,569**

Note: #Par, number of parameters; #Dim, number of dimensions; AIC, Akaike information criterion; 2PL_10, two-parameter logistic model with 10 quadrature points; 2PL_20, two-parameter logistic model with 20 quadrature points; BF_IB, bifactor model with item blocks as specific dimensions; 2O_IB, second-order model with item blocks as specific dimensions; BF_CP, bifactor model with comprehension processes as specific dimensions; 2O_CP, second-order model with comprehension processes as specific dimensions; 2D2PL, between-item two-dimensional two parameter logistic model with reading purposes as dimensions; 2DBF_IB, bifactor model with item blocks as specific dimensions and two general dimensions representing reading purposes. Models in bold are the best fitting of those considered.

the deviance and AIC. According to the AIC, the bifactor model with specific factor corresponding to item blocks was the preferred model. A closer inspection of the item parameter estimates may reveal in what sense this model provided a better fit to the data.

In Figures 24.5(1) through 24.5(10), the item loadings on the specific dimensions are plotted against the item loadings on the general dimension, separately for each item block. It can be seen that the loadings on the specific dimensions are smaller than the loadings on the general dimension, but many of them are still substantially different from zero. This explains why the bifactor model provides a better fit than the unidimensional model. The results vary to some degree across item blocks. For the item block labeled "Antarctica" (Figure 24.5(6)), all loadings on the specific dimensions are close to zero, except one item that has an outlying estimated value of 3.04. For four item blocks (see Figures 24.5(3), 24.5(6), 24.5(9), and 24.5(10)), the loadings on the specific dimensions are negative for some items, and positive for other items, indicating both negative and positive conditional dependencies given the general dimension. For the six other item blocks, all loadings on the specific dimension are larger than zero, indicating that all conditional dependencies are positive for the items in these item blocks.

Second, it can be seen in Figures 24.5(1) through 24.5(10) that for most item blocks the loadings on the specific dimension are not proportional to the loadings of the general dimension. Were this to be the case, the dots within each figure would (approximately) form a straight line. The lack of proportionality within each item block is a violation of the assumption of the second-order model, as the second-order model is a bifactor model in which the loadings on the specific dimension are constrained to be proportional to the loadings on the general dimension within each item block. This explains why the bifactor model provides a better fit to the data than the second-order model.

For the bifactor model with the comprehension processes constituting the specific dimensions, the loadings on the specific dimensions are closer to zero than was the case for the bifactor model with item blocks as specific dimensions. The mean of the loadings on the specific dimensions across all items was 0.30 for the former and 0.36 for the latter. This mean was heavily influenced by some outlying estimated values for the bifactor model with comprehension processes as specific dimensions. The median value of the loadings on the specific dimensions amounted to 0.15 for this model, which is less than half of the median for the bifactor model with item blocks as specific dimensions (0.32). These results may explain why the bifactor with the item blocks as specific dimensions provided a better fit than the bifactor model with the comprehension processes as specific dimensions. In terms of model fit, it was more important to incorporate item cluster effects that stem from the item blocks than to incorporate item cluster effects related to comprehension processes. Another indication that taking into account the comprehension processes does not have a substantial impact is that the deviance for the second-order model with comprehension processes as first-order

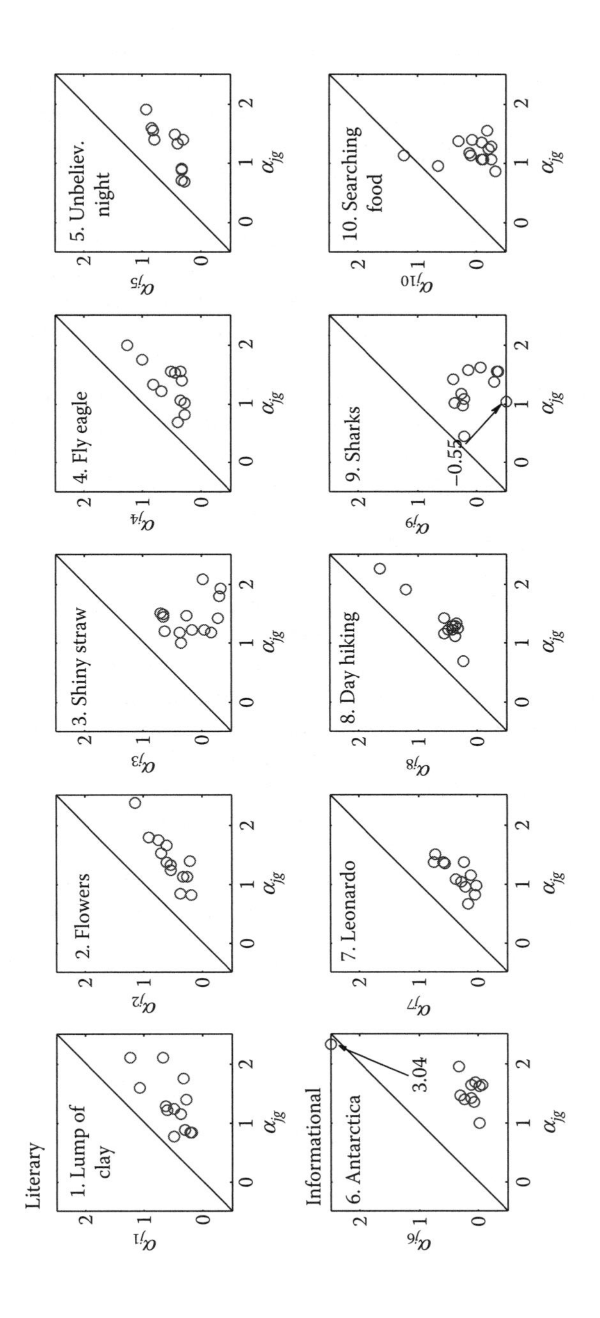

FIGURE 24.5
Scatter plots of the loadings on the general dimension versus the loadings on the specific dimensions for the bifactor model with item blocks as specific dimensions.

dimensions is a little higher than that of the unidimensional model with 20 quadrature points (due to slight imprecision in evaluating the log likelihood with fewer quadrature points for the higher-dimensional model).

From the first set of estimated models, it is apparent that the item blocks do constitute a separate source of individual differences. In contrast, the four comprehension processes do not seem to constitute separate dimensions, but rather are blended together into one overall dimension.

In a second stage, two models were formulated that incorporated correlated dimensions for reading purposes. One of the models was a between-item two-dimensional model, similar to the models currently in operational use. The other model incorporated separate dimensions for item blocks in addition to the two correlated dimensions for reading purposes. Because the first set of results indicated that a bifactor structure was preferred to a second-order structure, no proportionality restrictions were imposed on the loadings of the item block dimensions. A similar model was proposed by Cai (2010). The model is presented in Figure 24.6. In the figure, θ_{iI} represents reading to acquire and use information and θ_{iL} represents reading for literary experience. Because of the undirected edge between θ_{iL} and θ_{iI}, representing the fact that both dimensions are correlated, the graph is no longer a directed graph but a chain graph.

In principle, both models can be estimated efficiently by carrying out computations in two-dimensional spaces. However, technically, when the latent variables are assumed to be normally distributed, the model is estimated through a Cholesky decomposition of the covariance matrix of the latent variables. As a consequence, one of the correlated reading purpose factors, say θ_{iL}, is reformulated as a weighted sum of the other purpose factor, say θ_{iI}, and an independent residual factor, say $\theta_{iL\ \mathrm{res}}$. For the bifactor model, the items loading on θ_{iL} are now loading on three dimensions: the item block factor, θ_{iI}, and the residual purpose factor, $\theta_{iL\ \mathrm{res}}$.

The two-dimensional model provided a better fit than the unidimensional model, but not as good as the bifactor model with item blocks as specific

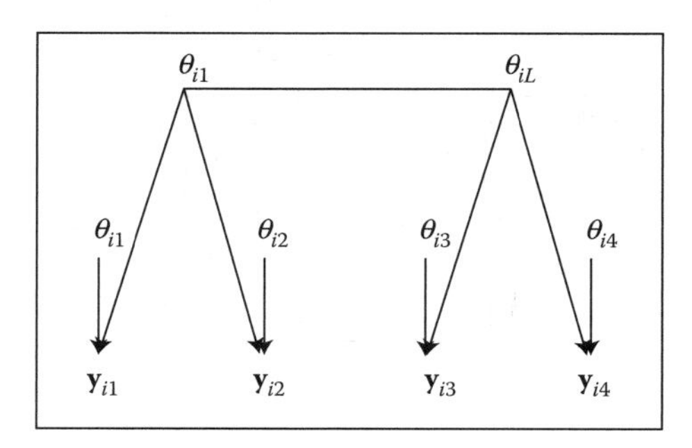

FIGURE 24.6
Chain graph of a bifactor model with a general factor for each reading purpose.

dimensions (Table 24.1). The estimated correlation between the two purposes of reading was .91. A visual inspection of the scatter plot of the estimates of the loadings of the two-dimensional model versus the estimated loadings of the unidimensional two-parameter logistic model revealed that the estimated loadings were very similar for both models.

After taking item blocks into account through the incorporation of specific dimensions, the model with a separate dimension for each of the two reading purposes provided a better fit than the corresponding model with only a single general dimension according to the AIC. The correlation between the two reading purposes was estimated as .93. A visual inspection of the scatter plots revealed that the estimated loadings were very similar to the estimates of the loadings for the bifactor model with one general dimension. This was the case for both the general dimensions and the specific dimensions corresponding to item blocks. The median loadings on the specific dimensions was 0.23, which is about one-third lower than the median of the loadings on the specific dimensions for the bifactor model with a single general dimension (0.32). Overall, the results indicate that the effect of item blocks can be attributed in part but not entirely to the fact that item blocks are clustered within reading purposes. On the other hand, the high correlations between the two reading purposes in the two models presented in this section, and the fact that the item loadings were very similar between the models with one general dimension and the models with a dimension for each reading purpose, do indicate that the two purposes of reading do not constitute substantially different sources of individual differences. Rather, they are blended together into one overall dimension for reading.

Acknowledgment

The research reported here was supported by the Institute of Education Sciences, U.S. Department of Education, through Grant R305D110027 to Educational Testing Service. The opinions expressed are those of the authors and do not represent views of CTB/McGraw-Hill, Educational Testing Service, the Institute of Education Sciences, or the U.S. Department of Education.

References

Akaike, H. 1973. Information theory and an extension of the maximum likelihood principle. In: B.N. Petrov and F. Csáki (Eds.), *2nd International Symposium on Information Theory*, pp. 267–281. Tsahkadsov, Armenia, USSR.

Adams, R. J., Wilson, M., and Wu, M. L. 1997. Multilevel item response models: An approach to errors in variables regression. *Journal of Educational and Behavioral Statistics, 22*, 47–76.

Besag, J. and Green, P. J. 1993. Spatial statistics and Bayesian computation. *Journal of the Royal Statistical Society, Series B, 55*, 25–38.

Bock, R. D. and Aitkin, M. 1981. Marginal maximum likelihood estimation of item parameters: An application of the EM algorithm. *Psychometrika, 46*, 443–459.

Bradlow, E. T., Wainer, H., and Wang, X. 1999. A Bayesian random effects model for testlets. *Psychometrika, 64*, 153–168.

Breslow, N. E. and Clayton, D. G., 1993. Approximate inference in generalized linear mixed models. *Journal of the American Statistical Association, 88*, 9–25.

Cai, L. 2010. A two-tier full-information item factor analysis model with applications. *Psychometrika, 75*, 581–612.

Chib, S. 1996. Calculating posterior distributions and modal estimates in Markov mixture models. *Journal of Econometrics, 75*, 79–97.

Cowell, R. G., Dawid, A. P., Lauritzen, S. L., and Spiegelhalter, D. J. 1999. *Probabilistic Networks and Expert Systems.* New York: Springer.

De Boeck, P. and Wilson, M. (Eds.) 2004. *Explanatory Item Response Models. A Generalized Linear and Nonlinear Approach.* New York: Springer.

Fox, J. P. and Glas, C. A. W. 2001. Bayesian estimation of a multilevel IRT model using Gibbs sampling. *Psychometrika, 66*, 271–288.

Foy, P. and Kennedy, A. M. 2008. *PIRLS 2006 User Guide for the International Database.* Chestnut Hill, MA: TIMSS & PIRLS International Study Center, Boston College.

Fischer, G. H. 1973. Linear logistic test model as an instrument in educational research. *Acta Psychologica, 37*, 359–374.

Gibbons, R. D. and Hedeker, D. 1992. Full-information item bi-factor analysis. *Psychometrika, 57*, 423–436.

Gibbons, R. D., Bock, R. D., Hedeker, D., Weiss, D. J., Segawa, E., Bhaumik, D. K., Kupfer, E. F., Grochocinski, V. J. and Stover, A. 2007. Full-information item bifactor analysis of graded response data. *Applied Psychological Measurement, 31*, 4–19.

Goldstein, H. and Rasbash, J. 1996. Improved approximations for multilevel models with binary responses. *Journal of the Royal Statistical Society, Series A, 159*, 505–513.

Jensen, F. V. 1996. *An Introduction to Bayesian Networks.* London, UK: University College London Press.

Jöreskog, K. G. 1994. On the estimation of polychoric correlations and their asymptotic covariance matrix. *Psychometrika, 59*, 381–389.

Lauritzen, S. L. 1996. *Graphical Models.* Oxford, UK: Clarendon Press.

Li, D., Oranje, A. and Jiang, Y. 2009. On the estimation of hierarchical latent regression models for large-scale assessments, *Journal of Educational and Behavioural Statistics, 34*, 433–463.

Lin, X. and Breslow, N. E. 1996. Bias correction in generalized linear mixed models with multiple components of dispersion. *Journal of the American Statistical Association, 91*, 1007–1016.

Mullis, I. V. S. and Kennedy, A. M., Martin, M. O., and Sainsbury, M. 2006. *PIRLS 2006 assessment framework and specifications.* (2nd Edition Tech. Rep.). Boston College, Chestnut Hill, MA: TIMSS & PIRLS International Study Center.

Martin, M. O., Mullis, I. V. S., and Kennedy, A. M. 2007. PIRLS 2006 technical report. (Tech. Rep.). Boston College, Chestnut Hill, MA: TIMSS & PIRLS International Study Center.

Masters, G. N. 1982. A Rasch model for partial credit scoring. *Psychometrika, 47*, 149–174.

Mislevy, R. J. 1985. Recent developments in the factor analysis of categorical variables (ETS Research Rep. No. RR-85–24). Princeton, NJ: ETS.

Mislevy, R. J. 1987. Exploiting auxiliary information about examinees in the estimation of item parameters. *Appied Psychoogica Measurement, 11*, 81–9.

Mislevy, R. J., Beaton, A. E., Kaplan, B., and Sheehan, K. M. 1992. Estimating population characteristics from sparse matrix samples of item responses. *Journal of Educational Measurement, 29*(2), 133–161.

Molenberghs, G. and Verbeke, G. 2005. *Models for Discrete Longitudinal Data.* New York: Springer.

Muthén, B. 1984. A general structural equation model with dichotomous, ordered categorical, and continuous latent variable indicators. *Psychometrika, 49*, 115–132.

Muthén, B., du Toit, S. H. C., and Spisic, D. 1997. Robust inference using weighted least squares and quadratic estimating equations in latent variable modeling with categorical and continuous outcomes. Unpublished manuscript.

Pinheiro, P. C. and Bates, D. M. 1995. Approximations to the log-likelihood function in the nonlinear mixed-effects model. *Journal of Computational and Graphical Statistics, 4*, 12–35.

Rabe-Hesketh, S., Skrondal, A., and Pickles, A. 2004. Generalized multilevel structural equation modeling. *Psychometrika, 69*, 167–190.

Rabe-Hesketh, S., Skrondal, A., and Pickles, A. 2005. Maximum likelihood estimation of limited and discrete dependent variable models with nested random effects. *Journal of Econometrics, 128*, 301–323.

Rabiner, L. R. 1989. A tutorial on hidden Markov models and selected applications in speech recognition. *Proceedings of the IEEE, 77*, 257–286.

Rasch, G. 1960. *Probabilistic Models for Some Intelligence and Attainment Tests.* Copenhagen, Denmark: Danish Institute for Educational Research.

Raudenbush, S. W., Yang, M.-L., and Yosef, M. 2000. Maximum likelihood for generalized linear models with nested random effects via high-order, multivariate Laplace approximation. *Journal of Computational and Graphical Statistics, 9*, 141–157.

Raudenbush, S. W. and Sampson, R. 1999. Ecometrics: Toward a science of assessing ecological settings, with application to the systematic social observation of neighborhoods. In: P. V. Marsden (Ed.), *Sociological Methodology*, pp. 1–41. Oxford: Blackwell.

Rijmen, F. 2009. *An Efficient EM Algorithm for Multidimensional IRT Models: Full Information Maximum Likelihood Estimation in Limited Time* (ETS Research Report No. RR-09-03), Princeton, NJ: ETS.

Rijmen, F. 2010. Formal relations and an empirical comparison between the bi-factor, the testlet, and a second-order multidimensional IRT model. *Journal of Educational Measurement, 47*, 361–372.

Rijmen, F., Tuerlinckx, F., De Boeck, P., and Kuppens, P. 2003. A nonlinear mixed model framework for item response theory. *Psychological Methods, 8*, 185–205.

Rijmen, F., Vansteelandt, K., and De Boeck, P. 2007, October/2008, June. Latent class models for diary method data: Parameter estimation by local computations [E-published on October 4, 2007]. *Psychometrika, 73*(2), 167–182.

Rutkowski, L., Gonzalez, E., Joncas, M., and von Davier, M. 2010. International large-scale assessment data: Issues in secondary analysis and reporting. *Educational Researcher, 39*, 142–151.

Scott, S. L. 2002. Bayesian methods for hidden Markov models: Recursive computing in the 21st century. *Journal of the American Statistical Association, 97*, 337–351.

Skrondal, A. and Rabe-Hesketh, S. 2004. *Generalized Latent Variable Modeling: Multilevel, Longitudinal and Structural Equation Models*. Boca Raton, FL: Chapman & Hall/CRC.

Schilling, S. G. and Bock, R. D. 2005. High-dimensional maximum marginal likelihood item factor analysis by adaptive quadrature. *Psychometrika, 70*, 533–555.

Vermunt, J. K. 2003. Multilevel latent class models. *Sociological Methodology, 33*, 213–239.

von Davier, M. 2005. *A General Diagnostic Model Applied to Language Testing Data* (ETS Research Report No. RR-05–16). Princeton, NJ: ETS.

von Davier, M. Sinharay, S., Oranje, A., and Beaton, A. 2006. Marginal estimation of population characteristics: Recent developments and future directions. In: C. R. Rao and S. Sinharay (Eds.), *Handbook of Statistics (Volume 26: Psychometrics)*. Amsterdam: Elsevier.

von Davier, M. and Sinharay, S. 2007. An importance sampling EM algorithm for latent regression models. *Journal of Educational and Behavioral Statistics, 32*(3), 233–251.

von Davier, M. and Sinharay, S. 2010. Stochastic approximation for latent regression item response models. *Journal of Educational and Behavioral Statistics, 35*(2), 174–193.

Wainer, H., Bradlow, E. T., and Wang, X. 2007. *Testlet Response Theory and Its Applications*. New York, NY: Cambridge University Press.

Winick, D. M., Avallone, A. P., Smith, C. E., and Crovo, M. 2009. *Reading Framework for the 2009 National Assessment of Educational Progress*. National Assessment Governing Board: U.S. Department of Education. Retrieved May 27, 2010, from: www.nagb.org

Xu, X. and von Davier, M. 2006. *Cognitive Diagnosis for NAEP Proficiency Data*. Research Report, RR-06–08. ETS: Princeton, NJ.

Yung, Y.-F., Thissen, D., and McLeod, L. D. 1999. On the relationship between the higher-order factor model and the hierarchical factor model. *Psychometrika, 64*, 113–128.

Zwinderman, A. H. 1991. A generalized Rasch model for manifest predictors. *Psychometrika, 56*, 589–600.

Index

F

G

H

J

M

Q

R

S

Y